Handbook of Drying of Vegetables and Vegetable Products

Advances in Drying Science and Technology

Series Editor:
Arun S. Mujumdar

PUBLISHED TITLES

Handbook of Drying of Vegetables and Vegetable Products
Min Zhang, Bhesh Bhandari, Zhongxiang Fang

Computational Fluid Dynamics Simulation of Spray Dryers:

An Engineer's Guide
Meng Wai Woo

Advances in Heat Pump-Assisted Drying Technology
Vasile Minea

Handbook of Drying of Vegetables and Vegetable Products

Edited by
Min Zhang
Bhesh Bhandari
Zhongxiang Fang

CRC Press
Taylor & Francis Group
Boca Raton London New York

CRC Press is an imprint of the
Taylor & Francis Group, an **informa** business

CRC Press
Taylor & Francis Group
6000 Broken Sound Parkway NW, Suite 300
Boca Raton, FL 33487-2742

Printed on acid-free paper

International Standard Book Number-13: 978-1-4987-5386-9 (Hardback)

Library of Congress Cataloging-in-Publication Data

Names: Zhang, Min (Food consultant), editor. | Bhandari, Bhesh, editor. |
Fang, Zhongxiang, editor.
Title: Handbook of drying of vegetables and vegetable products / [edited by]
Min Zhang, Bhesh Bhandari, Zhongxiang Fang.
Description: Boca Raton : Taylor & Francis, a CRC title, part of the Taylor &
Francis imprint, a member of the Taylor & Francis Group, the academic
division of T&F Informa, plc, [2017] | Series: Advances in drying science &
technology | Includes bibliographical references and index.
Identifiers: LCCN 2016059178 | ISBN 9781498753869 (hardback : acid-free paper)
Subjects: LCSH: Vegetables--Processing. | Dried vegetables.
Classification: LCC TP443.H33 2017 | DDC 664/.8--dc23
LC record available at https://lccn.loc.gov/2016059178

Visit the Taylor & Francis Web site at
http://www.taylorandfrancis.com

and the CRC Press Web site at
http://www.crcpress.com

MIX
Paper from responsible sources
FSC
www.fsc.org
FSC® C013056

Printed and bound in Great Britain by
TJ International Ltd, Padstow, Cornwall

Contents

SECTION II Drying of Specific Vegetable Products

SECTION III Changes in Properties during Vegetable Drying

SECTION IV Others (Modeling, Measurements, Packaging, and Safety of Dried Vegetables and Vegetable Products)

Preface

Drying of vegetables and vegetable products is very important in many developing countries, such as China, Brazil, India, and Thailand, as this can reduce the waste of fresh produce since consumption is usually less than production in a given period, increase product value and job opportunities, and provide varieties of vegetable products to the market. This is also important in developed countries as consumers demand high-quality dried vegetables for both convenience and health benefits.

According to the data provided by the Food and Agriculture Organization of the United Nations, most of the vegetables in the world are produced by the developing countries. China alone produces about half of the world's vegetables. Unfortunately, the vegetable processing technology, including drying technology, is developing slowly. The aim of this handbook is to act as a handy tool for R&D researchers in the field of vegetable drying, a textbook for undergraduate and postgraduate students in universities, and a reference book for researchers in relevant fields, including agricultural engineering and chemical engineering. Twenty chapters are included in this book, and they are authored by experts from 12 different countries, representing the current global technologies and industrial status in this specific area.

The book comprises of four main sections: Drying Processes and Technologies, Drying of Specific Vegetable Products, Changes in Properties during Vegetable Drying, and Others (Modeling, Measurements, Packaging, and Safety of Dried Vegetables and Vegetable Products).

The actual contents of this handbook include: Main current vegetable drying technologies in hot air and freeze drying, and their combinations; Highly efficient vegetable drying technologies in microwave drying, radio frequency drying, infrared radiation drying, ultrasound assisted drying and their combinations; Smart drying and foam-mat drying; Drying of specific vegetable products like herbs, spices, vegetable snacks, edible flowers, and mushrooms; Properties or quality changes in pigments and nutrients, texture and aroma changes during drying; Modeling of dryer and microstructural and morphological changes during drying; Nondestructive measurement by optical sensing technology and by computer vision image analysis for vegetable drying; Novel package and microbiology safety of dried vegetable products.

Many research papers have been and are being published on these topics, indicating continuing interests in vegetable drying research and development in the industry. We hope this comprehensive handbook, which provides updated information on various aspects of vegetable and vegetable products drying, will be helpful for the readers and researchers engaged in product research and development in food industries and for academics in research institutions and universities.

We acknowledge and thank all of the contributing authors for their hard work. We would like to thank Prof Arun S. Mujumdar for his strong support and Dr Weiqiao Lv for his efficient assistance in communication with chapter authors during editing this book.

Min Zhang
Bhesh Bhandari
Zhongxiang Fang

Editors

Min Zhang performed two periods of postdoctoral research from 1992 to 1997. He is currently the key professor and PhD supervisor in the School of Food Science and Technology, Jiangnan University. He is the director of R&D, Center of Food Resource and Utilization Technology in Jiangnan University. His research areas focus on the mechanism, technology, and engineering of processing and storing fresh foods.

Professor Zhang has been approved as a master's supervisor and has supervised 105 master's students (of whom 96 graduated) since 1998. Since 2000, he has been approved as a PhD supervisor and has supervised 42 PhDs (of whom 31 graduated). Professor Zhang began to supervise postdoctoral researchers in 2004 and has supervised 6 postdocs. In 2003 and 2005, he was listed in *Who's Who in Science and Engineering*. In 2004, he was listed in the first batch of the National Personnel Training Project of the New Century Bai-Qian-Wan Project. In 2006, he was listed as a National Expert and was entitled to special government allowances. In 2005, he was invited as the scientific adviser in the food science by the International Foundation for Science (IFS). From 2007, he has been listed as early or late listed as the editorial member of three SCI journals (*Journal of Food Engineering*, IF2.576; *Drying Technology*, IF1.77; *International Agrophysics*, IF 1.142) and several domestic journals, such as *Journal of Food Science and Biotechnology*, *Journal of Food Safety and Food Quality*, and *Drying Technology and Equipment*. In August 2013, he was appointed as an honorary professor in the field of food science by the University of Queensland, Australia.

In recent years, Professor Zhang has actively promoted the industrialization of new quality control technology in fresh food processing and preservation and has established a longtime collaboration mechanism of Industry-Academia-Research (joint research institute) with more than 10 large- and medium-sized domestic and foreign food or equipment companies like Haitong Food Group Co. Ltd., PepsiCo Food Co., and Jiahao Food Co., which created obvious economic benefits for the enterprise.

Professor Zhang has published 26 monographs or translated books in national presses and 10 English book chapters, edited one international conference proceedings (English edition), and coedited one English academic book. He has published more than 240 SCI international journal papers as a corresponding author. About 116 applied invention patents from his group (with him as the main inventor) have already been authorized by the Chinese National Intellectual Property Bureau. He has also applied for 11 international patents, including patents from Germany, Japan, France, South Korea, and Switzerland. He was recognized as the fifth of the top ten outstanding patent inventors in Jiangsu Province (China) in 2012. The projects carried out by Professor Zhang have been granted more than 10 awards by government and professional associations, including an award from the national government (second prize for the National Scientific and Technological Progress Award, 2012), an award from the government of Jiangsu Province (first prize for the Province Scientific and

Technological Progress Award, 2009), two awards from China General Chamber of Commerce (first prize for the National Commerce Scientific and Technological Progress Award, 2009, 2011), an award from China National Light Industry Council (first prize for the National Light Industry Scientific and Technological Progress Award, 2007), and an award from China National Food Industry Association (first prize for the National Food Industry Scientific and Technological Progress Award, 2011).

Bhesh Bhandari has been associated with the University of Queensland for more than 22 years. His research is mainly focused on food materials science, including microencapsulation of food ingredients and glass transition–related issues in food processing and product systems. Various microencapsulation processes, such as spray drying, molecular encapsulation, cocrystallization, precipitation, and gel entrapment, have been investigated by his research team. Professor Bhandari's current research focus has been to relate the nanostructure of food systems to their bulk properties. He has developed a continuous method to produce microgel particles that can be used to encapsulate various functional ingredients and pharmaceutical drugs. He has performed a number of pioneering studies on the stickiness issues of food powders encountered during drying and handling. Recently, he developed a patented technique to produce an ethylene powder that can be used for fruit ripening as well as control of other physiological functions in plants. Professor Bhandari has authored more than 200 papers, including 30 book chapters. He has coedited *Food Materials Science and Engineering* and *Handbook of Food Powders*, which have been published recently. Professor Bhandari is an editor of *The Journal of Food Engineering.*

Zhongxiang Fang earned bachelor (1993) and MPhil (2003) degrees in agricultural processing and storage engineering at the China Agricultural University and a PhD (2007) in food science at Jiangnan University, China. He was a lecturer at Zhejiang University, China, from 2007 to 2009 and moved to Australia in 2009 as a postdoctoral research fellow at the University of Queensland. Dr. Fang was appointed as a Curtin Research Fellow in the Food Science and Technology Program, School of Public Health, Curtin University, in 2012. From 2016, he worked as a lecturer in food processing in the Faculty of Veterinary and Agricultural Sciences, The University of Melbourne. Dr. Fang has secured research funding of about 1.7 million AUD, published more than 100 peer-reviewed papers, and has 9 patents. He has (co-) supervised six PhD students and ten master's students. He has served as a reviewer for major journals in associated fields. His research interests include the effect of climate, environment, and food processing technology on food quality and safety; antioxidant activity and metabolism of plant polyphenols; food processing technology and food engineering; and encapsulation of food bioactives and bioactive films in food coating and packaging.

Contributors

Karim Allaf
Intensification of Industrial Eco-Processes
Laboratory of Engineering Science for Environment
University of La Rochelle
La Rochelle, France

Tamara Allaf
Director
ABCAR-DIC Process
La Rochelle, France

Romy Gleyse Chagas Barros
Laboratory of Flavor and Chromatographic Analysis
Federal University of Sergipe
São Cristóvão, Sergipe, Brazil

Diego Alvarenga Botrel
Department of Food Science
Federal University of Lavras
Lavras, Minas Gerais, Brazil

Serkan Boyar
Department of Agricultural Machinery and Technologies Engineering
Süleyman Demirel University
Isparta, Turkey

Huizhi Chen
School of Food Science and Technology
Jiangnan University
Wuxi, Jiangsu, People's Republic of China

Marina Denadai
Laboratory of Flavor and Chromatographic Analysis
Federal University of Sergipe
São Cristóvão, Sergipe, Brazil

Erkan Dikmen
Department of Energy Systems Engineering
Süleyman Demirel University
Isparta, Turkey

Robert H. Driscoll
Food Science and Technology Group
University of New South Wales
Sydney, New South Wales, Australia

Xu Duan
Food and Biology Engineering College
Henan University of Science and Technology
Luoyang, Henan, People's Republic of China

Sabri Erbaş
Department of Field Crops
Süleyman Demirel University
Isparta, Turkey

Regiane Victória de Barros Fernandes
Department of Food Science
Federal University of Lavras
Lavras, Minas Gerais, Brazil

Oliver Hensel
Department of Agricultural Engineering
University of Kassel
Witzenhausen, Germany

Min Huang
Key Laboratory of Advanced Process Control for Light Industry
Jiangnan University
Wuxi, Jiangsu, People's Republic of China

Sachin V. Jangam
Department of Chemical and Biomolecular Engineering
National University of Singapore
Singapore

Hao Jiang
College of Food Science and Engineering
Northwest A&F University
Yangling, Shaanxi, People's Republic of China

Xin Jin
Suzhou Key Laboratory of Green Chemical Engineering
Soochow University
Suzhou, Jiangsu, People's Republic of China

H. C. P. Karunasena
Department of Mechanical and Manufacturing Engineering
University of Ruhuna
Galle, Sri Lanka

S. J. Kowalski
Institute of Technology and Chemical Engineering
Poznań University of Technology
Poznań, Poland

Wunwisa Krasaekoopt
Faculty of Biotechnology
Assumption University
Hua Mak, Bangkok, Thailand

Weiqiao Lv
School of Food Science and Technology
Jiangnan University
Wuxi, Jiangsu, People's Republic of China

Haile Ma
School of Food and Biological Engineering
Jiangsu University
Zhenjiang, Jiangsu, People's Republic of China

Alex Martynenko
Department of Engineering
Dalhousie University
Truro, Nova Scotia, Canada

D. Mierzwa
Institute of Technology and Chemical Engineering
Poznań University of Technology
Poznań, Poland

Sabah Mounir
Food Science Department
Zagazig University
Zagazig, Egypt

and

Intensification of Industrial Eco-Processes
Laboratory of Engineering Science for Environment (LaSIE)
University of La Rochelle
La Rochelle, France

Arun Sadashiv Mujumdar
Department of Bioresource Engineering
McGill University
Montreal, Quebec, Canada

Narendra Narain
Laboratory of Flavor and Chromatographic Analysis
Federal University of Sergipe
São Cristóvão, Sergipe, Brazil

Maria Terezinha Santos Leite Neta
Laboratory of Flavor and Chromatographic Analysis
Federal University of Sergipe
São Cristóvão, Sergipe, Brazil

Zhongli Pan
School of Food and Biological Engineering
Jiangsu University
Zhenjiang, Jiangsu, People's Republic of China

and

Healthy Processed Foods Research Unit
West Regional Research Center, USDA-ARS
Albany, California

and

Department of Biological and Agricultural Engineering
University of California, Davis
Davis, California

Gomathi Rajkumar
Laboratory of Flavor and Chromatographic Analysis
Federal University of Sergipe
São Cristóvão, Sergipe, Brazil

Wijitha Senadeera
School of Mechanical and Electrical Engineering
University of Southern Queensland
Springfield, Queensland, Australia

Saravanan Shanmugam
Department of Pharmacy
Federal University of Sergipe
São Cristóvão, Sergipe, Brazil

George Srzednicki
Food Science and Technology Group
University of New South Wales
Sydney, New South Wales, Australia

Barbara Sturm
Department of Agricultural Engineering
University of Kassel
Witzenhausen, Germany

Ya Su
School of Food Science and Technology
Jiangnan University
Wuxi, Jiangsu, People's Republic of China

J. Szadzińska
Institute of Technology and Chemical Engineering
Poznań University of Technology
Poznań, Poland

Bei Wang
School of Food and Biological Engineering
Jiangsu University
Zhenjiang, Jiangsu, People's Republic of China

Shaojin Wang
College of Mechanical and Electronic Engineering
Northwest A&F University
Yangling, Shaanxi, People's Republic of China

Bengang Wu
School of Food and Biological Engineering
Jiangsu University
Zhenjiang, Jiangsu, People's Republic of China

Baoguo Xu
School of Food and Biological Engineering
Jiangsu University
Zhenjiang, Jiangsu, People's Republic of China

Farid Zerrouq
Research Group QHSE-RG of LCME Laboratory
Sidi Mohamed Ben Abdellah University
Fez, Morocco

Min Zhang
School of Food Science and Technology
Jiangnan University
Wuxi, Jiangsu, People's Republic of China

Qibing Zhu
Key Laboratory of Advanced Process Control for Light Industry
Jiangnan University
Wuxi, Jiangsu, People's Republic of China

Section I

Drying Processes and Technologies

1 Main Current Vegetable Drying Technology I
Hot Airflow Drying and Related Combination Drying

Weiqiao Lv and Min Zhang
Jiangnan University

CONTENTS

1.1 INTRODUCTION

Hot airflow drying (AD) is a traditional drying method in the vegetable dehydration industry, and the products are always called AD vegetables (Zhang and Chen 2007).

Because the moisture content of most fresh vegetables is more than 80%, they are regarded as highly perishable commodities (Sagar and Kumar 2010). Through the AD process, the vegetables are dried to enhance storage stability, minimize packaging requirement, and reduce transport cost. Currently, the AD process is the main drying method used in vegetable dehydration. In this process, the pretreated vegetables are subjected to hot airflow of 50°C–90°C (Demiray et al. 2013; Wu et al. 2015). Heat can be transferred from hot air to the vegetables, and when the heat is absorbed by the materials, two types of moisture diffusion occur. One process is external diffusion, in which moisture moves from the material surface to the dry medium. The other process is internal diffusion, in which the internal moisture moves to the material surface. These two diffusion processes develop at the same time until the moisture content decreases to the level where the materials can be stored safely (Cheng et al. 2015; Srikiatden and Roberts 2006). Because the materials in the hot airflow environment are exposed to oxygen and the temperature is always higher than 50°C, some biochemical reactions happen, such as oxidation of phenolic compounds catalyzed by oxidase, vitamin loss, etc. To improve the retention rate of nutrients, the physiological activity of enzymes should be inhibited by blanching or hot steam pretreatment. In practice, blanching or hot steam temperature, pretreating time, airflow speed, drying temperature, and drying time are critical factors that affect the nutrient retention and drying quality of the vegetables.

With the development of the drying industry, novel drying technology and processes have been developed recently. For example, microwave drying (MD), infrared drying, vacuum freeze-drying, and some combined drying technologies have been widely used in vegetable dehydration. However, as the main process in vegetable dehydration, AD is a well-established technology owing to its easily available equipment and facilities.

In addition, the AD process dries the surface of the materials first, which has an advantage in drying products with high moisture and high sugar content, such as tomatoes (Demiray et al. 2013) and pumpkins (Hashim et al. 2014). It is expected that the traditional method of the AD process will still be widely used for vegetable dehydration in the future. Therefore, the AD process and other combined drying technologies need to be studied further.

1.2 CHARACTERISTICS OF HOT AIRFLOW DRYING

1.2.1 Heat and Mass Transfer

In an AD process, heat and moisture transfer between vegetable materials and hot air occur simultaneously. Heat is transferred from hot air to the material surfaces, and then moves to the interior of the materials. The internal moisture of the materials diffuse to the surface and then evaporate into the environment. The driving force of heat transfer is the temperature difference, and the driving force of mass transfer is the moisture difference or the partial pressure difference of water vapor. The two processes have a close relationship, but in opposite directions. Compared with MD, the drying resistance is more in the AD process. However, this is an advantage for some vegetables with high moisture and sugar content in which the surface layer

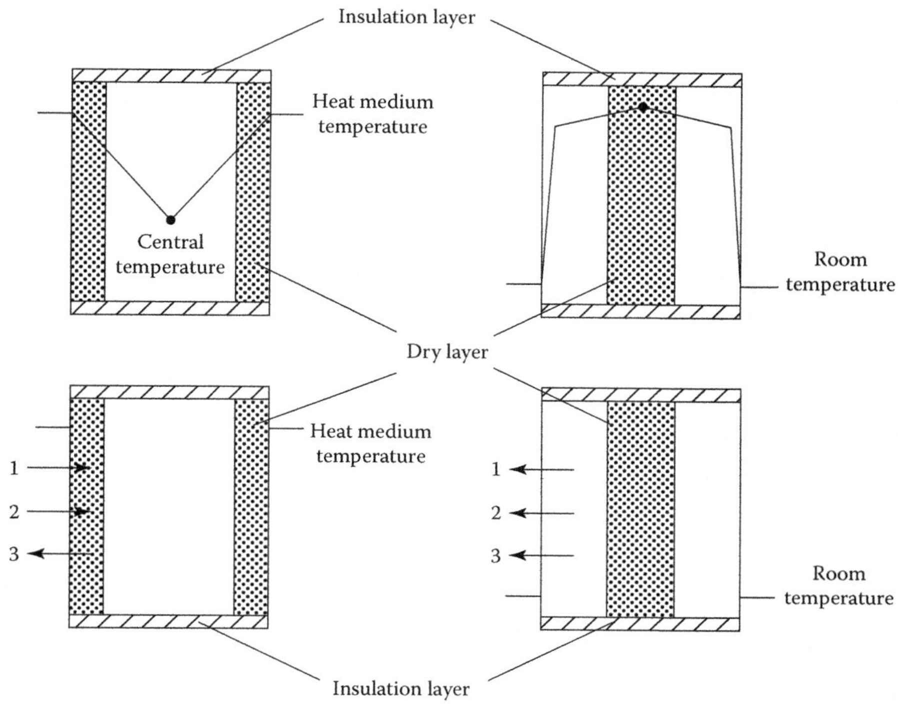

FIGURE 1.1 Model schematic drawing of AD process (left) and microwave drying process (right). 1, Direction of temperature gradient; 2, direction of heat conduction; 3, direction of moisture migration.

damage and adhesion could be avoided. The heat and water transfer diagram of AD and MD are shown in Figure 1.1.

In the AD process, the water on the surface of materials is dried by hot airflow. When the moisture content in the external part is lower than that in the internal part, difference occurs in the moisture concentration. Because of this, water diffuses from the region of high moisture content to the region of low moisture content. At the same time, the temperature in the external part is higher than that in the internal part, thus creating temperature difference. Because they occur in opposite directions, the two mechanisms are barrier factors to each other. When there is a moisture difference, moisture will be transferred to the surface; when there is a temperature difference, moisture will move in direction of the heat flow, in which condition the materials' drying speed becomes slow, or even stops, and always results in a hard surface on the surface. This phenomenon is mainly caused by very high temperature and fast airflow speed. Sometimes, this phenomenon is required to protect the shape of the vegetables, especially in the AD process combined with fluidized bed or spouted bed drying for vegetables with high moisture and high sugar content. Their surface layer can be dried first, and cannot be adhered together. This is the advantage of AD, more so for the drying of tomatoes, pumpkins, etc.

1.2.2 Drying Quality Evaluation

Consumers have certain expectations for the product quality of dehydrated vegetables, especially with regard to their visual appearance. They always tend to associate food product color and other visual properties with taste, hygiene, shelf life, and nutritional value; therefore, color and shape are regarded as the primary criteria in terms of consumer acceptance. In general, parameters such as temperature, mass of the materials, and airflow speed can be adjusted in the drying process, and thus the color and shape can be maintained within acceptable levels after drying.

Color is always measured by color difference meters. The L^*, a^*, b^* values of dehydrated vegetables are determined, where L^* (whiteness/darkness) is the lightness index and a^* (redness/greenness) and b^* (yellowness/blueness) are the saturation index (Chin and Law 2010; Chua et al. 2001). Sometimes, the color of the vegetables changes significantly in different detecting spots. In this case, fresh samples are always smashed into slurry and dried samples ground to powders so that color measurement can be conducted uniformly (Ding et al. 2012).

In the AD process, the temperature gradient direction is opposite to the direction of moisture migration. Therefore, if the temperature is high, the surface layer easily becomes hard, and this may lead to size shrink. Moreover, the nutrition loss of the vegetables is more under this condition. However, if the temperature is too low, the drying efficiency decreases, and some biological reactions occurring at high moisture content may result in low-quality products. In general, the AD process has a long drying time and slow moisture evaporation rate. As the water molecules migrate slowly, the molecular groups of the dry matter are not stimulated by a high-efficiency physical field, and compared with MD, the microstructure could be protected better in AD vegetables (Ding et al. 2012). Rehydration ability is another critical index to evaluate the quality of the dehydrated vegetables and is closely related to the microstructure of the products (Rahman 2003).

1.2.3 Energy Consumption

In the production of dehydrated vegetables, minimizing drying cost is one of the most highly regarded criteria. However, conditions that produce high-quality products using minimal costs are always difficult to achieve, and the optimum requirements of heat and mass transfer do not necessarily match with those for optimum quality. In terms of energy efficiency, reducing/increasing the temperature for the whole drying process is not very effective. In general, the drying parameters frequently vary at different stages with the decreasing moisture content in the AD process. The hot airflow can be recycled from the outlet air, and thus the energy efficiency can be improved. However, if the moisture content in the recycled hot air is too high, the water in the vegetables cannot be evaporated in time because of the saturated vapor pressure, and thus the drying efficiency could be reduced.

In the AD process, the vegetable materials may be dried with the hot air flowing on the surface, but in most cases the hot air flows through the entire material layer. A common feature of these dryers is their high energy consumption. There is worldwide concern over global warming, which is attributed to greenhouse gases produced by the combustion of fossil fuels, and there is increasing pressure to reduce energy consumption. As consumption of energy in the vegetable dehydration industry is high, highly energy efficient processes and equipment should be developed.

1.3 VEGETABLE AD PROCESS

1.3.1 Drying Process and Main Equipment

Dehydration of vegetables is not only helpful for storage and transport but also has many other benefits like prepared food or leisure food. Some dehydrated vegetables need to be rehydrated before cooking—these belong to the class of prepared food and include potatoes, cabbage, etc. Some dehydrated vegetables can be eaten as leisure food, such as edamame and carrot sticks, and these need to be blanched for a longer time before drying. In general, fresh vegetables can be processed to dehydrated products in the harvest season. With an efficient drying process, the dehydrated vegetables can maintain good color and nutrition at room temperature, which is welcomed by consumers. The drying procedures in the AD process include five steps as follows:

1. *Raw material selection*
   Fresh vegetables selected for industrial drying should be free of disease and insect pests. Some vegetables, such as pumpkin and wax gourd, have a hard and waxy skin that affect moisture diffusion during drying, and therefore this skin needs to be removed before processing.
2. *Cutting and blanching*
   The selected raw materials should be cut into slices, dices, or strips according to the product requirements. The cut vegetables are blanched in boiling water for 1–3 min or using hot steam for a few seconds to deactivate the oxidative enzyme activities. However, for some heat-sensitive vegetables, such as oyster mushroom and spinach, blanching is not most of the time necessary.
3. *Cooling and draining*
   The blanched vegetable slices, dices, or strips should be cooled in cold water. It is better to use flow water so that the material temperature can be decreased faster. The prepared materials can be drained by air or by a centrifugal method.
4. *Hot-air drying*
   According to the varieties and characteristics of the vegetables, drying temperature, airflow speed, and drying time can be set at different levels. In the AD process, the materials are placed in a drying house or in a hot airflow-drying machine. In the drying house, the materials move in one layer, and the hot air flows in the backward direction. Meanwhile, the materials can also be placed in two layers or multilayers, and the airflow can be set to forward and backward directions for different layers. In most cases, the temperature of the drying house is about 50°C. The drying machine is always different from the drying house; it has a small volume and the temperature can be adjusted more precisely. To improve energy efficiency,

the materials are always placed in multilayers, and thus the hot air with unsaturated water vapor can be recycled. Hot airflow temperature is generally controlled and maintained between 50°C and 90°C.

5. *Packaging*
   Dehydrated vegetables should be of good quality with regard to color, shape, rehydration ability, nutrition content, etc. and should also meet the hygiene requirements. The dehydrated vegetables can be packed in plastic bags and sealed.

At present, some different types of AD equipment include tunnel-type dryer, box-type dryer, conveyor dryer, fluidized bed dryer, drying room, etc. The main factors influencing AD quality are the hot-air temperature, airflow speed, and thickness of the materials. Some common hot AD equipment are discussed in the following sections.

1.3.2 Box-Type Vegetable Dryer

As the main AD equipment, the box-type vegetable dryer is widely used in the industry. It has a simple structure, which is shown in Figure 1.2. In this equipment, the vegetables are dried in the box via the hot air from the air inlet. In practice, the drying processes are organized by different stages under different temperatures. With the decrease in mass, the materials in different boxes can be put together. The box-type vegetable dryer has low equipment investment and ensures quality drying. The heat source can be steam, coal, fuel, or other combined energy sources, and it is economical and suitable for use in small-scale industry processing. However, the energy efficiency is low when the heat from the wet air passing though the materials is exhausted. As the hot air passes over the materials only once, the airflow is much drier than that in the recycled process. It is more suitable for vegetables with low moisture content and leafy vegetables, such as peanuts, green beans, cabbage, etc. In order to improve the energy efficiency, multilayer AD vegetable dryers and mesh belt AD equipment are also widely used for vegetable dehydration.

FIGURE 1.2 Box-type vegetable dryer.

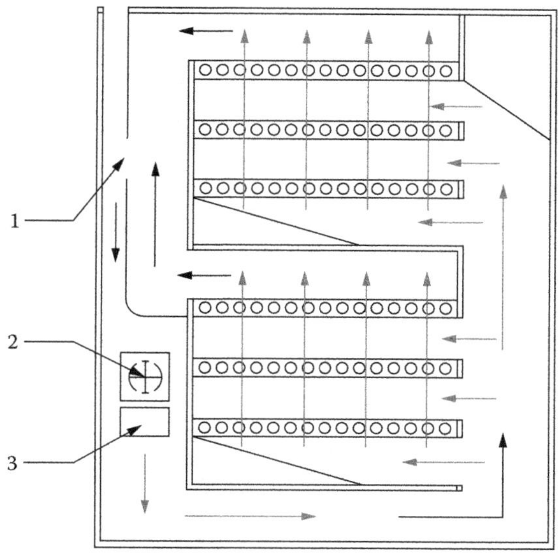

FIGURE 1.3 Multilayer AD vegetable dryer. 1, Humid-air outlet; 2, circulating centrifuge; 3, steam heater.

1.3.3 Multilayer AD Vegetable Dryer

A multilayer AD vegetable dryer has more than one layer of materials, and a schematic structure is shown in Figure 1.3. In this equipment, hot airflow passes through three layers of prepared vegetables. The heat due to this process can be absorbed three times in one cycle. A part of the used hot air is exhausted, but most of the used hot air can be pumped to the steam heater again, for reuse in the next cycle. As steam can carry more heat, the reused hot air with some moisture content is enough to make the water evaporate; thus, the multilayer AD vegetable dryer has higher drying efficiency than the box-type dryer.

However, a disadvantage of this dryer is that the drying speed of the materials in different layers vary. Hot air is not only the carrier of heat but also the carrier of water in vegetables. The materials in the lower layer are easy to dry, but the upper-layer materials that are dried at low temperature and high moisture content have a low drying speed. Therefore, multilayer AD vegetable dryers cannot be installed with too many layers. However, different recycle systems can be designed, and every recycle system can have no more than three layers. Figure 1.3 shows two recycle systems in a dryer, and each system has three layers, so drying uniformity can be achieved.

1.3.4 Mesh Belt AD Dryer

Vegetable drying in a multilayer AD vegetable dryer is still a batch operation. So, mesh belt AD equipment have been developed to meet the requirements of a continuous drying process. The adjacent conveyor belts have opposite moving directions, and the materials can be conveyed from the upper layer to the lower layer, where the

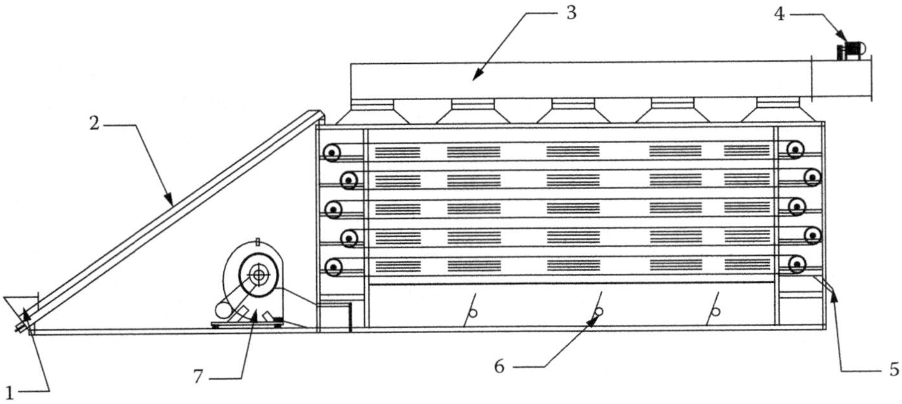

FIGURE 1.4 Mesh belt AD equipment. 1, Feed hopper; 2, automatic feeding machine; 3, moisture outlet pipe; 4, moisture pumping fan; 5, discharge port; 6, air-regulating plate; 7, hot air blower.

hot air passes through the mesh belts from the bottom to the top. The drying process and the dryer structure are shown in Figure 1.4.

In mesh belt AD equipment, the pretreated materials are automatically fed on the top. With the drying process, the materials move to the lower conveyor belt. The materials with high moisture content are dried in the upper layers, where the moisture content of the hot air is also high; thus, more heat can be transmitted to the materials and the drying efficiency can be improved. At the same time, materials with low moisture content moving in the lower layers will be dried by low moisture content but high temperature hot air, and this moisture concentration and temperature difference also have a positive effect on the AD process. Air-regulating plates are installed at the bottom of the drying chamber, where the hot air enters; thus, the uniformity of airflow in different parts can be adjusted. Because the utilization efficiency of hot air is relatively high, the used air can be exhausted directly. In general, mesh belt AD equipment has not only achieved continuous drying but also has a high drying efficiency.

1.3.5 Fluidized Bed Dryer

In a typical fluidized bed drying system, hot air flows through the materials quickly, and the fluidized bed is connected with a vibrating motor (Figure 1.5). With the action of mechanical vibration and hot airflow, the weight of the materials can be balanced in a fluidized state. Fluidized drying is always regarded as one of the best AD processes for granular materials, and its has the following advantages:

1. Because the airflow with high temperature and low moisture content can contact with granular materials sufficiently, and most of the hot airflow can be recycled, the drying rate and energy efficiency are high.
2. The equipment is relatively simple, requires low capital investment, and is easy to maintain.

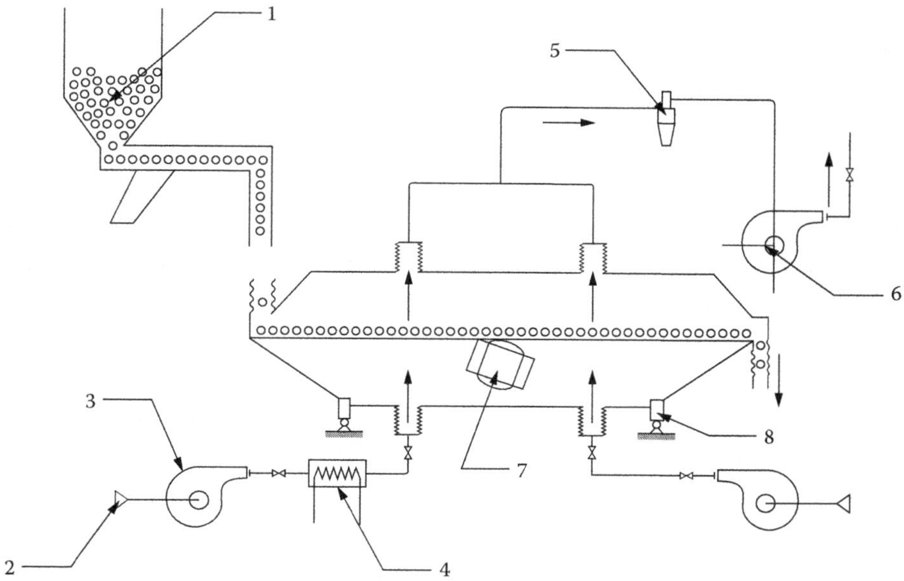

FIGURE 1.5 Hot airflow fluidized bed dryer. 1, Raw materials; 2, filter; 3, fan; 4, heat changer; 5, cyclone separator; 6, exhaust fan; 7, vibrator motor; 8, spring.

3. The drying process is easy to control, and the vegetable materials' drying uniformity is stable. It is applied in drying most of the granular vegetables, such as potato dices, apple dices, carrot dices, etc.

1.4 AD-RELATED COMBINATION DRYING

At present, the AD process is the main method used in vegetable dehydration. Owing to high drying temperature and long drying time, the AD product quality including color, texture, flavor, and nutrition could be altered significantly. In addition, other high-efficiency physical fields such as microwave, infrared, and radio frequency (RF) drying have their own limitations. It is much difficult to obtain dehydrated vegetables from single high-efficiency physical field drying in industry. Therefore, the combination of AD with these new drying techniques has the potential to not only improve the product quality but also to save energy, and so it is widely applied in the vegetable drying industry.

1.4.1 AD Combined with Microwave

In vegetable dehydration, AD has the characteristics of long drying time, high energy consumption, and relatively large amount of nutritional loss. MD has many advantages, such as strong penetrability, high drying efficiency, low nutrition loss, etc. However, because of the cavity's reflection and different penetration depth in vegetable materials, uneven drying and local burn are common problems in MD (Clark and Sutton 1996; Drouzas and Schubert 1996).

Ginger is a traditional herbal medicine and a typical vegetable used as a condiment in Asian countries. Because the dehydration process not only extends ginger's storage time but also provides good quality and flavor, the dehydrated ginger slices have a huge market value and vast development space (Yudthavorasit et al. 2014). An experiment has been completed to compare the drying uniformity of ginger slices with MD and AD, and the dehydrated samples are shown in Figure 1.6.

The MD process leads to uneven heating with obvious local burns (Figure 1.6), and the AD product leads to uniform yellow color with no brown burns. The AD product is better than the MD product; however, AD takes a longer time, and in the final stage the energy efficiency is lower.

While studying the different drying processes on lettuce slices, Roknul et al. reported that the product quality of hot air-assisted RF drying was the best, when compared with AD and MD methods. The result indicates that microwave has the disadvantage of uneven drying, which was in agreement with the results reported for microwave and hot-air dehydration of ginger slices.

In general, the burn phenomenon in MD always occurs at the final drying stage. Controlling the quality of products and final moisture content is another research focus of the process. Different drying methods have a different influence on the ginger tissue structure, which affect products' rehydration capability (Bondaruk et al. 2007; Han et al. 2010). It was reported that the microstructure and flavor of dehydrated ginger slices (dehydrated by AD and MD) was significantly different (Lv 2015).

On one hand, the microstructure of AD dehydrated ginger was maintained better than the ginger from the MD process. In the AD process, moisture migrating speed is slow when the water molecules pass through the cell wall and evaporate. The tissue structural damage is minimized and the cell framework and the starch granules are clearly visible. However, for the MD process, certain polar molecule

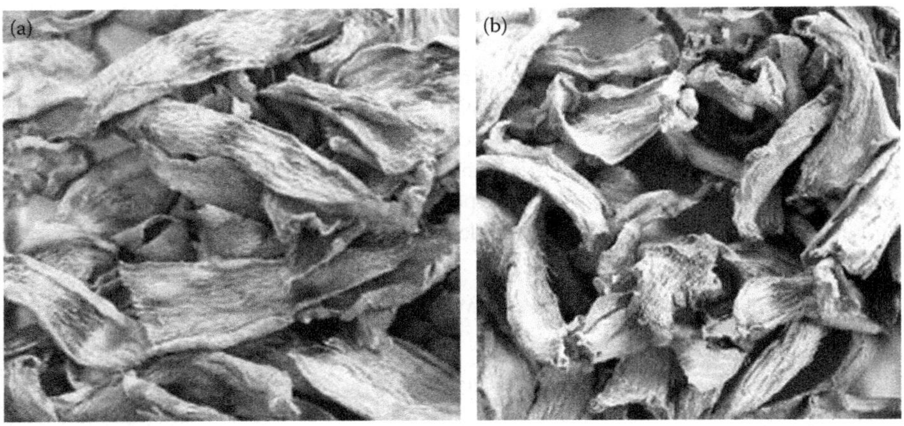

FIGURE 1.6 Dehydrated ginger slices by MD and AD processes. (a) MD with the microwave power of 1.8 W/g. (b) AD with the hot airflow at 75°C.

hydroxyl and carboxyl groups, which come from cellulose, hemicellulose, pectin, and other polysaccharides, always occur in a stable state and significantly interfere with this process; when the cell wall skeleton is mixed with starch granules under the intense impact, damage of the cell's regular structure is evident. With the information that the rehydration ratios from AD are much better than those from MD, it can be concluded that there is a strong relationship between the gingers' rehydration capacity turning poor and the structures being damaged in the drying process.

On the other hand, the cell wall of the vegetables is damaged by MD, and then the flavor can be released from the cells. Therefore, hot air microwave combined drying in different stages could also be designed; for example, the AD process may be applied in the first stage to fix the vegetable shape, or in the last stage to avoid overheating, which is more flexible than the simultaneous AD and MD combination of drying.

Rehydration ability is another important index to evaluate the quality of dehydrated vegetables, and disordered microstructure always decreases the rehydration ratios (Rahman 2003). The rehydration radios of the dried ginger slices from AD at a temperature of 75°C and MD with the microwave power of 0.9 W/g are shown in Table 1.1.

It can be seen that in the same soaking condition, the rehydration ratios of MD are weaker than those of AD. In different soaking temperatures, the rehydration ratio differences from MD are much more obvious than AD. Therefore, the MD product is much more sensitive to soaking temperature than the AD product.

For different soaking temperatures of 40°C, 60°C, and 80°C, there is a high rehydration speed in the beginning stage. With the development of the process, the samples absorb much more water when rehydration speed decreases. In the final stage, with moisture content in the tissue becoming close to saturation, rehydration speed decreases gradually, and at last the rehydration ratio is stabilized at a fixed level. If the soaking temperature was higher, the product's rehydration ability is stronger. Analysis shows that rehydration ability of the ginger following MD is poorer than that following AD, this is due to the damage to the tissue structure caused by MD. With the increase in the soaking temperature, the water molecules can penetrate to the internal side of the tissue structure quickly; when the influence caused by irregular organization structure is weakened, the difference in the rehydration ratios becomes small.

TABLE 1.1
Rehydration Ratios of Dried Ginger Slices from Different Drying Processes

Soaking Processes (°C)	Air Drying (75°C)	Microwave Drying (0.9 W/g)
40	5.12 ± 0.12	4.18 ± 0.13
60	5.18 ± 0.13	4.36 ± 0.13
80	5.34 ± 0.13	4.83 ± 0.10

Hot-air microwave fluidized drying is a novel drying process that combines the AD and microwave fluidized drying processes simultaneously. A schematic drawing of the microwave fluidized drying test device is shown in Figure 1.7.

The device is composed of six magnetrons, which can feed microwave energy independently. Through mechanical vibration, the vegetable samples will be dried in a fluidized state. This design is aimed to ease local overheating and materials burnt in MD, which is mainly caused by the microwave electromagnetic field's uneven distribution and the energy penetrating dissipation. Carrot dices were dried using the hot-air microwave fluidized drying process, and the product was obtained with good color, flavor, and uniform quality (Han et al. 2014). Reyes et al. (2007) also reported that the drying efficiency and quality of potato slices were improved by hot-air microwave fluidized drying (Reyes et al. 2007).

Some other researchers also reported drying of vegetables by synchronous combination of AD with MD. For example, if the microwave power was added when the carrot slices are dried in the AD process, the drying time could be decreased by 25%–90%, and if the microwave power was less the drying quality was good (Prabhanjan et al. 1995). AD and MD combined drying of *Agaricus bisporus* synchronously not only shortened the processing time but also improved the product quality and the rehydration ratios (Riva et al. 1991). With regard to drying of garlic,

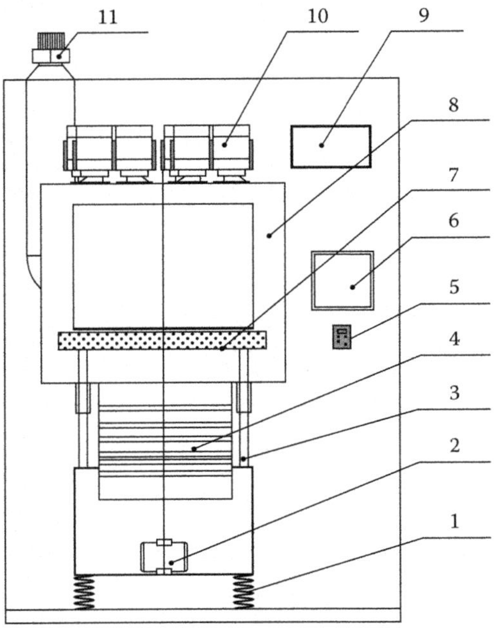

FIGURE 1.7 Schematic drawing of a microwave fluidized drying test device. 1, Damping spring; 2, vibration motor; 3, vibration frame; 4, electric heater; 5, frequency converter; 6, touch screen; 7, material tray; 8, microwave drying chamber; 9, view port; 10, magnetron; 11, induced draft fan.

compared with AD alone, the drying time of the combined AD and MD process was reduced by 80%–90% (Sharma and Prasad 2001).

In the AD process, the microwave can be added at the initial stage, middle stage, and final stage, respectively. At the initial stage, the temperature increases rapidly, when the moisture migration channel opens. At the middle stage, the moisture outside is dried by hot air and the moisture inside can be dehydrated quickly by microwave. At the final stage, with hot airflow, the rest moisture is difficult to dry, when microwave can improve drying efficiency and quality. MD intermittently with the drying time and tempering time ratio as 12min:4min first, and hot AD consecutively, good-quality dehydrated red jujube was obtained (Liu et al. 2013). In addition, some intermittent microwave hot-air combined drying methods have also been used in drying gingers (Lv et al. 2015a; Zhao et al. 2014).

1.4.2 AD Combined with Heat Pump

AD combined with heat pump process is also named heat pump drying. A typical heat pump dyer diagram is shown in Figure 1.8, which has two working subsystems, where 1-2-3-4-1 is a heat pump medium cycle subsystem, and A-B-C-D-E-A is an air cycle subsystem. The heat pump subsystem contains a condenser, compressor, evaporator, and expansion valve, which supplies the heat to the low moisture air before coming into the drying chamber, and releases the heat in high moisture content air in the drying chamber. The air system comprises the drying chamber, exhaust air, cold water, and heat pumper, which take the moisture away under the action of the heat pump subsystem.

The hot air from the drying chamber has high moisture content, and it can be coagulated in the medium's evaporator. The rest of the air with low moisture content will be heated in the medium's condenser and is then prepared for drying of the materials. The heat pump drying of vegetables has a relatively lower drying temperature than that of normal AD, which is generally at 40°C–80°C and suitable

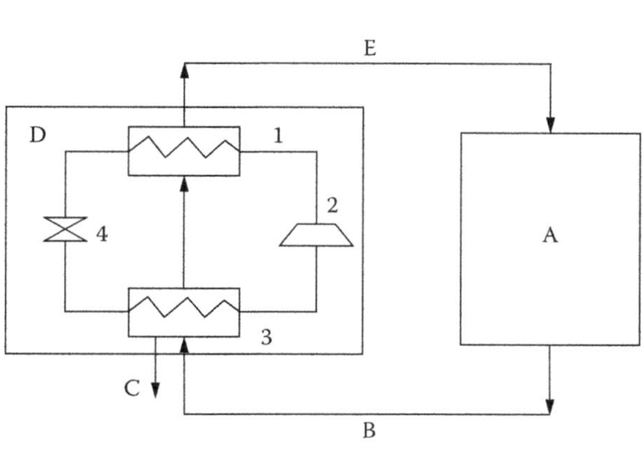

FIGURE 1.8 Diagram of a heat pump dryer. A, Drying chamber; B, exhaust air; C, cold water; D heat pumper; E hot air. 1, Condenser; 2, compressor; 3, evaporator; 4, expansion valve.

for heat-sensitive vegetables. Heat pump drying has many advantages, such as (1) high energy efficiency; (2) good product quality, especially for temperature-sensitive materials; and (3) easy-to-control drying condition. However, there are also some disadvantages of this drying method, which include the following: (1) initial capital investment is high; (2) refrigerant is likely to leak; (3) compared with normal convective AD, the structure is complex; and (4) the equipment needs more space.

There is no energy loss in a cycle of heat pump drying; thus, the energy efficiency is high. Heat pump drying process has been used in drying of garlic, ginger slices, and onion slices with a drying temperature not higher than 75°C and drying time not more than 8 h (Boonnattakorn et al. 2004; Chua et al. 2001). The hot airflow's thermal efficiency is more than 85%, and the products' quality is good. The heat pump drying technology has also been applied in drying cabbage, and when compared with normal AD, energy saved was 40%–50% (Yang et al. 2013).

In fact, the heat pump dryer has been widely used in drying of high value-added products. For a convective hot airflow dryer, if a refrigeration cycle was added, the rest heat can be absorbed by the refrigeration cycle, and the moisture from the rest hot air can be separated; thus, the normal hot airflow dryer can be modified to a heat pump dryer. In addition, the heat pump dryer can also be combined with far infrared, microwave, over heat steam, and solar energy drying processes to further improve drying efficiency.

1.4.3 AD Combined with Infrared Drying

Infrared drying (IRD) has high drying efficiency and is pollution free. However, its penetrability of the vegetable materials is poor; therefore, the drying rate could be different in different drying areas. At present, IRD is commonly combined with AD for vegetable drying.

Infrared light is an electromagnetic wave, and the wavelength range is 0.72–1000 μm, which is between the wavelength of microwave and visible light. The infrared can be divided into near infrared (0.72–2.5 μm), middle infrared (2.5–25 μm), and far infrared (25–1000 μm). In general, the penetrability depth of the infrared is short, which only focuses on the thin layer of the surface. The penetrability of near infrared is weaker than that of middle infrared, and the penetrability of far infrared is the weakest, and its energy only focuses on the surface of the vegetables. In addition, the moisture content of the materials will affect the penetrability of the infrared radiation, the higher the moisture content, the stronger the penetrability. Therefore, in an IRD process, the penetrability decreases with the decreasing moisture content of the vegetables. For most vegetable materials, the infrared absorption wavelength is about 0.72–30 μm, and the wavelength at other ranges is difficult to be absorbed. In order to control the drying efficiency and product quality, the infrared wavelength range is always selected at 0.72–9.6 μm in the industry, which belongs to the middle infrared and near infrared.

In the drying process of onion, it was found that the efficiency of combined AD and IFD is higher when compared with either of the techniques alone (Praveen Kumar et al. 2005). It was also suggested that AD could decrease the loss of infrared energy (Afzal and Abe 2000). In one of our team researches, the sensory quality and chemical quality of the dried mushroom product by IRD was better than that by AD

(Wang et al. 2015). Combined IRD and AD also reduced the AD time of mushroom chewing tablets, and the product flavor was better than that of either of the individual methods (Wang 2014).

1.4.4 AD Combined with RF

RF is a type of high-frequency electromagnetic wave, with a frequency ranging from 3 kHz to 300 MHz. When RF penetrates into the wet materials, the movement of the water molecules is dramatically stimulated, leading to conversion of the electric energy into heat energy, and thus the drying process is achieved. Compared with MD, radio frequency drying (RFD) can result in a more uniform drying of vegetable materials. However, because of the high cost of RFD, it is always combined with AD in the drying industry.

When the RF power is well controlled, good quality can be achieved in AD combined with RFD. For example, if the material is hot air-dried at a temperature above 90°C, a hard case could be formed on the surface of the material. However, when combined with RFD by 10% of the whole drying energy, the hard case would not be formed on the material surface even if the temperature rises to 200°C (Wang et al. 2014). AD combined with RFD can also be used in the pretreatment of fried potato chips (Marra et al. 2009).

1.5 ASSESSMENT OF DRYING FACTORS IN AN AD PROCESS

1.5.1 Effect of Pretreatment

1. *Physical pretreatments*

   To improve the product quality of dried vegetables, some physical pretreatments could be applied:

   a. *Blanching*

      It is well-known that enzymatic browning is one of the main biochemical factors that influences the quality of dried products. In fact, the enzyme is very sensitive to the hot and humid environment. Because the temperature in the AD process is generally below 80°C, the enzyme activity could not be inactived completely, and therefore pretreatment is necessary. Protein will be denatured in blanching, and the enzyme activity will be lost; therefore, blanching before the AD process is an effective and necessary pretreatment.

      The vegetables are immersed in boiling water for 1–3 min and then cooled with cold water (Beveridge and Weintraub 1995; Reis et al. 2008). With the influence of physical parameters (such as the boiling time, etc.), the cell membrane's permeability will be changed greatly, so that the resistance of the cell membrane for moisture evaporation is also changed. At the blanching temperature of 100°C, the cell's ability to control the water state is lost, and the ability to prevent the moisture's evaporation is weakened. In this way, blanching pretreatment is helpful for vegetable materials' efficient dehydration. However, blanching has its disadvantage in the AD process. With the increase of cell permeability, rehydration ability will be decreased, and the rehydration quality may turn bad. Therefore, blanching should be controlled in a short time, and according to the nature of the raw materials, the detailed parameters should be adjusted.

b. *Hot steam treatment*

In hot steam treatment, the vegetable materials are heated by hot steam (100°C–200°C) for 1–3 min and then cooled by cold water (Qin et al. 2011). Through instantaneous pretreatment by high temperature steam, the enzyme inactivation in vegetables will be lost. After hot steam treatment, enzymatic browning is avoided; thus, the quality of the fresh vegetables is stable in dehydration, including the flavor, color, etc.

c. *Other physical treatments*

Although blanching and hot steam treatment are simple processes and easy to operate, some negative influences occur, such as shape shrinkage, low rehydration ratio, etc. Some new technologies are applied in the pretreatment of vegetables before dehydration, such as microwave, ultrasonic, or air impingement blanching technology. The application of these new technologies can change the vegetables' tissue structure and moisture state, which causes the moisture to dehydrate easily, and improves the products' quality (Kostaropoulos and Saravacos 1995; Mothibe et al. 2011). Comparing the effects of heat treatment and ultrasonic treatment on the Vitamin C (VC) content and sensory quality of dehydrated carrot, results show that ultrasonic treatment had better sensory quality (Gamboa-Santos et al. 2013).

2. *Chemical pretreatment*

With the addition of some chemical reagents, the vegetables' color and rehydration ability could be maintained after drying. Some typical chemical pretreatment methods are as follows:

a. *Sulfur treatment*

By pretreating with sulfite solution or sulfur dioxide, enzymatic browning and nonenzymatic browning can be inhibited, and microbial and insect damage can also be minimized (Ong and Liu 2010).

b. *Metal ion treatment*

With the addition of metal ions, some metal complexes will be formed in the cells, which is helpful to protect the vegetables' color. For example, cupric and zinc ions have been successfully used for protecting the green color in beans and balsam pears (Canjura et al. 1999).

c. *Salt and alkali solution treatment*

The vegetables are soaked in salt or alkali solution before dehydration, which not only reduces the dissolved oxygen and inhibits the enzymatic browning but also improves the structure of the rehydrated vegetable product.

In addition, edible alcohols and sugars such as glucose, sucrose, and glycerol can also be added to the vegetables before drying to improve the structure and flavor of the dried product (Gong et al. 2009).

1.5.2 Effect of Water State

Because of the interaction between water and dry matter in vegetables, the physical and chemical properties of the materials always vary during the AD process. The water in vegetables can be divided as free water (FW), immobilized water (IW), and bond water (BW) (Trout 1988). FW is the main state in fresh vegetables. It is always distributed in cell capillaries with good fluidity; thus, it can easily be evaporated during the AD process. IW is adsorbed by dry matter, and it cannot flow freely. Compared with FW, IW is much more stable, and it can be dehydrated until the FW decreases to a certain level. BW is tightly combined with organic molecules, and it cannot be dried under the low temperature conditions of AD. Generally, the water is difficult to be removed at the final AD stage, which is mainly caused by two factors. On one hand, the water from the internal part of the vegetable materials is difficult to move to the surface, on the other hand, the IW and BW are hard to dehydrate. Recently, some new methods have been used for water state analysis, such as infrared spectroscopy, low-field nuclear magnetic resonance (NMR), and magnetic resonance imaging (MRI), which could be useful in the analysis of drying rate and drying efficiency in an AD process.

In NMR, through magnetic field, some nuclear magnetic energy levels of hydrogen protons can be formed when pulse electromagnetic waves are fed in the perpendicular direction of the magnetic field; thus, some of the hydrogen protons in the low-energy state will jump to a high-energy state, and this is when resonance happens (Chen et al. 2006). If these pulse electromagnetic waves stopped, the hydrogen proton would go back to the original low-energy state. The time of the return process is accepted as relaxation time. When the energy of an activated hydrogen proton is transmitted to the environment, the time cost in the process is called longitudinal relaxation time, T1. At a certain distance, two activated hydrogen protons with the same frequency but different movements tend to change their directions to the same level, and the time taken for this process is called as transverse relaxation time, T2. T2 is much more sensitive than T1; therefore, T2 is used to assess the moisture state. When the combined force between water and dry matter is weak, T2 is long. Conversely, when the combined force between water and dry matter is strong, T2 is short. Therefore, the water state of FW, IW, and BW can be identified from NMR experiments (Zhang et al. 2012).

At present, there are many reports about moisture analysis with NMR in the field of AD (Clark et al. 1997). NMR and MRI were used to research the moisture transfer characteristics in carrots' AD process by Xu et al., and it was found that the moisture profile was irregularly contracted to the central direction, when a multidimensional, unsteady mass transfer process was shown (Xu et al. 2013).

MRI is an extension of NMR. It can provide spatial information about the distribution of protons in the material space and is applied to reflect the uniformity of moisture's spatial distribution of the vegetable materials in each stage of an AD process. Figure 1.9 shows the MRI of ginger slices at different MD stages (Lv et al. 2015b), which indicates that the moisture is distributed much less in dark zones, and at the final stage the materials were dried completely.

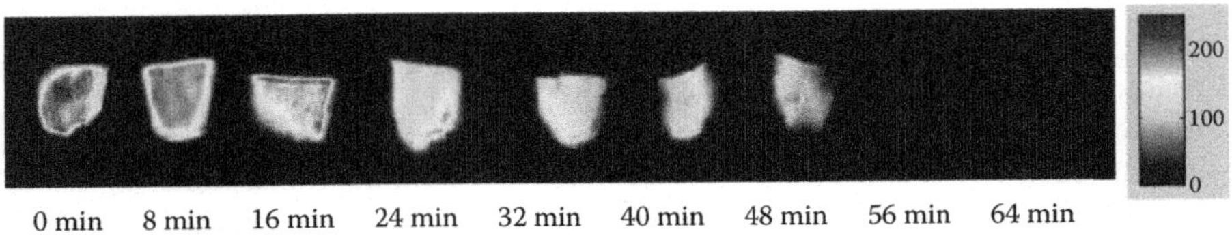

FIGURE 1.9 Moisture's distribution in the microwave drying process.

1.5.3 Effect of Airflow

In vegetables' AD process, the moisture diffusion process contains both liquid and gas migration. With high moisture content, the moisture migrates in liquid, and with low moisture content, the rest moisture can evaporate in the internal part and migrates to the environment in gas. If the airflow temperature was low and moving speed was slow, and the vegetable materials with high moisture content stayed for a long time in the warm temperature, some biological changes occur and the drying quality is likely to turn bad. However, if the drying rate is too high under a very hot and quick airflow, crusts always form on the surface layer of the vegetables. To minimize the loss of nutrients, controlling the drying speed in a high level is effective, but owing to the equilibrium theory of internal and external diffusion, a hardening and cracking phenomenon cannot be avoided. Therefore, the temperature of airflow and vegetables should be monitored in the AD process. A thermal infrared imager can collect the temperature distribution information visually (Liu et al. 2015). According to the data provided by the thermal infrared imager and the sensory quality of the materials, the thickness of the materials, the hot air temperature, wind speed, and other parameters should be adjusted accordingly to improve the drying quality.

1.5.4 Effect of Vegetable Varieties

The physical structure and chemical composition of vegetable varieties are different; therefore, the drying temperature and airflow speed should be adjusted accordingly. In addition, if the vegetables are cut into dices, slices, and strips, to increase the surface area, the drying rate can also be increased. In AD equipment, the mass of vegetable materials per unit area of the drying tray should also be controlled for the best drying process.

1.6 CONCLUDING REMARKS

As a main vegetable drying method, AD is a mature technology with easily available equipment and widely acceptable dried products. In addition, the AD process has the advantage in drying materials with high moisture and sugar content. At present, the most commonly used AD vegetable dryers include box-type dryer, multilayer dryer, and mesh belt dryer. In order to improve the drying efficiency and product quality, some other techniques such as microwave, infrared, RF, etc. have been combined with AD for drying of vegetables. New emerging detection and monitoring technologies, including NMR and MRI systems and thermal infrared imagers, have been used in monitoring the quality changes in vegetables during the drying process.

Effective pretreatment is also critical to improve the vegetable product drying quality, where blanching and hot steam treatment can inhibit the enzymatic browning, and chemical pretreatment such as sulfur treatment, metal ion treatment, and salt and alkali solution treatment can protect the chlorophyll content and improve the product sensory quality.

In the future, AD processes will be widely used in high-efficiency physical fields for drying vegetables, and the drying process will be monitored by a smart system. At the same time, good pretreatment and storage methods need to be developed when the AD vegetable industry is upgraded.

REFERENCES

Afzal, T and Abe, T. Simulation of moisture changes in barley during far infrared radiation drying. *Computers and Electronics in Agriculture* 26(2), 137–145 (2000).

Beveridge, T and Weintraub, S. Effect of blanching pretreatment on color and texture of apple slices at various water activities. *Food Research International* 28(1), 83–86 (1995).

Bondaruk, J, Markowski, M and Błaszczak, W. Effect of drying conditions on the quality of vacuum-microwave dried potato cubes. *Journal of Food Engineering* 81(2), 306–312 (2007).

Boonnattakorn, R, Phoungchandang, S, Leenanoon, B, Khajarern, S and Khajarern, J. Comparative study of garlic powder processing by heated air and dehumidifier heat pump drier. *Food Journal (Thailand)* 35(1), 35–38 (2004).

Canjura, F L, Watkins, R H and Schwartz, S J. Color improvement and metallo-chlorophyll complexes in continuous flow aseptically processed peas. *Journal of Food Science* 64(6), 987–990 (1999).

Chen, W J, Lin, X Y, Ruan, R S, He, C Y, Zhu, R B and Liu, Y H. Study on quickly and non-destructive estimate the moisture content of food using NMR. *Food Research and Development* 27(4), 125–127 (2006).

Cheng, L S, Fang, S and Ruan, M L. Influence of blanching pretreatment on the drying characteristics of cherry tomato and mathematical modeling. *International Journal of Food Engineering* 11(2), 265–274 (2015).

Chin, S K and Law, C L. Product quality and drying characteristics of intermittent heat pump drying of *Ganoderma tsugae* murrill. *Drying Technology* 28(12), 1457–1465 (2010).

Chua, K J, Mujumdar, A S, Hawlader, M N A, Chou, S K and Ho, J C. Batch drying of banana pieces—Effect of stepwise change in drying air temperature on drying kinetics and product colour. *Food Research International* 34(8), 721–731 (2001).

Clark, C, Hockings, P, Joyce, D and Mazucco, R. Application of magnetic resonance imaging to pre- and post-harvest studies of fruits and vegetables. *Postharvest Biology and Technology* 11(1), 1–21 (1997).

Clark, D E and Sutton, W H. Microwave processing of materials. *Annual Review of Materials Science* 26(1), 299–331 (1996).

Demiray, E, Tulek, Y and Yilmaz, Y. Degradation kinetics of lycopene, β-carotene and ascorbic acid in tomatoes during hot air drying. *LWT-Food Science and Technology* 50(1), 172–176 (2013).

Ding, S, An, K, Zhao, C, Li, Y, Guo, Y and Wang, Z. Effect of drying methods on volatiles of Chinese ginger (*Zingiber officinale* Roscoe). *Food and Bioproducts Processing* 90(3), 515–524 (2012).

Drouzas, A and Schubert, H. Microwave application in vacuum drying of fruits. *Journal of Food Engineering* 28(2), 203–209 (1996).

Gamboa-Santos, J, Soria, A C, Pérez-Mateos, M, Carrasco, J A, Montilla, A and Villamiel, M. Vitamin C content and sensorial properties of dehydrated carrots blanched conventionally or by ultrasound. *Food Chemistry* 136(2), 782–788 (2013).

Gong, Z, Gao, L, An, J, Zhang, M, Mujumdar, A and Sun, J. Effects of predrying and vacuum impregnation with nano-calcium carbonate solution on strawberries, carrots, corn, and blueberries. *Drying Technology* 28(1), 36–41 (2009).

Han, Q H, Xie, S J, Li, S J, Ma, J W, Yin, Q and Wang, Y. Multiple-sources microwave combining with hot-air fluidized drying test device. *Transactions of the Chinese Society for Agricultural Machinery* 2, 35 (2014).

Han, Q H, Yin, L J, Li, S J, Yang, B N and Ma, J W. Optimization of process parameters for microwave vacuum drying of apple slices using response surface method. *Drying Technology* 28(4), 523–532 (2010).

Hashim, N, Daniel, O and Rahaman, E. A preliminary study: Kinetic model of drying process of pumpkins (*Cucurbita moschata*) in a convective hot air dryer. *Agriculture and Agricultural Science Procedia* 2, 345–352 (2014).

Kostaropoulos, A and Saravacos, G. Microwave pre-treatment for sun-dried raisins. *Journal of Food Science* 60(2), 344–347 (1995).

Liu, X D, Xu, H D, Sun, T K, Zhang, S J and Huang, X Q. Optimization of microwave-hot air drying of Chinese jujubes. *Food Science* 10, 022 (2013).

Liu, Z, Zhang, M and Wang, Y. Drying of restructured chips made from the old stalks of *Asparagus officinalis*: Impact of different drying methods. *Journal of the Science of Food and Agriculture* 96(8), 2815–2824 (2015).

Lv, W. Analysis on the drying process and quality of ginger slices in microwave fluidized drying, PhD thesis. China Agricultural University (2015).

Lv, W, Han, Q, Li, S, Zhang, X, Xu, T and Sun, L. Thermal dynamic and physical qualities of ginger (*Zingiber officinale*) slices in intermittent microwave fluidized drying. *International Agricultural Engineering Journal* 24(1), 39–46 (2015a).

Lv, W, Han, Q, Li, S, Zhou, H, Zhao, D and Zhu, W. Analysis of microwave drying's model and water removing mechanism about ginger slices. *Transactions of the Chinese Society for Agricultural Machinery* 46(4), 233–237 (2015b).

Marra, F, Zhang, L and Lyng, J G. Radio frequency treatment of foods: Review of recent advances. *Journal of Food Engineering* 91(4), 497–508 (2009).

Mothibe, K J, Zhang, M, Nsor-atindana, J and Wang, Y C. Use of ultrasound pretreatment in drying of fruits: Drying rates, quality attributes, and shelf life extension. *Drying Technology* 29(14), 1611–1621 (2011).

Ong, P K and Liu, S Q. Flavor and sensory characteristics of vegetables. In Nirmal K S. (eds) *Handbook of Vegetables and Vegetable Processing*, Chapter 3. Hoboken, NJ: Blackwell Publishing Ltd. 59–82 (2010).

Prabhanjan, D, Ramaswamy, H and Raghavan, G. Microwave-assisted convective air drying of thin layer carrots. *Journal of Food Engineering* 25(2), 283–293 (1995).

Praveen Kumar, D G, Umesh Hebbar, H, Sukumar, D and Ramesh, M. Infrared and hot-air drying of onions. *Journal of Food Processing and Preservation* 29(2), 132–150 (2005).

Qin, S, Wen, X, Shen, T and Xiang, L. Thin layer drying characteristics and quality evaluation of steam blanched chrysanthemum. *Transactions of the Chinese Society of Agricultural Engineering* 27(6), 357–364 (2011).

Rahman, M S. A theoretical model to predict the formation of pores in foods during drying. *International Journal of Food Properties* 6(1), 61–72 (2003).

Reis, F R, Masson, M L and Waszczynskyj, N. Influence of a blanching pretreatment on color, oil uptake and water activity of potato sticks, and its optimization. *Journal of Food Process Engineering* 31(6), 833–852 (2008).

Reyes, A, Cerón, S, Zúñiga, R and Moyano, P. A comparative study of microwave-assisted air drying of potato slices. *Biosystems Engineering* 98(3), 310–318 (2007).

Riva, M, Schiraldi, A and Di Cesare, L. Drying of *Agaricus bisporus* mushrooms by microwave-hot air combination. *Lebensmittel-Wissenschaft und Technologie* 24(6), 479–483 (1991).

Sagar, V and Kumar, P S. Recent advances in drying and dehydration of fruits and vegetables: A review. *Journal of Food Science and Technology* 47(1), 15–26 (2010).

Sharma, G P and Prasad, S. Drying of garlic (*Allium sativum*) cloves by microwave–hot air combination. *Journal of Food Engineering* 50(2), 99–105 (2001).

Srikiatden, J and Roberts, J S. Measuring moisture diffusivity of potato and carrot (core and cortex) during convective hot air and isothermal drying. *Journal of Food Engineering* 74(1), 143–152 (2006).

Trout, G R. Techniques for measuring water-binding capacity in muscle foods—A review of methodology. *Meat Science* 23(4), 235–252 (1988).

Wang, L P. The function and processing research on barley green tablets and barley green-mushroom tablets, Master thesis. Jiangnan University (2014).

Wang, H C, Zhang, M and Adhikari, B. Drying of shiitake mushroom by combining freeze-drying and mid-infrared radiation. *Food and Bioproducts Processing* 94, 507–517 (2015).

Wang, Y, Zhang, L, Johnson, J, Gao, M, Tang, J, Powers, J R and Wang, S. Developing hot air-assisted radio frequency drying for in-shell macadamia nuts. *Food and Bioprocess Technology* 7(1), 278–288 (2014).

Wu, Z H, Li, W, Lijuan, Z, Jianfang, S and Qing, L. Drying characteristics and product quality of *Lycium barbarum* under stages-varying temperatures drying process. *Transactions of the Chinese Society of Agricultural Engineering* 31(11), 287–293 (2015).

Xu, J, Xu, G, Zhang, X, Gu, Z, Zhang, S and Li, H. Moisture transport in carrot during hot air drying using magnetic resonance imaging. *Transactions of the Chinese Society of Agricultural Engineering* 29(12), 271–276 (2013).

Yang, Z, Zhu, E, Zhu, Z, Wang, J and Li, S. A comparative study on intermittent heat pump drying process of Chinese cabbage (*Brassica campestris* L. spp) seeds. *Food and Bioproducts Processing* 91(4), 381–388 (2013).

Yudthavorasit, S, Wongravee, K and Leepipatpiboon, N. Characteristic fingerprint based on gingerol derivative analysis for discrimination of ginger (*Zingiber officinale*) according to geographical origin using HPLC-DAD combined with chemometrics. *Food Chemistry* 158, 101–111 (2014).

Zhang, M and Chen, G. Research progress on the energy reduce of open-type air drying process of vegetables. *Drying Technology and Equipment* 6, 003 (2007).

Zhang, X, Zhu, S, Huang, J, Xu, G, Xu, J and Li, H. Analysis on internal moisture changes of carrot slices during drying process using low-field NMR. *Transactions of the Chinese Society of Agricultural Engineering* 28(22), 282–287 (2012).

Zhao, D, An, K, Ding, S, Liu, L, Xu, Z and Wang, Z. Two-stage intermittent microwave coupled with hot-air drying of carrot slices: Drying kinetics and physical quality. *Food and Bioprocess Technology* 7(8), 2308–2318 (2014).

2 Main Current Vegetable Drying Technology II

Freeze-Drying and Related Combined Drying

Xu Duan
Henan University of Science and Technology

CONTENTS

2.1 INTRODUCTION

Drying alters the characteristics of vegetable products and leads to loss of juiciness and lignified tissue, both of which result in a harder or chewier product. Other typical quality-loss attributes include case hardening, wherein the outer layers of the product are overdried in the process of removing moisture from the less accessible inner core of the product along with product shrinkage, both of which are issues associated with convective air-drying. As a result, the present demand for high-quality dehydrated vegetable products in the market requires dried vegetables to be maintained at a very high level to preserve the nutritional and organoleptical properties of the initial fresh product. Moreover, new technologies for drying of new products, higher capacities, better quality and quality control, reduced environmental impact, higher energy efficiency and lower cost, and safer operation are needed. Freeze-drying (FD) produces the highest quality food product obtained by any drying method. Despite unmatched advantages, FD has always been considered as the most expensive operation for manufacturing a dehydrated product owing to high energy consumption and high operation and maintenance costs.

The analysis of energy requirements for the conventional drying methods and FD have shown that the basic energy required to remove 1 kg of water is almost double for FD than for the conventional drying.

The FD process involves the following three stages: (1) the freezing stage, (2) the primary drying stage, and (3) the secondary drying stage. In the freezing stage, the foodstuff to be processed is cooled down to a temperature at which all the materials are in a frozen state. In the primary drying stage, the frozen solvent is removed by sublimation; this requires the pressure of the system (freeze dryer) at which the product is dried to be less than or near to the equilibrium vapor pressure of the frozen solvent. Because the water usually exists in a combined state, the material must be cooled below 0°C to keep the water in the frozen state. For this reason, during the primary drying stage, the temperature of the frozen layer is usually maintained at −10°C or lower at absolute pressures of about 2 mmHg or less.[1] As the ice sublimes, the sublimation interface, initiated at the outside surface (see Figure 2.1) recedes, and a porous shell of dried material remains. The latent heat of sublimation can be conducted through the layer of dried material and through the frozen layer. The vapor is transported through the porous layer of dried material. During the primary drying stage, some of the sorbed water (nonfrozen water) in the dried layer may be desorbed. The desorption process in the dried layer could affect the amount of heat that appears at the sublimation interface, and therefore it could affect the velocity of the moving sublimation front (interface). The time at which there is no more frozen layer (i.e., there is no more sublimation interface) is taken to represent the end of the primary drying stage. The secondary drying stage involves the removal of water that remains unfrozen (this is termed sorbed or bound water). The secondary drying stage starts at the end of the primary drying stage, and the desorbed water vapor is transported through the pores of the material that is dried.[2]

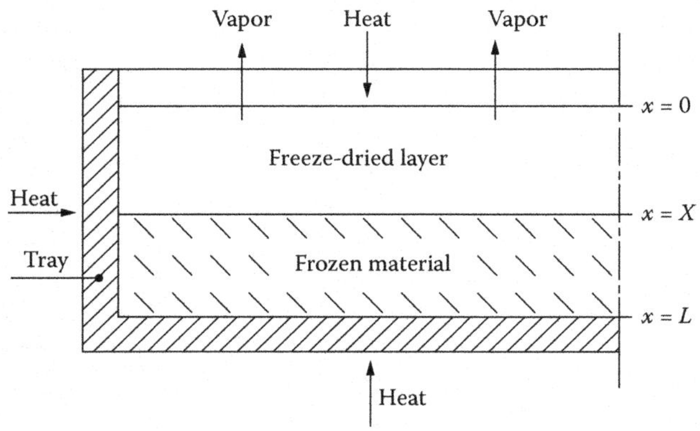

FIGURE 2.1 Diagram of a material on a tray during freeze-drying. The variable x denotes the position of the sublimation interface (front) between the freeze-dried layer and the frozen material.

2.2 PROCESS PARAMETERS AFFECTING THE PERFORMANCE AND PRODUCT QUALITY

Among the main parameters affecting the quality of freeze-dried products are process temperatures. For example, increasing the heating plate temperature certainly reduces costs associated with energy consumption during FD, but this could lead to product deterioration. Volume reduction due to FD is minimal if operating pressures and temperatures are appropriate. However, if the temperature is not appropriate, collapse may occur, leading to the sealing of capillaries, which in turn leads to less dehydration and puffiness. Thus, in the FD process, drying temperatures seem to have a significant impact on final product quality.[3]

Most food materials have a eutectic form. If the lowest eutectic temperature of a material is exceeded during the primary drying stage of FD, then melting in the frozen layer (Figure 2.1) may occur. The melting at the sublimation interface, or any melting that occurs in the frozen layer, can cause gross material faults such as puffing, shrinking, and structural topologies filled with liquid solution.

The maximum allowable temperature in the frozen layer is determined by both structural stability and product stability (e.g., product bioactivity) factors; that is, the maximum value of the temperature in the frozen layer during the primary drying stage must be such that the drying process is conducted without loss of product property (e.g., bioactivity) and structural stability. For structural stability, the same phenomena, as in the case of the primary drying stage, have to be considered in the second drying stage: Collapse, melting, or dissolution of the solid matrix can occur. Because many products are temperature-sensitive, it is necessary to control product stability by limiting the value of the temperature during the secondary drying process, and then the final moisture content is checked before the end of the cycle. In the secondary drying stage, the bound water is removed by heating the product under vacuum. The following product temperatures are usually employed: (1) between 10°C and 35°C for heat-sensitive products and (2) 50°C or more for less-heat-sensitive products.[4]

Besides this traditional temperature control strategy, some reports also introduced the glass transition concept into the FD temperature control. The change from the glassy to the rubbery state occurs as a second-order phase transition at a temperature known as the glass transition temperature (T_g). The breakdown of structure above a "collapse" temperature during FD is considered to be a major contributor in deteriorating the quality of product. Other related processes leading to quality loss can happen in several amorphous systems. It has also been reported that when temperature of some processes exceeds T_g, the quality of foodstuffs is extremely altered.[5–8]

In effect, above T_g, the viscosity drops considerably to a level that facilitates deformation. To better understand the collapse phenomena, the progression of the product temperature during FD could be compared to its glass transition temperature at comparable moisture contents (as shown in Figure 2.2) for the case of strawberries. Throughout the process of air-drying, the product temperature is above T_g, which demonstrates that collapse and poor quality occurs while air-drying. On the other hand, product FD temperatures are below T_g only at the end of the process.[9] Nevertheless, comparing temperatures at the average moisture content of the product

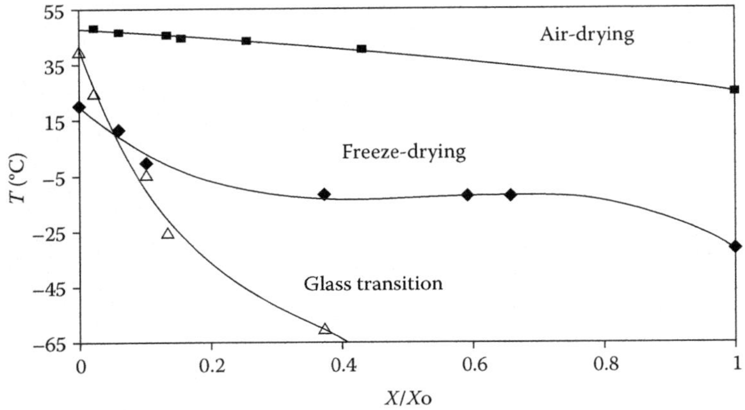

FIGURE 2.2 Progression of temperatures during drying (at 50°C) and freeze-drying (at 20°C) of strawberries as compared to its glass transition temperature. (From Ratti, C. *J. Food Eng.*, 49(4), 311–319, 2001. With permission.)

during FD would lead to a misunderstanding of the problem because, in this particular process, the solid is divided into dry and frozen regions separated by a receding front. The portion of the solid that is in contact with the highest temperatures for long periods of time is thus the amorphous dry matrix, which has a low moisture content and T_g corresponding closely to that of the dry solids. The previous explanation and the fact that collapse in a time–temperature phenomenon suggests that the glass transition temperature of dry solids (T_{gs}) would be a crucial optimization parameter for the FD process. This parameter may also serve as a useful tool for the choice of the most appropriate materials to be freeze dried. Caparino et al.[10] reported that a higher T_{gs} for a freeze-dried tomato when compared to an air-dried tomato is attributed to the structural differences between freeze-dried and air-dried products.

The glass transition temperature of biological materials is strongly plasticized by water. This behavior can be shown in a glass curve or in a state diagram where phase transition temperatures (glass transition and melting) are plotted against moisture content. Figure 2.3 presents the state diagram obtained for freeze-dried pineapple.[11] For water activities lower than 0.90, the plasticizing effect of water on the glass transition temperature was evident, with a great reduction in T_g caused by increasing moisture content. Nevertheless, at water activities higher than 0.90, the glass transition curve exhibited a discontinuity, with a sudden increase in glass transition temperatures that approached a constant value. This would configure a "practical glass curve," and the constant value of T_g obtained at high moisture contents would be the glass transition of the maximally freeze-concentrated amorphous matrix, T_g', represented in Figure 2.3 by the horizontal arrow at −51.6°C. The state diagram plotted in Figure 2.3 also shows the onset ice melting temperature (T_m) from samples that exhibited water freezing. As expected, melting temperature increased with increasing water content, and the intersection of the melting and glass transition curves occurred at T_g'.

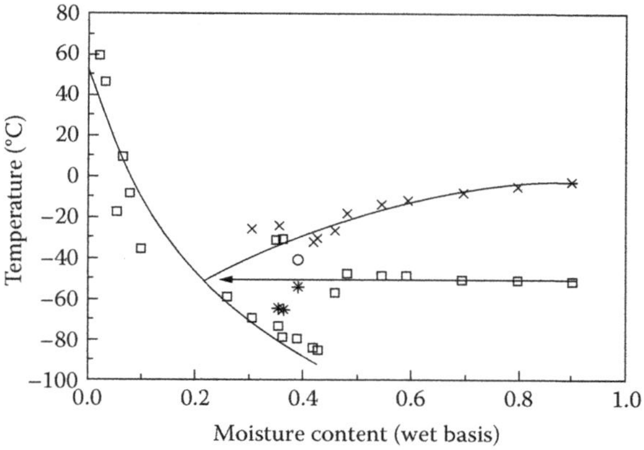

FIGURE 2.3 State diagram of freeze-dried pineapple (without annealing: □, T_g and ×, T_m; with annealing: *, T_g and ○, T_m). (From Telis, V.R.N., Sobral, P.J.A., *LWT-Food Sci. Technol.*, 34(4), 199–205, 2001. With permission.)

As a result, at the beginning of drying, when product presents high moisture content, the difference between sample temperature and T_g' is high, resulting in a higher rate of shrinkage. As the drying process continues, the difference between sample temperature and T_g decreases, rapidly reducing the rate of volume change.[11] At low moisture content, T_g increases, allowing the material to pass from rubbery to glassy state: The dried material became more rigid, decreasing significantly the extension of shrinkage. Therefore, glass transition temperature of the freeze-dried product could also be used as an optimization parameter for determining the optimal heating plate temperature of a particular food product.

2.3 COMBINED TECHNOLOGIES TO IMPROVE ENERGY EFFICIENCY

It has been reported that the energy spent in the FD process is significant when dealing with low-value raw materials. FD should, therefore, be regarded as a preservation process for high-value materials. Especially, as compared to air-drying, FD costs are four to eight times higher.

The FD process consists of four main operations: freezing, vacuum, sublimation, and condensing. Sublimation takes almost half of the total energy of the process, and vacuum and condensation shares are practically equal (25%).[9] Therefore, it can be concluded that improvements in classical vacuum freeze-drying should be addressed to assist sublimation by the improvement of heat transfer, to reduce the vacuum requirements by shortening drying times, and to avoid condensers.

To reduce energy consumption in FD, the optimization process, combined drying methods, and some assisted treatments were studied during past decades. Boss et al.[12] developed a dynamic mathematical model that well-represents the process of FD. The FD process was optimized to achieve lower energy consumption, and thus good product quality was also obtained. There are also other optimization methods such as quality-by-design, response surface method, and time-scale modeling.

2.3.1 Ultrasound-Assisted Freeze-Drying

Schössler et al.[13] developed a contact ultrasound system to improve FD efficiency (Figure 2.4). In this system, temperature generation by sonication can be kept under control by the FD process or by intermittent application of power ultrasound. Studies dealing with ultrasound-assisted drying have shown that water removal due to ultrasonic waves can be carried out at lower temperatures and in less time. Consequently, ultrasound-assisted FD is a technique that can prove beneficial for both, the temperature-reduced drying of sensitive foodstuff as well as for shortened FD processes with the potential to reduce processing costs. Pisano et al.[14] also confirmed that power ultrasound could increase the rate of water removal based on mechanical actions on both gas–solid interfaces and product structure during the FD process. In addition, the application of high-intensity ultrasonic energy can induce cavitation, which enhances the removal of water that is strongly linked to the solid product.

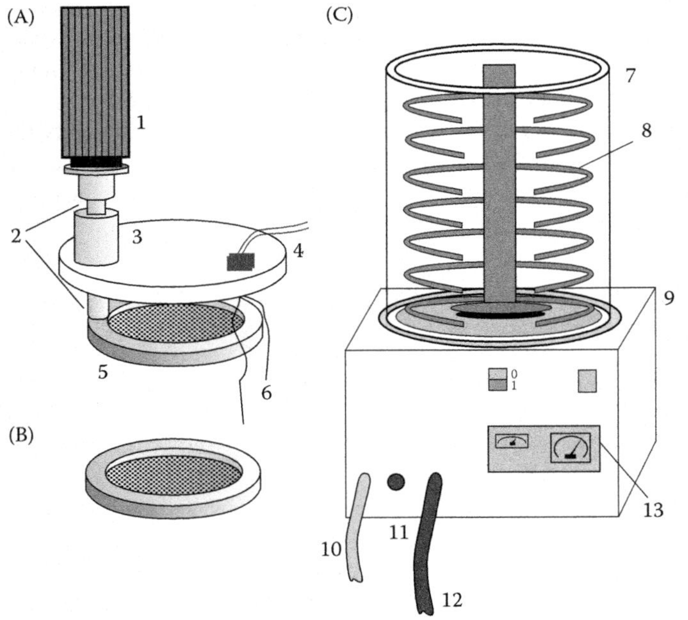

FIGURE 2.4 Laboratory scale system for contact ultrasound-assisted freeze-drying. (A) Acrylic lid with ultrasonic system, (B) screen for control samples, (C) freeze-dryer. 1, Ultrasound processor UIP1000; 2, Sonotrodes BS2d34; 3, vibration-free flange; 4, acrylic lid; 5, ultrasound drying screen; 6, thermocouples; 7, drying chamber; 8, rack; 9, freeze-dryer; 10, water outlet; 11, vacuum regulation; 12, vacuum tube; 13, display. (From Schössler, K. et al., *Innovat. Food Sci. Emerg. Technol.*, 16, 113–120, 2012. With permission.)

Unfortunately, so many attempts to modify and optimize the whole drying process of FD still cannot significantly reduce the energy consumption. Some researchers also tried to develop combined drying based on FD (FD combined with hot air, microwave, far-infrared radiation, and ultrasound systems, for instance). Although energy consumption could be significantly decreased, the product quality deteriorates because most of the water was not removed by sublimation. Moreover, these combined dryings need to be carried out in different dryers, resulting in an inconvenience in production. As a result, alteration of heating mode or drying pressure of FD should be a better choice for the probable decrease in cost.

2.3.2 Microwave Freeze-Drying

A simple microwave FD apparatus is shown in Figure 2.5. An independent polypropylene drying cavity was set up in a rectangle resonant cavity, which could effectively avoid the corona discharge at the vacuum condition. The pressure of the drying cavity was set at 10 Pa to 30 kPa (absolute pressure). The power of the microwave could be adjusted continually. The core temperature of materials was detected by the optic fiber sensor. The surface temperature of materials was detected by infrared thermometer. To avoid nonuniform distribution of the microwave field, three magnetrons were placed at different angles. In fact, all the drying process takes place under vacuum environment by sublimation; so the quality of microwave freeze drying (MFD) product does not differ from that of conventional FD product. The difficulties found in heat transmission in FD disappear with microwave (MW) heating. In the MFD systems, energy is directly absorbed by the water molecules for sublimation within the food material, without being affected by the dry zone. As a result, the microwave heating offers a good opportunity to increase the drying rate

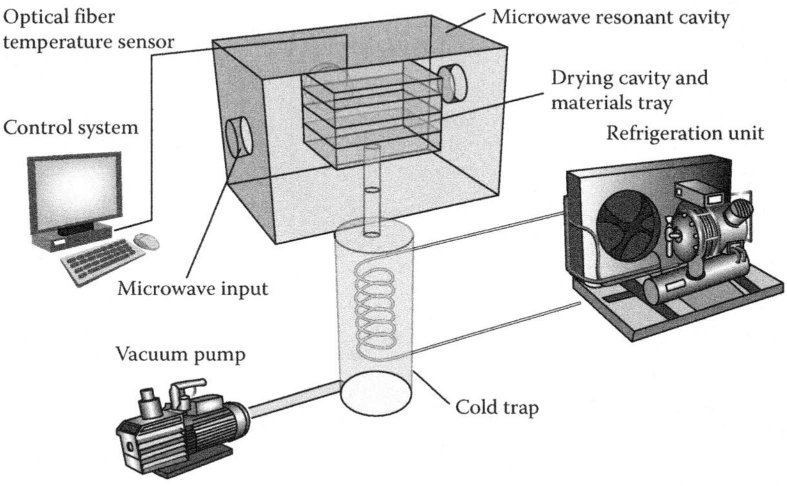

FIGURE 2.5 Schematic diagram of microwave freeze-dryer. (From Duan, X. et al., *Int. J. Agric. Biol. Eng.*, 8(1), 91–97, 2015.)

in FD. Experiments and numerical predictions all showed that the drying rate was significantly increased, and the drying cost reduced with microwave heating.

Duan et al.[15] reported that the MFD process could be divided into two main phases: the primary drying stage and the secondary drying stage. Unlike traditional FD, in the MFD process, the primary drying stage was relatively short and the temperature rose quickly. In the secondary drying stage, the temperature rose more quickly. This is a characteristic of microwave heating. Some research results showed that MFD can reduce the drying time by 40% as compared to FD and can provide similar quality. Apart from accelerating the drying rate, some researches show that the MFD process can cause a reduction in the microbial content of the product.

Duan et al.[16] developed a control strategy of MFD based on the dielectric property of materials when they investigated MFD apple slices. The drying process had a clear effect on the critical discharge microwave power (Figure 2.6). It was found that corona discharge readily took place at the initial and end stage. During the middle stage of drying, i.e., from moisture content of 70%–30%, the critical power was relatively high. As a result, microwave power should be carefully controlled at the initial and end stage to avoid corona discharge. This phenomenon can be explained because the two stages have much lower loss factor, leading to lower microwave energy dissipation. On the basis of the earlier studies, generally, higher moisture content in the samples leads to higher loss factor. However, the sample temperature curve showed that the temperature of samples at the initial stage was below −15°C, which led to low loss factor. Therefore, Figure 2.6 implied that corona discharge took place readily when the loss factor of samples was low. As a result, microwave power should be carefully controlled according to the changes of dielectric property of apple during MFD. Based on the dielectric property of samples, a changed microwave-loading scheme can lead to perfect product quality and can greatly reduce drying time.

Ren et al.[17] investigated the influence of different microwave-loading programs on the process of MFD of button mushrooms. They demonstrated that the difference

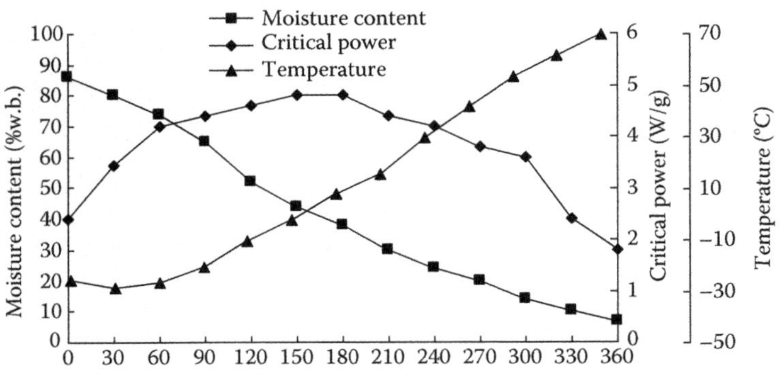

FIGURE 2.6 Variation of critical discharge microwave power during the MFD process. (From Duan, X. et al., *Drying Technol.*, 30(5), 535–541, 2012. With permission.)

between the temperature of the samples and T_g governs the rate of the mobility matrix. To reduce the difference between product temperature and T_g, after the moisture content was below 50% w.b., a step-down microwave power-loading scheme was applied. The step-down microwave-loading scheme based on the glass transition temperature significantly improved the product quality and did not increase the drying time.

Figure 2.7 provides the temperature and moisture content profiles of the samples under different microwave-loading schemes. The results show that microwave power has a significant effect on the drying time, and higher microwave power could obviously reduce the drying time. It was shown that MFD had two drying stages (sublimation and desorption), but the sublimation stage of MFD was short and the temperature of samples rose fast. Besides, there was no obvious boundary between the sublimation and desorption stages. As shown in Figure 2.7, in the sublimation stage, different microwave power had an insignificant effect on the temperature of the samples, but the water loss rate of the materials exhibited an obvious change in variety under different microwave power-loading programs. The possible reason was that most of the free water sublimated in this phase, and a great deal of frozen water in mushrooms absorbed microwave energy. As a result, relatively high microwave power could be applied in this phase to improve drying rate. From Figure 2.7, it was found that when the moisture content dropped below 50% w.b., fixed microwave power scheme (1 and 2 kW) led to rise in temperature rate.

According to the reports of Levi and Karel, the difference between the temperature of samples and T_g governs the rates of mobility matrix. The mobility of the solid matrix is closely related to its physical state: High mobility corresponds

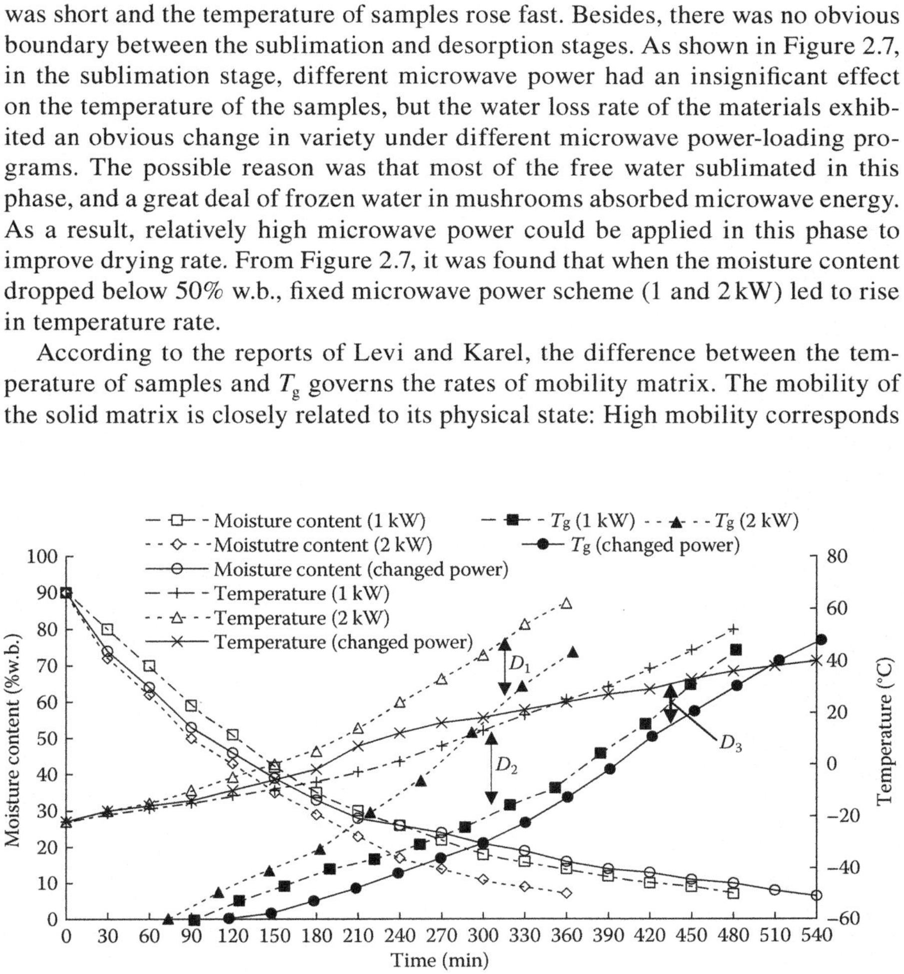

FIGURE 2.7 Temperature and moisture content profiles of the samples under different microwave-loading schemes during the MFD process. (From Ren, G.Y. et al., *Drying Technol.*, 33(2), 169–175, 2015. With permission.)

to viscoelastic behavior typical of a rubbery state, whereas low mobility corresponds to elastic behavior of a glassy state. Figure 2.7 also showed the comparison of the glass transition temperature and product temperature under different microwave power-loading programs throughout the MFD process. As can be seen from Figure 2.7, product temperature was higher than the T_g up to a moisture content of 20% w.b., showing that the material is in the rubbery state during most of the drying process. At the beginning of drying, moisture content was high, and the difference between product temperature and T_g (D_1, D_2, and D_3) was large, resulting in quality deterioration. As the MFD process progressed, the difference decreased, which led to a more stable matrix of the product. At the end stage of MFD, T_g increased to a high level. Therefore, when the microwave power was controlled precisely allowing the material to pass from rubbery to glassy state, the dried material became more rigid, significantly decreasing the extension of shrinkage.

From Figure 2.7, it can be found that T_g of samples gradually rises when the moisture content is dropped. According to the changing tendency of T_g, a changed microwave-loading scheme was conducted. To reduce the difference between product temperature and T_g, after the moisture content was below 50% w.b., a step-down microwave power-loading scheme was applied. As shown in Figure 2.7, glass state of the samples sustained for about 40 min. According to the glass transition theory, long duration in glass state will be helpful for reducing shrinkage and collapse of dried product. As a result, the step-down microwave power loading is useful. On the other hand, too low microwave power means a longer drying time, which would lead to more expensive operation cost.

Lombrana et al.[18] used a microwave on–off cycle strategy with simultaneous up–down modification of pressure as an acceptable power regulation system when magnetrons, whose field strength exceeds the breakdown voltage of gas, were used. In this case, chamber pressure became a convenient control parameter to avoid plasma discharge and subsequent melting of product.

Wang et al.[19] improved drying uniformity of MFD by introducing pneumatic pulse agitation in a laboratory system (Figure 2.8). Results showed that pulse-spouted bed mode resulted in dried stem lettuce slices with lower discoloration, a more uniform and compact microstructure, higher rehydration capacity, as well as greater hardness after rehydration over shorter drying time relative to those obtained in a steady spouting condition.

The MFD process also was optimized by some mathematical models. Wang and Shi developed a sublimation–condensation model based on an unsteady state model. Based on this model, Wu et al. developed a conjugate heat-and-mass transfer model of porous media with dielectric cores. Tao et al. studied a conjugate heat-and-mass transfer model of cylindrical porous media with cylindrical dielectric cores. However, the microwave absorption capability of materials was treated as a constant value, resulting in relatively high error. Witkiewicz and Nastaj developed a one-dimensional MFD model, taking into account temperature dependency of material loss factors and a variety of electric field strengths during the process. As a result, the dielectric property of materials is important for MFD process simulation and control.

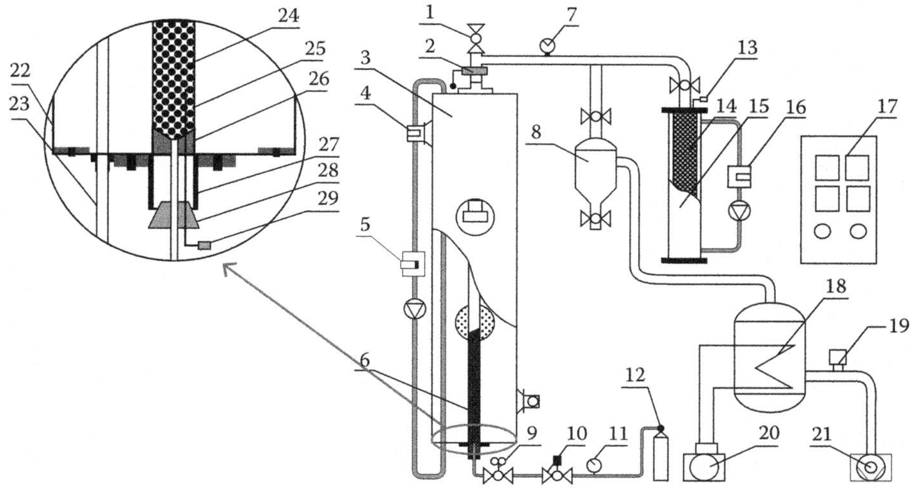

FIGURE 2.8 Schematic diagram of a freeze-drying (FD) system for Microwave-Assisted Pulse-Spouted Bed Freeze-Drying (PSMFD), MFD, and FD. 1, Feeding ball valve; 2, plate valve with 3-mm diameter hole; 3 and 22, microwave heating cavity; 4, magnetron; 5 and 16, circulating water unit; 6, drying chamber for MFD and PSMFD; 7 and 11, pressure gauge; 8, solid–gas separator; 9, gas flow electromagnetic valve; 10, gas flow adjustable valve; 12, nitrogen gas source; 13 and 29, fiber optic temperature sensor; 14 and 24, sample; 15, drying chamber with a jacket for FD; 17, control panel; 18, vapor condenser; 19, vacuum pressure transducer; 20, refrigerator unit; 21, vacuum pump unit; 23, water load pipe; 25, Teflon tube; 26, gas distributer; 27, fixed unit for drying chamber holder; 28, silicon rubber stopper. 4 (×4). (From Wang, Y.C. et al., *Food Bioprocess Technol.*, 6(12), 3530–3543, 2013. With permission.)

2.3.3 Atmospheric Freeze-Drying

The diffusion of water vapor from the drying boundary through the dried shell occurs by vapor pressure gradient, rather than by the absolute pressure on the system. Hence, it is possible to freeze-dry at atmospheric pressure if the partial pressure of water vapor in the drying medium is kept low enough to provide a mass transfer driving force for water vapor transfer from the frozen sample. This drying method can be called atmospheric freeze-drying (AFD). AFD products show similar characteristics of rehydration kinetics and hygroscopicity compared to the vacuum freeze-drying.[20]

The most important characteristic of AFD is convective drying at temperatures below the freezing point of the product. Compared with vacuum freeze-drying in FD, the temperature is higher, typically in the range of –3°C to –10°C. This is because of the physical properties of humid air, as lower air temperature reduces the ability to remove moisture. Furthermore, low air temperature also requires more energy and therefore reduces the specific moisture extraction rate (SMER).

The advantages of the AFD process, in comparison to vacuum freeze-drying, are as follows: (1) low initial investment cost as expensive vacuum auxiliary equipment could be eliminated, (2) the process could be designed as a continuous system with higher productivity and lower operating cost, (3) the application of a heat pump system and different process temperature elevating modes in AFD would decrease energy consumption and drying time, and (4) inert gas drying environment such as nitrogen or helium can be applied to minimize the product degradation caused by oxidation.[20]

Most of the literature deals with AFD in fluid-bed dryers and spray freeze-dryers. Figure 2.9 shows a fluid-bed dryer with heat pump system, which can be used to carry out AFD.

Despite the promises of low energy consumption and better product quality, certain problems still exist in the AFD process, which has limited its practical implementation. In AFD process, drying periods are very long, ice thawing and product shrinkage affect the drying rate, and diffusion of water leads to poor quality of the end product.[21] As a result, the temperature of samples should be controlled to maintain a relatively high drying rate and good product quality without causing melting of ice.

A wet material is constituted by solute and water that interacts and changes progressively during the drying process. Because the freezing point depression resulted in a decrease in moisture content, the drying process has to be designed individually for each product. Therefore, it becomes vital to control the AFD temperature according to the freezing point of the material depression during drying.[22]

Duan et al.[23] developed a drying strategy under a changed air temperature program based on glass transition of apple cubes. Figure 2.10 provides the results for the changed air temperature program experiment. The results show that the overall drying time is reduced to 34 h, which results in a significant improvement in the drying efficiency. From the beginning till 6 h of the drying procedure, moisture content of samples decreased to about 60% (w.b.), and at –5°C air temperature it could be ensured that most of the water was not thawed during high sublimation rate. When the moisture content of the samples reached below 60%, the freeze point decreased considerably; thus, air temperature of –10°C was used until the moisture content dropped to about 35% (w.b.). According to Claussen's report,[20] the freeze point would decrease to about –25°C during this period. Furthermore, the glass transition temperature could be also very low (below –30°C). Figure 2.10 shows that –10°C air temperature can ensure the temperature of the material to be located in a range between –20°C and –12°C, which could lead to some water thawing. Although much lower air temperature could be used to avoid water thawing, extremely low inlet air temperature means a lower evaporation temperature, which would lead to more expensive operation cost.

From Figure 2.10, it can be seen that T_g of samples rises quickly when the moisture content is below 35% (w.b.). According to the changed tendency of T_g, a step-up program was conducted. Under this drying strategy, a similar drying rate was maintained compared with the beginning stage. As a result, the overall drying time was greatly reduced.

Based on experimental tests reported in this chapter, the inlet air temperature has a significant effect on product quality and drying efficiency during the AFD process. Low air temperature can ensure good product quality but cost too long a drying time.

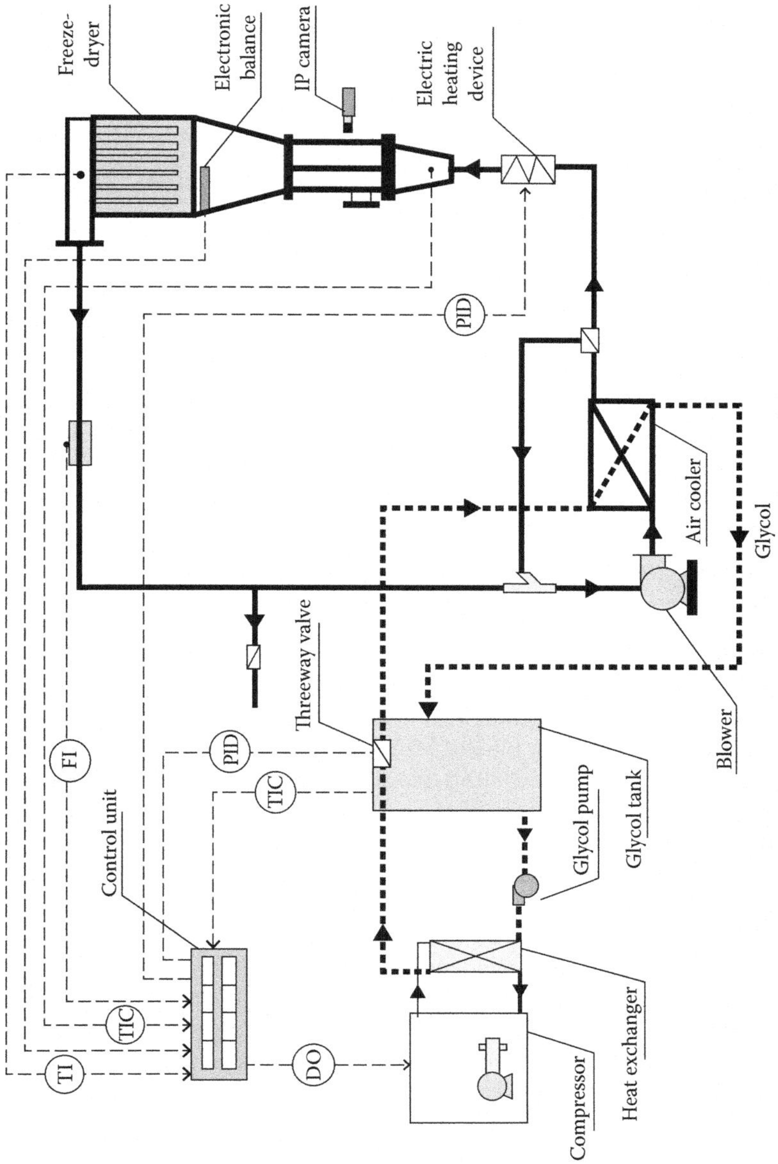

FIGURE 2.9 The heat pump freeze-drying unit and components. (From Li, S. et al., *Drying Technol.*, 25(7–8), 1331–1339, 2007. With permission.)

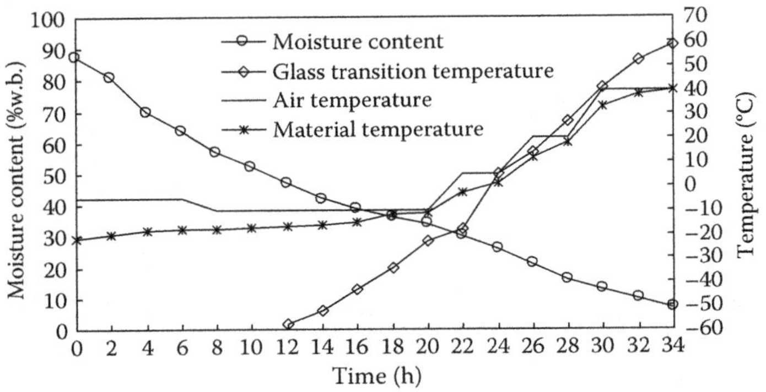

FIGURE 2.10 Effect of changed temperature program based on glass transition of apples on atmospheric freeze-drying process. (From Duan, X. et al., *Bioproducts Process.*, 91(4), 534–538, 2013. With permission.)

After the moisture content of apple decreases below 35% (w.b.), the glass transition temperature increases quickly. As a result, high air temperature could be adopted during the end stage of AFD. A step-up temperature program during the AFD process can reduce the drying time by almost half on the premise of maintaining product quality.

2.3.4 Comparison of the Effect of Different Freeze-Drying Methods on Product Quality

As shown in Figure 2.11, for mushrooms, the drying rate of AFD was the slowest when followed by FD. MFD could lead to a very rapid drying rate. Drying curves shown in Figure 2.11 show that the drying time for the AFD and FD process was 24h and 15h, respectively, whereas the drying time of MFD was 8h. This suggests that microwave application can strengthen heat and mass transfer, resulting in an increase in the drying rate, as expected. This phenomenon is in agreement with many reports about microwave drying methods. The drying curve of AFD showed an evident two-stage tendency because the air temperature was applied at two levels (−10°C and 40°C). The most important method of AFD is convective drying at temperatures below the freezing point of the product. Nevertheless, lower air temperature reduces the ability to remove moisture. As a result, generally, a step-up temperature program is performed in the AFD process in order to improve drying rate. Compared to FD, the temperature is higher. Moreover, a food product undergoing AFD has a freezing point depression owing to reduction in solvent fraction while increasing the solute concentration. As a result, ice thawing occurs more easily in AFD than in traditional FD. Therefore, low air temperature should be applied when the moisture content of the samples is high. To reduce the drying time, air temperature can be increased after most of the water is removed.

Table 2.1 shows the result of rehydration ratio, *L*-value, shrinkage ratio, bulk density, and Vc preservation rate of the products dried by the three different FD methods. It can be seen that the AFD products have lower *L*-value than that of MFD and FD products, which means AFD leads to more browning. The possible reason is that

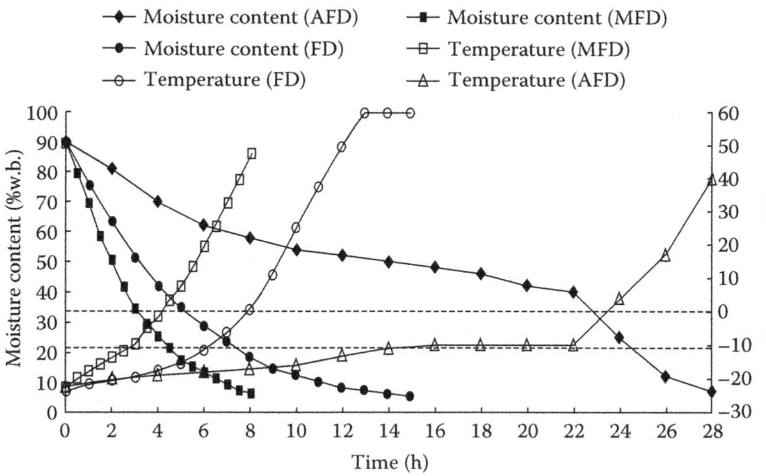

FIGURE 2.11 The profiles of product temperature and moisture content during different freeze-drying processes. (From Duan, X. et al., *Int. J. Agric. Biol. Eng.*, 8(1), 91–97, 2015.)

TABLE 2.1
Effects of Different Freeze-Drying Methods on the Dried Product Quality

Drying	*L*-value	Bulk Density (g·cm^{-3})	RR	SR	Vc Preservation Rate (%)
MFD	60.28 ± 2.42b	0.20 ± 0.02c	4.66 ± 0.24b	0.68 ± 0.24a	72.42 ± 2.21c
AFD	42.46 ± 2.38c	0.38 ± 0.02a	3.42 ± 0.28c	0.51 ± 0.28b	84.64 ± 3.18a
FD	62.26 ± 2.22a	0.22 ± 0.01b	5.38 ± 0.42a	0.69 ± 0.22a	80.68 ± 2.46b

Source: Duan, X. et al., *Int. J. Agric. Biol. Eng.*, 8(1), 91–97, 2015.

AFD is carried out at atmospheric condition resulting in some oxidation reaction. Besides, it can be seen that the bulk density of AFD samples is the highest, followed by that of FD and MFD products, respectively. Microwave application leads to reduced shrinkage and thus provides a more porous structure in a vacuum environment. This can be used to explain that the bulk density of MFD products was lower than that of FD products. The bulk density of AFD products is higher than MFD and FD samples because AFD also leads to a more significant shrinkage than that of FD and MFD (Table 2.1). On the other hand, the vitamin C content of AFD products was higher than that of the FD samples. The possible reason is that most of the AFD process is carried out at low temperature, thereby preventing vitamin C degradation, which is a heat-sensitive component. From Table 2.1, it can be found that the rehydration ratio of AFD samples is the lowest, and FD can lead to the highest rehydration capability. It is well known that the more the water is removed by sublimation, the better the form and structure can be retained during drying. Therefore, it is likely that

AFD treatment leads to more water thawing, which is removed by evaporation rather than sublimation, resulting in the destruction of the porous structure.

Texture is considered as one of the most important criteria concerning consumer acceptance of dehydrated foods. Moreover, it can also reflect the quality of drying products. For example, higher hardness and lower crispness generally imply more shrinkage and deformation. The hardness of AFD samples was the highest, and its crispness was the lowest. The possible reason is that AFD leads to a higher shrinkage, which agrees with the result of the shrinkage test mentioned earlier. Moreover, from the scanning electron microscope (SEM) pictures (Figure 2.12), it can be observed that FD and MFD result in a clear porous structure, which accounts for the low hardness of FD and MFD products. Thus, it can be concluded that shrinkage and porous structure have dominant effects on the texture of dried products.

From the earlier discussion, it can be concluded that atmospheric freeze drying (AFD) leads to a stronger browning reaction, shrinkage, bulk density, and lesser rehydration capability compared with freeze drying (FD) and microwave freeze drying (MFD) treatments. The only advantage of AFD product is its higher Vc preservation rate. According to some reports, the co-melting temperature of vegetables is about –10°C. From Figure 2.11, it can be observed that about 55% of total water was removed below –10°C during the MFD process, and about 60% of the total water was removed below –10°C during the FD process. However, only less than 50% of water was removed below –10°C during the AFD process, which indicates that more than 50% of water was removed by evaporation rather than by sublimation during the AFD process, resulting in more deterioration of color and structure. This is an obvious disadvantage of AFD, which needs to be resolved in the future.

As shown in Figure 2.13, even though processing time is longer in AFD than FD and MFD, AFD has the lowest energy consumption. This is because AFD is carried out at atmospheric condition, and the vacuum pump and cold trap are cancelled. Moreover, the heat pump can make full use of latent and sensible heat from the moisture in the air. Compared to FD, MFD consumes low energy. Its lower energy consumption is attributed to the reduction in drying time and the decline in working time of the vacuum system and the refrigeration system.

Figures 2.14 and 2.15 exhibited that white mushroom and hawthorn products under different FD treatments could help to maintain good shape and color.

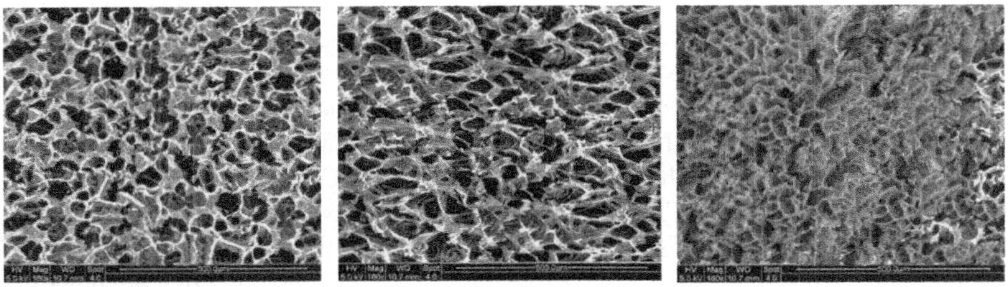

(A) MFD (B) FD (C) AFD

FIGURE 2.12 Effects of different drying methods on the microstructure of dried products. (From Duan, X. et al., *Int. J. Agric. Biol. Eng.*, 8(1), 91–97, 2015.)

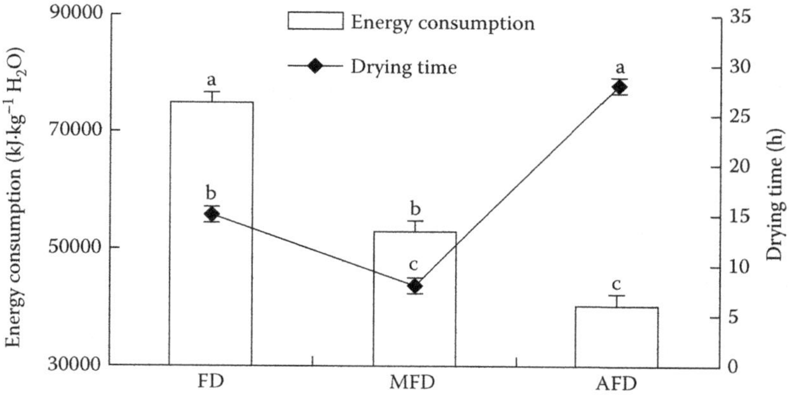

FIGURE 2.13 Effects of different freeze-drying methods on energy consumption. (a, b, c) Different letters indicate a significant difference ($p < .05$). (From Duan, X. et al., *Int. J. Agric. Biol. Eng.*, 8(1), 91–97, 2015.)

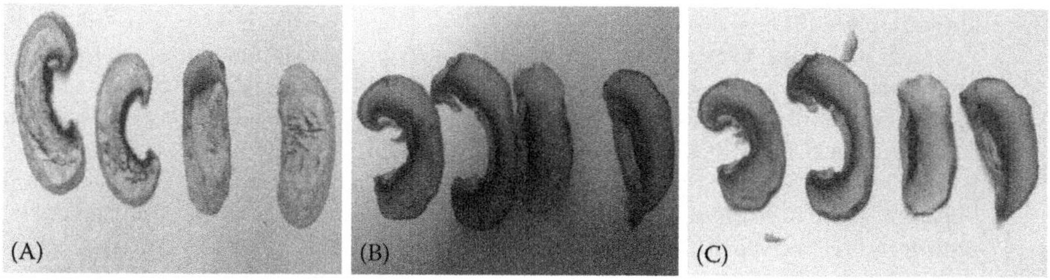

FIGURE 2.14 Pictures of white mushroom under different freeze-drying methods. (A) MFD; (B) AFD; (C) FD.

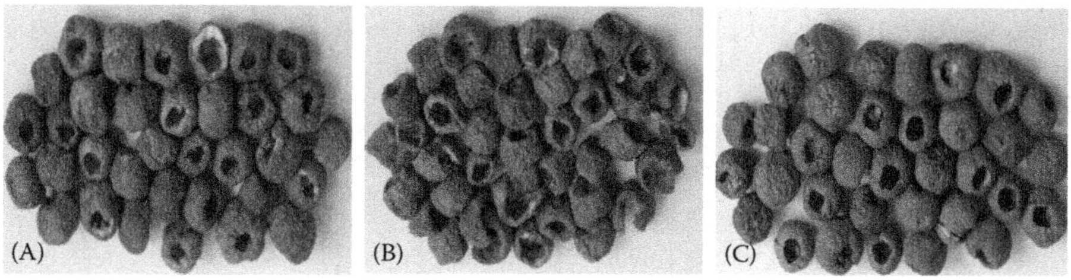

FIGURE 2.15 Pictures of hawthorn under different freeze-drying methods. (A) MFD; (B) AFD; (C) FD.

2.4 CONCLUSION

Different sublimation drying methods such as FD, MFD, and AFD have different effects on the quality and total drying time of the dehydrated mushrooms. FD, MFD, and AFD may lead to different temperature profiles resulting in different quantities of sublimated water. By comparing the drying curves, it can be found that about 55% of total water is sublimated during the MFD process, and about 60% of the total water is sublimated during the FD process. However, less than 50% of water is sublimated during the AFD process, which indicates that the AFD process could lead to more ice thawing, resulting in greater deterioration of color and structure. Compared with FD and AFD, MFD could be relatively of better quality and drying efficiency. As a result, MFD can be used to replace traditional FD, and AFD should also be popularized because of its low energy consumption.

REFERENCES

1. Liapis, A.I.; Bruttini, R. A theory for the primary and secondary drying stages of the freeze-drying of pharmaceutical crystalline and amorphous solutes: Comparison between experimental data and theory. *Separations Technology* 1994, 4(3), 144–155.
2. Sadikoglu, H.; Liapis, A.I.; Crosser, O.K. Optimal control of the primary and secondary drying stages of bulk solution freeze drying in trays. *Drying Technology* 1998, 16(3–5), 399–431.
3. Freire, F.B.; Vieira, G.N.; Freire, J.T.; Mujumdar, A.S. Trends in modeling and sensing approaches for drying control. *Drying Technology* 2014, 32(13), 1524–1532.
4. Liapis, A.I.; Bruttini, R. Freeze drying. In *Handbook of Industrial Drying*, 4th edition, Mujumdar, A.S Ed. CRC Press, Boca Raton, FL, 2014; 262–271.
5. Islam, M.I.U.; Sherrell, R.; Langrish, T.A.G. An investigation of the relationship between glass transition temperatures and the crystallinity of spray-dried powders. *Drying Technology* 2010, 28(3), 361–368.
6. Anglea, S.A.; Karathanos, V.; Karel, M. Low-temperature transitions in fresh and osmotically dehydrated plant materials. *Biotechnology Progress* 1993, 9(2), 204–209.
7. Levi, G.; Karel, M. Volumetric shrinkage (collapse) in freeze-dried carbohydrates above their glass transition temperature. *Food Research International* 1995, 28(2), 145–151.
8. Peleg, M.; Chinachoti, P. On modeling changes in food and biosolids at and around their glass transition temperature range. *Critical Reviews in Food Science and Nutrition* 1996, 36(1–2), 49–67.
9. Ratti, C. Hot air and freeze-drying of high-value foods: A review. *Journal of Food Engineering* 2001, 49(4), 311–319.
10. Caparino, O.A.; Sablani, S.S.; Tang, J.; Syamaladevi, R.M.; Nindo, C.I. Water sorption, glass transition, and microstructures of refractance window- and freeze-dried mango (Philippine "Carabao" Var.) powder. *Drying Technology* 2013, 31(16), 1969–1978.
11. Telis, V.R.N.; Sobral, P.J.A. Glass transitions and state diagram for freeze-dried pineapple. *LWT-Food Science and Technology* 2001, 34(4), 199–205.
12. Boss, E.A.; Filho, R.M.; de Toledo, E.C.V. Freeze drying process: Real time model and optimization. *Chemical Engineering and Processing: Process Intensification* 2004, 43(12), 1475–1485.
13. Schössler, K.; Jäger, H.; Knorr, D. Novel contact ultrasound system for the accelerated freeze-drying of vegetables. *Innovative Food Science and Emerging Technologies* 2012, 16, 113–120.

14. Pisano, R.; Fissore, D.; Barresi, A.A. (2014). Intensification of freeze-drying for the pharmaceutical and food industries. In *Modern Drying Technology*, Vol. 5; Tsotsas, E., Mujumdar, A.S., Editors. Wiley-VCH Verlag GmbH & Co. KGaA, Weinheim, Germany, 2014.
15. Duan, X.; Liu, W.; Ren, G.Y.; Liu, W.C.; Liu, Y.H. Comparative study on the effects and efficiencies of three sublimation drying methods for mushrooms. *International Journal of Agricultural and Biological Engineering* 2015, 8(1), 91–97.
16. Duan, X.; Ren, G.Y.; Zhu, W.X. Microwave freeze drying of apple slices based on the dielectric properties. *Drying Technology* 2012, 30(5), 535–541.
17. Ren, G.Y.; Zeng, F.L.; Duan, X.; Liu, L.L.; Duan, B.; Wang, M.M.; Zhu, W.X. The effect of glass transition temperature on the procedure of microwave–freeze drying of mushrooms (*Agaricus bisporus*). *Drying Technology* 2015, 33(2), 169–175.
18. Lombrana, J.I.; Zuazo, I.; Ikara, J. Moisture diffusivity behavior during freeze drying under microwave heating power application. *Drying Technology* 2001, 19(8), 1613–1627.
19. Wang, Y.C.; Zhang, M.; Mujumdar, A.S.; Mothibe, K.J. Microwave-assisted pulse-spouted bed freeze-drying of stem lettuce slices–effect on product quality. *Food and Bioprocess Technology* 2013, 6(12), 3530–3543.
20. Claussen, I.C.; Ustad, T.S.; Strommen, I.; Walde, P.M. Atmospheric freeze drying—A review. *Drying Technology* 2007, 25(6), 947–957.
21. Li, S.; Stawczyk, J.; Zbicinski, I. CFD model of apple atmospheric freeze drying at low temperature. *Drying Technology* 2007, 25(7–8), 1331–1339.
22. Duan, X.; Liu, W.C.; Ren, G.Y.; Yang, X.T.; Liu, Y.H. Atmospheric freeze drying of garlic slices based on freezing point depression. *International Journal of Agricultural and Biological Engineering* 2015, 8(4), 133–139.
23. Duan, X.; Ding, L.; Ren, G.Y.; Liu, L.L.; Kong, Q.Z. The drying strategy of atmospheric freeze drying apple cubes based on glass transition. *Food and Bioproducts Processing* 2013, 91(4), 534–538.

3 Highly Efficient Vegetable Drying Technology I

Microwave and Radio Frequency Drying of Vegetables

Hao Jiang and Shaojin Wang
Northwest A&F University

CONTENTS

3.1 INTRODUCTION

Microwave (MW) and radio frequency (RF) drying can be attributed to dielectric drying. MW/RF waves are parts of the electromagnetic spectrum. The dielectric energy absorption in foods primarily involves two mechanisms: dipolar relaxation and ionic conduction (Figure 3.1). These interactions are carried out under the electric field of RF and MW. Water in food is often the primary component responsible for dielectric heating/drying. Owing to their dipolar nature, water molecules attempt to follow the electric field as they rotate at very high frequencies. Such rotations of the water molecules produce heat, which is expressed as the dipolar relaxation. Ions, such as those present in salty food, migrate under the influence of the electric field, thereby generating heat. This is the second major mechanism of heating using MW and RF energy (ionic conduction).[1,2] Both the dielectric constant and the dielectric loss factor can measure the ability of the material to interact with the electric field of MW/RFs. The capacitance meter has been used to estimate the dielectric properties, but now vector network analyzer (VNA) and the open-ended coaxial-line probe are used for easy and effective measurement of dielectric properties. As dielectric-related new technologies are developing, MW/RF drying is becoming increasingly popular in food processing industries.

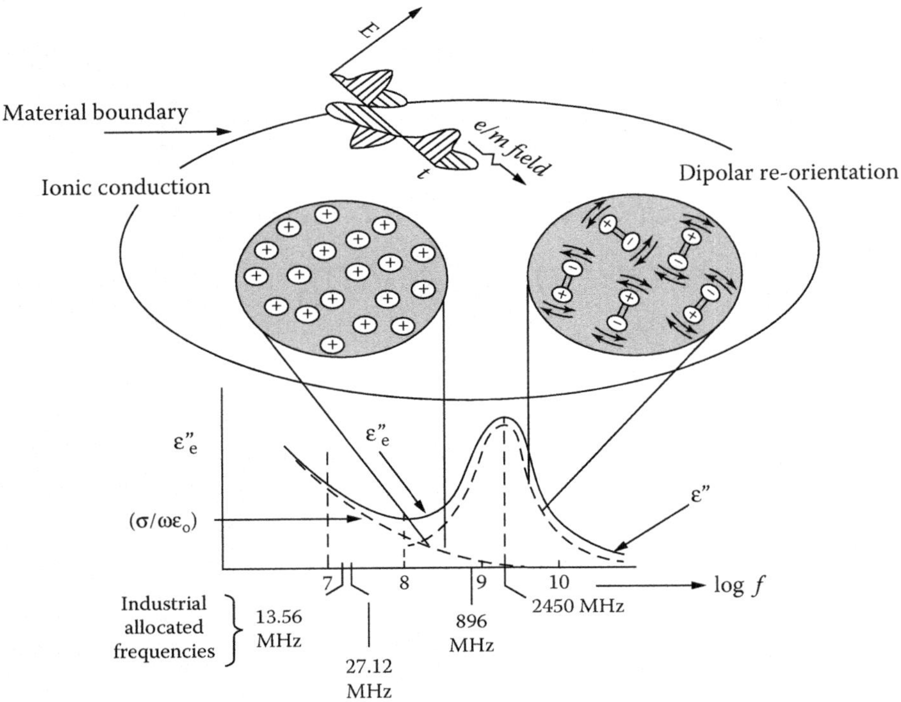

FIGURE 3.1 A schematic diagram depicting the dipolar and ionic loss mechanisms and their contributions to the dielectric properties as a function of frequency. Some commonly used RF and microwave frequencies are noted (the 896-MHz frequency is used in the UK whereas 915-MHz is used in the US). The dashed lines are contributions due to individual mechanisms, and the solid line stands for the combined effect. (From Metaxas, A.C., *Foundations of Electroheat*, John Wiley & Sons, Chichester, UK, 1996. With permission.)

Humans receive benefit from vegetables with high nutrition, high water content, colorful appearance, and tasteful flavor. These characteristics also complicate the process of vegetable preservation. Most vegetables are perishable and can be stored for only a short time. The reasons for quality loss include spoilage caused by moisture, molds, microorganisms, and vermin. Although temperature controlling is an effective method for vegetable preservation, adequate cold chain storage facilities are not available in most counties/regions (more than 50%).[3] However, drying exhibits interesting characteristics including low cost, easy operation, and high efficiency, making it a good method for vegetable preservation. In addition to food preservation, drying also plays an important role in the processing of some snacks, such as potato chips.

The market for dehydrated food is important for most of the countries worldwide. For instance, the world raisin production, mainly produced in the USA (297,557 tons) and Turkey (190,000 tons), was about 500,000 tons and valued at more than US$125 million in 2000 (Free Alongside Ship online data, 2000). The growth in the popularity of convenient foods in many Asian countries has stimulated increasing demand for high-quality dehydrated vegetables and fruits. This trend is expected to continue and even accelerate over the next decade in all emerging economic zones of the world. However, drying is an energy-consuming and labor-intensive process, because 15% energy is consumed worldwide in drying every year.[4] Another fact is that most of the popular drying methods are still solar drying or hot air drying, which can hardly offer high-quality dried products. In recent decades, drying-related researches mostly focus on process simulation, energy saving, and product quality enhancement. As emerging drying techniques, MW/RF drying shows remarkable advantages. The main advantages of MW/RF drying include the following: (1) Process speed is increased; (2) Uniform heating may be obtained throughout the material. Although that is not always true, often the self-moisture balance heating effect does produce uniform heating, avoiding the large temperature gradients that occur in conventional heating systems; (3) High efficiency of energy conversion is achieved; (4) Better and more rapid process control is provided; (5) Floor space requirements are usually reduced; (6) Selective heating may occur; (7) Product quality may be improved; and (8) Desirable chemical and physical effects may be produced.[5,6]

The moisture content and components vary for different types of vegetables. Researches on MW/RF vegetable drying should accurately focus on each type of vegetable. Although nonuniform heating leads to some difficulties during MW/RF food drying, it is proving to be a potential method for vegetable drying.

3.2 MICROWAVE DRYING

Compared with traditional drying methods (solar or hot air drying), MW drying offers significantly accelerated drying rates and acceptable quality of dried products. Although the nonuniform temperature distribution in MW heating has not been well resolved, it is still widely used in industrial drying. Szadzińska et al.[7] studied the effect of MW-assisted convective drying. By analyzing the data, they found that this drying method shortened the drying time significantly, reduced the energy consumption, and improved the quality factors. As compared to ultrasonic-assisted convective drying, MW-assisted convective drying showed superiority on drying rate (Figure 3.2). The heat-and-mass transfer of canola seeds using MW-assisted convective drying was studied by Hemis et al.[8] The results indicated that during reduced drying rate period, the use of MW energy resulted in faster drying (Figure 3.3), whereas the predicted results of drying rate showed that a MW power density of 0.5–1 W/g and a controlled flow of convective hot air at 60°C with a low relative humidity gave better results. Sarimeseli[9] used coriander leaves as the material to study the effect of MW drying. It was found that the moisture content was affected by the MW power input, and thus when the power input was increased the drying time of the leaves was significantly reduced from 14 to 4 min (Figure 3.4). It was also found that raising the power output resulted in shortened drying times, which can be maximally shortened to 75% total drying time with the accepted quality. Similar results can be found on various materials, including onion,[10] spinach,[11] pistachios,[12] and pumpkin slices.[13] To investigate the effect of relative humidity on MW drying of carrot, Pu et al.[14] developed a new MW drying system with humidity control, and the relationship between the humidity and the drying rate as well as the product quality were discussed. The results showed that adding balance water was a more direct and effective method to adjust the convection air's

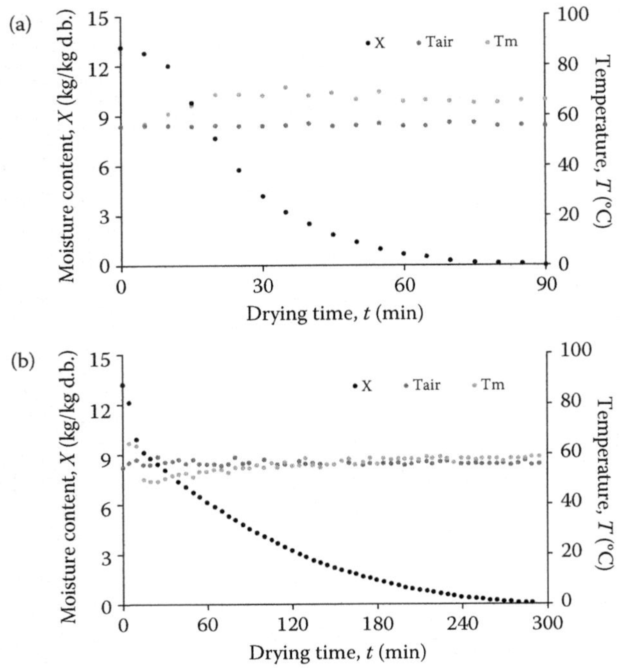

FIGURE 3.2 The drying curves and the temperature profiles: (a) Convective-microwave drying with power of 100 W, (b) convective drying assisted with microwave with power of 100 W for the first 10 min. X indicates the drying curves; T indicates the air temperature; and T_m indicates the material temperature. (From Szadzińska, J. et al., *Ultrason. Sonochem.*, 2017, 34, 531–539. With permission.)

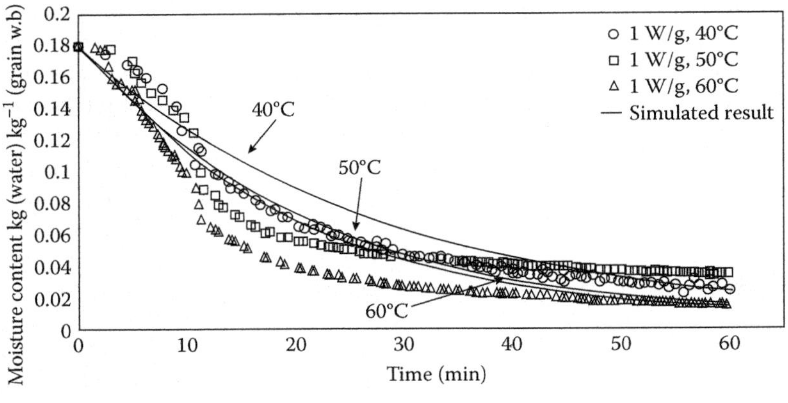

FIGURE 3.3 Comparison between simulated and experimental results of moisture loss from canola seeds during drying by a coupled system of microwave-assisted hot air using 1 W/g of density and three inlet air temperatures: 40°C, 50°C, and 60°C. (From Hemis, M. et al., *Biosyst. Eng.*, 139, 121–127, 2015. With permission.)

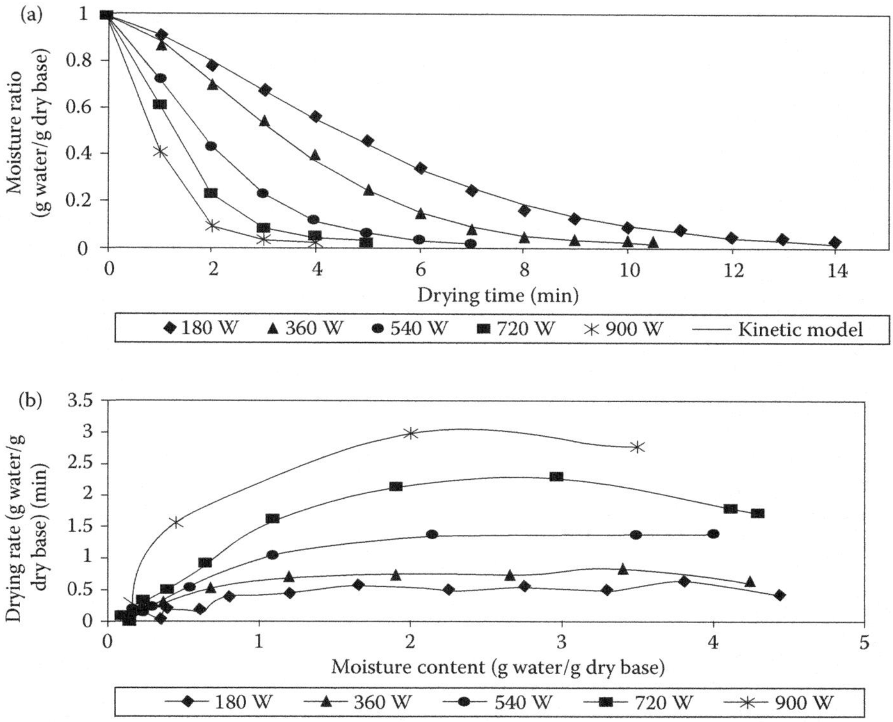

FIGURE 3.4 (a) Drying and (b) drying rate curves of coriander leaves during the microwave drying process. (From Sarimeseli, A. et al., *Energy Convers. Manag.*, 52, 1449–1453, 2011. With permisson.)

relative humidity, especially in the initial drying process; the color and rehydration were, therefore, improved, but the dehydration rate decreased. Holtz et al.[15] investigated the energy efficiency with different material (swede, potato, bread, and lightweight concrete) properties and drying rates. The results indicated that the energy consumption was approximately the same for all materials at intermediate to high moisture content; however, the specific energy consumption increased when drying was pursued into the hygroscopic region (low moisture content). Those evidences indicated that the MW drying effect was influenced by MW power, shape, and initial moisture content of materials. It is necessary to conduct experimental exploration before MW drying is applied on special materials.

It is believed that MW drying can enhance the quality of dried products compared with that of traditionally dried products. Lombraña et al.[16] analyzed the impact of internal vapor pressure and temperature on structure of dried samples. The moisture sorption in the dried layer was three times lower than that under atmospheric pressure, because of structural collapse and subsequent occlusion of moisture inside the product in this last case. The higher level of heat and wider temperature oscillations associated with the temperature control in this case would explain the poorer quality in the dried product. Arslan and Özcan[10] compared the color and preservation

TABLE 3.1
Comparison between Microwave Output Powers for Color Parameters during Spinach Drying

Microwave Power (W)	*L*	*a*	*b*	*C*	α°
Fresh	41.98 ± 0.99	−14.83 ± 0.82	16.69 ± 0.53	22.33 ± 0.56	131.62 ± 0.82
90	31.74 ± 1.71	−8.94 ± 0.45	12.12 ± 0.37	15.06 ± 0.39	126.41 ± 1.93
160	33.64 ± 1.03	−9.80 ± 0.78	13.26 ± 0.64	16.49 ± 0.82	126.47 ± 1.08
350	32.24 ± 1.16	−10.32 ± 0.75	12.91 ± 0.97	16.53 ± 1.00	128.64 ± 1.30
500	35.99 ± 0.39	−11.28 ± 0.89	14.50 ± 0.65	18.37 ± 0.35	127.88 ± 1.34
650	35.58 ± 2.20	−10.45 ± 0.72	13.30 ± 1.13	16.91 ± 1.12	128.16 ± 0.92
750	36.00 ± 1.02	−10.86 ± 0.39	14.51 ± 1.03	18.12 ± 0.66	126.81 ± 1.92
850	34.15 ± 1.16	−9.53 ± 0.19	13.43 ± 0.72	16.47 ± 0.58	125.36 ± 1.17
1000	29.90 ± 2.25	−6.93 ± 0.16	8.92 ± 0.90	11.30 ± 0.80	127.84 ± 1.87

L, brightness; *a*, redness; *b*, yellowness; *C*, chroma; α°, hue angle.
Source: Alibas Ozkan, I. et al., *J. Food Eng.*, 78, 577–583, 2007. With permission.

rate phenolics of onion dried by solar, hot air, and MW. The results indicated that low MW power (210 W) could retain phenolics of the product better than sun and oven drying. MW drying of onion at 210 W showed the best color; however, high MW power (700 W) obtained the darkest color. Alibas Ozkan et al.[11] figured out that the color values closest to that of fresh spinach were obtained in the drying processes using the energy levels of 750, 650, and 500 W (Table 3.1). Higher MW power is good for color protection, which was also agreed upon by Chua and Chou.[17] Their results showed that ascorbic acid values of the spinaches dried at energy levels of 160 W (with 25.67 mg/100 g) and 350 W (25.70 mg/100 g) were found to be lower than those dried at higher energy levels (500, 650, 750, 850, and 1000 W were 42.86 ± 1.61, 43.57 ± 1.24, 43.09 ± 1.88, 42.68 ± 1.50, and 41.79 ± 2.89 mg/100 g, respectively). In brief, there is no doubt that MW would offer better quality of dried products compared with solar- and hot air-dried samples, which was in accordance with the results of Łechtánska et al.,[18] Mujumdar et al.,[4,19,20] and Koné et al.[21]

3.3 RF DRYING

The major difference between MW and RF heating is the frequency (wavelength) (Figure 3.5).[22] Both MW (300 MHz–300 GHz) and RF waves (3 kHz–300 MHz) are part of the electromagnetic spectrum that results in heating of dielectric materials by induced molecular vibration as a result of dipole rotation or ionic migration. They have been credited with volumetric heat generation resulting in rapid heating of foodstuff. However, owing to the lower frequency levels, RF waves have a larger penetration depth than MW and hence could find better application in larger-size foods. The heating efficiency of MW would be superior than RF because of the higher frequency. Besides, MW and RF heating are both classified as dielectric

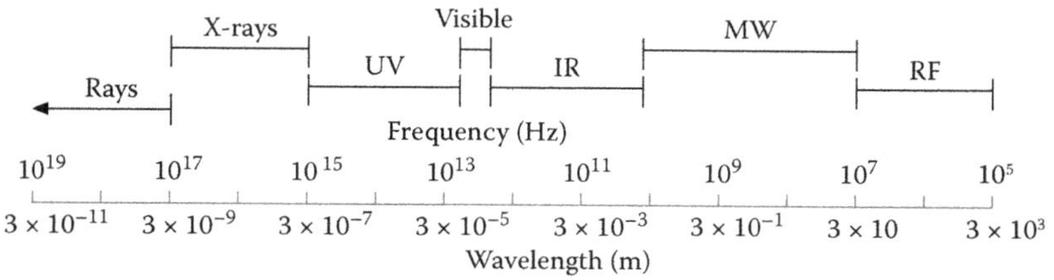

FIGURE 3.5 The electromagnetic spectrum. (From Marra, F. et al., *J. Food Eng.*, 91, 497–508, 2009. With permission.)

heating methods. It is generally accepted that ionic depolarization tends to be the dominant heating mechanism at the lower frequencies encountered in the RF range, whereas both ionic depolarization and dipole rotation can be dominant loss mechanisms at frequencies relevant to MW heating (i.e., 400–3000 MHz) depending upon the moisture and salt content within a product.[23] Therefore, dissolved ions in the RF range are more important for heat generation than the water dipoles.[24] In short, samples with high ionic concentration are good for RF drying.

Similar to MW heating, the major obstacle for RF technology to be commercially applicable is its nonuniform heating. Different factors, such as the sample's dielectric properties, size, shape, components, position between the RF electrodes, and the electrode configuration may affect temperature uniformity in an RF-treated food product. Related studies revealed that the dielectric properties of agricultural products and foods were functions of frequency, moisture content, and temperature.[25,26] Zhu et al.[27] considered that the moisture content and temperature had strong effects on permittivity values. The penetration depth decreased with increasing frequency, moisture content, and temperature at the RF range. It should be noted that the dielectric properties in the RF range are quite different from those in the MW range. Zhang et al.,[28] found that both dielectric constant and loss factor decreased sharply with increasing frequency over the RF range (10–300 MHz) but gradually over the MW range (300–4500 MHz) when using peanut kernels. Besides, both dielectric constant and loss factor in the RF range increased with increasing moisture content and temperature. The rate of increase was more at higher temperature and moisture levels than at lower levels. Penetration depth decreased with increasing frequency, moisture content, and temperature.[27]

Being an emerging heating technology, there is limited literature on RF drying of vegetables. Orsat et al.[29] tried to treat ready-to-eat carrot chips by RF energy. After treating for 2 min, the quality of the RF-treated samples was better than either of the control samples (chlorinated water) or hot water-treated carrot samples. Furthermore, RF heating treatments may reduce levels of initial competitive microflora, which allowed pathogens of lactic acid bacteria to proliferate. Ildikó et al.[30] used white mustard as the material to investigate the effect of RF heat treatment on its nutritional and colloid-chemical properties. The data obtained from this study proved that upon a mild heating treatment, the important nutritional compounds in white mustard were not damaged by RF treatment. Moreover, the myrosinase enzyme was successfully inactivated, while the product had a bland flavor. When the emulsion stability as well

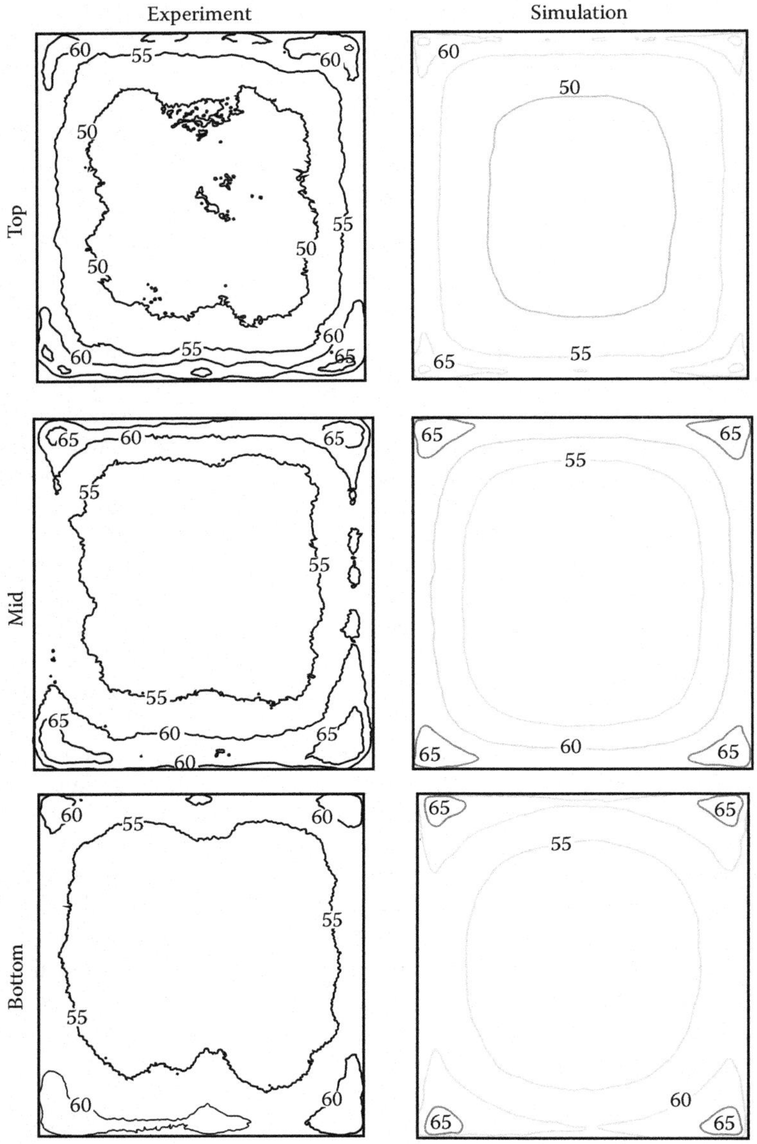

FIGURE 3.6 Experimental and simulated temperature distributions (°C) of dry soybeans in top, middle, and bottom layers (20, 40, and 60 mm from the bottom of sample) placed in a polypropylene container (300 × 220 × 60 mm^3) on the bottom electrode, after 6 min RF heating with an initial temperature of 25°C and a fixed electrode gap of 120 mm. (From Huang, Z. et al., *Biosystem. Eng.*, 129, 34–47, 2015. With permission.)

as the oil and water holding capacity of mustard flour was increased, no significant changes were observed in the emulsion activity based on the effect of RF heat treatment. Huang et al.[31] tried to simulate the RF treatment using the finite element-based commercial software—COMSOL. Dry soybeans packed in a rectangular plastic container were used to determine the heating uniformity and validate the simulation model using a 27.12 MHz, 6 kW RF system. Both simulated and experimental results showed similar heating patterns in RF-treated soybeans, in which corners and edges were more heated and the temperature values were higher in the lower part of the container (Figure 3.6). In addition, simulated results showed that smaller top plate area, placing the sample in the middle of two electrodes, and surrounding the container with similar dielectric constant sheets provided better RF heating uniformity. Further, Huang et al.[32] and Yu et al.[33] also devoted themselves to improve the uniformity of RF heating/drying of vegetables. Because of the better penetrability, RF drying is also applied on some nut kernels, such as peanuts,[28] walnuts,[34] and macadamia nuts.[35,36]

3.4 MICROWAVE/RF-RELATED HYBRID DRYING

Although MW/RF heating can readily deliver energy to generate heat in the moist portion of foods, one of its major drawbacks is the inherent nonuniformity of the electromagnetic field within a microwave cavity.[37] Cohen and Yang[38] and Clark[39] reported that the excessive temperatures created by more than 500 W MW power along the edges and corners of products may lead to overheating and irreversible drying out, resulting in possible scorching and development of off-flavors. To overcome the drawback of MW/RF drying, it is desirable to develop hybrid drying.

Microwave-assisted vacuum drying, or MW vacuum drying, (MVD) is a technology that not only has the advantages of MW heating (rapid heating, high efficiency, good controllability, and sanitation)[40] but also lowers the boiling point of water caused by the vacuum environment, thus improving the energy efficiency and reducing the formation of burned spots on the surface of the final products.[1] Furthermore, this new processing technology can raise the expansion ratio and improve the texture of the finished products.[37] Zielińska et al.[41] studied multistage combined drying (heat pump and MW vacuum drying) to process green peas, and investigated the quality factors, including the drying kinetics, moisture diffusivities, microstructure, and physical parameters of green peas. The results showed that MVD increased the drying rate; meanwhile, MVD samples were characterized by a structure with minimal changes with respect to fresh samples. It is a good choice to set the MVD as the final stage of hybrid drying. Figiel[42] investigated the drying kinetics and quality of MVD dehydrated garlic cloves and slices. Whole garlic cloves, halved cloves, and sliced cloves were subjected to MVD with power levels of 240, 480, and 720 W, respectively. It was found that higher microwave power resulted in increasing drying rate (Figure 3.7). Furthermore, the MW drying of the garlic samples with the vacuum made the color brighter, shifting it toward red and blue, and the best retention of volatile oils was observed in garlic slices dehydrated by MW at 720 W. Li et al.[43] used garlic as the material and tried to optimize the drying process by adjusting the MVD drying parameter dynamically. The results showed that the optimal MVD condition was drying for 3 min under the MW output power of 376.1 W, then 282.1 W for 3 min, followed by 188.0 W for 9 min, and finally for 3 min under the

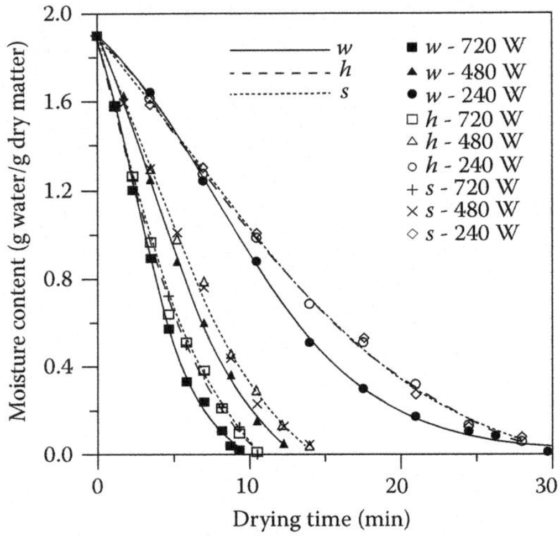

FIGURE 3.7 Drying curves for whole garlic cloves (*w*), halves (*h*), and slices (*s*) under vacuum drying condition at microwave powers of 240, 480, and 720 W. (From Figiel, A., *J. Food Eng.*, 94, 98–104, 2009. With permission.)

output power 94.0 W. After drying, the thiosulfinates retention was 90.2%, which was as good as the product prepared by freeze-drying (FD). Potato,[44,45] mushroom,[46–48] and edamames[49,50] are also chosen as the materials to prepare dried samples by MVD.

MW combined with FD (MW freeze-drying, MFD) is an emerging drying technology. A conventional freeze-dryer had the added capability of allowing MW to be introduced within the drying chamber. FD is used as a gentle dehydration method for heat-sensitive food, pharmaceuticals, and biological materials. It is well known for its ability to retain the quality of products better (color, shape, aroma, texture, biological activity, etc.) than any other drying methods, because of its low processing temperature and almost noninvolvement of oxygen in the process. However, FD is an expensive and lengthy dehydration process because of low drying rate, which leads to relatively small throughputs and high capital and energy costs generated by refrigeration and vacuum systems.[1] Mild MW can be absorbed by ice crystals without melting the ice; thus, FD combined with MW heating can accelerate the drying rate significantly. A typical MFD drying apparatus is shown in Figure 3.8. In the study by Jiang et al.[51], FD with heating plate power set at 400 W needed 600 min to finish drying, whereas MFD with 1 W/g MW power density needed only 360 min (Figure 3.9). After optimizing the process, the energy consumption of MFD was only 55% compared to that of FD. Abbasi and Azari[52] investigated the characteristics of the onion slices dried by MFD. The quality properties of the onion slices produced by MFD were completely comparable and competitive with the commercial freeze-drier, with more than 96% saving on processing time. Except for drying rate acceleration, Duan et al.[53] considered that MFD provided remarkable sterilization of dried cabbage. It should be noted that corona discharge and nonuniform heating were the inherent problems of MFD. Corona discharge could be alleviated by lowering the

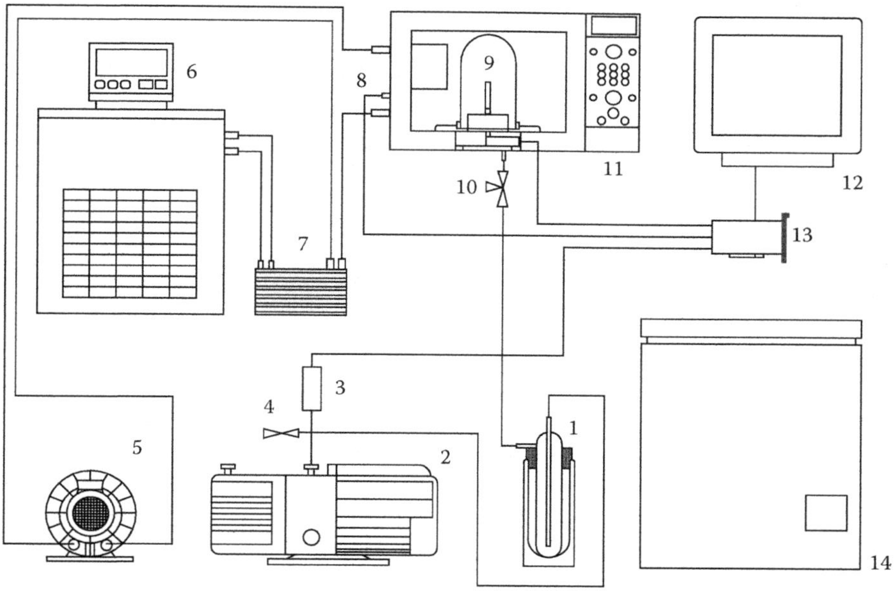

FIGURE 3.8 Schematic diagram of the microwave freeze-drying apparatus. 1, cool trap; 2, vacuum pump; 3, pressure sensor; 4, exhaust valve; 5, blower; 6, circulator; 7, heat exchanger; 8, infrared sensor; 9, drying chamber; 10, isolation valve; 11, microwave oven; 12, personal computer (PC); 13, data acquisition card; 14, freezer. (From Huang, J. et al., *J. Food Eng.*, 177, 80–89, 2016. With permission.)

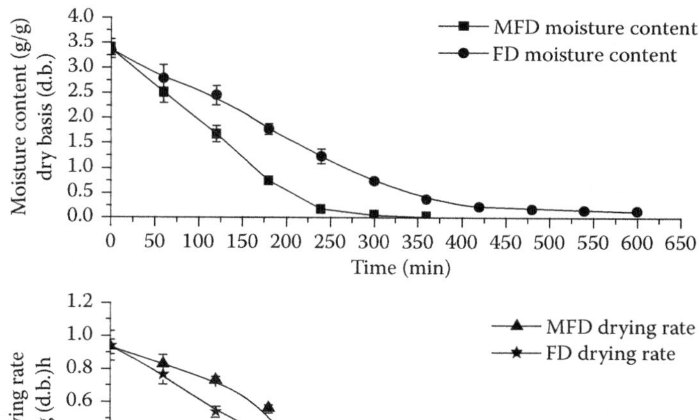

FIGURE 3.9 Drying and drying rate curves of freeze-drying/microwave freeze-drying samples. (From Jiang, H. et al., *Food Bioprod. Process.*, 91(4), 464–472, 2013. With permission.)[51]

vacuum degree. To solve the problem of nonuniformity, Wang et al.[54] improved the MFD dryer by adding the pulse-spouted bed unit. Intermittent air pumping made samples move in the drying cavity to alleviate the nonuniform heating. The results indicated that the samples dried by MW-assisted pulse-spouted bed freeze-dryer (PSMFD) showed better color, microstructure, texture, and rehydration compared to those dried by MFD dryer. Huang et al.[55] used MW air-spouted drying to process carrot cubes. It was observed that the carrot cubes processed by this technology offered better color; higher level of carotenoids and chlorophyll contents; lesser loss of terpenoids, alcohols, aldehydes, and more volatile compounds; as well as better rehydration capacity.

3.5 PRETREATMENT ON MW/RF DRYING

Pretreatment can alter the dielectric properties of samples, further accelerating the drying rate. It was reported that the components of materials, moisture content, temperature, and frequency could influence the dielectric properties.[27] The pretreatment, mainly blanching, ultrasound, and adding extrinsic dielectric core, would raise the dielectric constant and loss factor to accelerate the dielectric drying significantly.

Applying ultrasound in drying shortened the drying time and thus reduced the total energy consumption. Moreover, it can improve the product quality. The above-mentioned studies showed that the application of ultrasound might greatly accelerate the drying process without causing a noticeable increase in the material's temperature (1°C–2°C). For this reason, the ultrasound-assisted drying method is useful especially for temperature-sensitive food materials.[56] Two mechanisms of ultrasound transfer (contactless by the air and by contact materials direct under static pressure) between the transducer and the material being dried were tested. The results of these studies confirmed that although airborne ultrasounds accelerated the drying process in low temperature and low air velocity ranges, the contact methods were distinctly more efficient because the drying time can be reduced by about 65%–70%.[57] It is believed that ultrasound would cause loss in tissue coherence, break in cell membrane and wall, formation of microchannels, increase in porosity, etc., thereby changing the diffusion coefficient, as well as the dielectric properties of materials, and finally accelerating the drying rate.[58] This phenomenon can influence the properties of materials, including carrots,[59] green peppers,[60] and potatoes,[61] dried by MW/RF energy.

Traditional blanching is a heat pretreatment using hot water or steam, which is applied in the agro-food sector, particularly important in the processing of green vegetables. The main goal is to inactivate the enzymes involved in the spoilage of fresh vegetables or fruits.[62] The other purposes of blanching include reducing the microbial load of products so as to improve its conservation, softening tissues for an easier canning step, shortening cooking time, and eliminating intracellular air to prevent oxidation.[63] However, blanching influenced the structure of vegetables and fruits. It was reported after hot water blanching (95°C) that the electrical conductivity of lettuce was twice that of nontreated samples. Furthermore, the dielectric constant and loss factor of lettuce after hot water blanching were lower than those of nontreated samples because of loss of soluble solutes and ions.[64]

MW blanching is one of those technologies that seem to provide better nutrient retention than hot water or steam blanching.[65] MW blanching damages the plant tissue, which may modify the capillary-porous characteristic and water sorption capacity of the plant material. The intermolecular friction produced by MW heating may raise the internal cell pressure, rupture the cell wall, result in the loss of cell contents and organization, and finally change the dielectric properties of materials. Wang et al.[64] reported that MW blanching can raise the dielectric constant and loss factor (Figure 3.10). Jiang et al.[47] compared the effect of hot water and MW blanching on vacuum MW drying using mushroom slices. The results indicated that the vacuum MW drying duration in the blanched samples was ~17.5 min, which was reduced

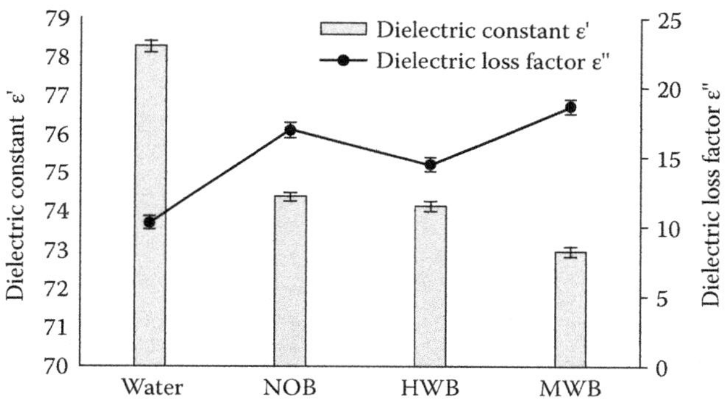

FIGURE 3.10 Influence of different blanching processes on the dielectric behaviors of stem lettuce cubes at 2450 MHz at 20°C. The scattered data are means of three replicates. NOB indicates nonblanching; HWB indicates hot water blanching; and MWB indicates microwave blanching. (From Wang, Y.C. et al., *J. Food Eng.*, 113, 177–185, 2012. With permission.)

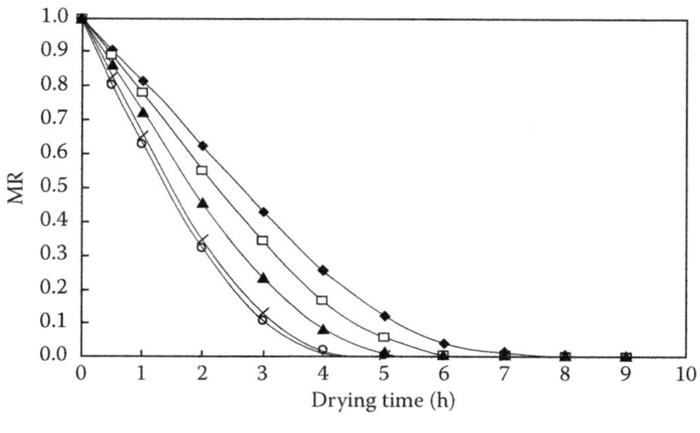

FIGURE 3.11 Effect of NaCl content on microwave freeze-drying drying rate. MR indicates moisture ratio. (♦: 1 g/100 g water, □: 3 g/100 g water, ▲: 5 g/100 g water, ×: 10 g/100 g water, o: 15 g/100 g water). (From Wang, R. et al., *LWT Food Sci. Technol.*, 43(7), 1144–1150, 2010. With permission.)[67]

by 22% compared to that of hot-water-blanched samples. The microstructure of the vacuum MW dried-treated MW blanching sample was more uniform than that of the hot water blanching sample. These results suggest that MW blanching improves the vacuum MW drying process when applied to mushroom before drying. Various foods were pretreated by MW blanching for different purposes.[66,67]

The ionic density can raise the loss factor; therefore, adding exogenous high dielectric cores is an effective method to improve the drying rate of MW/RF-related techniques. Wang et al.[68] found that the content of NaCl in the vegetable mixture for MFD drying altered the drying rate. The total drying time was reduced from 9 h to <5 h as the exogenous NaCl was added about 1 to 15 g/100 g (Figure 3.11). The dielectric properties of potato puree after adding NaCl are presented in Figure 3.12.

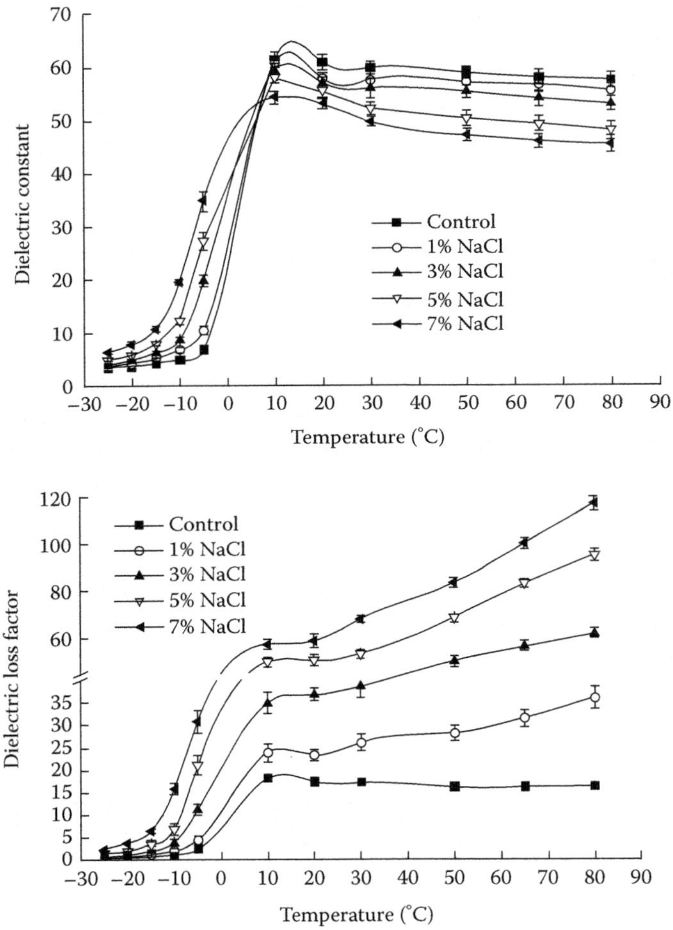

FIGURE 3.12 Dielectric of potato purees with different salt contents at different temperatures. (From Wang, R. et al., *J. Food Eng.*, 106(4), 290–297, 2011. With permission.)

3.6 FINAL REMARKS

Dielectric drying is efficient, but the nonuniform temperature distribution is the most important problem that hinders its applications in the developing industry. It can be concluded that the effect of dielectric drying depends on: (1) frequency of electromagnetic wave, (2) temperature of materials, (3) shape and thickness of materials, (4) moisture content of materials, (5) components of materials, and (6) ionic strength.

RF drying offers better drying uniformity and deeper penetration, but commercial applications could be developed based on the reduced equipment and operational cost and special treatment protocols for target vegetables. However, MW (2450 MHz) is still the most relevant technique. It has been proven that under laboratory scale, MW drying can offer higher-quality dried vegetable products but cost less than other drying methods (including hot air drying, FD, and contact drying). The main problems of MW drying in larger scale use include: (1) scale-up problems; (2) general lack of information, training, and equipment developed for the proposed application; (3) microwave leakage; (4) complexity of drying equipment; and (5) difficulty in controlling.

To solve the problem of nonuniform MW/RF drying, hybrid drying and pretreatment were reviewed and proposed in this chapter to enhance the quality of dried products. The discussion on the suitability of dielectric drying in vegetable processing is still open, and many new papers are delivered every year. Almost all of the published works have considered the laboratory-scale solutions; however, attempts on large-scale applications are frequently reported. This provides great potential for industrial applications of effective dielectric drying in the near future.

REFERENCES

1. Zhang, M., Jiang, H., Lim, R. X., Recent developments in microwave-assisted drying of vegetables, fruits, and aquatic products—Drying kinetics and quality considerations. *Drying Technology* 2010, 28(11), 1307–1316.
2. Metaxas, A. C., *Foundations of Electroheat*. John Wiley & Sons, Chichester, UK, 1996.
3. Garg, H. P., Prakash, J., *Solar Energy: Fundamentals and Applications*. Tata McGraw-Hill, New Delhi, 1997.
4. Mujumdar, A. S., *Handbook of Industrial Drying*. Taylor & Francis, Philadelphia, PA, 2007.
5. Athanasios, M., Georgios, L., Michael, K., A rapid microwave-assisted derivatization process for the determination of phenolic acids in brewer's spent grains. *Food Chemistry* 2007, 102, 606–611.
6. Datta, A. K., Anantheswaran, R. C., *Handbook of Microwave Technology for Food Application*. CRC Press, Cambridge, UK, 2001.
7. Szadzińska, J., Łechtańska, J., Kowalski, S. J., Stasiak, M., The effect of high power airborne ultrasound and microwaves on convective drying effectiveness and quality of green pepper. *Ultrasonics Sonochemistry* 2017, 34, 531–539.
8. Hemis, M., Choudhary, R., Gariépy, Y., Raghavan, G. S. V., Experiments and modelling of the microwave assisted convective drying of canola seeds. *Biosystems Engineering* 2015, 139, 121–127.

9. Sarimeseli, A., Microwave drying characteristics of coriander (*Coriandrum sativum* L.) leaves. *Energy Conversion and Management* 2011, 52, 1449–1453.
10. Arslan, D., Özcan, M. M., Study the effect of sun, oven and microwave drying on quality of onion slices. *LWT-Food Science and Technology* 2010, 43, 1121–1127.
11. Alibas Ozkan, I., Akbudak, B., Akbudak, N., Microwave drying characteristics of spinach. *Journal of Food Engineering* 2007, 78, 577–583.
12. Kouchakzadeh, A., Shafeei, S., Modeling of microwave-convective drying of pistachios. *Energy Conversion and Management* 2010, 51, 2012–2015.
13. Alibas, I., Microwave, air and combined microwave–air-drying parameters of pumpkin slices. *LWT-Food Science and Technology* 2007, 40(8), 1445–1451.
14. Pu, H., Li, Z., Hui, J., Raghavan, G. S. V., Effect of relative humidity on microwave drying of carrot. *Journal of Food Engineering* 2016, 190, 167–175.
15. Holtz, E., Ahrné, L., Rittenauer, M., Rasmuson, A., Influence of dielectric and sorption properties on drying behaviour and energy efficiency during microwave convective drying of selected food and non-food inorganic materials. *Journal of Food Engineering* 2010, 97, 144–153.
16. Lombraña, J. I., Rodríguez, R., Ruiz, U., Microwave-drying of sliced mushroom. Analysis of temperature control and pressure. *Innovative Food Science and Emerging Technologies* 2010, 11, 652–660.
17. Chua, J. K., Chou, K. S., A comparative study between intermittent microwave and infrared drying of bioproducts. *International Journal of Food Science and Technology* 2005, 40, 23–39.
18. Łechtánska, J. M., Szadzínska, J., Kowalski, S. J., Microwave- and infrared-assisted convective drying of green pepper: Quality and energy considerations. *Chemical Engineering and Processing* 2015, 98, 155–164.
19. Mujumdar, A. S., *Guide to Industrial Drying: Principles, Equipment and New Development*, 2nd edition. Colour Publications Pvt. Ltd, Mumbai, 2004.
20. Mujumdar, A. S., An overview of innovation in industrial drying: Current status and R&D needs. *Transport in Porous Media* 2006, 66(1–2), 3–18.
21. Koné, K. Y., Druon, C., Gnimpieba, E. Z., Delmotte, M., Duquenoy, A., Laguerre, J.-C., Power density control in microwave assisted air drying to improve quality of food. *Journal of Food Engineering* 2013, 119, 750–757.
22. Marra, F., Zhang, L., Lyng, J. G., Radio frequency treatment of foods: Review of recent advances. *Journal of Food Engineering* 2009, 91, 497–508.
23. Tang, J.M., Wang, Y.F., Chow Ting Chan, T.V., Radio frequency heating in food processing. In Gustavo V., Barbosa-Cánovas, G. V., Tapia, M. S., Cano, P. (eds) *Novel Food Processing Technologies*. Marcel Dekker, New York, 2004, pp. 501–524.
24. Ohlsson, T., Fundamentals of microwave cooking. *Microwave World* 1983, 4, 4–9.
25. Guo, W., Tiwari, G., Tang, J., Wang, S., Frequency, moisture and temperature dependent dielectric properties of chickpea flour. *Biosystems Engineering* 2008, 101(2), 217–224.
26. Jiao, S., Johnson, J., Tang, J., Tiwari, G., Wang, S., Dielectric properties of cowpea weevil, black-eyed peas and mung beans with respect to the development of radio frequency heat treatments. *Biosystems Engineering* 2011, 108(3), 280–291.
27. Zhu, X., Guo, W., Wu, X., Wang, S., Dielectric properties of chestnut flour relevant to drying with radio-frequency and microwave energy. *Journal of Food Engineering* 2012, 113, 143–150.
28. Zhang, S., Zhou, L., Ling, B., Wang, S., Dielectric properties of peanut kernels associated with microwave and radio frequency drying. *Biosystems Engineering* 2016, 145, 108–117.
29. Orsat, V. G. Y., Raghavan, G. S. V., Lyew, D., Radio-frequency treatment for ready-to-eat fresh carrots. *Food Research International* 2001, 34, 527–536.

30. Ildikó, S.-G., Klára, K. A., Marianna, T.-M., Ágnes, B., Zsuzsanna, M.-B., Bálint, C., The effect of radio frequency heat treatment on nutritional and colloid-chemical properties of different white mustard (*Sinapis alba* L.) varieties. *Innovative Food Science and Emerging Technologies* 2006, 7, 74–79.
31. Huang, Z., Zhu, H., Yan, R., Wang, S., Simulation and prediction of radio frequency heating in dry soybeans. *Biosystems Engineering* 2015, 129, 34–47.
32. Huang, Z., Marra, F., Wang, S., A novel strategy for improving radio frequency heating uniformity of dry food products using computational modeling. *Innovative Food Science and Emerging Technologies* 2016, 34, 100–111.
33. Yu, D., Shrestha, B. L., Baik, O., Temperature distribution in a packed-bed of canola seeds with various moisture contents and bulk volumes during radio frequency (RF) heating. *Biosystems Engineering* 2016, 148, 55–67.
34. Zhang, B., Zheng, A., Zhou, L., Huang, Z., Wang, S., Developing hot air-assisted radio frequency drying protocols for in-shell walnuts. *Emirates Journal of Food and Agriculture* 2016, 28, 459–467.
35. Wang, Y., Zhang, L., Gao, M., Tang, J., Wang, S., Temperature- and moisture-dependent dielectric properties of macadamia nut kernels. *Food and Bioprocess Technology* 2013, 6(8), 2165–2176.
36. Wang, Y., Zhang, L., Johnson, J., Gao, M., Tang, J., Powers, J. R., Wang, S., Developing hot air-assisted radio frequency drying for in-shell macadamia nuts. *Food and Bioprocess Technology* 2014, 7(1), 278–288.
37. Zhang, M., Tang, J., Mujumdar, A. S., Wang, S., Trends in microwave-related drying of fruits and vegetables. *Trends in Food Science and Technology* 2006, 17(10), 524–534.
38. Cohen, J. S., Yang, T. C. S., Progress in food dehydration. *Trends in Food Science and Technology* 1995, 6(1), 20–25.
39. Clark, D. E., Microwave processing of materials. *Annual Review of Materials Science* 1996, 26, 299–331.
40. Zhang, M., Li, C. L., Ding, X. L., Application of various drying methods in selenium-enriched vegetable dehydration. *Drying Technology* 2000, 18(4–5), 1139–1146.
41. Zielińska, M., Zapotoczny, P., Alves-Filho, O., Eikevik, T. M., Błaszczak, W., A multi-stage combined heat pump and microwave vacuum drying of green peas. *Journal of Food Engineering* 2013, 115, 347–356.
42. Figiel, A., Drying kinetics and quality of vacuum-microwave dehydrated garlic cloves and slices. *Journal of Food Engineering* 2009, 94, 98–104.
43. Li, Y., Xu, S., Sun, D., Preparation of garlic powder with high allicin content by using combined microwave–vacuum and vacuum drying as well as microencapsulation. *Journal of Food Engineering* 2007, 83, 76–83.
44. Quan, X., Zhang, M., Fang, Z., Liu, H., Shen, Q., Gao, Z., Low oil French fries produced by combined pre-frying and pulsed-spouted microwave vacuum drying method. *Food and Bioproducts Processing* 2016, 99, 109–115.
45. Bondaruk, J., Markowski, M., Błaszczak, W., Effect of drying conditions on the quality of vacuum-microwave dried potato cubes. *Journal of Food Engineering* 2007, 81, 306–312.
46. Jiang, N., Liu, C., Li, D., Zhang, Z., Yu, Z., Zhou, Y., Effect of thermosonic pretreatment on drying kinetics and energy consumption of microwave vacuum dried *Agaricus bisporus* slices. *Journal of Food Engineering* 2016, 177, 21–30.
47. Jiang, N., Liu, C., Li, D., Zhou, Y., Effect of blanching on the dielectric properties and microwave vacuum drying behavior of *Agaricus bisporus* slices. *Innovative Food Science and Emerging Technologies* 2015, 30, 89–97.
48. Giri, S. K., Prasad, S., Drying kinetics and rehydration characteristics of microwave-vacuum and convective hot-air dried mushrooms. *Journal of Food Engineering* 2007, 78, 512–521.

49. Hu, Q. G., Zhang, M., Mujumdar, A. S., Xiao, G. N., Sun, J. C., Performance evaluation of vacuum microwave drying of edamame in deep-bed drying. *Drying Technology* 2007, 25(4), 731–736.
50. Hu, Q. G., Zhang, M., Mujumdar, A. S., Xiao, G. N., Sun, J. C., Drying of edamames by hot air and vacuum microwave combination. *Journal of Food Engineering* 2006, 77(4), 977–982.
51. Jiang, H., Zhang, M., Liu, Y., Mujumdar, A. S., Liu, H., The energy consumption and color analysis of freeze/microwave freeze banana chips. *Food and Bioproducts Processing* 2013, 91(4), 464–472.
52. Abbasi, S., Azari, S., Novel microwave–freeze drying of onion slices. *International Journal of Food Science and Technology* 2009, 44(5), 974–979.
53. Duan, X., Zhang, M., Mujumdar, A. S., Studies on the microwave freeze drying technique and sterilization characteristics of cabbage. *Drying Technology* 2007, 25(10), 1725–1731.
54. Wang, Y. C., Zhang, M., Mujumdar, A. S., Mothibe, K. J., Microwave-assisted pulse-spouted bed freeze drying of stem lettuce slices—Effect on product quality. *Food and Bioprocess Technologies* 2013, 6, 3530–3543.
55. Huang, J., Zhang, M., Adhikari, B., Yang, Z., Effect of microwave air spouted drying arranged in two and three stages on the drying uniformity and quality of dehydrated carrot cubes. *Journal of Food Engineering* 2016, 177, 80–89.
56. Musielak, G., Mierzwa, D., Kroehnke, J., Food drying enhancement by ultrasound—A review. *Trends in Food Science and Technology* 2016, 56, 126–141.
57. Gallego-Juárez, J. A., Riera, E., de la Fuente Blanco, S., Rodríguez-Corral, G., Acosta-Aparicio, V. M., Blanco, A., Application of high-power ultrasound for dehydration of vegetables: Processes and devices. *Drying Technology* 2007, 25(11), 1893–1901.
58. Kowalski, S. J., Pawłowski, A., Intensification of apple drying due to ultrasound enhancement. *Journal of Food Engineering* 2015, 156, 1–9.
59. Kroehnke, J., Radziejewska-Kubzdela, E., Musielak, G., Stasiak, S., Ultrasonic-assisted and microwave-assisted convective drying of carrot: Drying kinetics and quality analysis. In *Proceedings of the 5th European Drying Conference* 2015, Taylor & Francis, U.K.
60. Szadzińska, J., Łechtańska, J., Kowalski, S. J., Microwave and ultrasonic assisted convective drying of green pepper: Drying kinetics and quality. In *Proceedings of the 5th European Drying Conference* 2015, 391–398, Taylor & Francis, U.K.
61. Schössler, K., Thomas, T., Knorr, D., Modification of cell structure and mass transfer in potato tissue by contact ultrasound. *Food Research International* 2012, 49, 425–431.
62. Williams, D. C., Lim, M. H., Chen, A. O., Pangborn, R. M., Whitaker, J. R., Blanching of vegetables for freezing: Which indicator enzyme to choose. *Food Technology* 1986, 40, 130–140.
63. Barrett, D. M., Theerakulkait, C., Quality indicators in blanched, frozen, stored vegetables. *Food Technology* 1995, 49, 62–65.
64. Wang, Y. C., Zhang, M., Mujumdar, A. S., Mothibe, K. J., Roknul Azam, S. M., Effect of blanching on microwave freeze drying of stem lettuce cubes in a circular conduit drying chamber. *Journal of Food Engineering* 2012, 113, 177–185.
65. Dorantes-Alvarez, L., Jaramillo-Flores, E., González, K., Martinez, R., Parada, L., Blanching peppers using microwaves. *11th International Congress on Engineering and Food (ICEF11), Procedia Food Science* 2011, 1, 178–183.

66. Patriciaa, C. M., Bibiana, D. Y., José, P. M., Evaluation of microwave technology in blanching of broccoli (*Brassica oleracea* L. *var Botrytis*) as a substitute for conventional blanching. *11th International Congress on Engineering and Food (ICEF11) Procedia Food Science* 2011, 1, 426–432.
67. Wang, R., Zhang, M., Mujumdar, A. S., Effect of food ingredient on microwave freeze drying of instant vegetable soup. *LWT-Food Science and Technology* 2010, 43(7), 1144–1150.
68. Wang, R., Zhang, M., Mujumdar, A. S., Jiang, H., Effect of salt and sucrose content on dielectric properties and microwave freeze drying behavior of re-structured potato slices. *Journal of Food Engineering* 2011, 106(4), 290–297.

4 Highly Efficient Vegetable Drying Technology II

Infrared Radiation Drying and Related Combination Drying

Bengang Wu, Bei Wang, Haile Ma, and Baoguo Xu
Jiangsu University

Zhongli Pan
Jiangsu University, West Regional Research Center, USDA-ARS, and University of California, Davis

CONTENTS

4.1 INTRODUCTION

Food products, such as fish, meat, fruits and vegetables, and oils and fats consist of a number of nutritional components, such as proteins, lipids, carbohydrates, vitamins, aromas, pigments, antioxidants, and mineral compounds, which are beneficial to human health. However, a large amount of food is wasted every year in the world owing to the fact that these food products cannot be preserved for a long time under natural conditions. Therefore, different preservation methods including drying, freezing, frying, and cooking are needed to extend the shelf life of food products and to avoid wastage.

In the food industry, drying is one of the most common and energy-consuming unit operations with simultaneous heat and mass transfer. Drying occurs as a result of vaporization of water by supplying heat to wet food products (Mujumdar 2007; Sahni and Chaudhuri 2012). In food processing, drying is mainly used for preserving food products and extending their shelf life by reducing the moisture content to a low level (Shi et al. 2008). Based on the mechanisms of heat transfer, drying can be classified into direct (convection), indirect or contact (conduction), radiant (radiation), and dielectric or microwave (radio frequency) drying (Sahni and Chaudhuri 2012). Infrared (IR) drying is a type of electromagnetic radiant drying and is also called thermal radiation drying. Based on the action of IR wavelength, the radiation from a source interacts with the internal structure of the food products and thus increases its temperature and facilitates the evaporation of moisture in the food products (Ruiz Celma et al. 2008). During the IR drying process, a medium for transmission of energy from source to food products is not needed. When IR radiates to the surface of food products, a part of IR radiation may be reflected, absorbed, or transmitted. This depends on the nature of the IR radiation and the surface characteristics of the food products (Xu et al. 2014).

With an increase in consumer demand for healthy food and social awareness of sustainable technology, the modern food industry is always exploring sustainable innovative technologies that can produce high-quality safe products, enhance the processing efficiency, and reduce energy consumption. Therefore, drying processes should be designed to shorten the drying time and minimize energy and capital costs while maintaining high product quality (Xu et al. 2014). IR drying technology is an extremely important source of heat treatment in the food industry owing to its considerable advantages as described in the following sentences. Because of its excellent performance, IR drying is a promising technique for obtaining high quality of dried food products, such as apple slices (Toğrul 2005), grapes (Celma et al. 2009), peach (Wang and Sheng 2006), onion slices (Sharma et al. 2005a), carrot (Toğrul 2006), soybean (Niamnuy et al. 2012), kelp (Xu et al. 2014), wet olive husk (Ruiz Celma et al. 2008), etc. In addition, IR heating has also been widely used in the food industry for enzyme inactivation (Vishwanathan et al. 2013), peeling (Li et al. 2014a,b,c), and sterilization (Zhang et al. 2013) in recent years.

In this chapter, we provide a brief introduction on IR drying and the application of IR drying on vegetables. In addition, special processing technologies including enzyme inactivation, peeling, and sterilization are also described. IR hybrid drying technology and future trends and developments in IR drying of vegetables are also summarized.

4.2 BASIC INTRODUCTION OF IR AND IR DRYING

4.2.1 IR Radiation

Figure 4.1 shows the electromagnetic wave spectrum, which is composed of γ-rays, X-rays, ultraviolet, visible, infrared, microwave, radio and TV waves, and long-wave. IR is electromagnetic radiation in the wavelength range of 0.78–1000 μm. The American Test and Material Association has classified IR into three categories: near infrared (0.75–2.5 μm; NIR), mid infrared (2.5–25 μm; MIR), and far infrared (25–1000; FIR). IR radiation has both a spectral and directional dependence because of its electromagnetic properties. The spectral dependence is very important for the application of IR. This is because the IR energy from an emitter includes different wavelengths and each wavelength depends on several factors including temperature and emissivity of the emitter (Pawar and Pratape 2015). Electric heater and gas-fired heater are the two main types of IR heaters for commercial applications. Radiation of electric IR heater is emitted by passing an electric current through a resistance, resulting in temperature rise. The gas-fired IR heater emits radiation by ignition of a premixed air and fuel stream, initiating combustion on the burner surface (Seyed-Yagoobi and Wirtz 2007).

4.2.2 Mechanism of IR Drying

IR drying depends on the potential of IR radiation sources, which are IR radiation generators such as special electric lamps and ceramic or metallic panels heated by electricity or gas. The source interacts with the internal structure of the sample and thus increases its temperature and promotes moisture evaporation (Ruiz Celma et al. 2008). The distinctive feature of IR drying is that it can transmit the energy from the source to sample without a medium, and the sample can be regarded as the absorber of IR radiation to carry out the drying process. The major food constituents are water, protein, lipid, and sugar. Sandu (1986) reported that the water strongly absorbs the energy of IR radiation at an IR wavelength of 2.7–3.3, 6.0, and 12.5 μm. This is because the O–H bonds in water absorb the IR radiation energy and then rotate with the same frequency as the IR radiation. This transformation of IR radiation energy to rotational energy results in water evaporation. In addition, Sandu (1986) also reported that the wavelength values for absorption bands of proteins are at 3–4 and 6–9 μm, of lipids are at 3–4, 6, 9–10 μm, and of sugars are at 3 and 7–10 μm.

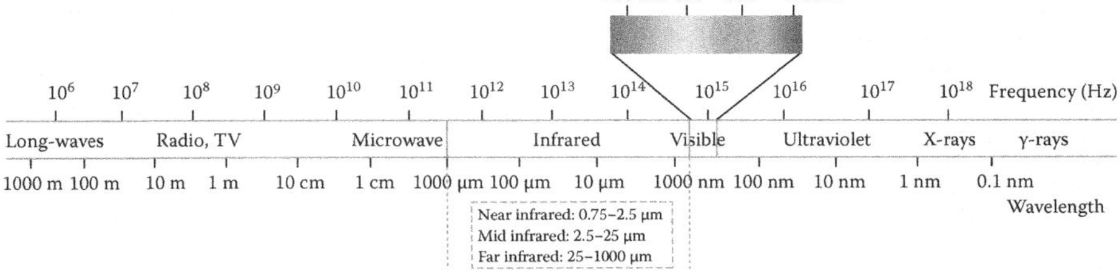

FIGURE 4.1 Electromagnetic wave spectrum.

Therefore, the matched degree between the sample and IR wavelength radiation to realize high absorption of IR radiation energy for the water inside the food sample is extremely important for enhancing the efficiency of drying.

4.2.3 Advantages of IR Drying

IR drying is regarded as one of the potential drying methods because of its intrinsic advantages such as high heat transfer rate, high drying rate, uniform temperature distribution, nutrient loss reduction, significant energy saving, and environment friendly.

High heat transfer rate: The transfer of the IR energy from heating source to the sample products does not require heating of the surrounding air. Thus, the temperature of the inner layers of the sample is higher than that of the surrounding air. As a result, the drying of the sample takes place from inner to outer layers via both radiation and convection thermal phenomena. This leads to a high rate of heat transfer (Ruiz Celma et al. 2008).

High drying rate: High heat transfer rate of IR results in reduction of drying time and increase of drying rate.

Uniform temperature distribution: Selecting an appropriate IR radiation wavelength band for food products is favorable for uniform temperature and energy distribution over the product surface.

Nutrient loss reduction: It is possible to generate a selective IR radiation in specific bands by using specific types of ceramics. The generation of specific IR radiation can significantly penetrate into the sample and efficiently affect the water on a molecular level. This method facilitates the drying process and reduces nutrient loss of the dried product (Bazyma et al. 2006).

Significant energy saving: IR energy is transferred directly to the sample according to the laws of optics. Thus, there is no gas resistance for heat flux as in the case of convection. This results in significant energy saving.

Environment friendly: The IR irradiation is harmless to humans (Bazyma et al. 2006). In addition, compared to fossil fuels, it does not pollute the environment and is also renewable.

4.3 APPLICATION OF IR DRYING ON VEGETABLES

4.3.1 Vegetable Drying by Mid-Near IR

Water has significant energy absorbance at an IR wavelength of 3.0, 4.7, and 6.0 μm (Nakamura 1969). Therefore, when mid-near IR radiates the surface of vegetables, the energy could be absorbed, reflected, or penetrated. Wang (2014) studied the feasibility of mid-infrared drying (MIRD) of Shiitake mushroom. The dehydration characteristics were studied systematically and the drying kinetics model was established, in comparison with several different drying techniques. Based on the IR absorption properties of the mushrooms, the feasibility of MIRD at the wavelength of 2.4–3.0 μm on mushrooms was confirmed. Factors such as heating temperature, air velocity, radiation distance, and loading coefficient were studied as well. The results showed that there was only a deceleration stage without constant stage. High temperature and air velocity, short radiation distance, and low loading lead to longer drying rate and shorter drying time. The optimum drying conditions for the Shiitake mushroom were heating temperature of 60°C, air velocity of 1.4 m/s, radiation distance of 14 cm, and loading of 2.0 kg/m (Wang 2014).

Kocabiyik et al. (2014) studied the physical and nutritional properties of tomatoes dried with short-IR radiation. A comprehensive analysis of dried products was performed based on their drying kinetics, drying time, specific energy consumption, shrinkage, rehydration ratio, color, and vitamin C and lycopene content. The results showed that drying time was prolonged with increasing air velocity, whereas it was shortened with increasing IR radiation intensity. The lowest energy consumption occurred at the air velocity of 1.0 m/s and at the IR radiation intensity of 2640 W/m^2. Use of short IR radiation resulted in the production of good quality tomatoes. Such as, shrinkage ratio varied between 0.139 to 0.203 and rehydration ratio varied between 2.14 to 3.40 for all the drying conditions, respectively. ΔE values varied from 5.70 to 13.06. The contents of vitamin C and beta-carotene in IR-dried tomatoes were decreased by 2%–51% and 5%–51%, respectively, whereas there was a significant increase in lycopene content varying between 50% and 529%. Overall, it was observed that IR drying of tomatoes provided good nutrient retention and consumed a low amount of energy. Therefore, IR radiation could be recommended in drying slices of tomatoes, in terms of nutritional quality of the product and to save energy.

Xu et al. (2014) studied the temperature and quality characteristics of IR radiation-dried kelp at different peak wavelengths (2.4, 3.0, 5.0, and 6.0 μm). Results showed that the drying time of IR-6.0 was the shortest, followed by IR-3.0, IR-5.0, and IR-2.4. However, the drying time for kelp between IR-6.0 and IR-2.4 was not very remarkable. In addition, temperature distribution of the IR-2.4 was found to be more uniform than the others, and the quality of the IR-2.4-dried kelp was the highest. Thus, the results indicated that near-IR drying can be also used as a method for drying of kelp. Wang and Zhang (2014) investigated the application of mid-IR drying in the preparation of mushroom chewing tablets. The results showed that mid-IR drying enhanced the drying rate for mushroom chewing tablets and preserved the temperature-sensitive materials in a better way by thermogravimetric analysis when compared to hot air drying. In addition, the total sensory score, flavor, and texture properties of mid-IR drying samples were better than those of hot air drying samples.

4.3.2 Vegetable Drying by Far IR

Dehydrated onions are generally dried with convection heating, which is inefficient and expensive. Gabel et al. (2006) compared the drying and quality characteristics of onion by catalytic infrared (CIR) heating and forced air convection (FAC) heating. In general, both CIR with and without air recirculation had shorter drying times, higher maximum drying rates, and greater drying constants than FAC at moisture contents greater than 50% (d.b.). Dried onion quality was similar in both CIR and FAC drying methods at 60°C and 70°C measured by pungent degradation. The product browning may be caused by the higher surface heat flux of the CIR heating and longer process times of FAC drying. A better product color (whiter and less yellow) was obtained for onions dried at lower temperatures for CIR and higher temperatures for FAC. The aerobic plate counts and coliform counts were not significantly different for either the product treated by CIR or the product treated by FAC drying. However, it was significantly lower for yeast and mold counts for the product dried by the CIR than those dried by the FAC. It is recommended that CIR should be used in the early stages of onion drying and FAC should be used in the later stages (Gabel et al. 2006).

Shi et al. (2008) studied the drying and quality characteristics of fresh and sugar-infused blueberries dried with IR radiation. IR drying tests were conducted at four product temperatures (60°C, 70°C, 80°C, and 90°C) to evaluate the drying rate and the color and texture of the finished product. The experimental data of moisture changes during IR drying were modeled with eight different models. The Thompson model showed the best fit to all experimental data. The IR drying produced firmer-texture products with much reduced drying time compared to hot air drying. For example, IR drying conserved drying time by 44% at 60°C.

China is ranked No. 1 in terms of the amount of vegetables exported in world trade, and the majority of them are leafy vegetables. Leafy vegetables are perishable, and since their nutritional value reduces faster postharvest, they should be processed within 24 h of harvest to maintain good quality. Currently, drying treatment of vegetables is the most effective method to extend their shelf life. Okamoto et al. (2012) studied far-IR drying of komatsuna. The authors determined quality changes (such as surface color and L-ascorbic acid, beta-carotene, and lutein contents) and energy consumption during the process. L-Ascorbic acid residual ratio after far-IR drying was significantly ($p < .05$) greater than that after hot air drying for identical drying rate constants. Energy consumption was 17% with far-IR drying less than that with hot air drying. Surface color or beta-carotene and lutein contents did not show any negative effects on far-IR drying and hot air drying. These results suggest that far-IR drying, along with hot air drying, is useful for drying of komatsuna.

Root vegetables are generally storage organs with a high carbohydrate concentration in the form of starch. Root vegetables are another important group of vegetables in international vegetable trade. Wu et al. (2014a) studied the drying and quality characteristics and models of carrot slices under catalytic infrared (CIR) heating. The influence of drying parameters on drying rate, time, surface color change, and rehydration ratio was determined by a two-factor factorial experiment design. Carrot slices with thicknesses of 3, 5, and 7 mm were dried at radiation distance of 26, 32, and 38 cm. It was observed that the time required decreased and the drying rate increased quite significantly as slice thickness decreased and/or the radiation distance decreased. The best processing parameters for CIR drying of carrot were radiation distance of 32 cm and slice thickness of 3 or 5 mm (Wu et al. 2014a).

Sharma et al. (2005b) studied the drying of onion slices by thin-layer IR radiation. A laboratory-scale IR-convective dryer was developed and single-layer drying of onion slices was carried out at different IR powers, air temperatures, and air velocities. Effective moisture diffusivity was significantly influenced by air temperature and infrared power. The drying time was reduced by about 2.25 times on increasing IR power from 300 to 500 W, air velocity from 1.0 to 1.5 m/s, and air temperature from 35°C to 45°C. The rehydration ratio of the dehydrated onion slices was found to be in the range of 4.5 and 5.3.

4.4 SPECIAL PROCESSING TECHNOLOGY ON IR DRYING OF VEGETABLES

4.4.1 Simultaneous IR Blanching and Drying of Vegetables

Blanching is an important procedure for the processing of most fruits and vegetables. Blanching also helps to control microbial population and keeps color stable for further processing. IR heating used to achieve a simultaneous blanching and dehydration effect is an innovative method, which is simpler and saves more energy than traditional methods like hot water blanching. During the IR heating, the radiation energy with specific wavelengths penetrates into products and heats water or desired components directly to achieve the purpose of blanching and dehydration. In the meantime, the IR energy does not heat the air and medium, and the energy transfer is highly efficient (Pan and Atungulu 2010).

The advantage of the simultaneous IR blanching and dehydration is that this technique uses no water or steam, contrary to the conventional methods of hot water blanching and steam blanching (Pan and Atungulu 2010; Pan and McHugh 2006). These conventional methods of blanching have many drawbacks such as low energy efficiency, quality deterioration, and environmental problems.

Wu et al. (2014) investigated the effects of various processing parameters on carrot slices exposed to IR radiation heating for achieving simultaneous blanching and dehydration. The influence of processing parameters on moisture reduction, drying rate, residual peroxidase (POD) activities, surface color change, and vitamin C retention was determined by a three-factor factorial design. Thin slices and/or high surface temperature resulted in faster inactivation of enzymes and quicker moisture removal compared to the thick slices and/or low surface temperature.

The quality of carrot slices at different heating times with IR dry-blanching, including residual POD, moisture reduction, and overall color change is shown in Figure 4.2. The process that produced 1 log reduction in POD activity resulted in moisture reduction from 40.2% to 88.8%, retention of vitamin C from 56.92 to 77.34 g/100 g, and overall color change (ΔE) from 3.17 to 5.13. It was concluded that simultaneous IR blanching and dehydration could be used as an alternative to produce high-quality blanched and partially dehydrated fruits and vegetables.

Zhu et al. (2010) investigated the effects of various processing parameters on apple slices exposed to IR radiation heating in a continuous heating mode for achieving simultaneous IR dry-blanching and dehydration (SIRDBD). The radiation intensity, slice thickness, and processing time were investigated. The influence of heating and drying rates, product temperature, moisture reduction, residual polyphenol oxidase (PPO), and POD activities and surface color change (ΔE) was determined by the three-factor factorial experiment. Thin slices and/or high radiation intensity quickly increased the product temperature and moisture removal as well as the inactivation of PPO and POD when compared with thick slices and/or low radiation intensity. The fractional conversion and the first-order kinetics models fitted well for POD and PPO inactivation curves. The simple page model also performed well for describing drying behavior during SIRDBD. Surface color changes were mainly due to the increase in a value and decrease in L value, which

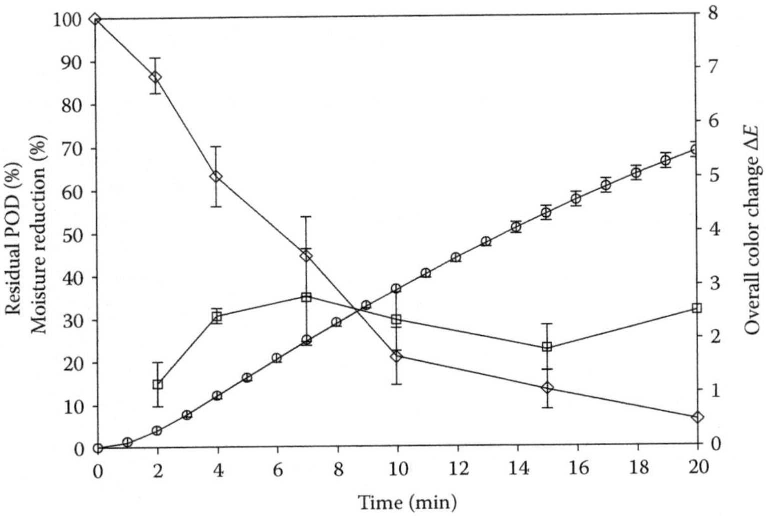

FIGURE 4.2 Quality of carrot slices at different heating times with infrared dry-blanching, ◇, residual POD; ○, moisture reduction; and □, overall color change ΔE. (From Wu, B. et al., *LWT-Food Sci. Technol.*, 57(1), 90–98, 2014b. With permission.)

corresponded to enzymatic browning that occurred during the process. It has been concluded that SIRDBD with continuous heating could be used as an alternative to the current processing methods for producing high-quality blanched and partially dehydrated fruits and vegetables.

4.4.2 Simultaneous IR Disinfestation, Sterilization, and Dehydration of Vegetables

Using heat treatment to achieve pasteurization or sterilization is a common method in food processing. However, high temperature may lead to a significant loss in the nutritional and sensory quality of food. For IR heating, quality changes can be reduced because of the lower total thermal energy supply. Up to 245% of energy reduction could be achieved by IR compared to conventional heating methods (Afzal et al. 1999; Krishnamurthy et al. 2007). IR heating has been widely applied to various thermal unit operations to achieve dehydration, frying, and pasteurization (Krishnamurthy et al. 2010; Sakai and Hanzawa 1994). In recent years, IR heating for pathogen inactivation has gained more attention.

Zhang et al. (2013) studied dehydration of spinach by catalytic IR radiation to improve the product microbial stability. The effectiveness of IR radiation and subsequent heat insulation to reduce total bacterial population in dehydrated spinach was evaluated in comparison to high-temperature steam sterilization. In addition, the changes in color and chlorophyll content in dehydrated spinach before and after these treatments were analyzed.

Logarithm values of total bacterial count of dehydrated spinach under IR treatment are shown in Figure 4.3. The results showed that the effectiveness of IR radiation followed by heat insulation for sterilizing dehydrated spinach is similar to high-temperature steam sterilization and is helpful in maintaining its quality.

Use of IR heating for improving the microbial safety and processing efficiency of dry-roasted almonds was investigated (Yang et al. 2010). Almonds were roasted at 130°C, 140°C, and 150°C with three different methods: IR roasting, sequential infrared and hot air (SIRHA) roasting, and traditional hot air (HA) roasting. The heating rate and pasteurization efficacy of almonds under different roasting methods and temperatures were evaluated. When SIRHA roasting at different temperatures was used to produce medium roasted almonds, bacterial reductions were achieved with 38%, 39%, and 62% time saving compared to HA roasting at each temperature, respectively. The activation energies were 73.58, 52.15, and 67.60 kJ/mol for HA, IR, and SIRHA roasting, respectively. No significant difference ($p > .05$) was observed in the sensory quality of medium-roasted almonds processed with different roasting methods. The authors conclude that the SIRHA roasting is a new promising method for the production of dry-roasted pasteurized almonds.

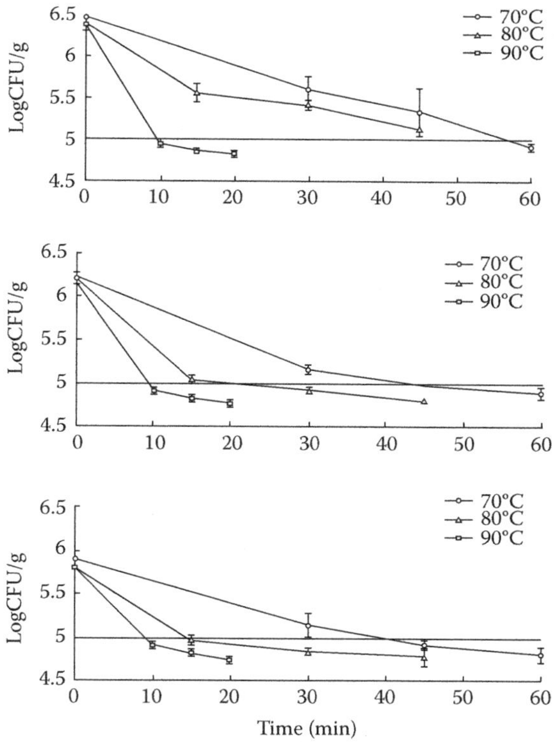

FIGURE 4.3 Logarithm values of total bacterial count of dehydrated spinach under different treatments. (From Zhang, X. et al., *Food Sci.*, 34(23), 133–137, 2013. With permission.)

4.4.3 IR Peeling of Vegetables

Literature (Li 2012; Pan et al. 2009) shows that in IR radiation heating can be used for peeling the tomato to eliminate the use of water, chemicals, and steam in the process. Rapid IR heating removed only a thin layer of tomato skin (<1 mm) while preserving the quality and nutritional values of the peeled tomatoes. Compared with conventional style peeling, IR dry-peeling of tomatoes showed a lower peeling loss and a firmer texture of peeled products while achieving a similar ease of peeling.

A pilot-scale IR dry-peeling system for tomatoes was designed and constructed by Pan et al. (2015). The system consisted of three major sections including the IR heating, vacuum, and pinch roller. The peeling performance of the system was examined under different operational conditions using tomatoes with different cultivars and sizes (Table 4.1).

Three lines of tomatoes were heated and processed at a residence time of 125 s, and 85% of the tomatoes were fully peeled, that is, a peeling yield of 82%, and an average thickness of 0.75 mm in the peeled tomato skin was achieved. Depending on tomato size, for achieving the same level of peeling percentage and yield, the required heating time was reduced to a range from 80 to 100 s, when tomatoes were loaded as a single line. The presence of the vacuum section could achieve cracks in 100% of the tomatoes after IR heating. The peeled products from IR heating had high firmness and appealing surface integrity, which indicated good quality. Because the dry-peeling is a chemical- and water-free process, residuals of tomato skins after IR peeling could be easily utilized as value-added by-products (Pan et al. 2015).

Wang et al. 2016 studied jujube peeling using novel IR radiation heating technology. The jujube fruits were heated using two electric IR emitters. Lye-peeled jujubes were used as a control. Heating with an IR intensity of 5.25 W/cm^2 at an emitter distance of 75 mm for 56 s were found as the optimum operating conditions, resulting in a peelability of 96%, peeling ease of 3.8, and moisture loss of 1.29% at the jujube

TABLE 4.1
Effect of Residence Time on Peeling Medium Size (50 mm) Tomatoes of cv. CXD 255 under Single-Line Heating

Residence Time (s)	Fully Peeled Percentage (s)		Peeling Yield (%)		Texture (N)		Temperature (°C)	
	Mean	Std	Mean	Std	Mean	Std	Mean	Std
60	76 a	8	85.7 a	4.8	10.9 a	2.2	95.9 a	2.3
80	80 a	8	81.0 b	5.1	8.8 b	2.8	97.0 a	3.2
100	88 b	5	81.6 b	1.0	8.6 b	2.0	105.5 b	1.8
120	94 c	7	76.7 c	3.6	8.2 b	2.6	113.0 c	2.6

Source: Pan, Z. et al., *Biosyst. Eng.*, 137, 1–8, 2015. With permission.

Note: Mean separation was analyzed via Duncan's Multiple Range Test. Means with a different letter in each column are significantly different at the .05 level.
Std, standard deviation.

surface temperature of 115°C. The IR-peeled jujube had significantly low peeling loss and color change compared to lye-peeled ones.

4.5 INFRARED HYBRID DRYING OF VEGETABLES

4.5.1 Combined IR and Freeze-Drying of Vegetables

Freeze-drying is widely used to obtain high quality dehydrated vegetables (Ratti 2001). However, freeze-drying is a slow and often low-throughput process, and the capital investment is high. Combining IR radiation with freeze-drying is a new method that has shown to have a great potential in industrial applications. Simultaneous IR freeze-drying is a two-step drying process of IR predehydration followed by freeze-drying.

Three major aspects of the dehydration process can be used to compare different techniques: the rate of dehydration, the quality and characteristic of the final product, and the economic and energy cost of the process. Some of the limitations of conventional dehydration technologies, such as air- or freeze-drying, can be addressed by combined IR and freeze-drying (Pan and Atungulu 2010).

Wang et al. (2014) studied the drying of shiitake mushroom by combining freeze-drying and mid-IR radiation. The effect of application of mid-IR drying before freeze-drying and after freeze-drying on drying time, color, rehydration ratio, apparent density, microstructure, and aroma compounds was measured. The results showed that the combination of freeze-drying followed by mid-IR drying saved 48% of the drying time compared to the freeze-drying method only, while the product quality was kept at an acceptable level. Mid-IR drying before freeze-drying was found to be inferior compared to mid-IR drying after freeze-drying as the former tended to produce products with a collapsed surface layer and poor rehydration capability. The combination of mid-IR drying with freeze-drying had a significant effect on aroma retention.

4.5.2 Combined IR and Hot Air Drying of Vegetables

Combining IR with hot air drying system is an efficient and rapid drying method compared to the usage of either of them separately. The combined system has advantages such as high quality of dried products and good drying characteristics, because heat and mass transfer takes place in a better way, which leads to reduction in drying time, decrease in energy consumption, and increase in energy efficiency. During drying, the temperature of the product is kept relatively low, and therefore IR is suitable for heat-sensitive products. Hebbar et al. (2004) showed that the energy and operation cost of combined drying model for many food and agricultural products were lower than conventional drying system. More and more researchers (Chen et al. 2015; Hebbar et al. 2004) had endeavored to find optimal operating conditions for combined IR and hot air drying to obtain improved products with high quality and low energy consumption.

Wanyo et al. (2011) studied on improvement of quality and antioxidant properties of dried mulberry leaves with combined far-IR radiation and air convection in

Thai tea processing. The authors observed a smaller decrease in L and b values in tea dried with combined far-IR radiation and hot air (FIR–HA) compared to commercial tea. FIR–HA tea was found to have a similar color like a fresh leaf, while the commercial tea had a darker color. A significant decrease in total phenolic acid content (TPC) and total flavonoid content (TFC) was found in hot air (HA)-dried commercial tea compared to fresh leaves, whereas TPC in FIR–HA-dried tea was significantly increased. Similar results were found in 1,1-diphenyl-2-picrylhydrazyl (DPPH) radical activities. However, the results were different for ferric-reducing antioxidant power (FRAP). Both the teas had lower FRAP values compared to that in fresh leaves. The TPC was increased in FIR–HA-dried samples compared to HA-dried tea, except for chlorogenic and syringic acids, which were found in larger amounts in HA-dried commercial tea. The results demonstrated that FIR–HA is a suitable drying method for mulberry tea, preserving its antioxidant properties and phenolic compounds (Wanyo et al. 2011).

4.5.3 Combined Infrared and Vacuum Drying of Vegetables

IR radiation can accelerate the drying process in food and agricultural materials. In order to obtain high-quality dried product, a lower drying temperature is preferred. Vacuum drying is one of the possible methods to achieve this goal, because under vacuum conditions moisture evaporates at a lower temperature because of its lower boiling point (Audebert and Temmar 1997). Drying under vacuum may also lead to dried products with high quality because of the lower drying temperature and lower level of oxidative reactions (Wu et al. 2007).

Ghaboos et al. (2016) studied the combined IR-vacuum drying of pumpkin slices. The results showed that the drying parameters such as IR radiation power, vacuum pressure, and slice thickness significantly affected the drying kinetics and various qualities of the dried pumpkin. The effective moisture diffusivity increased with power and ranged between 0.71 and $2.86 \times 10^{-9}\,m^2/s$. The β-carotene content of dried pumpkins was decreased with increase in IR radiation power and vacuum pressure, whereas ΔE values increased with increase in IR radiation power. In addition, the optimum condition (power: 238 W, pressure: 5 kPa and thickness: 5 mm) for combined IR-vacuum drying of pumpkin slices was obtained.

4.5.4 Combined Infrared and Microwave Drying of Vegetables

Microwave drying, which is characterized by low energy loss and high heating efficiency, has been reported for drying of fruits and vegetables for many years. However, it is well known that the problem of microwave drying is moisture accumulation at food surface, which results in a soggy surface instead of a crispy one (Aydogdu et al. 2015; Pawar and Pratape 2015). Therefore, the method to remove the surface moisture by additional heat source or by applying vacuum condition to increase the surface temperature needs to be adopted. IR drying with the advantages of high drying rate, energy saving, and uniform temperature distribution can be used to solve the problem of microwave drying.

Aydogdu et al. (2015) investigated the effects of microwave-infrared combination drying (MICD) on the quality of eggplants. They found that MICD had a significantly higher drying rate than that of hot air drying. In addition, the eggplants dried by MICD had a more porous structure than those dried by hot air drying, resulting in lower shrinkage and higher rehydration ratio for MICD eggplants. Therefore, MICD can be used for drying of eggplants, owing to its high drying rate and product quality. Saengrayap et al. (2015) studied the effect of far-infrared radiation-assisted microwave-vacuum drying (FIR-MVD) on drying characteristics and quality of red chilli. Different microwave powers of 100, 200, and 300 W under absolute pressures of 21.33, 28.00, and 34.66 kPa and different far-infrared powers of 100, 200, and 300 W were investigated. The results showed that the drying rate could be accelerated by an increase in microwave and IR power with a decrease in absolute pressure. In addition, the drying rate of FIR-MVD was higher than that of MVD only. Compared to MVD chilli, higher qualities (such as color changes, texture, rehydration ability, and shrinkage) of FIR-MVD chilli were also found. Eventually, the optimum condition for drying chilli at the microwave power of 300 W, under the absolute pressure of 21.33 kPa and with the applied FIR power of 300 W was also obtained. Sumnu et al. (2005) studied the effects of microwave-IR combination drying and hot air drying on the drying rate and quality attributes of carrots. The results exhibited that the drying time of MICD was reduced to 98% of hot air drying time. In addition, the MICD carrots had less color change and higher rehydration capacity than hot air drying carrots.

4.5.5 Infrared-Assisted Deep-Fried Dehydration of Vegetables

Bingol et al. (2012, 2014) used IR blanching as a pretreatment to produce lower-calorie deep-fried French fries. It was observed that complete inactivation of polyphenol oxidase enzyme could be achieved in 3 min for 9 mm French fries. The IR blanching-treated French fries had significantly lower oil content compared to the untreated sample. The result showed that following IR dry-blanching, the samples were fried at 146°C, 160°C, and 174°C for 1, 3, 5, and 7 min. At the end of 7-min frying, compared to unblanched samples, dry-blanched samples had 37.5%, 32%, and 30% less total oil at the frying temperatures of 146°C, 160°C, and 174°C, respectively. The final moisture contents of unblanched and dry-blanched samples were between 50% and 60% after 7-min frying. The sensory evaluation revealed that panelists preferred the IR dry-blanched French fries in terms of taste, texture, color, and appearance.

4.6 CONCLUSION

IR radiation heating technology is a new environment-friendly heating technology with the advantages of high efficiency and energy saving. IR can be used in vegetable drying, blanching, peeling, and can hybrid other drying technologies. Recently, the new IR heating technology has made significant progress in vegetable processing. It could be predicted that the further development and application of IR-related processing technology and equipment will help to achieve sustainable agricultural and food processing and produce high-quality, high-value, and healthy vegetables and foods.

REFERENCES

Afzal, T. M., Abe, T., and Hikida, Y. (1999). Energy and quality aspects during combined FIR-convection drying of barley. *Journal of Food Engineering*, 42(4), 177–182.

Audebert, P. and Temmar, A. (1997). Comparison of experimental results with analytical solutions and two-dimensional model of oak drying in an evacuated kiln. *Drying Technology*, 15(5), 1633–1652.

Aydogdu, A., Sumnu, G., and Sahin, S. (2015). Effects of microwave-infrared combination drying on quality of eggplants. *Food and Bioprocess Technology*, 8(6), 1198–1210.

Bazyma, L. A., Guskov, V. P., Basteev, A. V., Lyashenko, A. M., Lyakhno, V., and Kutovoy, V. A. (2006). The investigation of low temperature vacuum drying processes of agricultural materials. *Journal of Food Engineering*, 74(3), 410–415.

Bingol, G., Wang, B., Zhang, A., Pan, Z., and McHugh, T. H. (2014). Comparison of water and infrared blanching methods for processing performance and final product quality of French fries. *Journal of Food Engineering*, 121(1), 135–142.

Bingol, G., Zhang, A., Pan, Z., and McHugh, T. H. (2012). Producing lower-calorie deep fat fried French fries using infrared dry-blanching as pretreatment. *Food Chemistry*, 132(2), 686–692.

Celma, A. R., L6pez-Rodríguez, F., and Blázquez, F. C. (2009). Experimental modelling of infrared drying of industrial grape by-products. *Food and Bioproducts Processing*, 87(4), 247–253.

Chen, Q., Bi, J., Wu, X., Yi, J., Zhou, L., and Zhou, Y. (2015). Drying kinetics and quality attributes of jujube (*Zizyphus jujuba* Miller) slices dried by hot-air and short- and medium-wave infrared radiation. *LWT-Food Science and Technology*, 64(2), 759–766.

Gabel, M. M., Pan, Z., Amaratunga, K. S. P., Harris, L. J., and Thomson J. F. (2006). Catalytic infrared dehydration of onions. *Journal of Food Science*, 71(9), E351–E357.

Ghaboos, S. H., Ardabili, S. M., Kashaninejad, M., Asadi, G., and Aalami, M. (2016). Combined infrared-vacuum drying of pumpkin slices. *Journal of Food Science and Technology*, 53(5), 1–9.

Hebbar, H. U., Vishwanathan, K. H., and Ramesh, M. N. (2004). Development of combined infrared and hot air dryer for vegetables. *Journal of Food Engineering*, 65(4), 557–563.

Kocabiyik, H., Yilmaz, N., Tuncel, N. B., Sumer, S. K., and Buyukcan, M. B. (2014). Drying, energy, and some physical and nutritional quality properties of tomatoes dried with short-infrared radiation. *Food and Bioprocess Technology*, 8(3), 516–525.

Krishnamurthy, K., Khurana, H. K., Soojin, J., Irudayaraj, J., and Demirci, A. (2007). Infrared heating in food processing: An overview. *Comprehensive Reviews in Food Science and Food Safety*, 7(1), 2–13.

Krishnamurthy, K., Tewari, J. C., Irudayaraj, J., and Demirci, A. (2010). Microscopic and spectroscopic evaluation of inactivation of *Staphylococcus aureus* by pulsed UV light and infrared heating. *Food and Bioprocess Technology*, 3(1), 93–104.

Li, X. (2012). A study of infrared heating technology for tomato peeling: Process characterization and modeling. Dissertations and Theses, Gradworks Inc.

Li, X., Pan, Z., Atungulu, G. G., Wood, D., and McHugh, T. H. (2014a). Peeling mechanism of tomato under infrared heating: Peel loosening and cracking. *Journal of Food Engineering*, 128(1), 79–87.

Li, X., Pan, Z., Atungulu, G. G., Zheng, X., Wood, D., Delwiche, M., and McHugh, T. H. (2014b). Peeling of tomatoes using novel infrared radiation heating technology. *Innovative Food Science and Emerging Technologies*, 21(4), 123–130.

Li, X., Zhang, A., Atungulu, G. G., Delwiche, M., Milczarek, R., Wood, D., Williams, T., McHugh, T. H., and Pan, Z. (2014c). Effects of infrared radiation heating on peeling performance and quality attributes of clingstone peaches. *LWT-Food Science and Technology*, 55(1), 34–42.

Mujumdar, E. A. S. (2007). Book review: Handbook of industrial drying, third edition. *Drying Technology*, 25(25), 1133–1134.

Nakamura, A. (1969). Application of infrared radiation in food processing. *Journal of Symbolic Logic*, 34(3), 304.

Niamnuy, C., Nachaisin, M., Poomsa-ad, N., and Devahastin, S. (2012). Kinetic modelling of drying and conversion/degradation of isoflavones during infrared drying of soybean. *Food Chemistry*, 133(3), 946–952.

Okamoto, S., Orikasa, T., Kuwajima, G., Komoda, T., Saito, J., Yano, T., Muramatsu, Y., Koide, S., Shiina, T., and Tagawa, A. (2012). Application of far-infrared for drying of komatsuna. *Nippon Shokuhin Kagaku Kogaku Kaishi*, 59(9), 465–472.

Pan, Z. and Atungulu, G. G. (2010). *Infrared Heating for Food and Agricultural Processing*. Boca Raton, FL: CRC Press.

Pan, Z., Li, X., Bingol, G., McHugh, T. H., and Atungulu, G. (2009). Development of infrared radiation heating method for sustainable tomato peeling. *Applied Engineering in Agriculture*, 25(6), 935–941.

Pan, Z., Li, X., Khir, R., El-Mashad, H. M., Atungulu, G. G., McHugh, T. H., and Delwiche, M. (2015). A pilot scale electrical infrared dry-peeling system for tomatoes: Design and performance evaluation. *Biosystems Engineering*, 137, 1–8.

Pan, Z. and McHugh, T. H. (2006). Novel infrared dry blanching (IDB), infrared blanching, and infrared drying technologies for food processing. US Patent 20060034981.

Pawar, S. B. and Pratape, V. M. (2015). Fundamentals of infrared heating and its application in drying of food materials: A review. *Journal of Food Process Engineering*, doi: 10.1111/jfpe.12308.

Ratti, C. (2001). Hot air and freeze-drying of high-value foods: A review. *Journal of Food Engineering*, 49(4), 311–319.

Ruiz Celma, A., Rojas, S., and Lopez-Rodríguez, F. (2008). Mathematical modelling of thin-layer infrared drying of wet olive husk. *Chemical Engineering and Processing: Process Intensification*, 47(9–10), 1810–1818.

Saengrayap, R., Tansakul, A., and Mittal, G. S. (2015). Effect of far-infrared radiation assisted microwave-vacuum drying on drying characteristics and quality of red chilli. *Journal of food science and technology*, 52(5), 1–12.

Sahni, E. K. and Chaudhuri, B. (2012). Contact drying: A review of experimental and mechanistic modeling approaches. *International Journal of Pharmaceutics*, 434(1–2), 334–348.

Sakai, N. and Hanzawa, T. (1994). Applications and advances in far-infrared heating in Japan. *Trends in Food Science and Technology*, 5(11), 357–362.

Sandu, C. (1986). Infrared radiative drying in food engineering: A process analysis. *Biotechnology Progress*, 2(3), 109–119.

Seyed-Yagoobi, J. and Wirtz, J. W. (2007). An experimental study of gas-fired infrared drying of paper. *Drying Technology*, 19(6), 1099–1112.

Sharma, G. P., Verma, R. C., and Pathare, P. (2005a). Mathematical modeling of infrared radiation thin layer drying of onion slices. *Journal of Food Engineering*, 71(3), 282–286.

Sharma, G. P., Verma, R. C., and Pathare, P. B. (2005b). Thin-layer infrared radiation drying of onion slices. *Journal of Food Engineering*, 67(3), 361–366.

Shi, J., Pan, Z., McHugh, T. H., Wood, D., Hirschberg, E., and Olson, D. (2008). Drying and quality characteristics of fresh and sugar-infused blueberries dried with infrared radiation heating. *LWT-Food Science and Technology*, 41(10), 1962–1972.

Sumnu, G., Turabi, E., and Oztop, M. (2005). Drying of carrots in microwave and halogen lamp–microwave combination ovens. *LWT-Food Science and Technology*, 38(5), 549–553.

Toğrul, H. (2005). Simple modeling of infrared drying of fresh apple slices. *Journal of Food Engineering*, 71(3), 311–323.

Toğrul, H. (2006). Suitable drying model for infrared drying of carrot. *Journal of Food Engineering*, 77(3), 610–619.

Vishwanathan, K. H., Giwari, G. K., and Hebbar, H. U. (2013). Infrared assisted dry-blanching and hybrid drying of carrot. *Food and Bioproducts Processing*, 91(2), 89–94.

Wang, B., Venkitasamy, C., Zhang, F., Zhao, L., Khir, R., and Pan, Z. (2016). Feasibility of jujube peeling using novel infrared radiation heating technology. *LWT-Food Science and Technology*, 69, 458–467.

Wang, H. (2014). *Mid-Infrared Drying Shiitake Mushrooms and Its Hybrid Study*. Wuxi: Jiangnan University.

Wang, H. C., Zhang, M., and Adhikari, B. (2014). Drying of shiitake mushroom by combining freeze-drying and mid-infrared radiation. *Food and Bioproducts Processing*, 94, 507–517.

Wang, J., and Sheng, K. (2006). Far-infrared and microwave drying of peach. *LWT-Food Science and Technology*, 39(3), 247–255.

Wang, L., and Zhang, M. (2014). Application of intermediate-wave infrared drying in preparation of mushroom chewing tablets. *Drying Technology*, 32(15), 1820–1827.

Wanyo, P., Siriamornpun, S., and Meeso, N. (2011). Improvement of quality and antioxidant properties of dried mulberry leaves with combined far-infrared radiation and air convection in Thai tea process. *Food and Bioproducts Processing*, 89(1), 22–30.

Wu, B., Ma, H., Pan, Z., Wang, J., Qu, W., and Wang, B. (2014a). Drying and quality characteristics and models of carrot slices under catalytic infrared heating. *International Agricultural Engineering Journal*, 23(2), 70–79.

Wu, B., Pan, Z., Qu, W., Wang, B., Wang, J., and Ma, H. (2014b). Effect of simultaneous infrared dry-blanching and dehydration on quality characteristics of carrot slices. *LWT-Food Science and Technology*, 57(1), 90–98.

Wu, L., Orikasa, T., Ogawa, Y., and Tagawa, A. (2007). Vacuum drying characteristics of eggplants. *Journal of Food Engineering*, 83(3), 422–429.

Xu, B., Zhang, M., and Bhandari, B. (2014). Temperature and quality characteristics of infrared radiation-dried kelp at different peak wavelengths. *Drying Technology*, 32(32), 437–446.

Yang, J., Bingol, G., Pan, Z., Brandl, M. T., McHugh, T. H., and Wang, H. (2010). Infrared heating for dry-roasting and pasteurization of almonds. *Journal of Food Engineering*, 101(3), 273–280.

Zhang, X., Wen-Juan, Q. U., Hai-Le, M. A., Pan, Z. L., and Ben-Gang, W. U. (2013). Effectiveness of catalytic infrared radiation in the sterilization of dehydrated spinach. *Food Science*, 34(23), 133–137.

Zhu, Y., Pan, Z., McHugh, T. H., and Barrett, D. M. (2010). Processing and quality characteristics of apple slices under simultaneous infrared dry-blanching and dehydration with intermittent heating. *Journal of Food Engineering*, 97(1), 8–16.

5 Highly Efficient Vegetable Drying Technologies III

Ultrasound-Assisted Drying

S. J. Kowalski, D. Mierzwa, and J. Szadzińska
Poznań University of Technology

CONTENTS

5.1 INTRODUCTION

In the literature, the positive influence of ultrasound on the drying efficiency of biological materials like fruits and vegetables has been reported.[1–8] Some new and symptomatic experimental results based on modern equipment have also been reported, which seemed to be promising to food drying technology.[9–13]

Biological products like fruits and vegetables need very sublime drying methods, as they are very sensitive to high temperatures. Common drying methods (e.g., convective [CV] hot air drying) lead to strong degradation of valuable ingredients of biological products.[14–16] For this reason, new drying methods have been developed in the past few decades, such as CV, microwave, infrared, vacuum, and, recently, CV drying with ultrasound enhancement.[3,7,14] The positive outcomes of such hybrid drying methods may contribute to crucial changes in the drying technology of biological materials through

- Essential lowering of the drying temperature
- Reduction of energy consumption and its losses to the environment
- Shortening of the drying time
- Improvement of dried product quality through preservation of their valuable components (vitamins and minerals) and their appearance (color, and shape)

The new hybrid drying technologies seem to be eco-friendly. Recently new drying technologies utilize dryers equipped with ultrasonic setup, and these are used in the industry to increase productivity and decrease energy consumption.

This chapter focuses on recognizing the mechanism of ultrasound action on biological tissue, which leads to intensification of moisture removal from the dried products.[17,18] Based on research studies, it is found that high-frequency ultrasound causes periodical changes in pore pressure ("vibration effect") and insignificant "heating effect" in the dried material owing to absorption of ultrasonic energy.[18,19] Presumably, these phenomena contribute to the significant acceleration of liquid streaming from the interior of the tissue toward the surface, where it evaporates.

Both experimental and theoretical research regarding drying enhancement by ultrasound is described in this chapter. It is stated that power ultrasound can cause a high-frequency vibration of air near the dried material (the micro "vibration effect"), which leads to turbulence in the evaporation zone, and in this way promotes the heat and mass transfer processes. Meanwhile, a part of the ultrasonic energy is directly absorbed and leads to rise in temperature ("heating effect") in the dried material, which increases water vapor pressure in the evaporation zone. The combination of these two effects in addition to the "synergistic effect" in an ultrasound-assisted drying process will be discussed in this chapter to understand the mechanism of drying enhancement by ultrasound.[20–24]

5.2 DRYING TECHNOLOGIES FOR VEGETABLES

Fresh fruits and vegetables belong to a group of materials with low microbial stability due to high moisture content. Drying is one of the most popular processing methods to prevent food spoilage and decay. Dehydration of food materials leads to reduction in water activity and protection against bacteria, yeasts, and molds, thereby safeguarding these products and enabling their storage at ambient temperature. The main advantage of the drying process is that it enhances the shelf life of fruits and vegetables, reduces their volume and weight, and decreases the costs of storage and transportation. Biological food processed by drying has a wide range of applications, e.g., as a spice; as an addition to herbal teas, instant products, and muesli; as well as a healthy snack in the form of fruit bar and vegetable chips.

However, there are still disadvantages related to the drying process. This operation is extremely time and energy consuming, and both of these factors result in the high cost of this process. Another important factor is the effect of long-term exposure to high temperature, which directly influences the product quality. The value of color, shape, taste, flavor, nutrient content, and many other quality parameters may undergo significant changes. Because of the aforementioned problems, it is necessary to search for new and innovative drying methods that will allow the obtainment of better quality dried products and lead to a reduction of drying time and energy consumption.

Conventional drying techniques, like CV drying, are still extensively used. Hot air drying is commonly used owing to unquestionable advantages such as simple apparatus and a very well-known drying mechanism. However, the CV dryers should be operated under milder conditions while processing the heat-sensitive biological products to protect these materials against over-heating, shrinkage, discoloration, and hardening. However, on the other hand, a long-lasting hot air drying gives rise to low drying performance and high operating costs. Therefore, to overcome limitations of conventional dryers, some emerging drying technologies and new advancements in drying technology, such as hybrid drying, should be developed.

Hybrid mode combines different drying techniques such as CV drying, microwave (MW) drying, ultrasound drying, etc. These methods are characterized by different mechanisms of energy supply. In CV drying, the energy is transferred by the drying agent (air) and is used to evaporate the moisture from the material surface and to diffuse moisture from the inside of the material to the surface. Thus, the absorption of MW radiation leads to generation of heat in the entire material. Thus, the body temperature becomes higher inside the material when compared to its surface, which intensifies the heat and mass transfer. A completely different situation occurs during ultrasound drying. High-power ultrasound (20–500 kHz) causes various phenomena in the dried material, such as "vibration" and "heating" effects. Absorption of acoustic energy contributes to the rise in temperature of the material, which in turn, leads to increase in water vapor pressure in the dried body. High-frequency vibration of the air causes microvibrations and air turbulences near the material surface, which enhance the heat and mass transfer process.[25] Moreover, when a high-intensity ultrasonic energy travels through the body, it causes rapid contractions and expansions of the material tissue ("sponge effect"). This creates microscopic channels through which moisture dehydration is easier.[1]

Integration of various drying methods based on the theoretical knowledge allows to eliminate drawbacks and utilizes the advantages of combined techniques. The synergistic effect of such an action manifests itself most often as improvement in quality, reduction in drying time, and savings of energy. Experimental investigations have shown that appropriate and skillful use of convection and microwaves in a process significantly increases its efficiency. As shown by Kowalski and Mierzwa,[26] nonconventional methods, such as CV and MW drying, accelerate the drying rate of bell pepper almost 20-fold and shorten the drying time by 95%.

MW drying also improves the quality of dried products. Air drying followed by MW final drying showed that total color change (ΔE) in banana was reduced by more than three times, compared to the pure hot air and MW drying methods separately. Furthermore, in hybrid drying of cherries, the lowest value of water activity after application of MW in the falling drying period was observed. Another benefit of MW-related drying is energy saving. Combined hybrid drying is a cost-effective alternate system and is proven to minimize energy requirements.[27–29]

The first description on the intensification of heat and mass transfer by ultrasound was given just a few decades ago. Recently, high-power ultrasound has become an efficient tool for large-scale commercial applications such as in the drying processes. García-Pérez et al.[30] demonstrated that ultrasound application in air drying reduced the drying time of orange peel by up to 45%, significantly improved both the effective moisture diffusivity and the mass transfer coefficient, and reduced the total energy consumption by 20%. Similarly, Sabarez et al.[10] found a reduction in energy consumption and an increase in production throughput after the application of ultrasonic energy in the hot air drying process of apples. Moreover, many other researchers[7,31,32] showed that the ability of ultrasound to improve drying efficiency was greater at low temperatures, high ultrasonic power level, and low airflow rate. Furthermore, there are a lot of reports showing the positive effects of ultrasound on the final product quality. Power ultrasound is considered as a valuable tool to obtain high-nutritive dehydrated products than pure CV drying, because of its ability to retain higher vitamin C and β-carotene content. In turn, Gamboa-Santos et al.[33] showed that CVUS drying is an adequate procedure to obtain dried strawberry samples with high quality and appropriate microbiological stability. Application of airborne ultrasound in CV drying reduces the total color change ΔE (up to 11%) in comparison to the air-dried product.[34]

One of the important aspects of process engineering, including drying processes, is mathematical modeling. It reflects the actual processes carried out in the experiment and facilitates their designing. By mathematical modeling, it is possible to choose a suitable drying technique for heat-sensitive fruits and vegetables and to optimize (using computer simulation) the hybrid drying programs developed by experimental data.

This chapter describes ultrasound as an energy source that enhances the drying processes in biological materials. The study aims at a more profound recognition of the interaction between ultrasonic waves and the tissues of fruits and vegetables, which may contribute to an intensification of moisture removal during their drying. Absorption of acoustic energy causes heating and structural changes in the dried material because of rapid compressions and decompressions. The research hypothesis is based on the expectation that ultrasound may enhance moisture removal from the fruits and vegetables during drying owing to "vibration effect," "heating effect," and "synergistic effect." CV-ultrasound-assisted drying tests were conducted in a new hybrid dryer equipped with an ultrasonic generator. The effects of ultrasound on the drying process are based on a mathematical model and assessed quantitatively.

5.3 ULTRASOUND ACTION ON BIOLOGICAL TISSUE

5.3.1 Airborne Ultrasound

In the dryer with the ultrasound system, the ultrasound is delivered to the dried material through the air present in the drying chamber. The specially profiled ultrasound cylindrical plate is forced to vibrate by the piezoelectric transducer, so that ultrasonic waves in the air are generated by the vibrating plate (Figure 5.1).

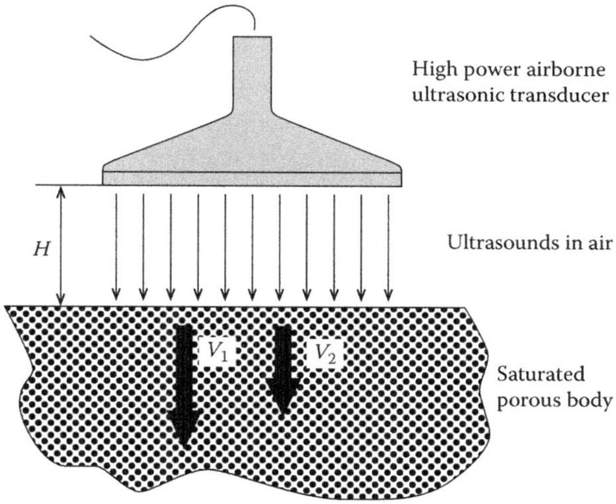

FIGURE 5.1 Ultrasound transmitted from the high-power ultrasonic transducer through air to the dried material.

To estimate the acoustic pressure acting on the dried material, the forced vibrations of the air particles need to be considered. A simple outline of ultrasonic wave propagation in air is described by using the following wave equation:[35]

$$\frac{\partial^2 u(x,t)}{\partial t^2} = c^2 \frac{\partial^2 u(x,t)}{\partial z^2} + F(x,t), \quad c = \sqrt{\kappa \frac{p}{\rho}}, \tag{5.1}$$

where u denotes the displacement of air particles in z direction, F is the force per unit mass, c is the speed of ultrasound in air, p and ρ denote the air pressure and density, and κ is the adiabatic index.

The solution for steady-state vibrations ($F = 0$) of the air particles takes the form

$$u(z,t) = \frac{\beta_0 \sin\alpha(H - z) + \gamma_0 \sin\alpha z}{\sin\alpha H} \sin\omega t, \tag{5.2}$$

where $\alpha = \omega/c$ is the wave number, H is the distance between the vibrating plate and the dried material, and β_0 and γ_0 are the vibration amplitudes at $z = 0$ and $z = H$, respectively.

The energy of an ultrasonic wave is equivalent to the maximum of kinetic energy of the vibrating air particles, which is

$$E(z) = \frac{1}{2}\rho\left(\frac{\partial u}{\partial t}\right)^2_{max} = \frac{1}{2}\rho\omega^2\left(\frac{\beta_0 \sin\alpha(H - z) + \gamma_0 \sin\alpha z}{\sin\alpha H}\right)^2 \left[\frac{\text{J}}{\text{m}^3}\right]. \tag{5.3}$$

The power of ultrasonic wave is determined as

$$I(z) = E(z)c = \frac{1}{2}\rho c\omega^2 \left(\frac{\beta_0 \sin\alpha(H-z) + \gamma_0 \sin\alpha z}{\sin\alpha H} \right)^2 \left[\frac{\text{W}}{\text{m}^2} \right]. \tag{5.4}$$

Assuming a perfect energy transition from the ultrasonic transducer (plate) to the air ($z = 0$), one can determine the incident ultrasonic power as

$$I(0) = \frac{1}{2}\rho c\omega^2 \beta_0^2 \left[\frac{\text{W}}{\text{m}^2} \right]. \tag{5.5}$$

Thus, the amplitude of air particle vibration close to the vibrating plate reads

$$\beta_0 = \sqrt{\frac{2I(0)}{\rho c\omega^2}} \quad [\text{m}]. \tag{5.6}$$

For an ultrasonic generator with a power of 200 W and plate of diameter $D = 0.4$ m, which gives $I(0) = 1592.36\,\text{W/m}^2$, and $\omega = 26\,\text{kHz}$, $c = 340\,\text{m/s}$, $\rho = 1.2\,\text{kg/m}^3$, $\beta_0 = 0.12\,\text{mm}$ is obtained.

The incident ultrasonic power that enters the dried material reads

$$I(H) = \frac{1}{2}\rho c\omega^2 \gamma_0^2 \left[\frac{\text{W}}{\text{m}^2} \right]. \tag{5.7}$$

The acoustic pressure p' expresses the difference between the total air pressure arising as the result of wave transition and the equilibrium pressure p_0 as follows:

$$p'(z,\ t) = p(z,\ t) - p_0 = \rho c \frac{\partial u}{\partial t} - p_0 = \rho c\omega \frac{\beta_0 \sin\alpha(H-z) + \gamma_0 \sin\alpha z}{\sin\alpha H} \cos\omega t \ [\text{Pa}]. \tag{5.8}$$

The acoustic pressure acting on the dried material surface reads

$$p'(H,t) = \rho c\omega\gamma_0 \cos\omega t = \rho c\dot{u}(H,t) \ [\text{Pa}], \tag{5.9}$$

where $\dot{u}(H,t)$ is the velocity of air particles at the interface of air-dried material.

In fact, a part of the ultrasound power is reduced on the way between the piezoelectric transducer and the dried material, so that $\beta_0 \geq \gamma_0$. If, however, the air is considered to be an elastic compressible medium that does not attenuate the ultrasonic wave, then $\beta_0 = \gamma_0$. Assuming the amplitude of air vibrations to be $\gamma_0 = \beta_0 = 0.12\,\text{mm}$, the maximal acoustic pressure is of the order

$$p'(H)_{\max} = \rho c w \gamma_0 = 1.2 \cdot 340 \cdot 26 \cdot 10^3 \cdot 0.12 \cdot 10^{-3} = 10608 \text{ Pa} = 10.61 \text{ kPa}. \tag{5.10}$$

The high-frequency vibration of air near the dried body surface causes a micro "vibration effect," i.e., air turbulence at the surface promotes the heat and mass transfer processes. Simultaneously, a part of the ultrasonic energy is absorbed by the dried material and contributes to an increase in temperature ("heating effect"), which causes an increase in water vapor pressure.

5.3.2 Ultrasound Incidence on Biological Material

The ultrasonic transducer causes vibrations of air particles, which propagate in the form of waves toward the dried material. The material expansion/compression cycle can be sinusoidal, mimicking that of ultrasonic waves. This section aims to show how the airborne vibration affects the dried material.[36] As an example, the structural and textural properties of apple tissue are assessed. Alternatively, for certain vibration amplitude sizes and acoustic pressures, the vibration expansion phase extends the pore size on the material surface and is then followed by violent compression, leading to a return to the smaller pore size.

Figure 5.2 presents a simple experimental setup used to test the effects of ultrasound incidence on biological material.

The cycles of expansion/compression of pores in the material owing to ultrasonic waves can be described according to the oscillating velocities (Equation 5.2, where $z = H$)

$$\dot{u}(H,t) = \gamma_0 \omega \cos \omega t, \tag{5.11}$$

where γ_0 is the vibration amplitude and ω is the ultrasound frequency.

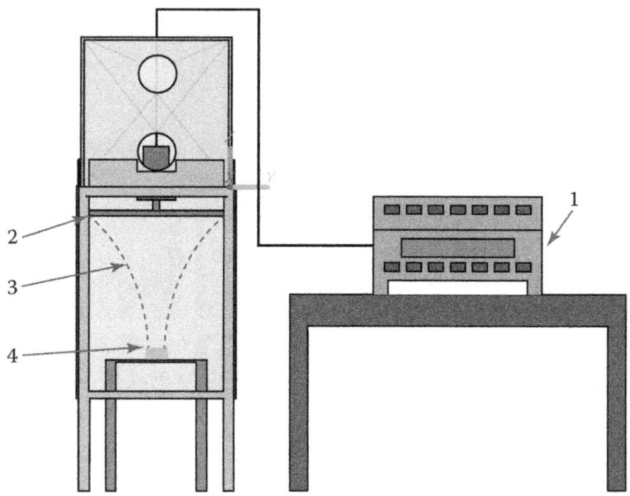

FIGURE 5.2 Experimental setup for testing the effects of ultrasound incidence on biological material: 1, ultrasound generator and amplifier; 2, disc transducer; 3, ultrasonic field outline; and 4, sample.

The acoustic pressure acting on the dried body surface reads

$$p'(H,t) = \rho c \omega \gamma_0 \cos \omega t \text{ [Pa]}, \tag{5.12}$$

where ρc is the impedance of the ultrasonic wave.

The impedance characterizes the dynamic properties of the air, i.e., it expresses the ratio of acoustic pressure to air particle velocity.

$$\rho c(H,t) = \frac{p'(H,t)}{\omega \gamma_0 \cos \omega t}. \tag{5.13}$$

The improvement in moisture transfer by ultrasound depends on pressure variations, oscillation velocities, microstreaming on the solid–gas interface and on the internal structure owing to a series of cyclic and rapid (>20 kHz) expansions and compressions. The mechanism known as the "sponge effect" is dependent on process variables such as temperature, air velocity, acoustic pressure, and dried product properties.

To analyze the effects of airborne ultrasound incidence on the apple sample, a visualization of the texture and porosity was carried out using the SEM microscopy. Cubic samples of raw apples (5 mm in dimension) were examined under the microscope to determine how ultrasound affects the structural parameter such as porosity.

From the observations of a number of microstructures and comparison of the microstructure in Figure 5.3, it is concluded that the incidence of ultrasound on apple tissue increases the pore dimensions in the decreased pressure zone and expands them in the lowered pressure zone (a kind of "sponge effect").

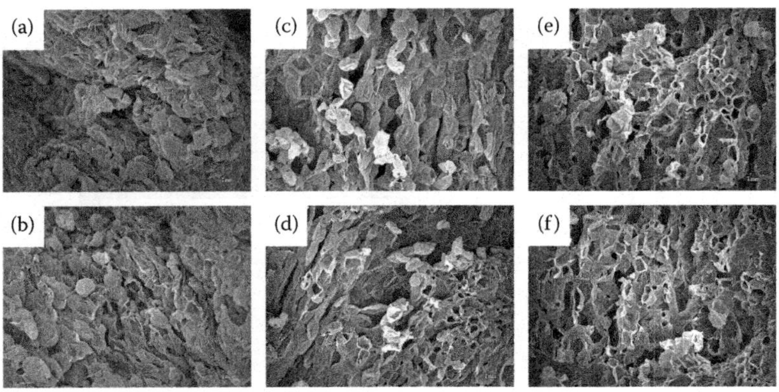

FIGURE 5.3 SEM images of apple tissue: (a, b) raw; (c, d) after 15 min of ultrasonic treatment with a power of 100 W; (e, f) after 15 min of ultrasonic treatment with a power of 200 W.

5.4 EXPERIMENTAL STUDIES

5.4.1 Equipment

The laboratory of the Department of Process Engineering at the Poznań University of Technology is equipped with a hybrid dryer constructed by the PROMIS-TECH company in Wrocław (Poland). The dryer is provided with ultrasonic systems supplied by the PUSONICS Company from Madrid (Spain) (Figure 5.4).

The hybrid dryer used for the experimental tests is equipped with three sources of energy: the hot air system, the magnetron, and the ultrasonic generator, which can work simultaneously or separately. The dryer has the capacity to work at an air temperature of up to 90°C, MW power of up to 500 W, and/or ultrasound power of up to 200 W with a frequency of 26 kHz. It is also possible to set a desired airflow through the scale pan with a velocity of up to 4 m/s.

A great advantage of such an experimental setup is the possibility of being able to measure all the process parameters online, i.e., drying time, material temperature, inlet air temperature, outlet air temperature, inlet air humidity, outlet air humidity, airflow, microwave power, ultrasound power, and energy usage. The dryer enables CV drying with different temperatures and velocities of the airflow, which can be set in advance with the programmer, and for CV drying enhanced with ultrasound and/or MW.

The following settings of the drying parameters were chosen:

- Air temperature: 40°C, 45°C, and 50°C (±1°)
- Air velocity: 0.72, 2, 3, and 4 m/s (±0.1 m/s)
- Ultrasound of frequency 26 kHz and power of 100 and 200 W

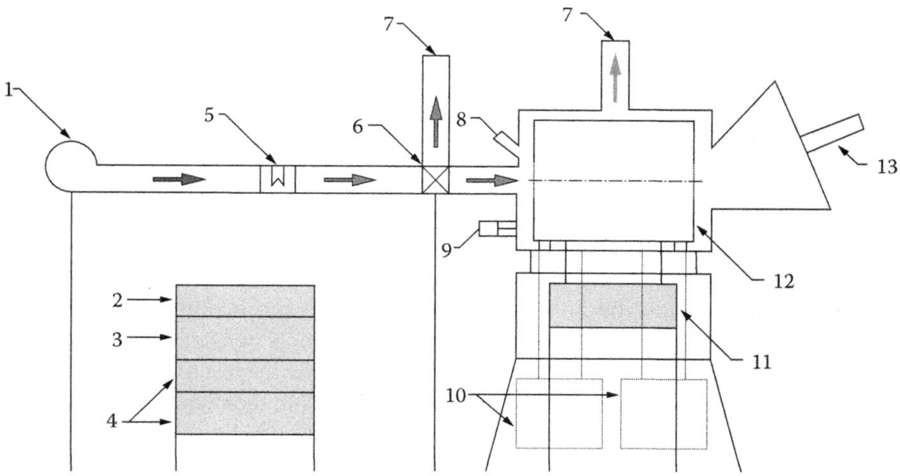

FIGURE 5.4 Hybrid dryer equipped with a convective heating system and ultrasound generator and magnetron: 1, computer (data acquisition unit); 2, hot air preparatory unit (heater + fan); 3, drying chamber; 4, ultrasonic transducer; 5, magnetron; 6, control unit; 7, balance; 8, ultrasonic amplifier and power unit; 9, drum drive; 10, microwave generator; 11, balance; 12, rotary drum; and 13, AUS ultrasound transducer.

5.4.2 Tested Material

Carrot (*Daucus carota* L.) is commonly found in areas of Europe, Asia, and North Africa and is mostly cultivated as a vegetable or fodder. As it is an edible root of high nutritional value and healing properties, it is widely used in the food industry. As a source of very high content of β-carotene, carrots possess attractive color and also valuable biological properties (provitamin A). Unfortunately, it is degraded when subjected to light, heat, and oxidizing agents. Dried carrots are mostly used as an ingredient in dry mixes for soups, sauces, and ready-meals, as well as healthy snacks. Because the β-carotene has a strong antioxidant activity, it can be used to treat degenerative diseases.[16,28]

The application of ultrasound in carrot drying is a relatively new technique that accelerates this process and improves the quality aspect of food products.[28] In these studies, the effort was made to find the best drying conditions for carrot from the kinetics and product quality point of view, combining ultrasonic pretreatment and CV drying.

First, hot air drying (AD) at a stable air temperature of 70°C was carried out. At the beginning of the process, a relatively high drying rate was observed. Therefore, dehydration occurred fairly rapidly, which affected the moisture distribution and caused shrinkage, rendering a rather bad quality to the dry material. The next drying tests at the same air temperature, but preceded by osmotic dehydration (OD) in aqueous fructose solutions of 40% for 2 h at 25°C and by OD with ultrasound, were performed to improve the kinetics and quality of carrots.

Figure 5.5 presents the drying kinetics for AD (Figure 5.5a), air drying preceded by osmotic dehydration (ODAD) (Figure 5.5b), and air drying preceded by ultrasound-assisted osmotic dehydration (USODAD) (Figure 5.5c).

The longest average drying time of 360 min was observed for the samples dried under stationary conditions. However, the carrots previously pretreated by osmosis achieved the final moisture content after 265 and 220 min.

Next, the hybrid drying tests of green pepper (*Capsicum annuum* L.), including CV drying as a reference, were carried out. The process parameters were as follows: air temperature T_{air} = 55°C, airflow velocity v = 2 m/s, ultrasound power = 100 and 200 W. The drying kinetics such as the drying curves, the air, and the material temperature curves were determined.

Figure 5.6a shows CV drying, which was carried out at constant air temperature of 55°C. In this case, the material temperature was below the air temperature during the drying process, and the final moisture content was reached after 759 ± 9 min. Owing to long-lasting hot AD, the dry material was strongly deformed and shrunken. As can be seen from Figure 5.6a, CV was found to be the longest drying process.

The second test concerned CV drying with an ultrasound enhancement of 100 to 200 W. The total drying time of green pepper samples was 514 ± 9 and 464 ± 5 min, respectively. In the hybrid process, the supply of energy resulted in the material temperature being close to air temperature in CVUS100 drying, and the heating effect occurred at the end of the process (Figure 5.6b). This phenomenon was different in CVUS200. Owing to higher ultrasound power, the pepper temperature increased after about 180 min of drying time (Figure 5.6c). In these two cases, the dried green pepper samples were less deformed.

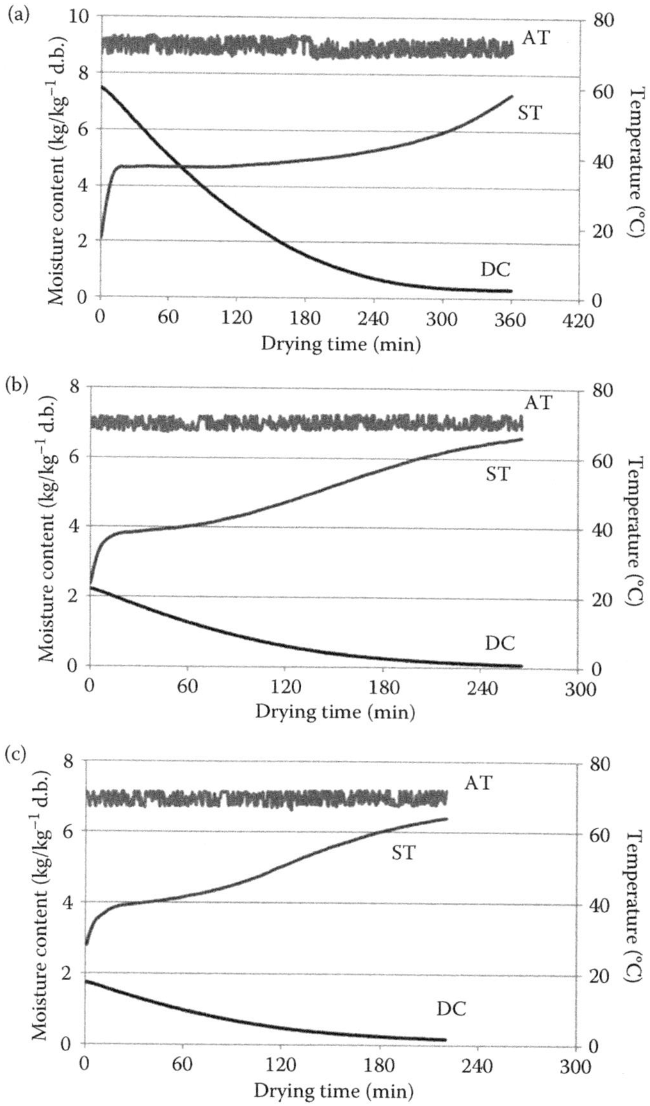

FIGURE 5.5 The drying curve (DC) and the temperature curves (AT, ST): (a) AD, (b) ODAD (40%), and (c) USODAD (40%).

Apart from the aforementioned vegetables, fruits such as apples (Ligol variety) and strawberries (*Fragaria ananassa*) were also used for drying experiments. They were purchased at the local market. Before the experiments, the apples were refrigerated at 4°C. Shortly before the drying tests, the apples were taken out of the fridge, and 9 or 16 sample slices of dimensions 40 × 20 × 5 mm (length × width × height)

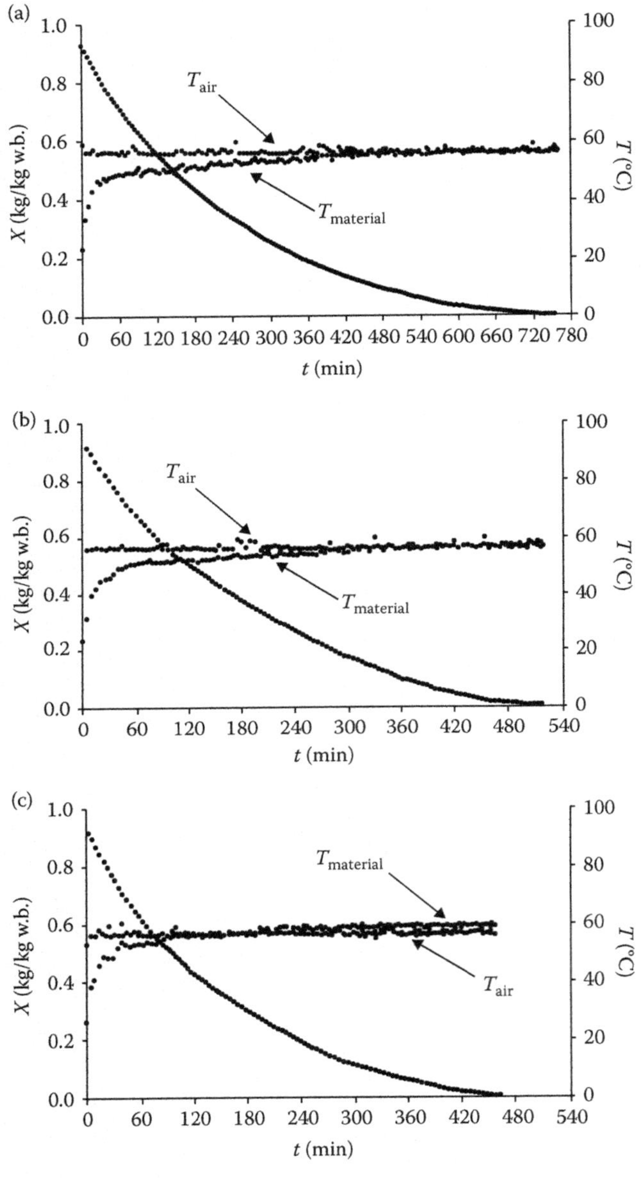

FIGURE 5.6 The drying curves and the temperature profiles: (a) CV, (b) CVUS100, and (c) CVUS200.

were cut and prepared for two different tests. The measurements of moisture mass reduction are attributed to a single sample slice, taken as an average from the 9 or 16 slices dried simultaneously in the individual drying tests. During each experiment, when preparing the samples, the initial moisture content of the apples was measured using a moisture analyzer to evaluate the approximate final mass of the processed material. Another more time-consuming method of moisture evaluation was used to precisely measure the initial moisture content of the apple samples. This was done by placing the apple sample in the CV dryer at 70°C for 24 h. After this period, the sample was weighed and the initial moisture content was estimated. It was specified that the average mass of a single slice amounted to 3.648 g, and its initial moisture content was 6.6 kg H_2O/kg d.b. The samples were placed on a scale pan as a monolayer and put into the dryer.

First, the experiments without ultrasound assistance were carried out at air temperature of 45°C with the airflow rate of 0.72 m/s through the scale pan. Next, the experiments with ultrasound assistance were carried out under the same conditions and with ultrasound power of 100 and 200 W.

Figure 5.7 presents the results of pure CV drying and of CV drying enhanced with ultrasound for apple samples in the form of rectangular slices of dimensions 40 × 20 × 5 mm.

The kinetics of pure CV drying at air temperature of 45°C and airflow of 0.72 m/s is shown by curve 1. Curves 2 and 3 represent the CV drying with ultrasound assistance (100 and 200 W).

Prior to each drying experiment, four fresh strawberry fruits of about 80 g with an initial water content of about 8.96 kg H_2O/kg d.b were deprived of stalks, washed in tap water, drained with blotting paper, and cut into eight halves. The samples were placed on a reticular pan in the form of a ring, with the skin side down, and then were dried to the final moisture content of about 0.01 kgH_2O/kg d.b.

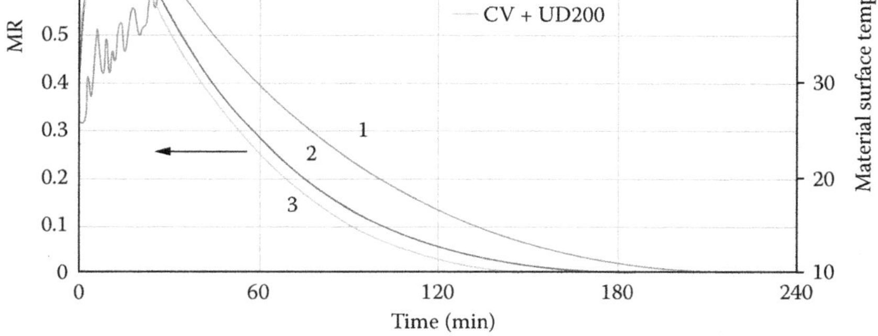

FIGURE 5.7 Drying curves and material temperature in the convective drying of apple samples: 1, pure convective drying; 2, convective drying with the assistance of ultrasound (100 W); and 3, convective drying with the assistance of ultrasound (200 W).

CV drying test of strawberries at air temperature of 50°C was carried out as a reference test (Figure 5.8).

The next series of experiments concerned the convective drying assisted with ultrasound (CVUS) (Figure 5.9). Application of an additional heat energy source in the form of high-power airborne ultrasound affected the moisture removal from strawberries and reduced the total drying time.

In CVUS, the overall drying time was about 600 min and was shorter by about 52%, as compared to CV drying. These results correspond to the results reported by Ortuño et al.[37] and Gamboa-Santos et al.,[33] where the application of ultrasound reduced drying time of orange peel by up to 48% and that of strawberries by up to 44%, compared to CV drying.

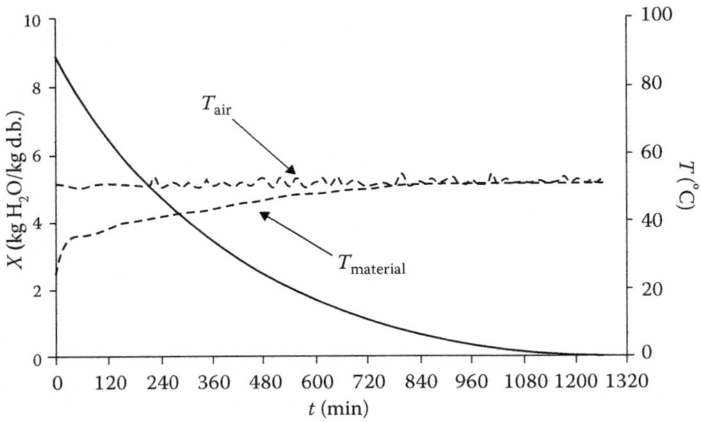

FIGURE 5.8 Convective drying of strawberries.

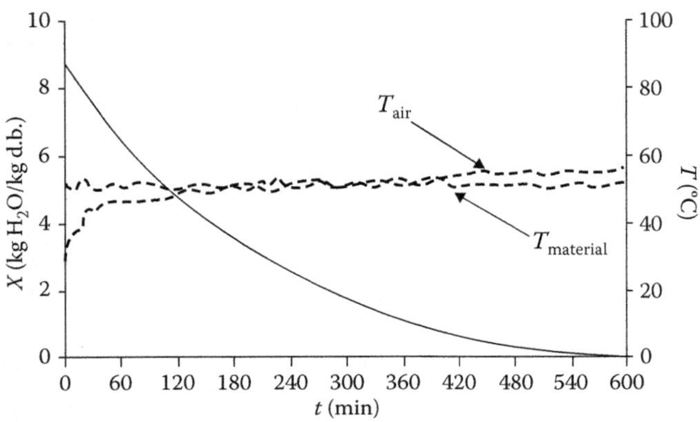

FIGURE 5.9 Convective drying of strawberries assisted with ultrasound.

5.5 MODELING

5.5.1 DRYING KINETICS

A mathematical model of drying kinetics is needed to assess the effectiveness of drying enhanced by ultrasound. For this, the equations of drying kinetics developed earlier by the author[38–41] are used. Drying kinetics expose the moisture content and temperature of the dried body as a function of time. The proposed mathematical model of drying kinetics enables numerical computation of the history of the drying curves and temperature of the drying material, which should have satisfactory adherence to that determined experimentally.

The mass and the energy balances constitute the basis for the construction of respective equations to determine the drying kinetics. The balance equations are of the form[40,41]

$$m_s \frac{dX}{dt} = -A_m J_m, \tag{5.14a}$$

$$m_s \frac{d}{dt}\left[(c_s + c_l X)T\right] = A_T J_T - A_m l J_m + \Delta Q, \tag{5.14b}$$

where m_s is the mass of dry body, X is the moisture content (dry basis), c_s and c_l are the specific heat for solid and liquid, J_m denotes the CV moisture flux, J_T is the convective heat flux, l is the latent heat of evaporation, A_m and A_T are the surfaces of mass and heat exchange, and ΔQ is the heat source due to ultrasound absorption.

To determine the rate relations for CV heat and mass fluxes, a boundary layer consisting of air and vapor mixture and temperature different from that of the surroundings is considered. The boundary layer of temperature T, vapor concentration y, and vapor chemical potential $\mu(T, y)$ is located close to the dried body. The ambient air is characterized by temperature T_a, vapor concentration y_a, and vapor chemical potential $\mu_a(T_a, y_a)$.[42,43]

Because the process of heat and mass transfer is irreversible, the entropy is produced during this process. The rate of entropy production per unit surface is analyzed. Assuming that the expression for entropy production is a positively defined quadratic form, there are sufficient conditions required for the heat and mass fluxes to satisfy this expression.[39,41,42] The sufficient conditions represent the coupled fluxes of heat and mass transfer. To fulfill the positively defined thermodynamic inequality, the phenomenological coefficients appearing in these rate equations have to be positive. In these considerations, the coupling Soret and Dufour effects[44,45] were assumed to be very small and therefore neglected in further considerations. Thus, the rate equations for heat and mass fluxes are reduced to the form[40,42,45]

$$J_T = h_T(T_a - T), \tag{5.15a}$$

$$J_m = L_{22}\left(\frac{\mu}{T} - \frac{\mu_a}{T_a}\right) \approx h_m \ln\frac{y}{y_a} = h_m \ln\frac{p_v|_{\partial B}}{p_{va}} = h_m \ln\frac{\phi|_{\partial B}\, p_{vs}(T)}{\phi_a\, p_{vs}(T_a)}, \tag{5.15b}$$

where h_T is termed the coefficient of CV heat transfer and h_m is termed the coefficient of convective vapor transfer.

The logarithm of the molar vapor concentration ratio in air (Equation 5.15b) constitutes the driving force for vapor transfer. This driving force was convenient to write using the relative air humidity $\varphi = (p_v/p_{vs})|_T$ and the vapor partial pressure for the saturated state p_{vs} as a function of temperature T.

The air relative humidity close to the surface of the drying sample $\varphi|_{\partial B}$ depends on the sample moisture content. One assumes the following form of air relative humidity at the sample surface as[46]

$$\phi|_{\partial B} = \begin{cases} 1 & \text{for} \quad X \geq X_{cr} \\ 1-(1-\phi_a)\dfrac{X_{cr}-X}{X_{cr}-X_{eq}} & \text{for} \quad X_{cr} \geq X \geq X_{eq} \end{cases}. \tag{5.16}$$

Parameters φ_a and T_a in Equations 5.15a and b denote the air relative humidity and temperature and are termed as the parameters of drying. The critical X_{cr} and equilibrium X_{eq} moisture contents in the dried sample are the parameters determined experimentally. Function $p_{vs}(T)$ is given in the literature in the form of tables.[47,48]

Thus, one can state, based on Equations 5.15b and 5.16, that the driving force responsible for moisture removal is constant for the saturated body ($X \geq X_{cr}$), obviously if the drying conditions are stable, and decreases for the unsaturated body ($X_{cr} \geq X \geq X_{eq}$), tending to air relative humidity φ_a for $X = X_{eq}$.

The governing equations describing the drying kinetics in the final form are

$$m_s \frac{dX}{dt} = -A_m h_m \ln \frac{\phi|_{\partial B}\, p_{vs}(T)}{\phi_a p_{vs}(T_a)}, \tag{5.17}$$

$$m_s \frac{d}{dt}\left[(c_s + c_l X)T\right] = A_T h_T (T_a - T) - A_m l h_m \ln \frac{\phi|_{\partial B}\, p_{vs}(T)}{\phi_a p_{vs}(T_a)} + a_U \chi_U P_U, \tag{5.18}$$

where a_U [–] denotes the dimensionless absorption coefficient of the ultrasonic wave, χ_U [–] is the dimensionless working efficiency of the ultrasonic transducer, and P_U [W] is the power of the ultrasonic generator.

It is assumed here that the thickness of the dried samples in the experimental tests is small, and so the acoustic energy attenuation is considered to be insignificant. The absorption coefficient of ultrasound a_U is determined by experiments. One assumes that the acoustic energy absorbed by the dried material is all converted into heat. Thus, the absorption coefficient of the ultrasonic waves can be calculated from the relation in Equation 5.18 for the adiabatic process, i.e., under the assumption that the CV heat and mass transfer are absent.

$$m_s \frac{d}{dt}\left[(c_s + c_l X)\mathrm{T}\right] = a_U \chi_U P_U. \tag{5.19}$$

The amount of acoustic energy absorbed by the dried material is estimated based on temperature variation in the dried material due to ultrasound action in comparison to drying without ultrasound assistance (Figures 5.5 through 5.7).

5.5.2 Equation of Evaporation

The phase transition of liquid into vapor takes place during drying. The equation of the evaporation curve relates the partial pressure of saturated vapor to the temperature of phase transition. Such a relation will be used further by estimating the effectiveness of drying assisted with ultrasound.

Based on the principles of irreversible thermodynamics, one can state that the rate of phase transition of liquid into vapor is proportional to the difference between the chemical potentials of liquid μ_l and vapor μ_v.[49] If the liquid and its vapor are in thermodynamic equilibrium, the chemical potentials of these two phases are equal ($\mu_l = \mu_v$). In thermodynamic equilibrium, the coexisting phases usually have the same pressures and temperatures. The Gibbs rule concerning the phase transition informs us that a system consisting of one component existing in two phases has one degree of freedom. Therefore, if the pressure of the system is changed on dp, then automatically the temperature of the phase transition T_{ph} will be changed on dT_{ph}, which leads to changes of chemical potentials from μ_l to $\mu_l + d\mu_l$ and from μ_v to $\mu_v + d\mu_v$, respectively. A new thermodynamic equilibrium is reached, but the continuity of the Gibbs function imposes an additional condition on chemical potential. Thus, one finds that the pressure of saturated vapor p_{vs} is related to the temperature of phase transition T_{ph} through *the equation of evaporation* in the form[49]

$$\ln p_{vs} = -\frac{l}{RT_{ph}} + \text{const}, \tag{5.20}$$

where l is the latent heat of evaporation and R is the gas constant for vapor.

By rearranging Equation 5.20 and assuming the initial (reference) condition to be p_{vs0} and T_{ph0}, one can determine the "const" and rewrite the equation of evaporation as follows:

$$p_{vs} = p_{vs0} \exp\left[\frac{l}{R}\left(\frac{1}{T_{ph0}} - \frac{1}{T_{ph}}\right)\right]. \tag{5.21}$$

Equation 5.21 shows that $p_{vs} > p_{vs0}$ when $T_{ph} > T_{ph0}$.

5.5.3 Estimation of Model Parameters

Based on data given in the textbooks,[47,48] one can state that the saturated vapor pressure p_{vs} is temperature dependent and can be approximated using the following function:

$$p_{vs}(T) = 9.61966 \cdot 10^{-4} T^4 - 1.08405264 \cdot 10^{-3} T^3 + 4.61325529 \cdot 10^{-2} T^2 - 2.77803513 \cdot 10^4 T + 6.29588464 \cdot 10^6. \tag{5.22}$$

The initial-value problem based on the set of kinetic Equations 5.17 and 5.18 was solved by the Adams–Bashforth non-self-starting multistep method. Selection of this method is motivated by good convergence and stability in long-term simulations.[50]

The method of model parameters estimation is based on the inverse problem, the concept of which follows from the solution of the direct problem applying optimization techniques. The solution of the inverse problem is compared directly with the curves of the experimental drying kinetics and the search is carried out to determine the best fit of the numerical and experimental curves. The kinetics consist of the drying curve and the temperature evolution curve. In the mathematical model (Equations 5.17 and 5.18), four parameters are introduced, which describe the drying process, namely, the effective coefficients of heat h_T and mass h_m convective transfer, the additional heat source $(a_U \chi_U P_U)$, as well as the critical moisture content X_{cr}. Therefore, the multiparameters and multiobjective optimization problem was formulated and solved. The objective function is defined as a sum of the squares of the normalized residuals of the experimental and numerical values of the moisture content and temperature.

$$f(h_m, h_t, a_U \chi_U P_U, X_{cr}) = \sum_{i=1}^{N} \left[\left(\frac{X_{num,i} - X_{exp,i}}{X_{max} - X_{min}} \right)^2 + \left(\frac{T_{num,i} - T_{exp,i}}{T_{max} - T_{min}} \right)^2 \right]. \quad (5.23)$$

The differences in the experimental maximal and minimal values of the moisture content and the temperature are used to get the same range of the residuals in defined objective function. Rosenbrock[51] optimization method is used for estimation of the model parameters. The method is then referred to optimization problem in which the objective function is inexpensive to compute, and the derivative either does not exist or cannot be computed efficiently. The simulations should be stopped when the improvement of error function is not observed.

5.6 DRYING EFFECTIVENESS ENHANCED WITH ULTRASOUND

5.6.1 Heating, Vibration, and Synergistic Effects

The mathematical model of drying kinetics presented earlier enables establishing the drying kinetics curves numerically. The coefficients of heat and mass transfer occurring in this model were estimated experimentally based on a series of tests on drying kinetics and by adjusting the theoretical to the experimental kinetic curves. In general, the CV heat and mass transfer coefficients h_m and h_T depend on both the moisture content and the temperature, although the influence of temperature in the range as it occurred in the drying tests can be considered as insignificant.

Good adherence of the theoretical and experimental drying curves allows to draw the increment of moisture mass Δm_{H_2O} removed from the dried sample in a given time, and to present the drying rate as a function of time. Figures 5.10 and 5.11 illustrate the time variation of moisture mass reduction and the drying rate for pure CV drying (Figure 5.10), and CV drying enhanced with ultrasound (200 W) (Figure 5.11).

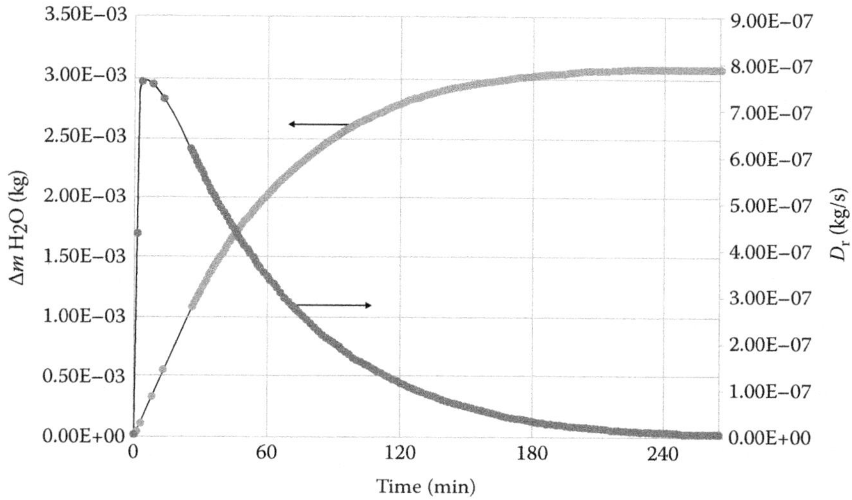

FIGURE 5.10 Moisture removal for pure convective drying: Δm—increment of moisture mass removed from the material in a given time period (dashed line), D_r—drying rate in a given time period (solid line).

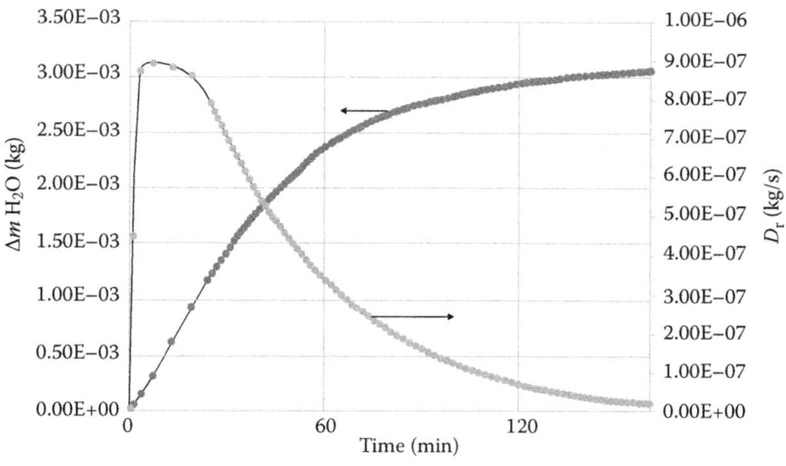

FIGURE 5.11 Moisture removal for ultrasound-assisted convective drying: Δm—increment of moisture mass removed from the material in a given time period (dashed line), D_r—drying rate in a given time period (solid line).

The dashed lines in Figures 5.10 and 5.11 present the amount of moisture mass removed from the dried material in a given time period, and the solid lines show the drying rate intensity as a function of time.

The drying rate D_r [kg/s] expresses the speed of moisture decrease in the drying material as a function of time (Equation 5.17).

$$D_r = m_s \frac{dX}{dt} = -A_m h_m \ln \frac{\phi|_{\partial B}\, p_{vs}(T)}{\phi_a p_{vs}(T_a)}. \tag{5.24}$$

The average drying rates $D_{r,ave}$ are determined as

$$D_{r,ave} = \frac{1}{t_e}\int_0^{t_e} D_r(t)\,dt, \tag{5.25}$$

where t_e is the drying time at which the moisture content reached equilibrium with surroundings X_e.

As it follows from Figure 5.6, the total drying time of pure CV drying lasted for about $t_e = t_e^{NU} = 235$ min, whereas in the CV drying with ultrasound enhancement only, $t_e = t_e^{U} = 185$ min for ultrasound power 100 W, and $t_e = t_e^{U} = 145$ min for ultrasound power 200 W.

The drying rate enhancement D_rE and the ratio of drying rate enhancement AD_rE are used to evaluate the effectiveness of the ultrasound-assisted drying. These are expressed as follows:

$$D_rE = D_{r,ave}^{U} - D_{r,ave}^{NU} = 20.971 - 12.940 = 8.031 \text{ [g/h]} \tag{5.26}$$

and

$$AD_rE = \frac{D_{r,ave}^{U} - D_{r,ave}^{NU}}{D_{r,ave}^{NU}} \cdot 100\% = \frac{8.031}{12.940} \cdot 100\% = 62.06\%. \tag{5.27}$$

The drying rate depends on the air contact area with the dried product, the mass transfer coefficient, and the driving forces. Owing to the "vibration effect" and "heating effect," the mass transfer coefficient and the material temperature in ultrasound-assisted drying are greater than those in drying without ultrasound. A part of the temperature rise in the dried body is due to the absorption of ultrasound, and the other part is caused by heat transfer enhancement due to the "vibration effect." Therefore, drying enhancement brought about by ultrasound can be expressed as

$$D_rE = D_{r,ave}^{U} - D_{r,ave}^{NU} = A_m(h_m + \Delta h_m)\ln \frac{\phi|_{\partial B}\, p_{vs}(T + \Delta T)}{\phi_a p_{vs}(T_a)} - A_m h_m \ln \frac{\phi|_{\partial B}\, p_{vs}(T)}{\phi_a p_{vs}(T_a)}, \tag{5.28}$$

where Δh_m represents the increase of mass transfer coefficients brought about by the "vibration effect," which is caused by the turbulence near the material surface, and ΔT represents the increase of material temperature because of ultrasound absorption ("heating effect"). The latter causes an increase in the saturated vapor pressure, and both these effects promote the heat and mass transfer processes.

Using the equation of evaporation 5.20, one can rewrite the drying enhancement 5.28 as follows:

$$D_r E = A_m h_m \frac{l}{R}\left(\frac{\Delta T}{T(T+\Delta T)}\right) + A_m \Delta h_m \ln \frac{\phi|_{\partial B}}{\phi_a} + A_m \Delta h_m \frac{l}{R}\left(\frac{1}{T_a} - \frac{1}{T+\Delta T}\right). \quad (5.29)$$

The first term on the right-hand side of Equation 5.29 contributes to the "heating effect," and the second term contributes to the "vibration effect." It is interesting to note that there is an extra third term, which is related to both the "heating effect" and the "vibration effect." This extra term can be defined as the "synergistic effect" which contributes to the drying efficiency.

This "synergistic effect," however, can be positive, provided that the heating effect is significant, i.e., when $T + \Delta T > T_a$. The temperature of the dried material becomes greater than the surroundings by intensive absorption of ultrasound or by additional volumetric heat supply, e.g., by microwave heating. Otherwise, the "synergistic effect" is negative.

Figure 5.12 represents the dependency of the drying time on different airflow velocities for pure CV drying and CV drying with ultrasound assistance at an air temperature of $T_a = 323$ K.

As can be seen in Figure 5.12, the course of material temperature for convective drying with ultrasound assistance differs significantly from that for pure CV drying at air temperature $T_a = 323$ K and different airflow velocities. For the processes with ultrasound enhancement, a disturbance at the boundary layer takes place, which intensifies the heat transfer. In such a case, the temperatures of the dried samples in CVUS drying with different airflow velocities reveal similar values, and the values are not as different as that for pure CV drying (Figure 5.12).

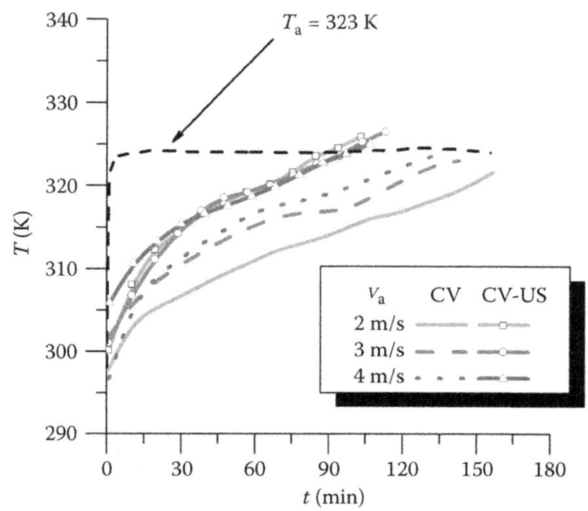

FIGURE 5.12 Temperature of the dried material vs. time for different air velocities at air temperature $T_a = 323$ K.

The "heating effect" due to the action of ultrasound is clearly seen in Figure 5.13. Here, the temperature of the sample in CV drying with ultrasound assistance reveals greater values than that for pure CV drying. One can notice the "synergistic effects" of ultrasound action in Figures 5.12 and 5.13, i.e., the increase of the sample temperature at some moment above the air temperature. This effect is even better seen for drying at a lower air temperature, i.e., 313 K, which may follow from the increase of air density and greater impedance, which have a positive influence on the ultrasound propagation.

Dividing the relation 5.29 by D_rE and multiplying it by 100%, one can write the contribution ratio (C) of the "heating effect" (T), "vibration effect" (v), and "synergistic effect" (s) to the drying enhancement as follows:

$$CD_rE_{\mathrm{T,eff}} = \frac{A_m h_m}{D_rE}\frac{l}{R}\left(\frac{1}{T} - \frac{1}{T+\Delta T}\right)\cdot 100\%. \tag{5.30a}$$

$$CD_rE_{\mathrm{v,eff}} = A_m \Delta h_m \frac{\ln \varphi|_{\partial B}/\varphi_a}{D_rE}\cdot 100\%. \tag{5.30b}$$

$$CD_rE_{\mathrm{s,eff}} = \frac{A_m \Delta h_m}{D_rE}\frac{1}{R}\left(\frac{1}{T_a} - \frac{1}{T+\Delta T}\right)\cdot 100\%. \tag{5.30c}$$

To assess quantitatively the impact of ultrasound on the drying process effectiveness involved in the "heating effect" and "vibration effect," it is necessary to estimate the increase of material temperature ΔT due to ultrasound absorption, and the increase of the mass transfer coefficient Δh_m due to turbulence on the dried material surface. The "synergistic effect" occurs, provided that the $T + \Delta T > T_a$.

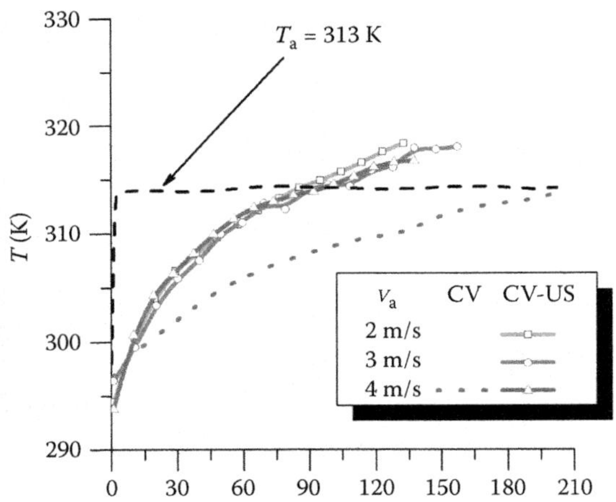

FIGURE 5.13 Temperature of the dried material vs. time for different air velocities at air temperature $T_a = 313$ K.

In the first drying test of apples, the air temperature was $T_a = 45°C$ and the relative humidity was $\varphi_a = 0.18$ on average. The material temperature was measured on the sample surface by a pyrometer. As is seen in Figure 5.6, the material temperature differs significantly from the air temperature only at the initial stage of drying, i.e., during the heating period. In the next period, these are close to each other.

Assuming that the difference between the air and material temperatures is 3°C, one can find the following vapor partial pressures for the saturated state:[47,48]

$p_{vs} = 0.009582$ MPa for air temperature $T_a = 45°C$
$p_{vs} = 0.008198$ MPa for material temperature $T = 42°C$ in pure CV drying
$p_{vs} = 0.008639$ MPa for material temperature $T + \Delta T = 43°C$ in CVUS drying.

By integrating the curves of the drying rate presented in Figures 5.12 and 5.13 and dividing the integral by the total drying time, one obtains the average drying rate for pure CV drying and for CVUS drying. The surface A_m of 16 samples through which the moisture transfer takes place and the average value of mass transfer coefficient h_m are estimated to be $A_m = 0.0256\,m^2$ and $h_m = 1.189\,kg/m^2{\cdot}h$ for pure CV drying and $h_m = 1.464\,kg/m^2{\cdot}h$ for CV drying enhanced with ultrasound, so that $\Delta h_m = 0.275\,kg/m^2{\cdot}h$.

Thus, the contribution ratio (C) of the "heating effect" (T) and "vibration effect" (v) together with the "synergistic effect" (s) gives the following values:

$$CD_rE_{T,eff} = \frac{A_m h_m}{D_r E} \ln \frac{p_{vs}(T+\Delta T)}{p_{vs}(T)} \cdot 100\% = \frac{256 \cdot 1.189}{80.31} \ln \frac{8639}{8198} \cdot 100\% \cong 17\%. \quad (5.31)$$

$$CD_rE_{v,eff} + CD_rE_{s,eff} = \frac{A_m \Delta h_m}{D_r E} \ln \frac{\phi|_{\partial B}\, p_{vs}(T+\Delta T)}{\phi_a p_{vs}(T_a)} \cdot 100\%$$

$$= \frac{256 \times 0.275}{80.31} \ln \frac{0.52 \times 8639}{0.18 \times 9582} \cdot 100\% \cong 83\%. \quad (5.32)$$

Based on the abovementioned calculations, one can state that the "vibration effect" contributes mostly to drying effectiveness with ultrasound assistance.

5.6.2 Energy Usage

One can also measure the electricity usage to estimate the energy efficiency *EE* in drying at air temperature of 45°C with and without ultrasound assistance. The total average energy consumption E_{ave} during an individual drying test can be determined as

$$E_{ave} = \frac{1}{t_e} \int_0^{t_e} E(t)\,dt, \quad (5.33)$$

where $E(t)$ is the current electric energy consumption used during drying in a given test, and $t_e = t_e^{NU}$ and $t_e = t_e^{U}$ denote the drying times at which the moisture content reached an equilibrium with surroundings X_{eq} in processes without the ultrasound assistance (NU) and with ultrasound assistance (U), respectively.

The energy savings ΔE and the ratio of energy efficiency EE can be used to evaluate the energetic effect of ultrasound-assisted drying. These are expressed as follows:

$$\Delta E = E_{ave}^{NU} - E_{ave}^{U} \text{ and } EE = \frac{E_{ave}^{NU} - E_{ave}^{U}}{E_{ave}^{NU}} \cdot 100\%. \qquad (5.34)$$

The drying setup used in this work for drying tests allowed to measure the total electric power used during each drying test. It was also possible to estimate the total amount of moisture mass removed from the 16 samples in each test. By dividing the total amount of consumed electric power per total moisture mass removed from the samples, one obtains the average usage of energy per unit moisture mass removed in each test.

$$E_{ave}^{NU} = 53.42 \text{ kWh/kg and } E_{ave}^{U} = 48.03 \text{ [kWh/kg]} \qquad (5.35)$$

for pure CV drying and for CVUS drying, respectively.

Thus, the energy savings ΔE and the average energy efficiency ratio EE amount, respectively, to

$$\Delta E = 53.42 - 48.03 = 5.39 \text{ kWh/kg and } EE = \frac{5.39}{53.42} \cdot 100\% = 10.08\%. \qquad (5.36)$$

The abovementioned data reveal a rather large energy usage per unit mass of removed moisture. The energy savings due to ultrasound enhancement of CV drying amounts merely to 10.08%. Such energy usage in the analyzed drying tests follows apparently from laboratory apparatus limitations and the experimental constraints. The volume of all 16 plaster samples was much smaller than the dryer volume of working chamber; so, an excessive energy consumption was needed to compensate the heat losses. Clearly, a larger load of the drying chamber will bring much better energy efficiency.

5.6.3 Effectiveness Dependent on Drying Parameters

Figure 5.14 illustrates the effect of air velocity on drying kinetics of the apple samples in CV and CVUS drying at air temperature $T_a = 323$ K and different air velocities (2, 3, and 4 m/s).

It is clearly visible that in the case of CV drying, an increase in air velocity results in acceleration of the process and shortening of the drying time, which amounts to 160 min at 2 m/s, 145 min at 3 m/s, and 130 min at 4 m/s (Figure 5.14). In the case of CVUS drying, the difference in air velocity has an insignificant influence on the drying rate. This shows that an increase in air velocity causes an awkward influence

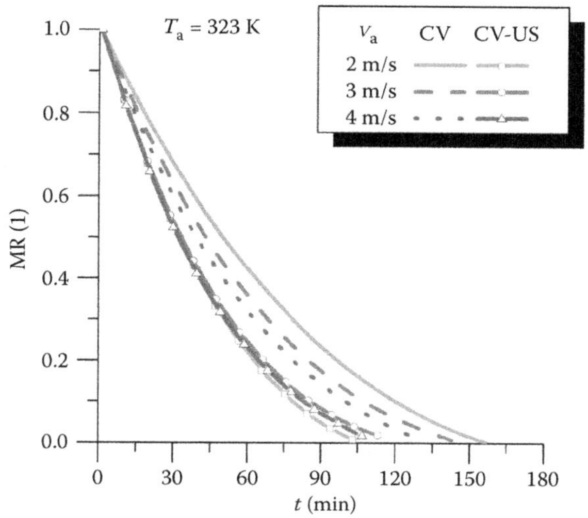

FIGURE 5.14 Drying curves of apple samples at different air velocities at temperature $T_a = 323\,K$.

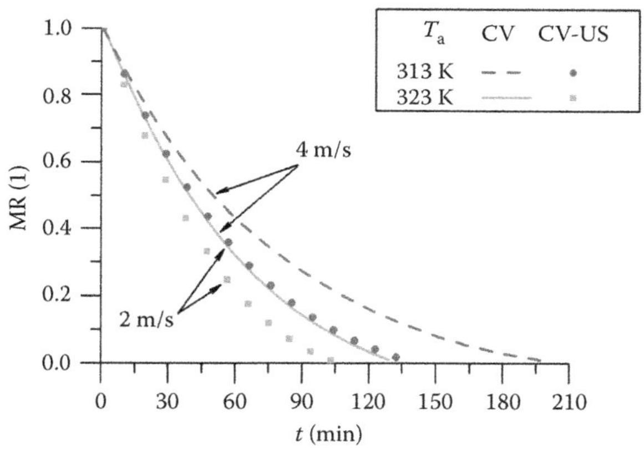

FIGURE 5.15 Drying curves of apple samples for CV and CVUS drying at different air velocities and temperatures.

on the ultrasound field and provokes difficulties in ultrasonic wave propagation. One can note that the fastest CVUS process amounting to 100 min was carried out with air velocity of 2 m/s, and the subsequent processes amounting to 105 min for 4 m/s, and 115 min for 3 m/s. This means that two factors have an impact on the drying rate, namely, ultrasound power and air velocity. Both of these factors intensify heat and mass transfer owing to gas turbulence at the boundary layer and thus reduce resistivity for the heat and mass transfer.

Figure 5.15 illustrates the effect of air velocity and temperature on drying kinetics in CV and CVUS drying of apple samples.

A comparison of the drying curves for the fastest CV and CVUS processes shows that the enhancement of CV drying with ultrasound not only lowers the drying temperature from 323 to 313 K but also reduces the air velocity from 4 to 2 m/s. Although the shortest drying time was gained in CVUS drying at a temperature of 323 K and air velocity of 2 m/s, the most effective process seems to be that conducted at a temperature of 313 K and air velocity of 2 m/s. This is because the kinetics of this process do not differ from that conducted at higher temperature of 323 K and at higher air velocity of 4 m/s. The positive effect may result in significant economical profit, because the drying times are almost the same for both of these processes, reached by much lower drying parameters from the energetic point of view.

Figure 5.16 presents the experimental validation of drying kinetics for the CV and the CVUS processes carried out at air velocity of 4 m/s and $T_a = 313$ K, as presented in Figure 5.14.

As is seen in Figure 5.13, the temperature of the drying material in the CVUS process at 90 min of drying time became greater than the temperature of the surrounding air. This means that the "synergistic effect" takes place after 90 min of drying. This effect can be quantitatively estimated by using data presented in Figures 5.12, 5.13, and 5.16. The drying kinetics presented in Figure 5.16 concern nine apple slices of dimensions 40 × 20 × 5 mm with an average mass of a single slice amounting to 3.648 g and an initial moisture content of 6.6 kg H_2O/kg d.b. The total mass of the nine slices amounts to 32.832 g, and the total mass of moisture is 28.512 g. Based on the drying curves in Figure 5.15, one can state that the total drying time amounted to 210 min for CV drying and 136 min for CVUS drying. Thus, the average drying rate enhancement D_rE in this drying process was

$$D_rE = D^{U}_{r,ave} - D^{NU}_{r,ave} = 10.679 - 8.146 = 2.533 \text{ [g/h]} \quad (5.37)$$

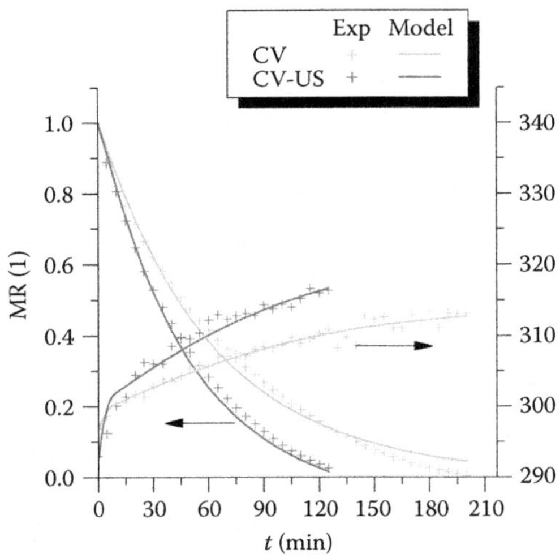

FIGURE 5.16 Drying curves of apple samples for CV and CVUS drying at air velocity of 4 m/s and $T_a = 313$ K.

To adjust the theoretical drying curves to the experimental ones, the CV mass transfer coefficient h_m has to be estimated. The average values of this coefficient were $h_m = 1.122\,kg/m^2 \cdot h$ for the CV drying and $h_m = 1.312\,kg/m^2 \cdot h$ for the CVUS drying. The difference between them is $\Delta h_m = 0.190\,kg/m^2 \cdot h$. An average difference of the material temperature between the CV and the CVUS drying is about 2°C, and the total surface of moisture exchange is $A_m = 0.0072\,m^2$. Assuming the difference between the air and the material temperatures to be 2°C, one can find the following vapor partial pressures for the saturated state:[47,48]

$p_{vs} = 0.007375$ MPa for air temperature $T_a = 40°C$
$p_{vs} = 0.006624$ MPa for material temperature $T = 38°C$ in pure CV drying
$p_{vs} = 0.008198$ MPa for material temperature $T + \Delta T = 42°C$ in CVUS drying.

The contribution ratio (C) of the "heating effect" (T), "vibration effect" (v), and "synergistic effect" (s) to drying with ultrasound assistance is as follows:

$$CD_rE_{T,eff} = \frac{A_m h_m}{D_rE} \ln \frac{p_{vs}(T+\Delta T)}{p_{vs}(T)} \cdot 100\% = \frac{72 \cdot 1.122}{25.33} \ln \frac{8198}{6624} \cdot 100\% \cong 68\%. \tag{5.38a}$$

$$CD_rE_{v,eff} = \frac{A_m \Delta h_m}{D_rE} \ln \frac{\phi|_{\partial B}}{\phi_a} \cdot 100\% = \frac{72 \cdot 0.190}{25.33} \ln \frac{0.25}{0.15} \cdot 100\% \cong 27\% \tag{5.38b}$$

$$CD_rE_{s,eff} = \frac{A_m \Delta h_m}{D_rE} \ln \frac{p_{vs}(T+\Delta T)}{p_{vs}(T_a)} \cdot 100\% = \frac{72 \cdot 0.190}{25.33} \ln \frac{8198}{7375} \cdot 100\% \cong 5\%. \tag{5.38c}$$

One can state that in the abovementioned process, the "heating effect" dominated over the "vibration effect." The reason might be the smaller capacity of the dried material in the drying chamber, i.e., 9 sample slices were subjected to drying instead of 16, as in the case of the previously discussed process.

The drying effectiveness of strawberries was evaluated based on the drying kinetics and material temperature during CVUS drying compared with CV drying (Figure 5.8). As it follows from the drying curve in Figure 5.8, the CV drying is a very long-lasting process, i.e., 1258 min, on average. During CV drying, the material temperature (T_m) was below the air temperature (T_a) for a long time, and only after about 800 min it reached 50°C. Even with rather low air temperature, the dried strawberries were characterized by a strongly shrunken surface, and their color darkened.

As can be seen in Figure 5.9, within the first 2 h, the value of material temperature was lower than air temperature, but after this period (i.e., 130 min) the temperature of the strawberry surface began to rise above the drying medium temperature. This is the so-called "heating effect." Because the increase in material temperature was marginal (up to 55°C), its effect on drying rate was regarded as negligible. The main reason for drying time reduction is "vibration effect," that arises because of vibration of air molecules with a high-frequency generated by the ultrasonic transducer near the dried body. Both effects resulting from the application of ultrasonic waves have definitely more powerful action (are compensated) in heat and mass transfer, i.e., "synergistic effect."

It was assumed that the heat and mass exchange occurs on the entire material surface, and the dried material undergoes linear volumetric shrinkage. Hence, the change in the surface dimension is a function of moisture content described by the following equation:

$$A_m = A_T = A(X) = [1 - \alpha_V (X_0 - X)]^{2/3} A_0, \tag{5.39}$$

where α_V and X_0 denote the volumetric shrinkage coefficient and the initial moisture content, respectively.

An averaged geometry for all samples was assumed in the simulations (Figure 5.17).

The proposed method for determining the model parameters is illustrated in the example of CV drying and CVUS drying of strawberry samples. Figures 5.18 and 5.19 present the comparison of the theoretical (Model) and experimental (Experiment) drying curves and the curves of material temperature evolution.

As it follows from Figures 5.18 and 5.19, a very good adjustment of the numerical drying curves as well as the material temperature curves to the experimental curve was achieved. Thus, the real drying processes have been successfully simulated with the proposed mathematical model of drying kinetics.

Figure 5.20 illustrates the variation of the drying rate in time for CV drying (a) and for CVUS (b).

As can be seen in Figure 5.20, at the very beginning of the CVUS process, the drying rate of the strawberries is about twice that of CV drying. Similarly, the

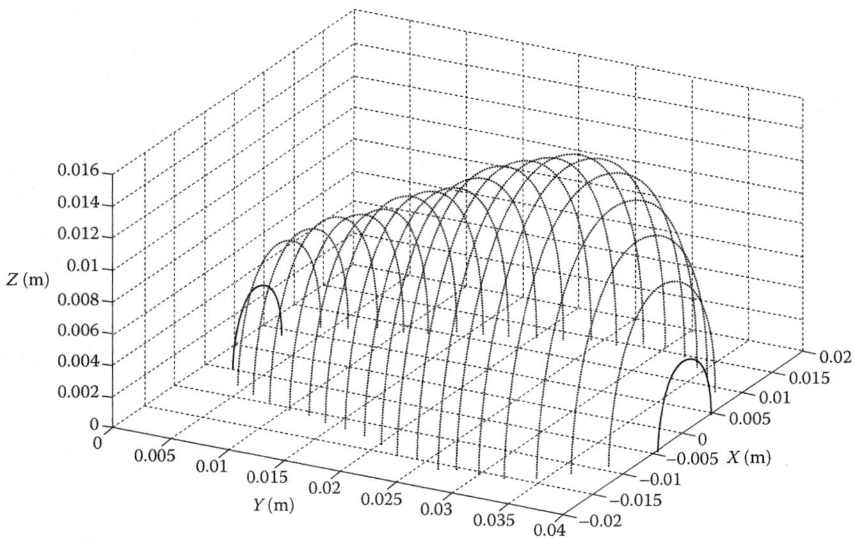

FIGURE 5.17 Strawberry sample geometry.

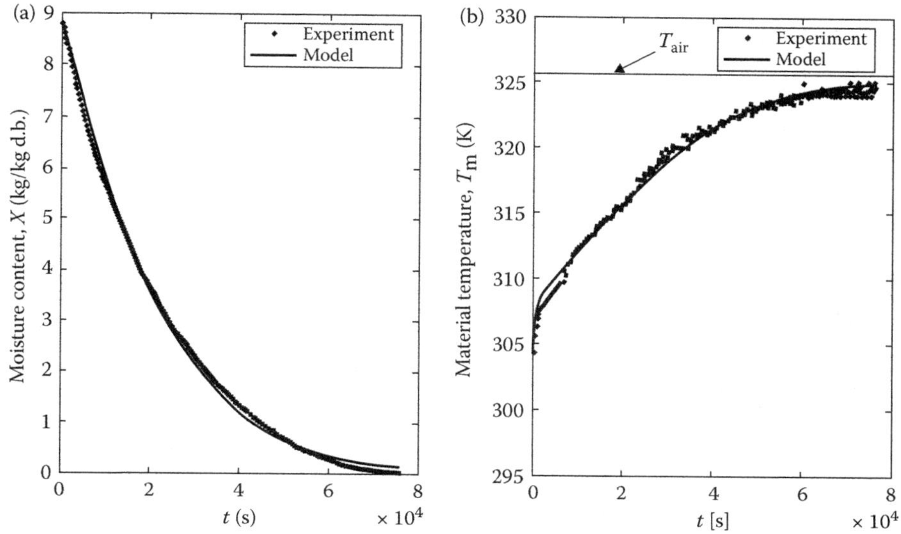

FIGURE 5.18 Comparison of CV theoretical and experimental tests: (a) drying curves and (b) material temperature.

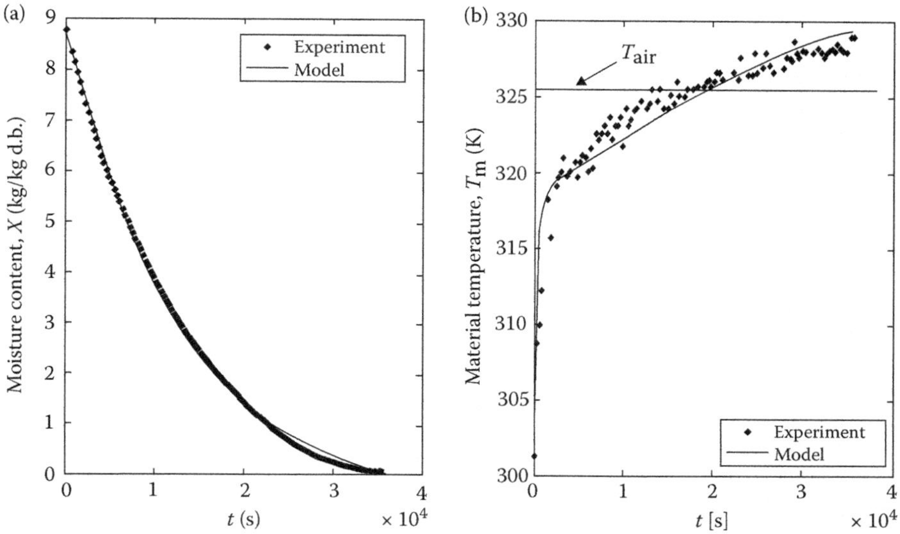

FIGURE 5.19 Comparison of CVUS theoretical and experimental tests: (a) drying curves and (b) material temperature.

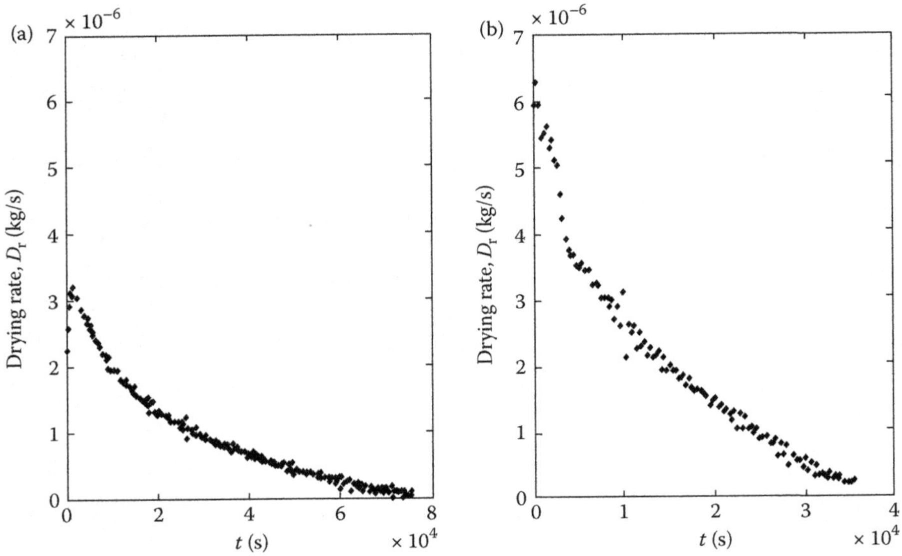

FIGURE 5.20 Comparison of drying rates: (a) CV drying and (b) CVUS drying.

total drying time of CVUS is twice shorter as compared with CV drying. The average drying time for CV drying is $t_e = t_e^{NU} = 1258\,\text{min}$ and that for CVUS drying is $t_e = t_e^{U} = 600\,\text{min}$. Thus, the average drying rates for the CV (NU) and the CVUS (U) processes calculated from Equation 5.13 amounted to

$$D_{r,ave}^{NU} = 14.90\ \text{g/h}, \quad D_{r,ave}^{U} = 27.16\ [\text{g/h}].$$

Drying rate enhancement D_rE and the ratio of drying rate enhancement AD_rE were used to evaluate the effectiveness of CVUS drying as follows:

$$D_rE = D_{r,ave}^{U} - D_{r,ave}^{NU} = 27.16 - 14.90 = 12.26\ [\text{g/h}] \tag{5.40}$$

$$AD_rE = \frac{D_{r,ave}^{U} - D_{r,ave}^{NU}}{D_{r,ave}^{NU}} \cdot 100\% = \frac{12.26}{14.90} \cdot 100\% = 82\%. \tag{5.41}$$

Taking into account the abovementioned results, one can state that ultrasound enhancement of strawberries drying significantly accelerated the drying process, as the ratio of the drying rate amounted to 82%. These results correspond to those reported by Kowalski and Pawłowski,[52] where the ratio of drying rate enhancement in apple drying amounted to 85.9%.

To describe the "heating effect" (T), "vibration effect," (v) and "synergistic effect" (s) quantitatively, i.e., the components of ultrasound action (C), a number of equations derived by Kowalski and Pawłowski[52] were used.

$$CD_rE_{T,eff} = \frac{A_m h_m}{D_r E} \ln \frac{p_{vs}(T_m + \Delta T_m)}{p_{vs}(T_m)} \cdot 100\% = \frac{20.8 \cdot 0.612}{12.26} \ln \frac{15741}{12960} \cdot 100\% = 20.17\%, \tag{5.42}$$

$$CD_rE_{v,eff} = \frac{A_m \Delta h_m}{D_r E} \ln \frac{\varphi|_{\partial B}}{\varphi_a} \cdot 100\% = \frac{20.8 \cdot 0.292}{12.26} \ln \frac{0.325}{0.075} \cdot 100\% = 72.64\%, \tag{5.43}$$

$$CD_rE_{s,eff} = \frac{A_m \Delta h_m}{D_r E} \ln \frac{p_{vs}(T_m + \Delta T_m)}{p_{vs}(T_a)} \cdot 100\% = \frac{20.8 \cdot 0.292}{12.26} \ln \frac{15741}{13613} \cdot 100\% = 7.19\%, \tag{5.44}$$

where the average surface of strawberry (A_m) was 0.0208 m^2, the average relative air humidity (φ_a) was 0.075, the h_m coefficients for CV and CVUS were 0.612 and 0.904 kg/m^{2}h, respectively, Δh_m was 0.292 kg/m^{2}h, and the vapor partial pressures p_{vs} for saturated state,[47,48] amounted to

p_{vs} = 13613 Pa for air temperature T_a = 52°C
p_{vs} = 12960 Pa for material temperature T_m = 51°C in the CV test
p_{vs} = 15741 Pa for material temperature $T_m + \Delta T_m$ = 55°C in the CVUS test.

Based on the abovementioned calculations, one can conclude that the "vibration effect" has contributed the most to the increase in drying efficiency.

5.7 CONCLUDING REMARKS

The examples of CV drying enhanced with ultrasound carried out on carrot, green pepper, apple, and strawberry samples revealed significant reduction of drying time with slight temperature elevation of the drying material (Figures 5.6, 5.7, and 5.9). In the case of apples, the drying time was shortened from 235 min for pure CV drying to 185 min for CV drying enhanced with ultrasound power of 100 W and to 145 min for ultrasound power of 200 W. In turn, the total drying time of strawberries was shortened from 1258 to 600 min when ultrasound was applied. This positive outcome confirms the successful application of ultrasound enhancement for drying of temperature-sensitive biological materials.

As was shown in Figures 5.3 and 5.4, the ultrasound transducer in the form of a disc generates air vibrations, which are transferred through the gas space as acoustic waves. These waves arrive at the dried material, create turbulence at the surface layer, and penetrate the dried material's interior. The medium-power ultrasound contributes to drying efficiency owing to "vibration effect" and "heating effect." From the observations of a number of microstructures and the comparison of the microstructure (Figure 5.3), it follows that the incidence of ultrasound on apple tissue increases the pore dimensions in the decreased pressure zone and expands them in the lowered pressure zone (a kind of "sponge effect").

Analytical assessment of the drying efficiency was enabled by the model of drying kinetics and the equation of evaporation given in this chapter. This model was validated by the experimental data. The calculated material coefficients allowed for estimating the contribution ratio of the "vibration effect" and "heating effect." It can be stated that drying efficiency increases mainly because of the special "vibration effect" and "heating effect" involved in the ultrasound. This is a very positive outcome as far as drying of heat-sensitive biological materials is concerned.

It is interesting to note that by CVUS, an extra effect may appear that contributes to the drying efficiency. This extra effect, termed as the "synergistic effect," can be positive, provided that the "heating effect" is significant, that is, when the temperature of the dried material becomes greater than the surrounding air. This may occur by intensive absorption of ultrasonic waves or by additional volumetric heat supply, e.g., by microwave heating.

One can expect that such emerging drying technology could be suitable for drying of biological materials and would find application in the industry, thus raising competitiveness by increasing productivity and decreasing energy consumption, thereby contributing to the concept of balanced development.

ACKNOWLEDGMENTS

This work was carried out as part of research project No. PBS1/A8/13/2012, sponsored by the National Centre for Research and Development of Poland.

REFERENCES

1. Galego-Juarez, J.A.; Rodríguez-Corral, G.; Gálvez-Moraleda, J.C.; Young, T.S. A new high-intensity ultrasonic technology for food dehydration. *Drying Technology* 1999, 17 (3), 597–608.
2. Galego-Juarez, J.A. Some applications of air-born power ultrasound to food processing. In: Powey, M.J.W.; Mason, T.J., eds. *Ultrasound in Food Processing*. Cornwall, UK: Blackie Academic and Professional; 1998.
3. Mulet, A.; Cárcel, J.; Benedito, J.; Rossell6, C.; Simal, S. Ultrasonic mass transfer enhancement in food processing. In: Welti-Chanes, J.; Velez-Ruiz, J.F.; Barbarosa-Cánovas, G.V., eds. *Transport Phenomena in Food Processing*. Boca Raton, FL: CRC Press; 2003.
4. Mason, T.J.; Riera, E.; Vercet, A.; Lopez-Buesa, P. Application of ultrasound. In: Sun, D.-W., ed. *Emerging Technologies for Food Processing*, 323–351. Amsterdam, the Netherlands: Academic Press; 2005.
5. García-Pérez, J.V.; Cárcel, J.A.; de la Fuente-Blanco, S.; Riera-Franco de Sarabia, E. Ultrasonic drying of foodstuff in a fluidized bed: Parametric study. *Ultrasonics* 2006, 44, e539–e543.
6. García-Pérez, J.V.; Rosselló, C.; Cárcel, J.A.; De la Fluente, S.; Mulet, A. Effect of air temperature on convective drying assisted by high power ultrasound. *Defect and Diffusion Forum* 2006, 258–260, 563–574.
7. de la Fuente-Blanco, S.; Sarabia, E.R.F.; Acosta-Aparicio, V.M.; Blanco, A.; Gallego-Juárez, J.A. Food drying process by power ultrasound. *Ultrasonics* 2006, 44, 523–527.
8. Gallego-Juárez, J.A.; Riera, E.; de la Fuente-Blanco, S.; Rodríguez-Corral, G.; Acosta-Aparicio, V.M.; Blanco, A. Application of high-power ultrasound for dehydration of vegetables: Processes and devices. *Drying Technology* 2007, 25, 1983–1901.

9. Frias, J.; Peñas, E.; Ullate, M.; Vidal-Valverde, C. Influence of drying by convective air dryer or power ultrasound on the vitamin C and β-carotene content of carrots. *Journal of Agricultural and Food Chemistry* 2010, 58, 10539–10544.
10. Sabarez, H.T.; Gallego-Juárez, J.A.; Riera, E. Ultrasonic-assisted convective drying of apple slices. *Drying Technology* 2012, 30(9), 989–997.
11. Bonazzi, C.; Dumoulin, E. Quality changes in food materials as influenced by drying processes. In: Tsotsas, E.; Mujumdar, A.S., eds. *Modern Drying Technology Volume 3: Product Quality and Formulation*, 1–20. Weinheim, Germany: Wiley-VCH Verlag GmbH & Co. KGaA; 2011.
12. Legay, M.; Gondrexon, N.; Le Person, S.; Boldo, P.; Bontemps, A. Enhancement of heat transfer by ultrasound: Review and recent advances. *International Journal of Chemical Engineering* 2011, Article ID 670108, 17 pages.
13. Pan, Y.K.; Zhao, L.J.; Dong, Z.X.; Mujumdar, A.S.; Kudra, T. Intermittent drying of carrot in a vibrated fluid bed: Effect on product quality. *Drying Technology: An International Journal* 1999, 17, 2323–2340.
14. Jayaraman, K.S.; Das Gupta, D.K. Drying of fruits and vegetables. In: Mujumdar, A.S., ed. *Handbook of Industrial Drying*, 606–631. Boca Raton, FL: CRC Press; 2006.
15. Ong, S.P.; Law, C.L. Hygrothermal properties of various foods, vegetables and fruits. In: Jangam, S.V.; Law, C.L.; Mujumdar, A.S., eds. *Drying of Foods, Vegetables and Fruits*, Vol. 1, 31–58. Singapore: TPR Group; 2010.
16. Kowalski, S.J.; Szadzińska, J.; Łechtańska, J. Non-stationary drying of carrot: Effect on product quality. *Journal of Food Engineering* 2013, 118 (4), 393–399.
17. Fernandes, F.A.N.; Gallão, M.I.; Rodrigues, S. Effect of osmosis and ultrasound on pineapple cell tissue structure during dehydration. *Journal of Food Engineering* 2009, 90, 186–190.
18. Kentish, S.; Ashokkumar, M. The physical and chemical effects of ultrasound, Chapter 1. In: Feng, H.; Barbosa-Cánovas, G.V.; Weiss, J, eds. *Ultrasound Technologies for Food and Bioprocessing (Food Engineering Series)*. New York: Springer; 2011. doi:10.1007/978-1-4419-7472-3_1.
19. Ashokkumar, M. The characterization of acoustic cavitation bubbles: An overview. *Ultrasonic Sonochemistry* 2011, 18, 864–872.
20. Ortuño, C.; García-Pérez, J.V.; Cárcel, J.A.; Femenia, A.; Mulet, A. Modelling of ultrasonically assisted convective drying of eggplant. In: *17th International Drying Symposium (IDS2010)*, Magdeburg, Germany, October 3–6, 2010.
21. Rodrigues, S.; Fonteles, T.V.; Leite, K.F.; Fernandes, F.A.N. Improvement on functionality of dried cashew apples subjected to ultrasonic treatment. In: *Eurodrying'2013*, Paris, October 2–4, 2013.
22. Cárcel, J.A.; García-Pérez, J.V.; Riera, E.; Simal, S.; Mulet, A. Convective drying intensification: Use of airborne ultrasound. In: *Eurodrying'2011*, Palma Balearic Island, Spain, 26–28 October 2011.
23. García-Pérez, J.V.; Cárcel, J.A.; Simal, S.; García-Alvarado, M.A.; Mulet, A. Ultrasonic intensification of grape stalk convective drying: Kinetic and energy efficiency. *Drying Technology* 2013, 31, 942–950.
24. Yao, Y. Using power ultrasound for regeneration of dehumidizers in desiccant air-conditioning systems: A review of prospective studies and explored issues. *Renewable and Sustainable Energy Reviews* 2010, 14, 1860–1873.
25. Kowalski, S.J.; Pawłowski, A. Intensification of apple drying due to ultrasound enhancement. *Journal of Food Engineering* 2015, 156, 1–9.
26. Kowalski, S.J.; Mierzwa, D. Hybrid drying of red bell pepper: Energy and quality issues. *Drying Technology* 2011, 129, 1195–1203.
27. Maskan, M. Microwave/air and microwave finish drying of banana. *Journal of Food Engineering* 2000, 44 (2), 71–78.

28. Kowalski, S.J.; Szadzińska, J.; Pawłowski, A. Ultrasonic-assisted osmotic dehydration of carrot followed by convective drying with continuous and intermittent heating. *Drying Technology* 2015, 33, 1570–1580.
29. Sagar, V.R.; Suresh Kumar, P. Recent advances in drying and dehydration of fruits and vegetables: A review. *Journal of Food Science and Technology* 2010, 47 (1), 15–26.
30. García-Pérez, J.V.; Ortuño, C.; Puig, A.; Cárcel, J.A.; Perez-Munuera, I. Enhancement of water transport and microstructural changes induced by high-intensity ultrasound application on orange peel drying. *Food and Bioprocess Technology* 2012, 5, 2256–2265.
31. García-Pérez, J.V.; Cárcel, J.A.; Riera, E.; Mulet, A. Influence of the applied acoustic energy on the drying of carrots and lemon peel. *Drying Technology* 2009, 27, 281–287.
32. Mulet, A.; Cárcel, J.A.; García-Pérez, J.V.; Riera, E. Ultrasound-assisted hot air drying of foods. In: Feng, H.; Barbosa-Cánovas, G.V.; Weiss, J., eds. *Ultrasound Technologies for Food and Bioprocessing*, Food Engineering Series, 511–534. New York: Springer; 2011.
33. Gamboa-Santos, J.; Montilla, A.; Soria, A.C.; Cárcel, J.A.; García-Pérez, J.V.; Villamiel, M. Impact of power ultrasound on chemical and physicochemical quality indicators of strawberries dried by convection. *Food Chemistry* 2014, 161, 40–46.
34. Kroehnke, J.; Musielak, G.; Boratyńska, A. Convective drying of potato assisted by ultrasound. *PhD Interdisciplinary Journal* 2014, 1, 57–65.
35. Śliwiński, A. *Ultrasounds and Their Application.* Warsaw, Poland: WNT; 2001 (in Polish).
36. Ozuna, C.; Alvarez-Arenas, T.G.; Riera, E.; Cárcel, J.A.; García-Pérez, J.V., Influence of material structure on airborne application in drying. *Ultrasonics Sonochemistry* 2014, 21, 1235–1243.
37. Ortuño, C.; Pérez-Munuera, I.; Puig, A.; Riera, E.; García-Pérez, J.V. Influence of power ultrasound application on mass transport and microstructure of orange peel during hot air drying. *Physics Procedia* 2010, 3, 153–159.
38. Kowalski, S.J. Ultrasonic waves in diluted and densified suspensions. *Ultrasonics* 2004, 43, 101–111.
39. Kowalski, S.J. *Thermomechanics of Drying Processes.* Heidelberg and Berlin: Springer; 2003.
40. Kowalski, S.J.; Pawłowski, A. Modeling of kinetics in stationary and intermittent drying. *Drying Technology* 2010, 28 (8), 1023–1031.
41. Kowalski, S.J.; Mierzwa, D. Influence of osmotic pretreatment on kinetics of convective drying and quality of apples. *Drying Technology* 2013, 31, 1849–1855.
42. Berry, R.S.; Kazakov, V.A.; Sieniutycz, S.; Szwast, Z.; Tsirlin, A.M. *Thermodynamic Optimization of Finite–Time Processes.* New York: John Wiley & Sons; 2000.
43. Coussy, O.; Eymard, R.; Lassabatere, T. Constitutive modeling of unsaturated drying deformable materials. *Journal of Engineering Mechanics* 1998, 124, 658–667.
44. Gumiński, K. *Thermodynamics of Irreversible Processes.* Warsaw, Poland: PWN-Polish Scientific Publishers; 1962 (in Polish).
45. Szarawara, J. *Chemical Thermodynamics.* Warsaw, Poland: WNT-Scientific Technological Publishers; 1985 (in Polish).
46. Kowalski, S.J.; Musielak, G.; Banaszak, J., Heat and mass transfer during microwave-convective drying, *AIChE* 2010, 56 (1), 24–35.
47. Strumiłło, Cz. *Fundamentals of the Theory and Technology of Drying*, 2nd edition. Warsaw, Poland: WNT-Scientific Technological Publishers, 1983 (in Polish).
48. Wiśniewski, S.; Wiśniewski, T.S. *Heat Exchange*, 4th edition. Warsaw, Poland: WNT-Scientific Technological Publishers, 1997 (in Polish).
49. Elwell, D.; Pointon, A.J. *Classical Thermodynamics.* Warsaw, Poland: WNT-Scientific Technological Publishers, 1976 (in Polish).

50. Krupowicz, A. *Numerical Methods of Initial Value Problems of Ordinary Differential Equations*. Warsaw, Poland: PWN, 1986 (in Polish).
51. Rosenbrock, H.H. An automatic method for finding the greatest or least value of a function. *The Computer Journal* 1960, 3 (3), 175–184.
52. Kowalski, S.J., Pawłowski, A. Intermittent drying: Energy expenditure and product quality. *Chemical Engineering Technology* 2011, 34 (7), 1123–1129.

6 Smart Drying Technology for Vegetable Products

Ya Su and Min Zhang
Jiangnan University

Arun Sadashiv Mujumdar
McGill University

CONTENTS

6.1 INTRODUCTION

Smart drying technology utilizes sensors to detect and monitor online various quality parameters of the material to proactively control and detect errors or deficiencies in dryer operation. It incorporates equipment, technologies, resources, and practices to save energy and promote environmental sustainability, and thus control the dryer operating conditions to make high-value products.[1–3] The monitored quality parameters generally include moisture content, color, shape, taste, and flavor and the conditions within the dryer such as pressure, velocity, temperature, and humidity to control the performance of the drying system. Consequently, smart drying technology includes not only dryers but also smart sensors, translators, and smart control systems to monitor the operating conditions and to improve drying product quality as well as to enhance the energy efficiency. Fresh food can be successfully manufactured into high-quality products through smart drying technology, which contains three parts: manufacturing process dryer, the probe, and online analysis. The probe monitors the drying process and sends messages for online analysis, which in turn controls the drying process according to the information received. The smart drying technology needs to be designed based on the knowledge of the specification of product being dried, such as local drying conditions and the quality parameters of the product after drying. Then the product quality is assured with optimal energy consumption and minimal environmental impact. It is also worth mentioning that the quality of foodstuff should be examined during the whole process and not only at the end point. Therefore, all unit operations need to be monitored and controlled during the entire process using appropriate instruments to assure the quality of the end products.[4] Drying is a complex, dynamic, nonlinear, uncertain, and often unpredictable operation because of simultaneous heat and mass transfer, potential for several physical transformations such as shrinkage, puffing, transformation, case-hardening, as well as potential for chemical and biochemical reactions.[5] This complexity becomes more severe in the process of food drying because of product heterogeneity, anisotropicity, and nonuniformity of drying conditions.[6] In recent years, with the rapid development of drying science and technology and designing of new equipment, the drying industry is making huge progress. However, the present drying technology has some disadvantages, such as limited application of smart automation, high energy and fuel consumption, etc. With the emergence of computer science in the early 1980s, computer-aided design was incorporated in many areas of science and engineering and in industrial drying.[7,8] About two decades ago, Professor Mujumdar[1–3] proposed the development of smart or intelligent dryers to make industrial drying more efficient and cost-effective. Because drying is a highly energy-intensive operation, which affects the quality of dried product decisively and also has an adverse effect on the environment through greenhouse gas emissions, it is necessary to make this a sustainable operation by utilizing the latest technologies including advanced computer hardware and software for reliable process control. With the recent advances in mathematical modeling of dryers and sophisticated sensors for real-time measurement of variables of interest in automatic control of dryers and robust control strategies, it is now feasible to design and operate smart dryers.

In recent years, considerable effort has been devoted to the development of smart drying technology.[9–12] Although most researchers are focusing on laboratory-scale research, application of smart drying technology in industrial-scale production is demanding. Interest in employing the most efficient methods to monitor the food quality during drying has increased.[13] In many industries, engineers use smart drying technology every day to monitor the production process and the quality of dried products. Recent developments in the area of smart drying technology for vegetables are discussed in this chapter. These include biomimetic systems, computer vision technology, microwave/dielectric spectroscopy, ultrasonic techniques, and control systems for the drying environment. Aghbashlo et al.[14] have reported the techniques used to monitor and control product quality in fluidized bed dryers. Burggraeve et al.[15] have reviewed the analytical techniques used for monitoring and controlling fluidized bed granulators. da Silva et al.[16] discussed facilities for monitoring and controlling coating and granulation processes. Ghasemi-Varnamkhasti et al.[10] reviewed biomimetic-based odor- and taste-sensing systems for food quality and safety characterization. Several publications have also investigated the applications of these techniques for analysis of processing and quality control of food[17] and control systems in drying technology.[18] Recent advances in smart drying of fresh vegetables are reviewed and the techniques that need to be monitored for quality parameters, such as moisture content, shape, odor, or other components in foods as well as the conditions in the dryer, are also discussed.

6.2 CATEGORIES OF SMART DRYING TECHNOLOGY

6.2.1 Biomimetic Systems

The appreciation of food is based on the perception of organoleptic properties such as odor, taste, and appearance. One of the major research areas is the development of monitoring and control methods concerning food stocks, their processing, as well as the quality of dried final products.[11] Among these methods, human sense-inspired sensor technologies containing odor-sensing systems (electronic noses) and taste-sensing systems (electronic tongues) are increasingly under consideration for application in several food processing operations such as drying, packaging, sorting, and storage.[10] The odor-, flavor-, and taste-sensing systems provide valuable information about the quality of the products by analyzing the compounds in the brain regions responsible for taste and smell. Combined with the use of appropriate computer software and data analysis methods, these sensing systems perform cost-efficient, multifunctional, and eco-friendly analysis for process control and for determining the quality of the dried foods.[9,12]

Electronic nose (e-nose), also known as artificial nose, is an electronic system that emulates the functions of the biological nose. The first stage for both biological and electronic noses is the interaction between the volatile compounds emitted by the drying material and the biological/electronic noses. One odorant is recognized by multiple odorant receptors and one odorant receptor can also be sensitive to multiple flavor-containing compounds. The next stage is the storage of the signals produced by the receptors in the brain or in a pattern recognition engine (learning stage), and later the recognition of one of the odors stored (classification stage).[4] Coupled with chemometric techniques, electronic noses can be applied in industrial drying as a reliable instrument for aroma monitoring during the drying process. This type of biomimetic system has three main components: a vapor delivery system, a sensor array, and a pattern recognition algorithm. It combines cross-reactive sensor arrays with pattern recognition algorithms to create robust odor-discrimination systems. A schematic diagram of an e-nose monitoring and control system is shown in Figure 6.1. Such an e-nose system can be used for food and beverage quality control as well as in medical diagnostics.[19] Table 6.1 shows some examples of the application of e-noses in industries. With their obvious advantages, biomimetic systems show great market potential, and numerous achievements have been reported in this field.[20–32]

Li et al. examined online monitoring of carrot volatiles with a zNose during microwave drying.[20,21] With a specially designed fuzzy logic controller, this system could retain carrot volatiles, avoid burning, maintain carrot color and overall appearance, shorten drying time, and save energy. A similar system was also designed and applied to monitor and control microwave drying of apple by Li et al.[20,21] Li et al.[22] have also developed another real-time, volatile-detection-assisted control system for microwave drying. In this system, a fast gas chromatograph (zNose) was used to detect volatiles emanating from the food samples during drying, and the detected volatile signals were integrated with a fuzzy logic algorithm to determine the drying temperature. Moreover, a phase controller was used to automatically and continuously adjust the microwave power, while a data acquisition unit was employed to integrate

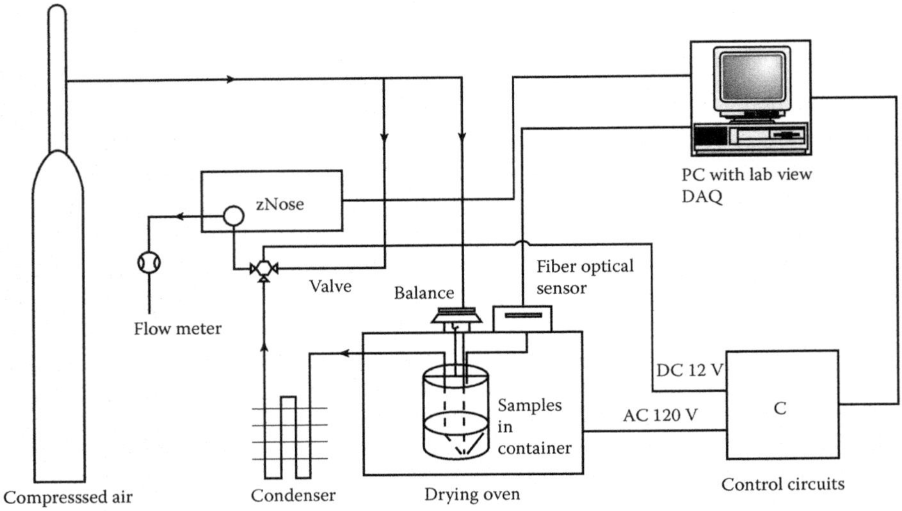

FIGURE 6.1 Schematic diagram of e-nose monitoring and control system.

the entire control. Their results showed that the system could successfully achieve the desired temperature, power, and volatile control. In a study by Raghavan et al., using an ultrafast gas chromatograph (zNose) to detect food aroma and a fuzzy logic algorithm to dynamically analyze the aroma signals to determine drying temperatures, a real-time aroma monitoring system to control a microwave drying process was designed.[23] In this system, an automatic phase controller was used to adjust the microwave power level to meet the temperature requirement. The application of the system showed that the newly developed control strategies could improve the quality of the dried products undergoing microwave drying in terms of aroma retention.

Yang et al.[24] researched the identification of coumarin-enriched Japanese green teas and their particular flavor using an e-nose. A new "absolute value expression" (AVE) method was developed and employed to characterize the tea flavors into a quality parameter to express them numerically. The e-nose successfully characterized the drying temperature-dependent trend of coumarin content during the manufacturing process, which was employed to identify green teas with particular flavor.[24] Infante et al. studied the effect of drying on lemon verbena aroma and infusion sensory quality using a trained sensory panel and an e-nose.[25] In addition, a novel e-nose based on miniaturized surface acoustic wave (SAW) sensor arrays coupled with solid-phase microextraction process (SPME) enhanced headspace analysis was studied by Barié et al.[26] With the SPME technique, this system was used for rapid determination of the volatile organic compounds for food quality monitoring.

The odor- and taste-sensing systems are also used for assessing the bitterness intensity of some active pharmacological ingredients,[27] measuring the ethanol content of wine synthetic matrices,[28] discriminating the grade of tea,[29] monitoring the fermentation process of black tea,[30] and detecting the optimum fermentation time for black tea manufacturing.[31–34] In some cases, e-noses can be used to reduce the

TABLE 6.1
Examples of the Application of Electric Noses

Application	Sample	Object of Investigation	Type of e-Nose	Method of Data Analysis	References
Food process monitoring	Apple	Monitoring and control in microwave drying	zNose (7100 Fast GC Analyzer)	Fuzzy logic control method	21
	Carrot	Monitoring and control in microwave drying	zNose (7100 Fast GC Analyzer)	Fuzzy logic control method	20
	Carrot	Real-time, volatile-detection-assisted control system was designed for microwave drying	zNose (7100 Fast GC Analyzer)	Fuzzy logic control method	22
Testing the shelf life of food	Lemon verbena	Determine the shelf life of lemon verbena aroma of different drying processes	e-Nose EOS 835	PCA	25
Evaluation of food	Japanese green teas	Identification of coumarin-enriched Japanese green teas and their particular flavor	e-Nose (FF-2A Fragrance and Flavor Analyzer)	PCA, CA absolute value expression (AVE)	24
	Apple	Rapid determination of volatile organic compounds in food quality monitoring	e-Nose SAW sensor SPME	PCA	26

amount of analytical chemistry that is performed in food production, especially when qualitative results are needed.[35] With the development of modern computational methods, biomimetic systems are becoming key elements of the most promising smart drying technologies.

The biomimetic systems are very useful in the food industry. However, the biomimetic systems also suffer from significant weaknesses, which limit their widespread application in food quality and safety assessment. For instance, their sensing ability is heavily affected by environmental factors, such as humidity (in the case of e-nose) and background noise, general drift caused by temperature, and sensor variations and poisoning.[10] Furthermore, the absorbability or catalysis of the sensitive materials of the sensors have a great influence on the detection ability of the sensors. Compared with the biological binding of specific odorants and tastants to the olfactory and taste receptor cells, biomimetic systems still have restrictions in sensitivity and specificity. Future work could be focused on using the e-nose for monitoring of various food drying processes and consequently automating of drying plants employing advanced intelligent controlling approaches.[4]

6.2.2 Computer Vision Technology

Appearance of dry foods is the most intuitive parameter that influences the quality of food and is also the first quality attribute of foods affecting consumers' acceptance. Rapid and objective measurement of food appearance is required in quality control for the commercial production of dried vegetables and is therefore a key area in the future development of smart drying technology.[36] Similar to biomimetic systems, computer vision technology is used for recognizing objects and extracting quantitative information from digital images in order to provide objective, rapid, noncontact, and nondestructive quality evaluation.[16,37] It is one of the latest techniques employed in the food drying process for real-time determination of physical properties of the products, including size, shape, and quality parameters. The schematic diagram of a computer vision system (CVS) is shown in Figure 6.2. Computer vision techniques fit very well in both simple and complex quality assurance or quality management systems in combination with charge coupled device (CCD) cameras, ultrasound and infrared (IR), and fusion of imaging technologies to capture a refined image or to extract useful insight from it.[15,38] Image acquisition, preprocessing, segmentation, extraction, and representation of the characteristic parameters are the main steps in image processing analysis.[39] The use of computer vision technology has been highly successful in solving food classification problems in the past, and it has continued this success in recent times.[40,41] Visual observability of a drying process is provided by computer vision technology using online image analysis and correlation of image attributes (area, color, and texture) with physical parameters of drying (moisture, quality). A relationship between area shrinkage and moisture content was used for online estimation of actual moisture content, and a relationship between color intensity and quality was used for online estimation of quality degradation of ginseng roots.[42] Computer vision is a promising technique being actively investigated for food appearance measurement, especially with the ability to provide a detailed characterization of color uniformity at the pixel-based level.

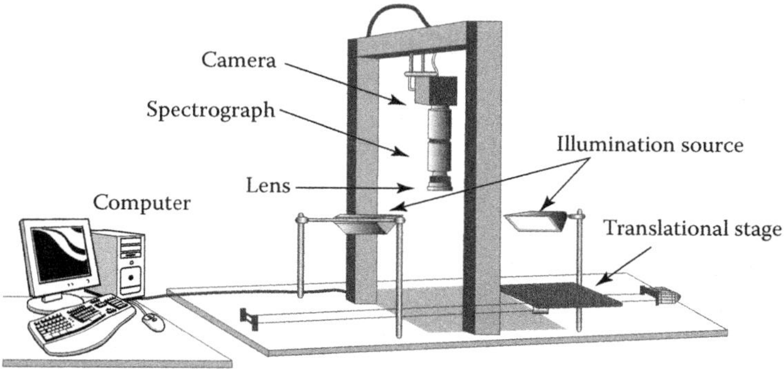

FIGURE 6.2 Schematic diagram of a computer vision system.

Machine vision and image analysis can provide surface or structural properties, while computer vision technology can quantitatively describe texture parameters, moisture content, and color parameters of the material. Sampson et al. assessed the ability of a low-cost, dual-view CVS in measuring the volume and color co-occurrence image of textural features of apple slices to determine the end point of drying by comparing physical texture parameters and moisture content.[43] Trinca et al. used a nondestructive method to evaluate lost moisture content (LMC) and moisture content (MC) of apples during slow drying, oven drying, and lyophilization using image processing and a multilayer neural network (NN) predictor. This technique showed high precision in controlling the quality parameters during the drying process.[44]

Computer vision technology and visual image systems are also analysis techniques suitable for quality inspection. They have been applied to evaluate, grade, and classify meat and other agricultural products, providing new ideas for the development of smart drying technology.[36–38,40,45,46] It was also reported that evaluation, grading, and classification of fish, vegetables, and other agricultural products by appearance, firmness, color, and other parameters could be achieved by computer vision technology,[47–51] offering technical support for the automatic control and hierarchical management and monitoring in the drying of vegetable products.

The computer vision technique is a real-time, nonintrusive, rapid, low-cost, efficient, repeatable, accurate, high-resolution, consistent, and objective inspection tool based on image analysis. It offers several advantages, such as simplicity of the experimental setup and ease of scale-up procedures, with applicability for nonintrusively obtaining a complete velocity vector field.[52] However, image processing necessitates massive computational efforts for data processing due to the very large amounts of data generated when compared with other techniques. Such a disadvantage would be diminished with further progress in computer technology. Moreover, the resolution of the computer imaging was also confined by the wavelength of light, the desired field of inspection, and the pixel density of the digital camera. Moreover, the images could also be influenced by the nonuniform illumination, variable solid density in the dryer, imperfections or dirt on the wall, background light, and photobleaching. Solid moisture content can affect the reflective attributes of the particles, particularly free surface water, and the captured image quality and characteristics. Most imaging techniques are limited to translucent media and two-dimensional images, apart from confocal laser scanning imaging, which is suitable for three-dimensional imaging.[53] Thus, the computer vision technique is inapplicable for large-scale drying processes in which opaque metal chambers are used. Therefore, improving the application scope would be both a chance and a challenge to the development of computer vision technology.[54–60]

6.2.3 Microwave Dielectric Spectroscopy

Microwave technology can be used in two models: microwave absorption mode and resonant mode. Since different moisture values of materials absorb microwave energy differently, the moisture content is measured in microwave absorption mode; according to the microwave resonance frequency changing with the moisture content of the material, the moisture content is measured in microwave resonant mode. The resonance frequency and the resonance bandwidth of the microwave field, produced by the induction of electromagnetic waves with frequencies ranging from 300 MHz to 300 GHz and wavelengths varying from 1.0 mm to 1.0 m, are affected during interaction with solid products and water molecules. Thus, by measuring the shift in microwave frequency and resonance attenuation simultaneously, moisture content and density of particles can be determined without influencing each other.[16] The microwave resonance technology utilizes the interaction between water molecules and changing electromagnetic fields. The measuring frequency of the employed stray field sensor is predetermined by the resonance wavelength of the microwave-inducing resonator. The resonance frequency depends on the geometries of the employed sensors. If the resonator is loaded with materials, an increase in the storage of electric field energy can be observed, which leads to a decrease in resonance frequency. The permittivity, which gets excited by the storage of energy, is significantly changed in relation to the water content. In addition, the wet material disposes the energy of the resonator, which results in an increasing width of the resonance waves. Because resonators respond very sensitively, a high accuracy of measurement is possible. While an increasing water burden leads to a decreasing resonance frequency, the frequency bandwidth broadens simultaneously. The broadening of the detected resonance frequency band is caused by both the product moisture and the material load in the focus of the sensor. By considering frequency and bandwidth simultaneously and comparing it to the unstressed resonator in air, two independent properties become available, which enable the determination of two product parameters—the density and the moisture. Thus, a density-revised moisture measurement, or a moisture-revised density measurement, can be obtained.[61] Microwave has a unique characteristic of penetrating detection and has the advantage of detecting internal moisture of the materials that have large size and complex surface features. With the applications of measurement probes, these parameters can be evaluated by microwave dielectric spectroscopy online.[62–64] The microwave dielectric spectroscopy gives information on the chemical relationship of compounds with their surroundings; thus, water content, state, and a_w can be studied using spectroscopic techniques,[64] with the analyzers and artificial intelligence techniques to transform the information and make adaptive decisions, thereby adjusting the drying conditions along with the drying process.

In the study by Cataldo, a procedure that is based on microwave reflectometry technique for *in situ* and continuous monitoring of the status of the hypertonic solutions used in industrial osmotic dehydration (OD) processing of fruit and vegetables was presented and validated.[65] With the OD process for tomatoes and a typical industrial application as the test, variation of the a_w and other parameters could be monitored, and specific calibration curves can be generated and subsequently adapted to monitor the real-time and *in situ* evolution of the process. It is demonstrated that the implementation of such a system would allow optimization of the production process, thus enhancing the overall quality of the final product and reducing the production costs.[65] According to the characteristics and requirements of the drying process for vegetable products, Castro-Giráldez et al. reported the application of microwave dielectric spectroscopy for controlling long-time OD of parenchymatus apple tissue.[66] Similar technology was also developed for OD of kiwifruit[67] and classification and adulteration control of vegetable oils.[68] It has been demonstrated that microwave dielectric spectroscopy, which has the advantage of being an objective and a rapid technique, is a good tool to control the surface a_w and can be considered for its novel rapid detection, better repeatability, and lower-cost online monitoring technology.

Microwave dielectric spectroscopy is a noncontact, nonchemical, and a very fast technique to determine moisture and density of materials in the drying process of vegetables. It is insensitive to environmental conditions (like vapor and dust), less sensitive to material buildup, safe at the low power levels used in measurements, and has low maintenance requirements. This technique provides more representative moisture data continuously, because the measurement is not limited to surface moisture and is independent from density or product load, color, and surface structure of solids.[14] The technique does not need sample withdrawal and provides in-process results because of its fast acquisition and processing times.[69] A chemometric method is not required if one-target information is needed, making it more appropriate for monitoring situations where sophisticated chemometric software and expert chemometricians are not available.[69] However, in order to extract multiple-target information from a large data set, multivariate data analysis techniques such as artificial intelligence techniques should be applied to transform the information and make adaptive decisions. Moreover, because the emitted microwave might slightly accelerate heat and mass transfer during drying, and its probe has a few intrusive parts, this technique is not always actually nonintrusive.[14] In addition, microwave dielectric spectroscopy requires expensive electronic components for applying high frequency, and it must be calibrated separately for different materials. It is difficult or even infeasible to attain an appropriate spatial resolution because of the relatively long wavelengths.[14] In conclusion, the commercial application of microwave dielectric spectroscopy still needs further research.

6.2.4 Ultrasonic Techniques

The applications of ultrasound in drying process, analysis, and quality control can be divided into low- and high-energy ultrasound based on the frequency range. Low-energy (low-power or low-intensity) ultrasound has frequencies higher than 100 kHz at intensities below 1 W·cm^2, which can be utilized for noninvasive analysis

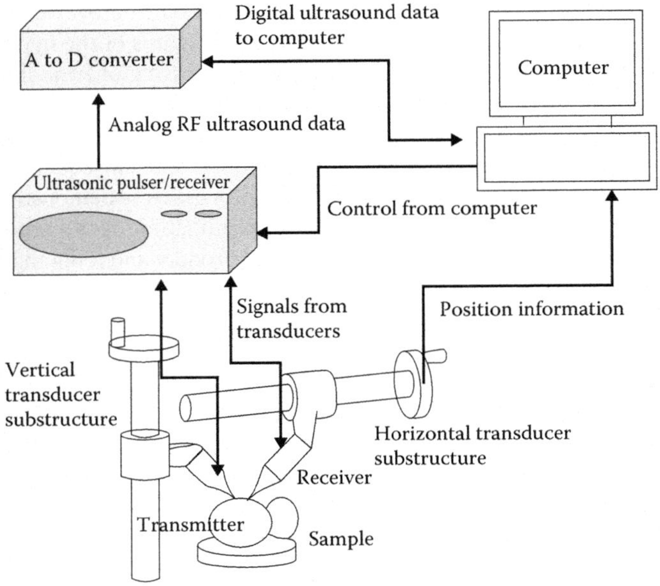

FIGURE 6.3 Schematic diagram of ultrasound technique.

and monitoring of various food materials during processing and storage to ensure high quality and safety.[17] Ultrasound technique is another emerging noninvasive, online acoustic measurement technique, which is fast and reliable for correlating specific quality-related indices and characteristics during the entire production process of vegetables.[17,70] The schematic diagram of ultrasound technique is shown in Figure 6.3.[17] The acoustic measurement concept is based on recording of passive acoustic emissions from systems that contain information that can be useful in their characterization. The mechanical structure of the tissue, its physicochemical quality indices, and every change in the quality attributes of the food affect the energy of the received signal. Its application spans various research and industrial monitoring of systems generating sound (vibrations). It is suitable for quality measurement in various products such as porous food products and vegetables.[71,72]

Ihunegbo et al. investigated the ultrasound technique for online monitoring and end point determination in fluidized bed drying. In his experiment, acoustic signals were acquired using four accelerometers mounted at different locations on the wall of the fluidized bed[73] (Figure 6.4). The final prediction results were satisfactory for monitoring the drying progress and for end point determination.[73] This online monitoring technique for the drying process can be applied in many relevant industries to improve the overall economics of material drying using this decisive end point determination approach.[70] It was reported that the ultrasound technique could also be applied to monitor the sugar content of vegetables.[74] This shows the potential for use of ultrasound for monitoring online process state by measuring the ultrasound propagation velocity and therefore the potential of ultrasound technology for the rapid inspection of the drying process of vegetables.

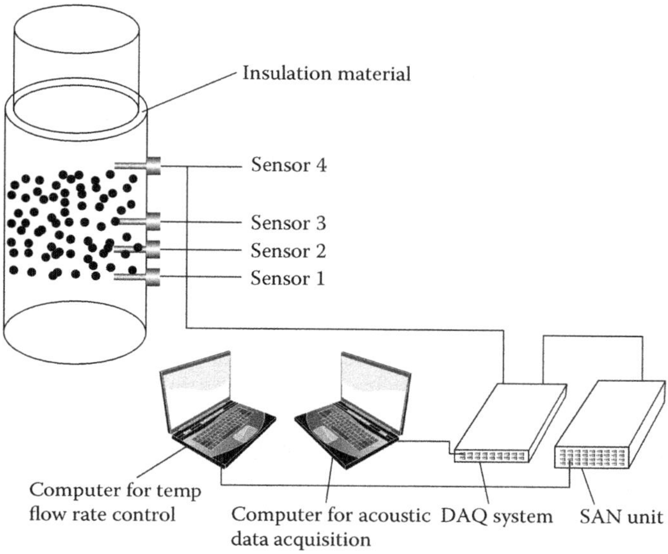

FIGURE 6.4 Schematic diagram of the experimental setup for the control and acoustic signal acquisition system.

However, ultrasound technique also has many drawbacks in the application of the drying process for vegetables. The main drawback of the ultrasound technique is that the relationship between signals and the quality parameters of drying food is indirect. It can be quite challenging to interpret the results to extract insight from the signals related to foods and the dryer attributes, which hinders the widespread application of ultrasound technique monitoring.[53] This problem could be overcome with statistical treatment of data to extract relevant information and decrease irrelevant information. On the other hand, ultrasound signals are not usually strong, and the sensitivity of the technique may be affected by the fluidizing airflow rate, construction material, and external uncontrollable and unintended interferences in the dryer,[75] necessitating the preprocessing of the spectra before the actual analysis. The ultrasound technique relies on highly specialized sensors and requires calibration with a reference technique. Therefore, the application of ultrasound technique in smart drying of vegetables needs further academic research.

6.2.5 Control Systems for Smart Drying Environment

In the drying process, environmental conditions such as pressure, temperature, and humidity have a great influence on the quality of the products, and at the same time, the parameters reflect information on the progress of the drying process.[76] Instead of detecting the quality parameters of the material itself, control systems of the drying environment monitor conditions such as pressure, temperature, and humidity in the process of drying, and thereby use this information to control the dryer operating conditions locally to yield high-quality products. The surrounding sensor systems not only contain smart sensors, but also translators, analyzers, and artificial intelligence techniques to transform the information and make adaptive decisions to change the conditions of the drying process. Today, most industrial dryers are equipped with varying levels of automatic control systems, involving very rapid drying or units that produce products within stringent quality specifications.[1,77] This system could sense the surroundings of the dryer and adjust the operating parameters consistent with the needs of product quality by drying timely and effectively; therefore, the control systems of the drying environment could be applied in various drying processes.

Correa-Hernando et al. developed an approach for smart sensors that makes use of the psychrometric properties of the air inside the drying chamber. These smart sensors characterize the drying process by monitoring of the surrounding environmental parameters, such as pressure, temperature, and humidity and adjust the operational conditions, thus meeting the inherent variability of the drying conditions during the day and over the year, and achieving a high agreement with experimental data at low cost.[78] A new way to use the instant controlled pressure drop (DIC) technology was adopted by Albitar et al. in the drying of onion. The DIC was performed as a blanching–steaming pretreatment of fresh cut onion, with a steam pressure of 0.2–0.5 MPa for 5–15 s, and an instant pressure drop toward a vacuum of 5 kPa. With this technology, the mass transfer rate is intensified and the total drying time is also reduced; furthermore, the sensory and nutritional attributes are preserved with a decontaminated end product.[79] Mounir et al.[80] studied the instant DIC process in manufacturing expanded granule powder of apple and onion. The results showed that the nutritional value of the product is partially preserved (vitamins) or even improved with more available flavonoids. The DIC technique can decontaminate the products very well, and end-quality attributes are better than those of the conventionally dried ones and are as good as freeze-dried samples.[80,81,82] Yang et al. proposed a method to handle the challenge of temperature control in a closed-loop heat pump dryer that can operate both in heating cycle and refrigeration cycle. Experimental results showed that the temperature fluctuations are reduced when using parallel conversion control with fast and stable response as compared to the fixed-frequency and control system.[83] Fissore et al. conducted the methods based on the pressure rise test for monitoring a freeze-drying process.[84] In another study, Fissore investigated online control of a freeze-drying process in vials. The operation was carried out at constant chamber pressure and shelf temperature, which were predicted by means of mathematical simulations, using a previously validated detailed model, thus allowing offline optimization of the process.[85]

Control system of the drying environment is sensitive, accurate, fast, robust, virtually nonintrusive, and relatively cheap and easy to implement in lab-, pilot-, and industrial-scale units, even under harsh conditions.[16] This technique can be considered to be a nonintrusive technique if the pressure transducer is flush-mounted at the vessel wall or if differential pressure measurements are applied[86]; thus, distortion of the flow around the point of measurement is avoided.[87] Meanwhile, signal processing is an essential tool for extracting information from the recoded environment parameter fluctuations related to the particle physical properties and the dryer hydrodynamics. However, this technique provides information only about changes of the environment index in the dryer, but it is useless for monitoring the quality parameters of vegetables inside the dryer and for ascertaining the location in the dryer in which variations in the dynamic behavior are taking place during drying. Unfortunately, few techniques are available for successful and satisfactory processing of environment signals to monitor physical properties of particles being processed, such as signal energy, average cycle time, dominant frequency, and attractor comparison tools. Moreover, these environment parameters need intrusive taps, and the transducer needs to be placed inside the process itself for industrial and experimental applications. This technique does not provide detailed information about height of drying media and clear knowledge during processing of very fine particles. Furthermore, to prevent fouling of the transducers with special materials, continuous back-flushing with pressurized air (which is expensive) or a mechanical scraper is necessary, which in turn diminishes the sensitivity of the probe,[88] and these control systems provide statistical information without resolving specific particle sizes. In addition, identification of the source of fluctuations among many simultaneously occurring phenomena is very difficult due to the extremely complex local flow structure through the dryer.[89] Therefore, control system of the drying environment needs to undergo further studies considering research and commercial values.

6.3 CLOSING REMARKS

A comprehensive definition for smart drying technology of vegetables is proposed in this chapter. Although many consumer dryers (e.g., domestic clothes dryers) are already marketed as smart dryers, most industrial dryers operating today cannot be classified as smart under this definition. Many dryers incorporate relatively simple control strategies that directly manage the operating conditions within a dryer by minoring them and indirectly manage the quality of the product. However, most of them do not monitor in real time the quality parameters due to either lack of suitable sensors or high cost. In the coming years, it is expected that the sensor technology could be more cost-effective for applications in the food processing industry. This will enable smart dryers to be developed cost-effectively as well. Apart from excellent quality control, smart drying can minimize energy consumption and lower environmental impact.

REFERENCES

1. Jumah, R.Y.; Mujumdar, A.S.; Raghavan, V.G.S. Control of industrial dryers. In *Handbook of Industrial Drying*, Mujumdar, A.S., Ed. CRC Press, Boca Raton, FL, 2006, 1161–1179.
2. Mujumdar, A.S. Research and development in drying: Recent trends and future prospects. *Drying Technology* 2004, 22(1–2), 1–26.
3. Mujumdar, A.S. Editorial: The role of *drying technology* in sustainable R&D and innovation. *Drying Technology* 2012, 30(13), 1387.
4. Ghasemi-Varnamkhasti, M.; Aghbashlo, M. Electronic nose and electronic mucosa as innovative instruments for real-time monitoring of food dryers. *Trends in Food Science and Technology* 2014, 38(2), 158–166. doi:10.1016/j.tifs.2014.05.004.

5. Aghbashlo, M.; Mobli, H.; Rafiee, S.; Madadlou, A. The use of artificial neural network to predict exergetic performance of spray drying process: A preliminary study. *Computers and Electronics in Agriculture* 2012, 88, 32–43.
6. Nazghelichi, T.; Aghbashlo, M.; Kianmehr, M.H.; Omid., M. Prediction of energy and exergy of carrot cubes in a fluidized bed dryer by artificial neural networks. *Drying Technology* 2011, 29(3), 295–307.
7. Gong, Z.X.; Mujumdar, A.S. Software for design and analysis of drying systems. *Drying Technology* 2008, 26(7), 884–894.
8. Damez, J.L.; Clerjon, S. Quantifying and predicting meat and meat products quality attributes using electromagnetic waves: An overview. *Meat Science* 2013, 95(4), 879–896.
9. Śliwińska, M.; Wiśniewska, P.; Dymerski, T.; Namieśnik, J.; Wardencki, W. Food analysis using artificial senses. *Journal of Agricultural and Food Chemistry* 2014, 62(7), 1423–1448.
10. Ghasemi-Varnamkhasti, M.; Mohtasebi, S.S.; Siadat, M. Biomimetic-based odor and taste sensing systems to food quality and safety characterization: An overview on basic principles and recent achievements. *Journal of Food Engineering* 2010, 100(3), 377–387.
11. Dymerski, T.M.; Chmiel, T.M.; Wardencki, W. Invited review article: An odor-sensing system—powerful technique for foodstuff studies. *Review of Scientific Instruments* 2011, 82(11), 111101.
12. Lorenz, J.K.; Reo, J.P.; Hendl, O.; Worthington, J.H.; Petrossian, V.D. Evaluation of a taste sensor instrument (electronic tongue) for use in formulation development. *International Journal of Pharmaceutics* 2009, 367(1–2), 65–72.
13. Mujumdar, A.S. An overview of innovation in industrial drying: Current status and R&D needs. *Transport in Porous Media* 2007, 66(1), 3–18.
14. Aghbashlo, M.; Sotudeh-Gharebagh, R.; Zarghami, R.; Mujumdar, A.S.; Mostoufi, N. Measurement techniques to monitor and control fluidization quality in fluidized bed dryers: A review. *Drying Technology* 2014, 32(9), 1005–1051.
15. Burggraeve, A.; Monteyne, T.; Vervaet, C.; Remon, J.P.; De Beer, T. Process analytical tools for monitoring, understanding, and control of pharmaceutical fluidized bed granulation: A review. *European Journal of Pharmaceutics and Biopharmaceutics* 2013, 83(1), 2–15.
16. da Silva, C.A.M.; Butzge, J.J.; Nitz, M.; Taranto, O.P. Monitoring and control of coating and granulation processes in fluidized beds: A review. *Advanced Powder Technology* 2014, 25(1), 195–210.
17. Awad, T.S.; Moharram, H.A.; Shaltout, O.E.; Asker, D.; Youssef, M.M. Applications of ultrasound in analysis, processing and quality control of food: A review. *Food Research International* 2012, 48(2), 410–427.
18. Dufour, P. Control engineering in drying technology: Review and trends. *Drying Technology* 2006, 24(7), 889–904.
19. Stitzel, S.E.; Aernecke, M.J.; Walt, D.R. Artificial noses. *Annual Review of Biomedical Engineering* 2011, 13, 1–25.
20. Li, Z.-F.; Raghavan, G.S.V.; Wang, N. Carrot volatiles monitoring and control in microwave drying. *LWT-Food Science and Technology* 2010, 43(2), 291–297.
21. Li, Z.; Vijaya Raghavan, G.S.; Wang, N. Apple volatiles monitoring and control in microwave drying. *LWT-Food Science and Technology* 2010, 43(4), 684–689.
22. Li, Z.; Vijaya Raghavan, G.S.; Wang, N.; Gariépy, Y. Real-time, volatile-detection-assisted control for microwave drying. *Computers and Electronics in Agriculture* 2009, 69(2), 177–184.
23. Raghavan, G.S.V.; Li, Z.; Wang, N.; Gariépy, Y. Control of microwave drying process through aroma monitoring. *Drying Technology* 2010, 28(5), 591–599.

24. Yang, Z.; Dong, F.; Shimizu, K.; Kinoshita, T.; Kanamori, M.; Morita, A.; Watanabe, N. Identification of coumarin-enriched Japanese green teas and their particular flavor using electronic nose. *Journal of Food Engineering* 2009, 92(3), 312–316.
25. Infante, R.; Rubio, P.; Contador, L.; Moreno, V. Effect of drying process on lemon verbena (*Lippia citrodora Kunth*) aroma and infusion sensory quality. *International Journal of Food Science and Technology* 2009, 45(1), 75–80.
26. Barié, N.; Bücking, M.; Rapp, M. A novel electronic nose based on miniaturized SAW sensor arrays coupled with SPME enhanced headspace-analysis and its use for rapid determination of volatile organic compounds in food quality monitoring. *Sensors and Actuators B: Chemical* 2006, 114(1), 482–488.
27. Legin, A.; Rudnitskaya, A.; Kirsanov, D.; Frolova, Y.; Clapham, D.; Caricofe, R.; Pardo, M.; Sberveglieri, G. Assessment of bitterness intensity and suppression effects using an electronic tongue. In *AIP Conference Proceedings*, Brescia (Italy), 2009, 271–274.
28. Macias Macias, M.; Agudo, J.E.; García Manso, A.; García Orellana, C.J.; González Velasco, H.M.; Gallardo Caballero, R. A compact and low cost electronic nose for aroma detection. *Sensors (Basel)* 2013, 13(5), 5528–5541.
29. Xiao, H.; Wang, J. Discrimination of Xihulongjing tea grade using an electronic tongue. *African Journal of Biotechnology* 2009, 8(24), 6985–6992.
30. Bhattacharyya, N.; Seth, S.; Tudu, B.; Tamuly, P.; Jana, A.; Ghosh, D.; Bandyopadhyay, R.; Bhuyan, M. Monitoring of black tea fermentation process using electronic nose. *Journal of Food Engineering* 2007, 80(4), 1146–1156.
31. Bhattacharyya, N.; Seth, S.; Tudu, B.; Tamuly, P.; Jana, A.; Ghosh, D.; Bandyopadhyay, R.; Bhuyan, M.; Sabhapandit, S. Detection of optimum fermentation time for black tea manufacturing using electronic nose. *Sensors and Actuators B: Chemical* 2007, 122(2), 627–634.
32. Wongchoosuk, C.; Lutz, M.; Kerdcharoen, T. Detection and classification of human body odor using an electronic nose. *Sensors* 2009, 9, 7234–7249.
33. Dymerski, T.; Gebicki, J.; Wardencki, W.; Namiesnik, J. Quality evaluation of agricultural distillates using an electronic nose. *Sensors (Basel)* 2013, 13(12), 15954–15967.
34. Schmitt, V.; Tan, T.T. Simplify quality control with electronic nose. *G.I.T Laboratory Journal* 2001, 1, 2–4.
35. Keller, P.E.; Kangas, L.J.; Liden, L.H.; Hashem, S.; Kouzes, R.T. Electronic noses and their applications. In *IEEE Northcon/Technical Applications Conference (TAC'95)*, Portland, OR, 1995.
36. Wu, D.; Sun, D.-W. Colour measurements by computer vision for food quality control: A review. *Trends in Food Science and Technology* 2013, 29(1), 5–20.
37. Sun, D.-W. *Computer Vision Technology for Food Quality Evaluation.* Elsevier Academic Press, San Diego, CA, 2008, 583.
38. Sun, D.-W. Computer vision technology in the food and beverage industries. *International Journal of Food Science and Technology* 2013, 48(7), 1556.
39. Dowlati, M.; de la Guardia, M.; Dowlati, M.; Mohtasebi, S.S. Application of machine-vision techniques to fish-quality assessment. *Trends in Analytical Chemistry* 2012, 40, 168–179.
40. Jackman, P.; Sun, D.-W. Recent advances in image processing using image texture features for food quality assessment. *Trends in Food Science and Technology* 2013, 29(1), 35–43.
41. Ahmad, R.; Tichadou, S.; Hascoet, J.-Y. New computer vision based Snakes and Ladders algorithm for the safe trajectory of two axis CNC machines. *Computer-Aided Design* 2012, 44(5), 355–366.
42. Martynenko, A.I. Computer-vision system for control of drying processes. *Drying Technology* 2006, 24(7), 879–888.

43. Sampson, D.J.; Chang, Y.K.; Rupasinghe, H.P.V.; Zaman, Q.U.Z. A dual-view computer-vision system for volume and image texture analysis in multiple apple slices drying. *Journal of Food Engineering* 2014, 127, 49–57.
44. Trinca, L.C.; Capraru, A.M.; Arotaritei, D.; Volf, I.; Chiruta, C. Monitoring methods and predictive models for water status in Jonathan apples. *Food Chemistry* 2014, 144, 80–86.
45. Jackman, P.; Sun, D.-W.; Allen, P.; Brandon, K.; White, A.M. Correlation of consumer assessment of *longissimus dorsi* beef palatability with image colour, marbling and surface texture features. *Meat Science* 2010, 84(3), 564–568.
46. Kamruzzaman, M.; ElMasry, G.; Sun, D.W.; Allen, P. Prediction of some quality attributes of lamb meat using near-infrared hyperspectral imaging and multivariate analysis. *Analytica Chimica Acta* 2012, 714, 57–67.
47. Zion, B.; Alchanatis, V.; Ostrovsky, V.; Barki, A.; Karplus, I. Classification of guppies' (*Poecilia reticulata*) gender by computer vision. *Aquacultural Engineering* 2008, 38(2), 97–104.
48. Quevedo, R.; Aguilera, J.M. Computer vision and stereoscopy for estimating firmness in the salmon (*Salmon salar*) fillets. *Food and Bioprocess Technology* 2008, 3(4), 561–567.
49. Riquelme, M.T.; Barreiro, P.; Ruiz-Altisent, M.; Valero, C. Olive classification according to external damage using image analysis. *Journal of Food Engineering* 2008, 87(3), 371–379.
50. Zion, B.; Alchanatis, V.; Ostrovsky, V.; Barki, A.; Karplus, I. Real-time underwater sorting of edible fish species. *Computers and Electronics in Agriculture* 2007, 56(1), 34–45.
51. Lee, D.-J.; Archibald, J.K.; Chang, Y.-C.; Greco, C.R. Robust color space conversion and color distribution analysis techniques for date maturity evaluation. *Journal of Food Engineering* 2008, 88(3), 364–372.
52. Liew, C.V.; Wang, L.K.; Heng, P.W.S. Development of a visiometric process analyzer for real-time monitoring of bottom spray fluid-bed coating. *Journal of Pharmaceutical Sciences* 2010, 99(1), 346–356.
53. Närvänen, T.; Seppälä, K.; Antikainen, O.; Yliruusi, J. A new rapid on-line imaging method to determine particle size distribution of granules. *AAPS PharmSciTech* 2008, 9(1), 282–287.
54. Marcone, M.F.; Wang, S.; Albabish, W.; Nie, S.; Somnarain, D.; Hill, A. Diverse food-based applications of nuclear magnetic resonance (NMR) technology. *Food Research International* 2013, 51(2), 729–747.
55. Ghosh, P.K.; Jayas, D.S.; Gruwel, M.L.H.; White, N.D.G. Magnetic resonance imaging studies to determine the moisture removal patterns in wheat during drying. *Canadian Biosystems Engineering* 2006, 48(7), 13–18.
56. Manzocco, L.; Anese, M.; Marzona, S.; Innocente, N.; Lagazio, C.; Nicoli, M.C. Monitoring dry-curing of S. Daniele ham by magnetic resonance imaging. *Food Chemistry* 2013, 141(3), 2246–2252.
57. Altan, A.; Oztop, M.H.; McCarthy, K.L.; McCarthy, M.J. Monitoring changes in feta cheese during brining by magnetic resonance imaging and NMR relaxometry. *Journal of Food Engineering* 2011, 107(2), 200–207.
58. Castell-Palou, A.; Rosselló, C.; Femenia, A.; Bon, J.; Simal, S. Moisture profiles in cheese drying determined by TD-NMR: Mathematical modeling of mass transfer. *Journal of Food Engineering* 2011, 104(4), 525–531.
59. Pérez-Palacios, T.; Caballero, D.; Caro, A.; Rodríguez, P.G.; Antequera, T. Applying data mining and computer vision techniques to MRI to estimate quality traits in Iberian hams. *Journal of Food Engineering* 2014, 131, 82–88.
60. Zhang, L.; McCarthy, M.J. NMR study of hydration of navy bean during cooking. *LWT-Food Science and Technology* 2013, 53(2), 402–408.
61. Buschmuller, C.; Wiedey, W.; Döscher, C.; Dressler, J.; Breitkreutz, J. In-line monitoring of granule moisture in fluidized-bed dryers using microwave resonance technology. *European Journal of Pharmaceutics and Biopharmaceutics* 2008, 69(1), 380–387.

62. Ramaswamy, H.; Tang, J. Microwave and radio frequency heating. *Food Science and Technology International* 2008, 14(5), 423–427.
63. Clerjon, S.; Damez, J.-L. Microwave sensing for an objective evaluation of meat ageing. *Journal of Food Engineering* 2009, 94(3–4), 379–389.
64. Castro-Giráldez, M.; Aristoy, M.C.; Toldrá, F.; Fito, P. Microwave dielectric spectroscopy for the determination of pork meat quality. *Food Research International* 2010, 43(10), 2369–2377.
65. Cataldo, A.; Cannazza, G.; De Benedetto, E.; Severini, C.; Derossi, A. An alternative method for the industrial monitoring of osmotic solution during dehydration of fruit and vegetables: A test-case for tomatoes. *Journal of Food Engineering* 2011, 105(1), 186–192.
66. Castro-Giráldez, M.; Fito, P.J.; Fito, P. Application of microwaves dielectric spectroscopy for controlling long time osmotic dehydration of parenchymatic apple tissue. *Journal of Food Engineering* 2011, 104(2), 227–233.
67. Castro-Giráldez, M.; Fito, P.J.; Dalla Rosa, M.; Fito, P. Application of microwaves dielectric spectroscopy for controlling osmotic dehydration of kiwifruit (*Actinidia deliciosa cv Hayward*). *Innovative Food Science and Emerging Technologies* 2011, 12(4), 623–627.
68. Cataldo, A.; Piuzzi, E.; Cannazza, G.; De Benedetto, E. Classification and adulteration control of vegetable oils based on microwave reflectometry analysis. *Journal of Food Engineering* 2012, 112(4), 338–345.
69. Corredor, C.C.; Bu, D.; Both, D. Comparison of near infrared and microwave resonance sensors for at-line moisture determination in powders and tablets. *Analytica Chimica Acta* 2011, 696(1–2), 84–93.
70. Mizrach, A. Ultrasonic technology for quality evaluation of fresh fruit and vegetables in pre- and postharvest processes. *Postharvest Biology and Technology* 2008, 48(3), 315–330.
71. Saeleaw, M.; Schleining, G. A review: Crispness in dry foods and quality measurements based on acoustic–mechanical destructive techniques. *Journal of Food Engineering* 2011, 105(3), 387–399.
72. Varela, P.; Chen, J.; Fiszman, S.; Povey, M.J.W. Crispness assessment of roasted almonds by an integrated approach to texture description: Texture, acoustics, sensory and structure. *Journal of Chemometrics* 2006, 20(6–7), 311–320.
73. Ihunegbo, F.N.; Ratnayake, C.; Halstensen, M. Acoustic chemometrics for on-line monitoring and end-point determination of fluidised bed drying. *Powder Technology* 2013, 247, 69–75.
74. Mizrach, A. Nondestructive ultrasonic monitoring of tomato quality during shelf-life storage. *Postharvest Biology and Technology* 2007, 46(3), 271–274.
75. Burggraeve, A.; Sandler, N.; Heinämäki, J.; Räikkönen, H.; Remon, J.P.; Vervaet, C.; De Beer, T.; Yliruusi, J. Real-time image-based investigation of spheronization and drying phenomena using different pellet formulations. *European Journal of Pharmaceutical Sciences* 2011, 44(5), 635–642.
76. Kim, T.H. Physical changes in maize (*Zea mays* L.) grains during postharvest drying. Doctor of Philosophy's thesis, Seed Technology, Department location, Paper 220, 2000.
77. Erenturk, K. Artificial intelligence techniques for food drying technology. In *1st International Symposium on Sustainable Development*, Sarajevo, June 9–10, 2009, 375–380.
78. Correa-Hernando, E.; Arranz, F.J.; Diezma, B.; Jiménez, H.T.; Robla, J.I.; Ruiz-García, L.; García-Hierro, J.; Barreiro, P.; Valero, C. Development of smart sensors for the supervision of a solar dryer: Agro-products dehydration application. In *International Conference on Agricultural Engineering*, Clermont-Ferrand, France, June 8–August 8, 2010.

79. Albitar, N.; Mounir, S.; Besombes, C.; Allaf, K. Improving the drying of onion using the instant controlled pressure drop technology. *Drying Technology* 2011, 29(9), 993–1001.
80. Mounir, S.; Besombes, C.; Al-Bitar, N.; Allaf, K. Study of instant controlled pressure drop DIC treatment in manufacturing snack and expanded granule powder of apple and onion. *Drying Technology* 2011, 29(3), 331–341.
81. Bosca, S.; Corbellini, S.; Barresi, A.A.; Fissore, D. Freeze-drying monitoring using a new process analytical technology: Toward a "Zero Defect" process. *Drying Technology* 2013, 31(15), 1744–1755.
82. Drăgoi, E.N.; Curteanu, S.; Fissore, D. On the use of artificial neural networks to monitor a pharmaceutical freeze-drying process. *Drying Technology* 2013, 31(1), 72–81.
83. Yang, Z.; Zhu, Z.; Zhu, E. Experimental research on parallel conversion control of drying temperature in a closed-loop heat pump dryer. *Drying Technology* 2013, 31(9), 1049–1055.
84. Fissore, D.; Pisano, R.; Barresi, A.A. On the methods based on the pressure rise test for monitoring a freeze-drying process. *Drying Technology* 2010, 29(1), 73–90.
85. Fissore, D.; Velardi, S.A.; Barresi, A.A. In-line control of a freeze-drying process in vials. *Drying Technology* 2008, 26(6), 685–694.
86. Chaplin, G.; Pugsley, T.; Winters, C. Application of chaos analysis to pressure fluctuation data from a fluidized bed dryer containing pharmaceutical granule. *Powder Technology* 2004, 142(2–3), 110–120.
87. Ruud van Ommen, J.; Sasic, S.; van der Schaaf, J.; Gheorghiu, S.; Johnsson, F.; Coppens, M.-O. Time-series analysis of pressure fluctuations in gas–solid fluidized beds: A review. *International Journal of Multiphase Flow* 2011, 37(5), 403–428.
88. Brennan, W.; Jacobson, M.; Book, G.; Briens, C.; Briens, L. Development of a triboelectric procedure for the measurement of mixing and drying in a vibrated fluidized bed. *Powder Technology* 2008, 181(2), 178–185.
89. Salehi-Nik, N.; Sotudeh-Gharebagh, R.; Mostoufi, N.; Zarghami, R.; Mahjoob, M.J. Determination of hydrodynamic behavior of gas–solid fluidized beds using statistical analysis of acoustic emissions. *International Journal of Multiphase Flow* 2009, 35(11), 1011–1016.

7 Foam-Mat Drying of Vegetables

Regiane Victória de Barros Fernandes and Diego Alvarenga Botrel
Federal University of Lavras

CONTENTS

7.1 INTRODUCTION

Reducing postharvest losses and ameliorating the poor distribution of food are major global challenges. Because vegetables are highly perishable, dehydrated forms present an alternative to the excess of production in the fresh-produce market besides offering the consumer and the food industry a sensorially differentiated product. This product, being less perishable, can be commercialized at any time of the year and may be used in the development of novel food formulations using vegetable powders, such as dehydrated goods and powdered seasonings and soups.

Vegetables and their derivatives in dry forms are important sources of energy and nutritional components such as vitamins, minerals (Sablani 2006), carbohydrates, fiber, and other nutrients. Moreover, vegetables are rich in bioactive compounds including tocopherols, flavonoids, phenolic acids, alkaloids, chlorophyll derivatives, and carotenoids, which possess substantial antioxidant properties and confer significant health benefits (Hudson 1990; Hall and Cuppett 1997; McDermott 2000; Hung and Duy 2012).

Removal of water from food products is one of the oldest methods of preservation. By reducing the moisture content of a foodstuff to very low levels, microbiological deterioration decreases, and the rates of other degradation reactions are also significantly reduced. In addition to enhancing preservation, dehydration reduces the weight and volume of foods, thereby increasing the efficiency of transport and storage. Compared to fresh forms, removal of water often results in a more convenient product for consumption. Hot air drying is widely used in the preservation of vegetables; however, a relative reduction in quality could result from this process. The most common problems associated with dehydrated products are excessive hardness, difficulty in rehydrating, undesirable color changes, and loss of flavor and nutrients. Stability is also an important factor to consider in the use of dried vegetables as ingredients, colorants, or even as a source of functional components in food and other applications. The stability of components presenting functional claims, during extended periods of time, is of great interest to industry and consumers.

Numerous methods exist for the dehydration of foods, from the most advanced, targeted to large-scale production, to the most simple, aimed at small producers (Baldwin et al. 1999). The choice of drying procedure depends on various factors, including product type, dryer availability, cost, and final product quality (Sagar and Kumar 2010), apart from the intended end use of the finished product (Raghavan and Orsat 2007). Although sun drying is still employed for certain fruits, such as prunes, apricots, and grapes, continuous drying processes, for example, tunnel, belt-through, fluidized-bed, and foam-mat drying, are mainly used for vegetable drying (Raghavan and Orsat 2007).

Air drying is considered as the most simple and economical process and is still widely used. The principal problems associated with this method include shrinkage caused by the collapse of cells upon loss of water, rehydration issues, unfavorable color, texture, and flavor changes, and loss of nutritional value (Mazza 1983). Hot air drying usually destroys cell structure, thereby taking more time for dehydration (Sagar and Kumar 2010). Moreover, various changes to the physical and chemical properties of these products can take place during storage. Many studies have been conducted to evaluate dried food rehydration characteristics and final product quality through the study of process variables including drying time, dryer type, and air temperature; type of material to be dried; and sample pretreatment (Cunningham et al. 2008).

Foam drying is an alternative method of hot air drying that has several advantages over the traditional process. The main advantage of foam-mat drying is the lower temperature and shorter drying time taken compared with nonfoamed materials in a comparable dryer, thereby reducing thermal degradation of the dried product (Fernandes et al. 2014a). This decreased drying time depends on the total contact surface exposed to the air and the specific heat and mass transfer characteristics of foamed materials (Ratti and Kudra 2006).

Research in the development of alternative drying processes results in a quality product with high added value and is of great interest. The demands of consumers for safe and nutritive foods with good sensory characteristics lead to the continual development of food processing technologies. Moreover, current interests in the consumption of healthy foods have challenged the food industry to design new products and ingredients with specific properties, paying special attention to those derived from fruits and vegetables (Villamiel et al. 2015).

7.2 FOAM-MAT DRYING

7.2.1 Foam-Mat Drying Concepts

The drying process offers several advantages, such as a considerably reduced water activity and weight, protection against degradation reactions, diminished microbial activity, energy savings as no refrigeration is required, and product availability throughout the year (Mayor and Sereno 2004; Fernandes et al. 2014a). Moreover, it is regarded not only as a preservation method but also as a means to increase the added value of foods.

Of the techniques used to obtain powdered food products, drying by foam layer (foam-mat drying) is notable, which consists of the transformation of liquid or semiliquid foods into stable foams through vigorous agitation and incorporation of air, with the aid of foaming agents and stabilizers (Kudra and Mujumdar 2001; Silva et al. 2008). A thin layer of the foamed material is then dried, before being disintegrated to yield a powder (Salahi et al. 2015). During this process, moisture content can be reduced to between 1% and 5% (Sangamithra et al. 2015a). The literature reports 80°C as the upper limit for the dehydration of vegetables, with 60°C being the temperature most commonly used for the dehydration of vegetables so as to keep the quality of the final product intact. These levels are principally determined according to the sensitivity of the food's components (Martinazzo et al. 2010).

The foam-mat drying method dates from 1917, when it was patented by Campbell (1917) for the drying of foamed evaporated milk, followed by other patents for the dehydration of egg albumin (Mink 1939, 1940; Ratti and Kudra 2006). During the 1990s, this relatively old technology received renewed attention due to the rapid drying times achievable, the potential to process difficult-to-dry materials, and the ability to retain volatiles that might be lost during dehydration of nonfoamed materials (Ratti and Kudra 2006).

Foam-mat drying is a relatively simple method that facilitates the removal of water from vegetables (Asokapandian et al. 2016). This process is considerably cheaper than freeze- and spray drying for the production of vegetable powders (Sangamithra et al. 2015a). Moreover, this method is suitable for heat-sensitive, sticky, or viscous materials that cannot be dried by spray drying (Kandasamy et al. 2014). Table 7.1 presents some examples of the foaming/stabilizing agents and air temperatures used in foam-mat drying of vegetables.

Whipping or beating can be carried out with various devices, such as manual and automatic blenders, vortex mixers, and homogenizers, which agitate the liquid to create an interface with the gas phase (Hardy and Jideani 2015). The volume of air incorporated into a foam generally increases with beating intensity (Arzhavitina and Steckel 2010; Hardy and Jideani 2015). Frequently, wet foams are dried using convection, by applying a flow of hot air over or through a relatively thin layer (3–10 mm) of foamed material (Cooke et al. 1976). This process primarily consists of three stages: changing the liquid consistency of the pulp or purée into a stable foam by the addition of foaming agents; drying the material in a thin layer; and pulverizing the dried materials (Travaglini et al. 2001; Silva et al. 2008). The preparation of stable foams plays a major role in foam-mat drying, with foam

TABLE 7.1
Foaming/Stabilizing Agents and Air Drying Temperatures Used in Foam-Mat Drying of Vegetables

Vegetable	Foaming/Stabilizing Agent	Air Drying Temperature (°C)	References
Bitter gourd, tomato, and cucumber	Egg albumin	50, 60, and 70	Chandrasekar et al. (2015)
Pumpkin	Glycerol monostearate	50 and 60	Das et al. (2015)
Yacon	Egg albumin/commercial emulsifier Emustab (monoglycerides, sorbitan monostearate, and polysorbate 60)	50, 60, and 70	Franco et al. (2015a)
Shallot	Cassava starch/ maltodextrin	80	Setyadjit and Sukasih (2015)
Tomato	Egg albumin/whey protein isolate/maltodextrin	50, 60, and 70	Sramek et al. (2015)
Potato	Glycerol monostearate	50, 60, and 65	Chakraborty et al. (2014)
Tomato	Egg albumin	45	Qadri and Srivastava (2014)
Tomato	Egg albumin	60 and 80	Fernandes et al. (2013a, 2014a)
Yam	Glyceryl monostearate	70	Falade and Onyeoziri (2012)
Tomato	Carboxyl methyl cellulose/ milk/egg white	65, 75, and 85	Kadam et al. (2012)
Tomato	Egg albumin	60, 65, and 70	Kadam and Balasubramanian (2011)
Cowpea	Glyceryl monostearate/ egg albumin	60	Falade et al. (2003)
Cowpea	Sodium palmitate	70	Olopade et al. (2003)

expansion and stability being the most important aspects (Pasban et al. 2014). The steps of this process are shown in Figure 7.1.

Foam-mat drying has an increasing demand and application in the commercial-scale dehydration of liquids for high-quality concentrate, including vegetables, fruit juices, tea, and coffee (Bag et al. 2011; Sangamithra et al. 2015a). This type of drying enables processing of hard-to-dry biomaterials such as tomato paste, as well as allowing the production of easily rehydratable materials that retain quality indicators such as color, aroma, texture, and nutritional value (Fernandes et al. 2014a). However, this technique may not necessarily result in products superior to those manufactured by spray drying. For example, the physical characteristics and reconstitution properties of foam-mat-dried soy extract were found to be better than those achieved using a conventional method, but were inferior to those obtained by spray drying (Akintoye and Oguntunde 1991). The end product of foam-mat drying is porous and can be easily reconstituted (Sangamithra et al. 2015a). However, the main drawback of this process is its limited throughput, caused by the small amount of foodstuff that can

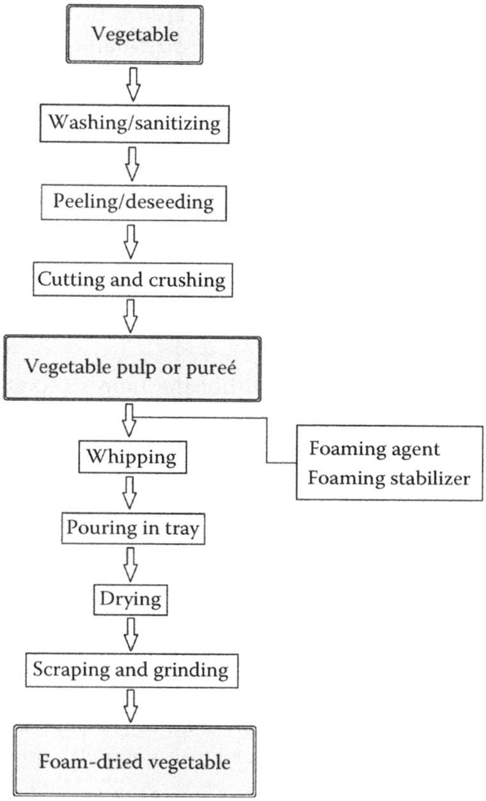

FIGURE 7.1 Flowchart of the foam-mat drying process.

be applied per unit area of the dryer surface, although this might be compensated by shorter drying times (Zbicinski et al. 2014).

7.2.2 Properties and Characteristics of Foam-Mat Drying

The success of foam-mat drying depends on the production of mechanically and thermally stable foams using suitable foaming agents. Mechanical stability is necessary to avoid collapse of the structure during the drying process. In general, foams that retain their structure for at least 1 h are considered mechanically stable. Based on studies conducted by Bates (1964), foams that do not collapse after the first minutes of the process can be considered structurally stable throughout the drying period.

Thermally stable foams retain their porous structure, which aids the reconstitution process. On the other hand, foam collapse increases drying time, makes rehydration more difficult, and changes the color, texture, flavor and nutritional value (Ratti and Kudra 2006). Foam formation is a significant step, understanding the factors contributing to its development or failure is crucial. The foam texture can be determined by the number, size, and distribution of bubbles (Hardy and Jideani 2015).

Foam density is commonly used to evaluate whipping properties. Larger volumes of air incorporated during whipping result in lower foam densities (Falade et al. 2003; Thuwapanichayanan et al. 2012), which decrease with increase in whipping time and speed. After this initial decrease, however, a slight increase in density can be noticed after 7 min of whipping (Fernandes et al. 2013a). Higher degrees of aeration result in thinner liquid between the bubbles, and mechanical deformation can cause the foam to rupture (Falade et al. 2003).

Generally, drying rates are comparatively higher in foamed pulps because of increased surface area at the liquid–gas interface, which facilitates rapid drying through internal moisture movement within the pulp (Rajkumar and Kailappan 2006). Figure 7.2 shows drying curves obtained for treatments with and without addition of a foaming agent, using the same drier type and temperature. Rapid drying is achieved by the capillary movement of water in the films, thereby separating liquid and gas in the foam bubbles. The foam structure makes for an extremely porous bed that is more prone to drying (Venkataraman 1996), and reduction of flavor loss by diffusion can be obtained due to the shorter drying period (Kerkhof 1994).

The physical characteristics of dried products relating to their ease of dispersion in an aqueous solution include bulk density, particle density and porosity, and instantanization properties (wetting, dispersibility, and solubility) (Fernandes et al. 2013b). Tapped density is an important factor affecting packaging, transport, and commercialization of powders, and this value can be useful in determining the weight of the material that fits into a container (Finney et al. 2002). High-density dry products

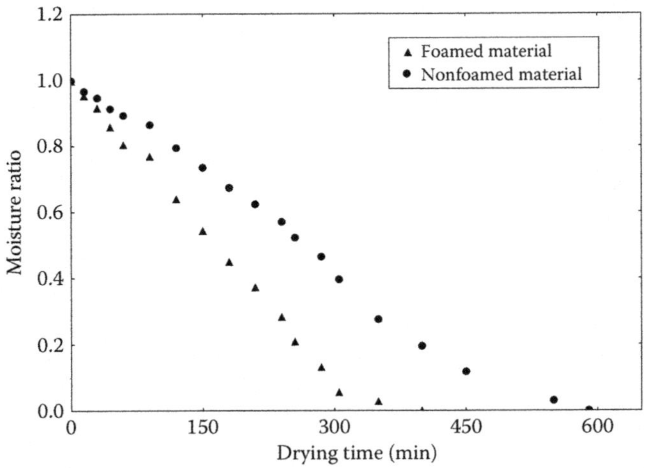

FIGURE 7.2 Drying profiles of tomato pulp dried at 60°C with and without application of the foaming agent albumin. (Adapted from Fernandes, R.V.B. et al., *Bioscience Journal*, 29, 816–825, 2013a.)

can be stored in smaller containers than low-density ones (Quispe-Condori et al. 2011). Particle density may be influenced by the temperature used for air drying, the size of particles produced (Fernandes et al. 2014b), and the foam's constituents (i.e., the nature of the vegetable and foaming agent). Another property of fundamental importance in foam-mat drying is porosity, which plays a key role in determining the reconstitution properties of the dried product. Dried product quality-control parameters such as the Carr Index and Hausner ratio, which evaluate powder flow, should also be considered (Fitzpatrick 2005). For instance, a high Hausner ratio indicates a more cohesive powder, less capable of flowing freely. Wettability, meanwhile, is characterized as the rehydration ability of the powder in water (Fernandes et al. 2014b). In addition, powders used as ingredients in the food industry should provide good solubility (Fernandes et al. 2014c,d), which is the last step in powder dissolution and is considered to be the decisive factor in the quality of such products (Jayasundera et al. 2011).

7.3 FOAMING AND STABILIZING AGENTS

Foams are generally regarded as dispersions of gas bubbles in aqueous, semiaqueous, or solid continuous phase and are commonly associated with detergents. The dispersed and continuous phases are referred to as the internal and external phases, respectively (Sangamithra et al. 2015a). Various substances are used to aid foam formation, including proteins, phospholipids (lecithin), and monoglycerides and diglycerides, or to stabilize the emulsions formed, such as gums (Fernandes et al. 2013a). The gas–liquid dispersions are represented by gaseous emulsions, in which a small volume of gas is dispersed in a large amount of liquid. The density of these "wet" foams ranges from 0.2 to 0.8 g/mL. However, the foam density used in foam-mat drying processes ranges from 0.3 to 0.6 g/mL. The final bubble size depends on agitator speed, the geometry of the apparatus, and the liquid's rheological properties (Arzhavitina and Steckel 2010; Hardy and Jideani 2015). Gas bubbles in low-density foams are polyhedral, in contrast to the spherical forms observed in denser foams. Regardless of their shape, bubbles typically have a defined size distribution and form a disordered structure (Figure 7.3). Highly heterogeneous bubble size distribution can lead to unstable foams, due to unbalanced forces at the liquid–gas interface (Ratti and Kudra 2006).

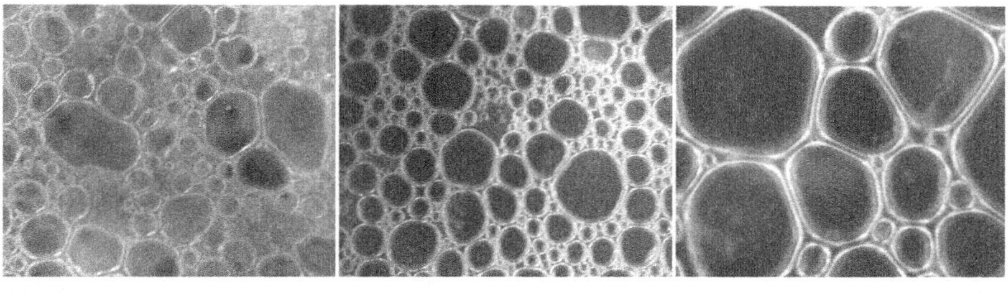

FIGURE 7.3 Structure of egg albumin foams.

7.3.1 Properties of Foaming Agents

Foaming agents are surfactants that reduce the surface tension between two liquids or between a liquid and a solid and facilitate foam formation (Sangamithra et al. 2015a). Proper understanding of the nature of a foam and its physical properties is important to help attain effective control over them (Foegeding et al. 2006). The volume of the foam formed by whipping depends on several factors, including whipping amplitude and frequency, container shape, and the volume, protein content, and temperature of the liquid used (Sangamithra et al. 2015a).

A foaming agent needs to be added to the product in order to aid the formation and stabilization of foams following the use of mechanical action for air incorporation. Increased drying time impairs product quality (Hart et al. 1963). Therefore, during drying, it is desirable that the foam remains stable and retains its typical open structure to facilitate moisture removal (Valenzuela and Aguilera 2013). The quality of the product would be compromised in case the foam structure collapses. Such a structure plays a fundamental role in moisture movement during subsequent drying and influences product quality. Activation of hydrophilic and hydrophobic groups of the foaming agents is essential to obtain stable structures as foam quality depends on emulsifier conformation at the interface (Ptaszek et al. 2015).

Foaming agents should have the following properties: the ability to stabilize foams effectively and rapidly at low concentrations; effective performance over the range of pH values found in various food production systems; and efficiency in the medium containing foam inhibitors such as fat, alcohol, or flavoring substances (Zayas 1997; Sangamithra et al. 2015a).

Stability usually refers to the ability of a foam to maintain some of its properties over time (e.g., volume, bubble size, and/or liquid content). Depending on the particular property and timescale of interest, the aforementioned characteristics can be of major importance (Denkov and Marinova 2006; Marinova et al. 2009). Following foam formation, several events leading to foam destruction occur, namely liquid drainage, bubble coalescence, and bubble disproportionation (Prud'homme and Khan 1996; Denkov 2004; Murray and Ettellaie 2004; Denkov and Marinova 2006).

Drainage (gravitational separation or syneresis) can be described as the rising of bubbles through the foam mass while the liquid drains through the lamellae and borders between the bubbles under the effect of gravity (Denkov 2004; Murray and Ettellaie 2004; Saint-Jalmes et al. 2005; Denkov and Marinova 2006). Bubble coalescence is the thinning and rupture of the thin liquid films separating two neighboring bubbles (or that between a bubble and the ambient atmosphere). The main factors that affect foam stability in relation to coalescence are the colloidal surface forces acting between film surfaces, the adsorption of surface-active molecules, the squeezing force pushing bubbles against each other (i.e., bubble capillary pressure), and film diameter (Denkov and Marinova 2006; Pereira et al. 2003; Denkov 2004; Marinova et al. 2009). Disproportionation is the process of bubble coarsening due to gas diffusion across the foam films, from smaller to larger bubbles (Prud'homme and Khan 1996; Weaire and Hutzler 1999). This phenomenon does not directly lead to a decrease in foam volume but exerts an indirect effect, because bigger bubbles join via larger, more easily ruptured films (Ivanov 1988).

Several foaming agents, such as modified soy protein (Bates 1964), soy protein isolate (Sankat and Castaigne 2004), and dehydrated egg albumin (García et al. 1988) can be used to produce dried food by the foam-mat drying process. Many types of foods, such as egg white, meat, and milk proteins, also can be converted into stable foam when whipped. These soluble proteins contribute to the formation and stability of the foam structure. High-molecular-weight polysaccharides may improve foam stability by increasing the interfacial viscoelasticity of lamellae (Muthukumaran et al. 2008). The stabilizer concentration should be optimized for maximum drying speed and efficiency (Azizpour et al. 2014).

7.3.1.1 Proteins

Proteins perform a vast array of functions in the food industry, in emulsification, foaming, encapsulation, viscosity enhancement, and gelation (O'Sullivan et al. 2016). Soluble proteins contribute to the formation and stability of foam structure. Proteins move through the aqueous phase and are spontaneously adsorbed at the air–aqueous interface, where viscoelastic films are subsequently formed (Thuwapanichayanan et al. 2008). Such adsorption results in a reduction of surface tension, improving foam formation (Prins 1988). Moreover, these viscoelastic films are typically resistant to the rupture and coalescence of the gas bubbles dispersed in the liquid phase (Karim and Wai 1999).

Whey protein is the main source of globular protein used in the food industry, owing to its emulsifying properties, gel- and foam-forming ability (Bernard et al. 2011), excellent nutritional quality, and inherent functional properties (Ezhilarasi et al. 2013), which also meet the demands of foam-mat vegetable drying. This protein is derived from the dairy industry, as a by-product of cheese production (Sangamithra et al. 2015a). It is commonly used as an ingredient in many food formulations, such as infant formula, supplements, and nutritional bars.

Studies have shown that whey protein and its hydrolysates exhibit antioxidant activity (Salami et al. 2010; Gad et al. 2011). Thus, whey protein has the potential to retard oxidative reactions in dried products. Furthermore, studies of interactions between proteins and volatile substances have shown that the former possess a high capacity to bind flavor compounds (Baranauskiene et al. 2006). This property is highly important when working with vegetables containing sensitive and volatile components. The greater solubility of whey protein isolate (90% protein) in water and its surface hydrophobicity account for its better foaming ability when applied to foam-mat drying (Abirached et al. 2012).

Egg albumin is a natural and easily available food-foaming agent with excellent foaming properties (Sangamithra et al. 2015b). This protein acts as an amphiphilic emulsifier and disperses within the continuous phase to stabilize foam. However, the use of commercially available egg albumin for foaming entails certain disadvantages, including limited pH range and ionic strength (Hardy and Jideani 2015). During the whipping process, egg albumin denature at the interface and interact with one another to form a stable interfacial film (Sangamithra et al. 2015a). Egg albumin has been used in various studies as an efficient foaming agent (Falade et al. 2003; Raharitsifa et al. 2006; Kadam et al. 2012; Fernandes et al. 2013a, 2014a).

Soybeans contain approximately 40% protein and 20% oil on an average dry-matter basis. By removing oil at lower temperatures, soy protein isolate is obtained (Nishinari et al. 2014). This represents a highly refined form of soy protein, with a minimum protein content of 90%. Owing to the functional properties, it is applied in gelation, viscosity, emulsification, foaming or whipping, and water binding (Daniel 2004; Asokapandian et al. 2016). The use of soy proteins as functional ingredients in food manufacturing is increasing because of their applications in human nutrition and health (Morales et al. 2015). In addition, soy protein has been applied as a foaming agent in foam-mat drying of tomatoes (Sharada 2013).

Gelatin is a denatured form of collagen, which is the most common protein in mammalian tissue, such as tendons, skin, and bones. In practice, large-scale manufacturing of gelatin uses the collagen present in pig and cattle tissues as the primary raw material (Rafieian et al. 2015). Gelatin is frequently used as a foaming agent because of its ability to form stable foams and gels (Zúñiga and Aguilera 2009), and is commonly employed as such to produce foods such as marshmallows and a type of premixed coffee beverage.

7.3.1.2 Carbohydrates

Gums are long-chain polymers of high molecular weight, capable of dispersing or dissolving in cold or warm water, producing a thickening or gelling effect (Zanaloni 1992). They are used in providing texture to food products, emulsion stability, viscosity control, crystallization, particle suspension, and inhibiting the release of water from processed foods (Glicksman 1982). Plant-based gums and colloids are used in the drying of food ingredients because they are edible and form part of the normal human diet (Arshady 1993). Both proteins and polysaccharides can be used as thickening agents, but the latter are preferred as they tend to have higher molecular weights and are extended even more, enabling their uses at much lower concentrations (McClements 1999).

Foam stability can also be improved through addition of both low-molecular-weight and high-molecular-weight compounds (polysaccharides) (Miquelim et al. 2010; Żmudziński et al. 2014). The addition of structure-forming hydrocolloids contributes to the generation of stable foams (Liszka-Skoczylas et al. 2014) and improves stability against coalescence (Miquelim et al. 2010). In the case of hydrocolloids, such as xanthan gum, gum arabic, or carrageenan, foam stabilization is a result of increased viscosity of the continuous phase with rheological properties typical for viscoelastic solids (Liszka-Skoczylas et al. 2014; Ptaszek et al. 2016).

Chemical modification of starches allows their uses as foaming agents by the addition of lipophilic functional groups. A potential alternative is the use of octenyl succinic anhydride (OSA)-derivatized starch. OSA starch is usually prepared in three stages: Granular starch is first derivatized with OSA in an aqueous alkaline medium below gelatinization temperature; second, gelatinization confers solubility to the esterified starch and reduces the viscosity of the solution; finally, the starch is degraded in an acid process. When modified with OSA, the partially hydrolyzed waxy starch gains a hydrophobic element in the form of octenyl groups, resulting in molecules of an amphiphilic nature (Sweedman et al. 2013). In addition, solubility needs to be modified because native starches are insoluble; however, this can be achieved by hydrolysis.

7.4 FOAM-MAT DRYING OF VEGETABLES

7.4.1 Preservation of Vegetables

Although the availability of vegetable products has increased for a growing proportion of the population, in many countries, unacceptable levels of losses are observed owing to the inadequacy of the techniques adopted from harvest to storage (Fernandes et al. 2014a). Postharvest losses vary greatly among commodities, production areas, and seasons. It is estimated that 20%–30% of vegetables and fruits produced do not reach the final consumer (Luengo and Calbo 2006), and this figure may be even higher, depending on the commodity. A considerable percentage of production is lost in the field as its below-optimal physical condition excludes to being sold in fresh form, although it is considered suitable for human consumption. This is the case, for example, with malformed and/or inconsistently sized products.

Many widely consumed vegetables lose their physical, chemical, and sensorial characteristics in a few days following harvesting, especially when kept in unfavorable environmental conditions. This is due to high transpiration rates, resulting in impaired appearance, including loss of brightness, shriveling and wrinkling of the skin, and diminished sensorial characteristics, particularly changes in texture. Processing of vegetables can contribute to waste reduction and convenience and time savings in daily food preparation, and is increasingly needed in the modern world (Alves et al. 2010). The excess of production during the harvest season and the high perishability of foods, combined with the absence and/or deficiency of appropriate handling techniques, transport, and storage, have generated great losses, which can be reduced by means of processing (Correia et al. 2008).

Because vegetables have a short shelf life due to their high water content and in most cases rely on seasonal production, it is necessary to employ preservation processes, of which dehydration is considered a highly viable option (Villamiel et al. 2015). Drying is also used in the preparation of raw materials for the development of new products in the food industry.

7.4.2 Vegetables Dried Using Foam-Mat Drying

Several studies have reported the foam-mat drying process to obtain dried vegetables. Franco et al. (2015a) evaluated the foam-mat drying of yacon by the addition of two foaming agents (egg albumin and the emulsifier/stabilizer Emustab [monoglycerides, sorbitan monostearate, and polysorbate 60]), using different air temperatures (50°C, 60°C, and 70°C) and layer thicknesses (0.5, 1.0, and 1.5 cm). The authors concluded that the foams produced by addition of egg albumin required less time to reach the equilibrium moisture content. In addition, higher air temperatures and reduced foam layer thicknesses reduced the processing time using both foaming agents. The results of the research may aid the design and optimization of foam-mat drying approaches for this high-nutritional-value vegetable.

Franco et al. (2015b) studied foams formed by the addition of egg albumin. By analyzing foam images, it was found that the greater the number of air bubbles present, the smaller their diameters, and this was affected by resting time (0, 10, or 20 min). In addition, foams formed with higher concentrations of foaming agent and longer whipping times showed the most desirable characteristics for the foam-mat drying. The addition of foaming agents to the yacon juice under investigation was a determining factor in the formation of foams and maintenance of their physical properties. This research demonstrated that higher concentrations of foaming agent and longer periods of air incorporation generated lower-density foams with higher overruns and air volume fractions, whereas those formed with emulsifier showed better stability.

Chakraborty et al. (2014) examined the optimization of conditions used for foam-mat drying of potato based on the functional properties of the final dried powder product. During mat preparation, a maximum foam expansion of 25% was attained by applying 10 min of magnetic stirring and using a glycerol monostearate solution. The optimum drying conditions identified in this work consisted of 60°C for 135 min using 2% glycerol monostearate as a foaming agent.

The effect of temperature, air velocity, and the addition of a foaming agent on the drying kinetics of Mexican hot salsa and their drying curve models have been studied (Escobedo-Avellaneda et al. 2013). Full drying times of 530, 340, and 180 min were observed at an air velocity of 2 m/s, while those of 380, 250, and 120 min were obtained at 4 m/s, at 45°C, 60°C, and 80°C, respectively. This work demonstrated that varying temperature and air velocity had a critical effect on the drying rate of food products. However, the addition of the foaming agent Tween 60 did not increase the drying rate but improved the appearance of Mexican hot salsa.

Roncheti et al. (2014a) obtained foam-mat drying curves for carrot at temperatures of 50°C, 60°C, and 70°C using a tray dryer. In preparing the juice, water was added to the carrots until they reached 50% by volume to assist in blender homogenization. The commercial emulsifier Emustab was used for foam production at a concentration of 2.5%. At all temperatures studied, the moisture ratio initially decreased rapidly before slowly declining as drying time increased. Henderson and Pabis, Page, and Newton models fitted to the experimental data satisfactorily. Roncheti et al. (2014b) also evaluated the foam-mat drying of beet, which showed similar results to those obtained using carrots, as previously discussed, with the Page model being best fitted to the experimental data.

The physicochemical and physical characteristics of papaya pulp and carrot processed by foam-mat drying at temperatures of 70°C, 60°C, and 50°C using foam layer thicknesses of 1.5, 1.0, and 0.5 cm, respectively, have been evaluated (Santiago et al. 2014). Using a 2:1 ratio of papaya pulp to carrot, foam-mat dehydration was carried out using 1% Emustab and a 0.5% preparation of a thickening agent based on carboxymethyl cellulose and guar gum. Higher solubility was obtained with the lower drying temperature (50°C), showing an increasing trend as foam layer thickness was decreased. Color brightness and red and yellow intensities were higher using the elevated drying temperature (70°C), which means that the samples had a more whitish appearance.

Falade et al. (2003) tested the foam-mat drying of cowpeas using glyceryl monostearate and egg albumin as foaming agents. Foam density was found to decrease as the concentrations of these foaming additives was increased in cowpea paste. Foams with suitable densities were obtained after 9 and 21 min of whipping with glyceryl monostearate and egg albumin, respectively. It was concluded that egg albumin-stabilized foams were too unstable to be dried.

Kadam et al. (2012) studied the influence of foam-mat drying parameters on the quality of tomato paste using carboxymethyl cellulose, egg albumin, and milk as foaming and stabilizing agents. The authors concluded that this technology showed great promise for the preservation of tomatoes in powdered form without considerable nutrient loss. This study showed that increasing the level of egg albumin as a foaming agent reduced the drying time up to a point, before leading to a decreasing trend (Kadam and Balasubramanian 2011). In addition, the drying of tomato juice at 60°C, 65°C, and 70°C took 510, 450, and 420 min, respectively, to reach the equilibrium moisture content, while the optimum egg albumin concentration was found to be 10%, with 5 min of whipping. Fernandes et al. (2014a) studied the drying process of tomato pulp, reported that the use of albumin for foam formation was responsible for reducing the time taken to reach the equilibrium moisture content during the drying process, at both temperatures under investigation. In this work, the equilibrium moisture content was achieved after 400 and 590 min at 60°C, and 180 and 240 min at 80°C, for foamed and nonfoamed treatments, respectively. The use of a foaming agent favored the production of powders with lower hygroscopicity, higher solubility, and lower density, which are important attributes for the reconstitution of dried materials. Figure 7.4 illustrates the tomato pulp powder produced by foam-mat drying.

Tomato pulp foam formation was optimized based on higher stability and lower density, thereby providing more expanded stable foam (Fernandes et al. 2013a). The authors concluded that an albumin concentration of 4.5% with 4.5 min of whipping time resulted in a more stable foam. The structure of the solutions with and without addition of the foaming agent can be seen in Figure 7.5. Using the aforementioned optimized conditions, the values for foam stability and density were 53.6% and 0.32 g/cm^3, respectively.

FIGURE 7.4 Tomato powder obtained by foam-mat drying.

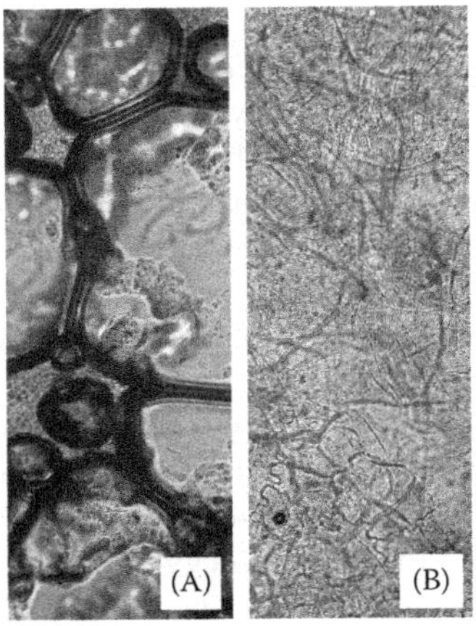

FIGURE 7.5 Images obtained by optical microscopy of tomato pulp with (A) and without (B) addition of 4.5% albumin after whipping for 4.5 min.

The quality attributes of foam-mat-dried okra have been evaluated previously (Falade and Omojola 2010) and compared with those resulting from other drying methods. Okra paste was observed to produce foams without the inclusion of foaming agents or stabilizers, probably due to the protein content of the seeds. These foams were stable during air drying for periods of 0.5–1.5 h. This investigation found that sun- and solar-dried okra demonstrated higher L^* and a^* color parameters compared to fresh, blanched, and foam-mat-dried forms. Fresh, frozen/thawed, and foam-mat-dried okra samples did not significantly differ in a sensory evaluation and received consistently high scores for overall color, taste, and aroma acceptability.

Pumpkin pulp was foamed using 3% and 4% glycerol monostearate as a foaming agent before being dried at 50°C and 60°C. Moisture diffusivity increased with drying temperature and glycerol monostearate concentration, with the latter having a significant effect on drying time. The study also demonstrated that pumpkin powder could be used in bakery products and confectionaries as a source of β-carotene and vitamin A (Das et al. 2015).

In a recent investigation, mixed vegetable juice (bitter gourd, tomato, and cucumber) was dried using the foam-mat method, egg albumin was used as the foaming agent (Chandrasekar et al. 2015). Foam properties including expansion, stability, and density were evaluated. The results showed that increased foaming agent concentration heightened the foam expansion and decreased stability and density.

The effects of cultivar (Efuru and Abuja) and drying method (cabinet and foam-mat) on the color, pasting, and sensory properties of instant pounded yam flours had been assessed (Falade and Onyeoziri 2012). Yams were foamed using glyceryl monostearate and air-dried to produce instant pounded yam flour with good sensory characteristics. The density of foams produced from yam pastes decreased from 0.97 g/cm^3 to between 0.362–0.538 and 0.308–0.507 g/cm^3 for Abuja and Efuru cultivars, respectively. Foam-mat-dried flour demonstrated higher L^*, ΔC, ΔE, and hue angle, but lower a^* and b^* values.

Qadri and Srivastava (2014) studied the effect of microwave treatment on the drying time and quality indices of dried tomato pulp using microwave-assisted foam-mat drying approaches. An enormous decrease in the drying time of tomato pulp–egg albumin foam was observed using the microwave-assisted process. The color, titratable acidity, and pH of the product were not affected. Retention of ascorbic acid was also more compared to that of samples dried with air convection foam-mat technique. This study indicates that microwave-assisted foam-mat drying could be a promising alternative to existing conventional drying methods applied to liquid foods in the dehydration industry.

ACKNOWLEDGMENTS

The authors thank CNPq (National Council for Scientific and Technological Development—Brazil) and FAPEMIG (Minas Gerais State Research Foundation) for their financial support.

REFERENCES

Abirached, C., Medrano, C.A., Araujo, A.C., Moyna, P., Añón, M.C., and Panizzolo, L.A. 2012. Comparison of interfacial and foaming properties of soy and whey protein isolates. *Journal of Food Science and Engineering* 2:376–81.

Akintoye, O.A. and Oguntunde, A.O. 1991. Preliminary investigation on the effect of foam stabilizers on the physical characteristics and reconstitution properties of foam-mat dried soymilk. *Drying Technology* 9:245–62.

Alves, J.A., Vilas Boas, E.V.B., Vilas Boas, B.M., and de Souza, É.C. 2010. Maintenance of the quality of fresh-cut products made up of pumpkin, carrot, chayote, and arracacha (*Peruvian carrot*). *Ciência e Tecnologia de Alimentos* 30:625–34.

Arshady, R. 1993. Microcapsules for food. *Journal of Microencapsulation* 10:413–35.

Arzhavitina, A. and Steckel, H. 2010. Foams for pharmaceutical and cosmetic application. *International Journal of Pharmaceutics* 394:1–17.

Asokapandian, S., Venkatachalam, S., Swamy, G.J., and Kuppusamy, K. 2016. Optimization of foaming properties and foam mat drying of muskmelon using soy protein. *Journal of Food Process Engineering* 39(6):692–701. doi:10.1111/jfpe.12261.

Azizpour, M., Mohebbi, M., Khodaparast, M.H.H., and Varidi, M. 2014. Optimization of foaming parameters and investigating the effects of drying temperature on the foam-mat drying of shrimp (*Penaeus indicus*). *Drying Technology* 32:374–84.

Bag, S.K., Srivastav, P.P., and Mishra, H.N. 2011. Optimization of process parameters for foaming of bael (*Aegle marmelos* L.) fruit pulp. *Food and Bioprocess Technology* 4:1450–58.

Baldwin, E.A., Burns, J.K., Kazokas, W., Brecht, J.K., Hagenmaier, R.D., Bender, R.J., and Pesis, E. 1999. Effect of two edible coatings with different permeability characteristics on mango (*Mangifera indica* L.) ripening during storage. *Postharvest Biology and Technology* 3:215–26.

Baranauskiene, R., Venskutonis, P.R., Dewettinck, K., and Verhé, R. 2006. Properties of oregano (*Origanum vulgare* L.), citronella (*Cymbopogon nardus* G.) and marjoram (*Majorana hortensis* L.) flavors encapsulated into milk protein-based matrices. *Food Research International* 39:413–25.

Bates, R.P. 1964. Factors affecting foam production and stabilization of tropical fruit products. *Food Technology* 18:93–96.

Bernard, C., Regnault, S., Gendreau, S., Charbonneau, S., and Relkin, P. 2011. Enhancement of emulsifying properties of whey proteins by controlling spray-drying parameters. *Food Hydrocolloids* 25:758–63.

Campbell, C.H. 1917. Drying milk. US Patent 1250427.

Chakraborty, S., Banerjee, S., and Mazumder, S. 2014. Functional properties based statistical optimization of foam mat drying parameters for potato (Kufri Chandramukhi). *Global Journal of Science Frontier Research* 14:1–10.

Chandrasekar, V., Gabriela, J.S., Kannan, K., and Sangamithra, A. 2015. Effect of foaming agent concentration and drying temperature on physiochemical and antimicrobial properties of foam mat dried powder. *Asian Journal of Dairy and Food Research* 34:39–43.

Cooke, R.D., Breag, G.R., Ferber, C.E.M., Best, P.R., and Jones, J. 1976. Studies of mango processing I. The foam-mat drying of mango (*Alphonso cultivar*) puree. *Journal of Food Technology* 11:463–73.

Correia, L.F.M., Faraoni, A.S., and Pinheiro-Sant'Ana, H.M. 2008. Efeitos do processamento industrial de alimentos sobre a estabilidade de vitaminas. *Alimentos e Nutrição* 19:83–95.

Cunningham, S.E., McMinn, W.A.M., Magee, T.R.A., and Richardson, P.S. 2008. Effect of processing conditions on the water absorption and texture kinetics of potato. *Journal of Food Engineering* 84:214–23.

Daniel, C. 2004. Soy protein concentrate. In *Soybeans as Functional Foods and Ingredients*, ed. Liu, K., 132–44. Champaign, IL: AOCS Publishing.

Das, S., Banerjee, S., and Pal, J. 2015. Mathematical modeling of foam-mat dried pumpkin pulp. *International Journal of Food and Nutritional Science* 4:50–55.

Denkov, N.D. 2004. Mechanisms of foam destruction by oil-based antifoams. *Langmuir* 20:9463–505.

Denkov, N.D. and Marinova, K.G. 2006. Antifoam effects of solid particles, oil drops and oil-solid compounds in aqueous foams. In *Colloidal Particles at Liquid Interfaces*, ed. Binks, B.P. and Horozov, T.S., 383–444. Cambridge, MA: Cambridge University Press.

Escobedo-Avellaneda, Z., Pérez-Pérez, C., Bárcenas-Pozos, M.E., Guerrero-Beltrán, J., and Welti-Chanes, J. 2013. Analysis of the drying process of Mexican hot salsa using the characteristic curve model. *Journal of Food Processing and Preservation* 37:441–48.

Ezhilarasi, P.N., Indrani, D., Jena, B.S., and Anandharamakrishnan, C. 2013. Freeze drying technique for microencapsulation of *Garcinia* fruit extract and its effect on bread quality. *Journal of Food Engineering* 117:513–20.

Falade, K.O., Adeyanju, K.I., and Uzo-Peters, P.I. 2003. Foam-mat drying of cowpea (*Vigna unguiculata*) using glyceryl monostearate and egg albumin as foaming agents. *European Food Research and Technology* 217:486–91.

Falade, K.O. and Omojola, B.S. 2010. Effect of processing methods on physical, chemical, rheological, and sensory properties of okra (*Abelmoschus esculentus*). *Food and Bioprocess Technology* 3:387–94.

Falade, K.O. and Onyeoziri, N.F. 2012. Effects of cultivar and drying method on color, pasting and sensory attributes of instant yam (*Dioscorea rotundata*) flours. *Food and Bioprocess Technology* 5:879–87.

Fernandes, R.V.B., Queiroz, F., Botrel, D.A., Rocha, V.V., de Souza, V.R., and de Lima, C.F. 2014a. Study of the albumin addition and drying temperature on the characteristics of tomato pulp powder. *Semina: Ciências Agrárias* 35:1267–78.

Fernandes, R.V.B., Borges, S.V., Botrel, D.A., and de Oliveira, C.R. 2014b. Physical and chemical properties of encapsulated rosemary essential oil by spray drying using whey protein–inulin blends as carriers. *International Journal of Food Science and Technology* 49:1522–29.

Fernandes, R.V.B., Borges, S.V., and Botrel, D.A. 2014c. Gum Arabic/starch/maltodextrin/inulin as wall materials on the microencapsulation of rosemary essential oil. *Carbohydrate Polymers* 101:524–32.

Fernandes, R.V.B., Marques, G.R., Borges, S.V., and Botrel, D.A. 2014d. Effect of solids content and oil load on the microencapsulation process of rosemary essential oil. *Industrial Crops and Products* 58:173–81.

Fernandes, R.V.B., Queiroz, F., Botrel, D.A., Rocha, V.V., de Lima, C.F., and de Souza, V.R. 2013a. Foam mat drying of tomato pulp. *Bioscience Journal* 29:816–25.

Fernandes, R.V.B., Borges, S.V., and Botrel, D.A. 2013b. Influence of spray drying operating conditions on microencapsulated rosemary essential oil properties. *Ciência e Tecnologia de Alimentos* 33:171–78.

Finney, J., Buffo, R., and Reineccius, G.A. 2002. Effects of type of atomization and processing temperatures on the physical properties and stability of spray-dried flavors. *Journal of Food Science* 67:1108–14.

Fitzpatrick, J.J. 2005. Food powder flowability. In *Encapsulated and Powdered Foods*, ed. Onwulata, C., 247–60. Boca Raton, FL: CRC Press.

Foegeding, E.A., Luck, P., and Davis, J. 2006. Factors determining the physical properties of protein foams. *Food Hydrocolloids* 20:284–92.

Franco, T.S., Perussello, C.A., Ellendersen, L.N., and Masson, M.L. 2015a. Effect of process parameters on foam mat drying kinetics of yacon (*Smallanthus sonchifolius*) and thin-layer drying modeling of experimental data. *Journal of Food Process Engineering*. doi:10.1111/jfpe.12264.

Franco, T.S., Ellendersen, L.N., Fattori, D., Granato, D., and Masson, M.L. 2015b. Influence of the addition of ovalbumin and emulsifier on the physical properties and stability of yacon (*Smallanthus sonchifolius*) juice foams prepared for foam mat drying process. *Food and Bioprocess Technology* 8:2012–26.

Gad, A.S., Khadrawy, Y.A., El-Nekeety, A.A., Mohamed, S.R., Hassan, N.S., and Abdel-Wahhab, M.A. 2011. Antioxidant activity and hepatoprotective effects of whey protein and *spirulina* in rats. *Nutrition* 27:582–89.

García, R., Leal, F., and Rolz, C. 1988. Drying of bananas using microwave and air ovens. *International Journal of Food Science and Technology* 23:73–80.

Glicksman, M. 1982. Functional properties of hydrocolloids. In *Food Hydrocolloids*, ed. Glicksman, M., 47–99. Boca Raton, FL: CRC Press/Taylor & Francis.

Hall, C.A. and Cuppett, S.L. 1997. Structure-activities of natural antioxidants. In *Antioxidant Methodology:* In Vivo *and* In Vitro *Concepts*, ed. Aruoma, O.I. and Cuppett, S.L., 2–29. Champaign, IL: AOCS Press.

Hardy, Z. and Jideani, V.A. 2015. Foam-mat drying technology: A review. *Critical Reviews in Food Science and Nutrition*. doi:10.1080/10408398.2015.1020359.

Hart, M.R., Graham, R.P., Ginnette, L.F., and Morgan, A.I. 1963. Foams for foam-mat drying. *Food Technology* 17:1302–04.

Hudson, B.J.F. 1990. *Food Antioxidants*. London: Elsevier Applied Science.

Hung, P.V. and Duy, T.L. 2012. Effects of drying methods on bioactive compounds of vegetables and correlation between bioactive compounds and their antioxidants. *International Food Research Journal* 19:327–32.

Ivanov, I.B. 1988. *Thin Liquid Films: Fundamentals and Applications*. New York: Marcel Dekker.

Jayasundera, M., Adhikari, B., Howes, T., and Aldred, P. 2011. Surface protein coverage and its implications on spray-drying of model sugar-rich foods: Solubility, powder production and characterization. *Food Chemistry* 128:1003–16.

Kadam, D.M. and Balasubramanian, S. 2011. Foam mat drying of tomato juice. *Journal of Food Processing and Preservation* 35:488–95.

Kadam, D.M., Wilson, R.A., Kaur, S., and Manisha. 2012. Influence of foam mat drying on quality of tomato powder. *International Journal of Food Properties* 15:211–20.

Kandasamy, P., Varadharaju, N., Kalemullah, S., and Maladhi, D. 2014. Optimization of process parameters for foam-mat drying of papaya pulp. *Journal of Food Science and Technology* 51:2526–34.

Karim, A.A. and Wai, C.C. 1999. Characteristics of foam prepared from starfruit (*Averrhoa carambola* L.) puree by using methylcellulose. *Food Hydrocolloids* 13:203–10.
Kerkhof, P.J.A.M. 1994. The role of theoretical and mathematical modelling in scale-up. *Drying Technology* 12:1–46.
Kudra, T. and Mujumdar, A.S. 2001. *Advanced Drying Technologies*. New York: Marcel Dekker, Inc.
Liszka-Skoczylas, M., Ptaszek, A., and Żmudziński, D. 2014. The effect of hydrocolloids on producing stable foams based on the whey protein concentrate (WPC). *Journal of Food Engineering* 129:1–11.
Luengo, R.F.A. and Calbo, A.G. 2006. *Embalagens para comercialização de hortaliças e frutas*. Brasília: Circular técnica/Embrapa Hortaliças.
Marinova, K.G., Basheva, E.S., Nenova, B., Temelska, M., Mirarefi, A.Y., Campbell, B., and Ivanov, I.B. 2009. Physico-chemical factors controlling the foamability and foam stability of milk proteins: Sodium caseinate and whey protein concentrates. *Food Hydrocolloids* 23:1864–76.
Martinazzo, A.P., Melo, E.C., Corrêa, P.C., and Santos, R.H.S. 2010. Mathematical modeling and quality parameters of lemon grass [*Cymbopogon citratus* (DC.) Stapf] leaf drying. *Revista Brasileira de Plantas Medicinais* 12:488–98.
Mayor, L. and Sereno, A.M. 2004. Modelling shrinkage during convective drying of food materials: A review. *Journal of Food Engineering* 61:373–86.
Mazza, G. 1983. Dehydration of carrots: Effects of pre-drying treatments on moisture transport and product quality. *Journal of Food Technology* 18:113–23.
McClements, D.J. 1999. *Food Emulsions: Principles, Practice and Technique*. Boca Raton, FL: CRC Press.
McDermott, J.H. 2000. Antioxidant nutrients: Current dietary recommendations and research update. *Journal of the American Pharmacists Association* 40:785–99.
Mink, L.D. 1939. Egg material treatment. US Patent 2183516.
Mink, L.D. 1940. Treatment of egg whites. US Patent 2200963.
Miquelim, J.N., Lannes, S.C.S., and Mezzenga, R. 2010. pH influence on the stability of foams with protein-polysaccharide complexes at their interfaces. *Food Hydrocolloids* 24:398–405.
Morales, R., Martínez, K.D., Ruiz-Henestrosa, V.M.P., and Pilosof, A.M.R. 2015. Modification of foaming properties of soy protein isolate by high ultrasound intensity: Particle size effect. *Ultrasonics Sonochemistry* 26:48–55.
Murray, B.S. and Ettellaie, R. 2004. Foam stability: Proteins and nanoparticles. *Current Opinion in Colloid and Interface Science* 9:314–20.
Muthukumaran, A., Ratti, C., and Raghavan, V.G.S. 2008. Foam-mat freeze drying of egg white and mathematical modeling Part I: Optimization of egg white foam stability. *Drying Technology* 26:508–12.
Nishinari, K., Fang, Y., Guo, S., and Phillips, G.O. 2014. Soy proteins: A review on composition, aggregation and emulsification. *Food Hydrocolloids* 39:301–18.
Olopade, A.A., Akingbala, J.O., Oguntunde, A.O., and Falade, K.O. 2003. Effect of processing method on the quality of cowpea (*Vigna unguiculata*) flour for akara preparation. *Plant Foods for Human Nutrition* 58:1–10.
O'Sullivan, J., Murray, B., Flynn, C., and Norton, I. 2016. The effect of ultrasound treatment on the structural, physical and emulsifying properties of animal and vegetable proteins. *Food Hydrocolloids* 53:141–54.
Pasban, A., Mohebbi, M., Pourazarang, H., and Varidi, M. 2014. Effects of endemic hydrocolloids and xanthan gum on foaming properties of white button mushroom puree studied by cluster analysis: A comparative study. *Journal of Taibah University for Science* 8:31–38.
Pereira, L.G.C., Johansson, C., Radke, C.J., and Blanch, H.W. 2003. Surface forces and drainage kinetics of protein-stabilized aqueous films. *Langmuir* 19:7503–13.

Prins, A. 1988. Principles of foam stability. In *Advances in Food Emulsions and Foams*, ed. Dickinson, E. and Stainsby, G., 91–122. New York: Elsevier Applied Science.

Prud'homme, R.K. and Khan, S.A. 1996. *Foams: Theory, Measurements, and Applications*. New York: Marcel Dekker.

Ptaszek, P., Kabziński, M., Kruk, J., Kaczmarczyk, K., Żmudziński, D., Liszka-Skoczylas, M., Mickowska, B., Łukasiewicz, M., and Banaś, J. 2015. The effect of pectins and xanthan gum on physicochemical properties of egg white protein foams. *Journal of Food Engineering* 144:129–37.

Ptaszek, P., Kabziński, M., Ptaszek, A., Kaczmarczyk, K., Kruk, J., and Bieńczak, A. 2016. The analysis of the influence of xanthan gum and apple pectins on egg white protein foams using the large amplitude oscillatory shear method. *Food Hydrocolloids* 54:293–301.

Qadri, O.S. and Srivastava, A.K. 2014. Effect of microwave power on foam-mat drying of tomato pulp. *Agricultural Engineering International: CIGR Journal* 16:238–44.

Quispe-Condori, S., Saldaña, M.D.A., and Temelli, F. 2011. Microencapsulation of flax oil with zein using spray and freeze drying. *LWT-Food Science and Technology* 44: 1880–87.

Rafieian, F., Keramat, J., and Shahedi, M. 2015. Physicochemical properties of gelatin extracted from chicken deboner residue. *LWT-Food Science and Technology* 64:1370–75.

Raghavan, G.S.V. and Orsat, V. 2007. Recent advances in drying of biomaterials for superior quality bioproducts. *Asia-Pacific Journal of Chemical Engineering* 2:20–29.

Raharitsifa, N., Genovese, D.B., and Ratti, C. 2006. Characterization of apple juice foams for foam-mat drying prepared with egg white protein and methylcellulose. *Journal of Food Science* 71:141–51.

Rajkumar, P. and Kailappan, R. 2006. Optimizing the process parameters for foam mat drying of totapuri mango pulp. *The Madras Agricultural Journal* 93:86–98.

Ratti, C. and Kudra, T. 2006. Drying of foamed biological materials: Opportunities and challenges. *Drying Technology* 24:1101–08.

Roncheti, E.F.S., Junqueira, M.S., Saraiva, S.H., Pizzani, L.A., and Rusciolelli, L.B. 2014a. Kinect and modelling of the foam mat drying in trays of carrot (*Daucus carota*). *Magistra* 26:962–67.

Roncheti, E.F.S., Schuina, G.L., García, M.V., Junqueira, M.S. and Saraiva, S.H. 2014b. Study of the foam mat drying kinect of beet (*Beta vulgarius* L.). *Magistra* 26:1694–8.

Sablani, S.S. 2006. Drying of fruits and vegetables: Retention of nutritional/functional quality. *Drying Technology* 24:123–35.

Sagar, V.R. and Kumar, P.S. 2010. Recent advances in drying and dehydration of fruits and vegetables: A review. *Journal of Food Science and Technology* 47:15–26.

Saint-Jalmes, A., Peugeot, M.-L., Ferraz, H., and Langevin, D. 2005. Differences between protein and surfactant foams: Microscopic properties, stability and coarsening. *Colloids and Surfaces A: Physicochemical and Engineering Aspects* 263:219–25.

Salahi, M.R., Mohebbi, M., and Taghizadeh, M. 2015. Foam-mat drying of cantaloupe (*Cucumis melo*): Optimization of foaming parameters and investigating drying characteristics. *Journal of Food Processing and Preservation* 39:1798–808.

Salami, M.; Moosavi-Movahedi, A.A.; Ehsani, M.R., Yousefi, R., Haertlé, T., Chobert, J.M., Razavi, S.H., Henrich, R., Balalaie, S., Ebadi, S.A., Pourtakdoost, S., and Niasari-Naslaji, A. 2010. Improvement of the antimicrobial and antioxidant activities of camel and bovine whey proteins by limited proteolysis. *Journal of Agricultural and Food Chemistry* 58:3297–302.

Sangamithra, A., Venkatachalam, S., John, S.G., and Kuppuswamy, K. 2015a. Foam mat drying of food materials: A review. *Journal of Food Processing and Preservation* 39:3165–74.

Sangamithra, A., Sivakumar, V., Kannan, K., and John, S.G. 2015b. Foam-mat drying of muskmelon. *International Journal of Food Engineering* 11:127–37.

Sankat, C.K. and Castaigne, F. 2004. Foaming and drying behaviour of ripe bananas. *LWT-Food Science and Technology* 37:517–25.
Santiago, V.M.S., Moura, R.L., Sousa, S.F., Figueirêdo, R.M.F. and Queiroz, A.J.M. 2014. Quality assessment of papaya pulp mixed with carrot powder. *Magistra* 26:324–28.
Setyadjit and Sukasih, E. 2015. Effect of addition of filler on the production of shallot (*Allium cepa* var. *ascalonicum* L.) powder with drum dryer. *Procedia Food Science* 3:396–408.
Sharada, S. 2013. Studies on effect of various operating parameters and foaming agents-drying of fruits and vegetables *International Journal of Modern Engineering Research* 3:1512–19.
Silva, A.S., Gurjão, K.C.O., Almeida, F.A.C., Bruno, R.L.A., and Pereira, W.E. 2008. Dehydration of tamarind pulp through the foam-mat drying method. *Ciência e Agrotecnologia* 32:1899–905.
Sramek, M., Schweiggert, R.M., Kampen, A.V., Carle, R., and Kohlus, R. 2015. Preparation of high-grade powders from tomato paste using a vacuum foam drying method. *Journal of Food Science* 80:E1755–62.
Sweedman, M.C., Tizzotti, M.J., Schäfer, C., and Gilbert, R.G. 2013. Structure and physicochemical properties of octenyl succinic anhydride modified starches: A review. *Carbohydrate Polymers* 92:905–20.
Thuwapanichayanan, R., Prachayawarakorn, S., and Soponronnarit, S. 2008. Drying characteristics and quality of banana foam mat. *Journal of Food Engineering* 86:573–83.
Thuwapanichayanan, R., Prachayawarakorn, S., and Soponronnarit, S. 2012. Effects of foaming agents and foam density on drying characteristics and textural property of banana foams. *LWT-Food Science and Technology* 47:348–57.
Travaglini, D.A., Aguirre, J.M., and Siqueira, E.T.F. 2001. *Desidratação de frutas*. Campinas, Brazil: CETEA/ITAL.
Valenzuela, C. and Aguilera, J.M. 2013. Aerated apple leathers: Effect of microstructure on drying and mechanical properties. *Drying Technology* 31:1951–59.
Venkataraman, P.S. 1996. Success and failures in mango pulp processing industries in Chitoor district of AP. *Indian Food Packer* 51:57–58.
Villamiel, M., Gamboa, J., Soria, A.C., Riera, E., García-Pérez, J.V., and Montilla, A. 2015. Impact of power ultrasound on the quality of fruits and vegetables during dehydration. *Physics Procedia* 70:828–32.
Weaire, D. and Hutzler, S. 1999. *The Physics of Foams*. Oxford, UK: Clarendon Press.
Zanaloni, E. 1992. Uso de gelificantes y espesantes en heladería. *Heladeria Panaderia Latinoamericana* 19:39–46.
Zayas, J.F. 1997. Foaming properties of proteins. In *Functionality of Proteins in Food*, ed. Zayas, J.F., 260–309. Heidelberg, Germany: Springer.
Zbicinski, I., Rabaeva, J., and Lewandowski, A. 2014. Drying of foamed materials. In *Modern Drying Technology: Process Intensification*, ed. Tsotsas, E. and Mujumdar, A.S., 163–90. Weinheim, Germany: Wiley-VCH Verlag GmbH & Co. KGaA.
Żmudziński, D., Ptaszek, P., Kruk, J., Kaczmarczyk, K., Rożnowski, W., Berski, W., Ptaszek, A., and Grzesik, M. 2014. The role of hydrocolloids in mechanical properties of fresh foams based on egg white proteins. *Journal of Food Engineering* 121:128–34.
Zúñiga, R.N. and Aguilera, J.M. 2009. Structure–fracture relationships in gas-filled gelatin gels. *Food Hydrocolloids* 23:1351–57.

Section II

Drying of Specific Vegetable Products

8 Drying of Herbs and Spices

Sachin V. Jangam
National University of Singapore

Arun Sadashiv Mujumdar
McGill University

CONTENTS

8.1 INTRODUCTION

Herbs are plants traditionally used for food, medicine, or perfume, whereas spices are substances used mainly for flavoring, coloring, and preservation of food. Although it is difficult to differentiate between them, in general, herbs come from the leafy part of the plant (even though other parts of some plants are used as herbs), while spices are the parts of the plant other than the leafy part. According to the American Spice Trade Association, spices are defined as any dried plant product used primarily for seasoning purposes. Herbs have attracted a lot of scientific attention in recent decades because of the health benefits they offer. The important properties are their biological activity for pharmacological use, antiaging and antioxidant properties, antimicrobial property, etc. The use of herbs has increased manyfold in pharmaceutical, cosmetic, and food industries owing to their aforementioned characteristics.

Medicinal herbs and spices have been traditionally used in many Asian countries such as China, India, Nepal, Malaysia, Indonesia, and Thailand; however,

Orphanides et al. (2016) pointed out the high import of herbs in the EU in 2012. A recent article by Narayanaswamy and Ismail (2015) provides an overview of a number of herbs found in Southeast Asia and their potential in the field of cosmetics. They provided the traditional cosmetic uses of a number of herbs. Further, they have categorized different herbs based on different properties such as antiaging, antiacne activity, antityrosinase activity, and other uses such as effects against dehydration and irritation. Xie et al. (2016) have provided a good overview of research progress of Chinese herbal medicine in cosmetics. Yasurin et al. (2015) have provided an overview of antimicrobial properties of common herbs and spices in Thai cooking. There are a number of other reviews available, which summarize applications of specific herbs and spices or variety of herbs and spices for specific applications (Zhu et al., 2004; Fan, 2005; Yarnell and Abascal, 2007; Jarald et al., 2008; Kumar et al., 2011; Arulselvan et al., 2014). There are a number of monographs and articles published by the World Health Organization (WHO) that provide scientific information on the safety, efficacy, and quality control/quality assurance of widely used medicinal plants (WHO monograph, 1999; Tilburt and Kaptchuk, 2008). In another document by the WHO, general guidelines and methodologies are provided on research and evaluation of traditional medicine (WHO, 2000).

The medicinal herbs are rarely consumed in fresh form; most herbs are consumed in a dried form. Drying plays an important role in extending the shelf life of many food products by removing moisture to a safe limit (water activity level below 0.8) so that no microbial growth takes place (Mujumdar, 2014; Jangam and Mujumdar, 2015). In the case of herbs and spices, dehydration is important for increasing the shelf life so that these products are available for consumption throughout the year. Drying also helps to minimize the storage and transportation volume and to reduce the packaging cost. Thermal drying involves application of heat at relatively high temperatures, which can destroy or denature various nutrients that are heat-sensitive. Moreover, drying of herbs and spices may also involve structural and physicochemical modifications—serious damage to flavor, loss of aroma, and color. Therefore, processing and handling of these products will preferably need treatment at lower temperature. Thermal drying is an energy-intensive operation as it involves evaporation of water. Recent concerns of energy usage and GHG emissions demand for energy-efficient processes to achieve a sustainable route (Xiao, 2015). Therefore, it is also necessary to minimize the energy use to produce dried products at competitive market price without compromising the product quality (Mujumdar, 2007; Mujumdar and Law, 2010; Mujumdar, 2014). All the aforementioned factors make thermal drying a complex unit operation. Some of the other key factors are the availability of a large number of dryer types and a wide range of herbs and spices to be handled. Most of the time, the assignment of selecting a proper dryer ends up with more than one dryer type, which needs to be tested based on various aspects.

This chapter begins with a review of some of the traditional drying techniques used for herbs and spices. The latter part of the chapter will focus on advanced drying methods used for herbs and spices over the past decade, with suggestions for future R&D in drying.

8.2 CONVENTIONAL DRYING METHODS

The traditional drying methods used for herbs and spices are natural drying in the sun or in a covered space. However, these drying methods are reasonable for processing of only small quantities of herbs or spices. The most commonly used industrial method for these products is convective air drying. However, drying temperature is the limiting factor that decides the quality of the dried product. The use of high temperatures can substantially degrade the properties such as color, texture, and nutritional content. In the case of most food products (including herbs and spices), drying is controlled by internal diffusion; therefore, using higher temperature results in overheating of the surface as the water diffusion rate is slow. This probably leads to overdrying of the outer surface, case hardening, and also loss of aroma (which is important for herbs and spices). On the contrary, using moderately lower temperatures results in longer drying times, which may cause degradation of product quality due to enzymatic and nonenzymatic reactions (Mujumdar, 2014). It is necessary to select proper operating conditions and/or drying methods in order to produce good quality dried herbs and spices, with minimal use of energy. There are several studies that report the use of fluidized bed drying method for several herbs and spices, because these dryers provide very good drying rates. Table 8.1 summarizes the use of conventional drying techniques for selected herbs and spices reported in recent decades. The techniques conventionally used include open sun drying, solar cabinet drying, other solar-integrated drying systems, fluidized bed drying, and microwave drying (MD). The selection of each of these techniques is done for several reasons such as for improvement of quality and enhancement of drying rates or the physical form of the product dried.

8.2.1 Recent Developments in Drying of Herbs and Spices

As discussed earlier, the two important aspects to consider for improvement in drying of herbs and spices are quality and energy consumption. A number of drying methods and variations have been proposed in recent years to address these aspects. In the following section, some of these drying methods have been discussed, which proposes new ideas for improvement.

8.2.2 Microwave Drying

While drying any material, it is very important to enhance the drying rate in order to prevent thermal damage of the product. At the same time, it is necessary to have a dried product with uniform moisture content for longer shelf life and also to get better market value for the product. The use of microwave (MW) is the most appropriate option to address these issues, because MW provides a volumetric heating that results in homogenous dried product (Mujumdar, 2014; Deepika and Sutar, 2015). The internal heat generation in MW drying rapidly generates vapors inside the material. This creates a pressure difference, resulting in mass transfer (Mujumdar, 2014). The higher the moisture content, the greater the pressure difference, which leads to very rapid drying without causing any surface overheating of the product.

TABLE 8.1
Drying of Herbs and Spices Using Conventional Dryers

Herbs/Spices	Research Outcome	References
Peppermint leaves	The effect of hot air temperature was studied, and the energy consumption and product characteristics were compared with dried product using MD at various MW power levels. Although hot air drying has lower energy efficiency, the essential oil yield was higher for hot air-dried product.	Torki-Harchegani et al. (2016)
Ginger and turmeric	Comparison of solar–biomass integrated drying system (IDS), FBD, electrical oven (EO), and open sun drying (OSD); IDS provided the best product qualities (texture and color attributes).	Borah et al. (2017)
Alfalfa	The effect of hot air temperature, hot air velocity, Alfalfa culms length, and culms physical pretreatment methods on quality parameters such as protein content and fiber content. The important parameters affecting product characteristics were hot air temperature and physical pretreatment.	Wang et al. (2015)
Sage, thyme, mint, and lemon balm	Comparison of hot air and oven drying. Air-dried herbs contained more phenolics, antioxidant activity, and flavonoids than oven-dried herbs	Rababah et al. (2015)
Kaffir lime (*Citrus hystrix* D.C.) leaves	FBD with sand as inert particles under different air velocities, mass ratio of leaves to sand at constant temperature. The use of inert particles enhanced the drying rate, although the rates decreased when higher mass ratios of leaves to sand were used. No significant loss of properties such as retention of essential oil, color, vitamin content was observed.	Tasirin et al. (2014)
Basil leaves	Comparison of oven drying, hot air drying, sun drying, ambient-air drying, and MW drying based on the nutritional characteristics. The MW drying and oven drying were found to show the best quality.	Danso-Boateng (2013)
Torch ginger (*Etlingera elatior*)	Response surface methodology (RSM) was used to understand the effect of hot air drying on physicochemical properties such as texture, color, and water activity. The optimized results were provided.	Juhari et al. (2012)
Peppermint	Performance evaluation of mixed-mode solar cabinet dryer (SCD) and comparison with open sun drying and cabinet drying; better product quality using SCD.	Eltawil et al. (2012)
Parsley leaves	Energy and exergy analysis of forced convection solar drying and open sun drying (natural convection) was carried out; solar drying was shown to be an effective alternative for hot air and open sun drying	Akpinar (2011)
Lemon myrtle plant (*Backhousia citriodora*) leaves	FBD at different temperatures to analyze color retention and retention of the principal volatile compound, citral, in dried products. Higher temperature caused more damage; use of hot water blanching did not show any improvement.	Buchaillot et al. (2009)

The combination of MW and vacuum drying can provide much higher mass transfer and better product qualities (Deepika and Sutar, 2015). The use of vacuum system also prevents oxidation of material. However, the vacuum-assisted MW dryers require large condensers and vacuum pumps. Therefore, MD is mainly used for initial or final stages of drying as the cost of operation can be quite high. Although MW can heat and dry the products quickly, rapid heating can be destructive. If the heating is too rapid and if the moisture cannot escape quickly, this may result in rupture of the material. In the case of herbs and spices, the use of MW may also result in unacceptable changes in other polar compounds present in these plant materials. Therefore, one has to cautiously select proper conditions for MD of herbs and spices. The use of MW has been practiced for drying of several herbs and spices (Deepika and Sutar, 2015; Xie et al., 2016). Table 8.2 shows some applications of MD for selected herbs and spices.

8.2.3 Freeze-Drying

Freeze-drying (FD), without doubt, provides the best product quality as the frozen product is not allowed to melt during drying. It prevents shrinkage and produces highly porous material with very good rehydration properties (Jangam and Mujumdar, 2015). The only limitation is high cost because of high energy usage in FD. Although there have been significant efforts to develop new techniques that can provide similar product quality, it is difficult to match the quality attributes of dried products that only FD can provide, such as rehydration, color, and texture. In the case of herbs and spices, FD helps to retain many active components, which is a key goal.

There have been few attempts to reduce the cost of FD by improving the drying rates using external heating (Woo and Mujumdar, 2010). This can be achieved either by using magnetic and electric fields, using atmospheric FD, or using MW in FD. A detailed review by Woo and Mujumdar (2010) on the use of electric and magnetic fields during freezing, and their application to FD, suggests that one can have a better control on the ice crystals and, ultimately, product quality. The use of microwave FD for a number of food products has been successfully demonstrated with better drying rates and product quality (Zhang et al., 2010). A detailed discussion on FD and advances can be found in the *Handbook of Industrial Drying* (Mujumdar, 2014).

This technique has been used for drying of several herbs such as medicinal ginger, pepper (Tambunan et al., 2001), spearmint leaves (Antal et al., 2011), curry leaves (Shivanna and Subban, 2013), basil, coriander, and turmeric (Nugboon and Intarapichet, 2015). Some comparative studies have been reported in Tables 8.1 and 8.2.

8.2.4 Heat Pump-Assisted Drying and Advances

Heat pump drying (HPD) is a comparatively older technique that has already been used for several applications; however, the HPD system has been improving as far as the design and application in different industrial sectors is concerned. In traditional dryers, the exhaust gas from a dryer is purged to the atmosphere, resulting in loss of both sensible and latent heat associated with the water vapor in it. The heat pump

TABLE 8.2
Applications of Microwave for Drying of Herbs and Spices

Herbs/Spices	Research Outcome	References
Ginger rhizome slices	Quality attributes of the dried samples were compared in terms of volatile compounds, 6, 8, 10-gingerols, 6-shogaol, antioxidant activities and microstructure for air drying (AD), freeze-drying (FD), infrared drying (IR), MD, and intermittent MW convective drying (IMCD); it was found that AD and IR drying could preserve volatiles, whereas FD, IR, and IMCD could retain maximum gingerols with better antioxidant activity.	An et al. (2016)
Basil, lovage, mint, oregano, parsley	Microwave convection drying was carried out to understand the effect on physicochemical properties. The highest retention of phenolic compound, good resistance to color degradation was found for lovage and parsley, whereas worse properties were found for basil, oregano, and mint.	Śledź et al. (2013)
Celak leaves	The effect of various drying treatments (sun, shade, oven, MW, and FD) on the essential oil yield, composition, and color characteristics of leaves was studied. Although MW drying showed low essential oil yield, it shortened drying time, provided high color quality, and increased major compounds of the leaves.	Rahimmalek and Goli (2013)
Curry leaves	Lutein content and β-carotene content of dried curry leaves were analyzed for various drying methods (e.g., MW, FD, IR, hot air, and shade drying); the results showed that the MW-processed leaves contain higher levels of lutein (99.4 mg/100 g) and β-carotene than other dryers.	Shivanna and Subban (2013)
Ginger	The effect of MW drying and silica gel drying was studied on the degree of dehydration and volatile components of ginger. MW-dried ginger had good amount of zingiberene and satisfactory drying characteristics. The authors also recommended the use of MW and silica gel drying to maintain the taste and appearance of fresh ginger	Huang et al. (2012)
Thai herbal teas	MW, FD, and oven drying (OD) and comparison based on antioxidant properties (AOP) and sensory properties. Although FD showed the highest AOP, the MW-dried tea had the highest scores for aroma, flavor, and overall acceptability.	Chan et al. (2012)
Marjoram herb; sweet basil	The influence of convective drying (CD), vacuum MD (VMD), and their combination (convective drying followed by VM finish-drying) on aroma compounds. The combination of these two drying methods was found to be the best option.	Calín-Sánchez et al. (2012, 2015)
Rosmarinus officinalis (rosemary)	VMD drying was carried out. The effect of vacuum and MW levels on aroma compounds and sensory quality was evaluated. Low vacuum level and MW levels were recommended to get the highest concentrations of volatiles and the best sensory quality	Calín-Sánchez et al. (2011)
Peppermint	Comparison of sun, oven, and MW drying to study the drying kinetics, nutritional, and color characteristics. MW oven drying shortened the drying time, revealed the highest phenolic content, and optimum color values.	Arslan et al. (2010)
Astragalus slices	Characterization of the microstructure of Astragalus slices dried by MW technique. The MW-dried sample shows much shorter drying time with increased number of pores of larger sizes and open structure on the surface layer of the matrix.	Yang et al. (2009)

(Continued)

TABLE 8.2 (*Continued*)
Applications of Microwave for Drying of Herbs and Spices

Herbs/Spices	Research Outcome	References
Basil leaves	Comparison of atmospheric pressure MD drying, air drying, and FD. Authors reported that the MD allowed a larger retention of chlorophyll pigments than air drying and FD (with or without blanching) and preserved the color of the raw basil. Microwave drying required a much shorter treatment and implied the simultaneous blanching of the material.	Di Cesare et al. (2003)

dryer was initially used mainly to recover this exhaust heat, although the design has evolved over the years to make use of other advantages that HPD offers, such as product quality, especially for application in drying of foods and other heat-sensitive products (Mujumdar, 2014). HPD consists of a refrigeration cycle to recover heat from dryer exhaust air, which is followed by reheating it to the required temperature before it enters back into the dryer. The details of the HPD system, its operation, and different designs can be found elsewhere (Jangam and Mujumdar, 2011). A simple schematic of HPD is shown in Figure 8.1, although the design can vary substantially depending on the purpose. It consists of two cycles—refrigeration and air cycle. The important advantages of HPD systems are to provide a wide range of operating conditions (humidity and temperature) and the capability to use inert gas as a drying medium. It is also easy to combine other modes of heat transfer that provides greater flexibility in HPD. Especially for heat-sensitive products such as herbs and spices with long drying times, HPD is very useful because drying can be safely carried out using relatively lower temperature dehumidified air. The use of inert gas can provide good physical properties as well as the possibility of allowing the closed HPD system

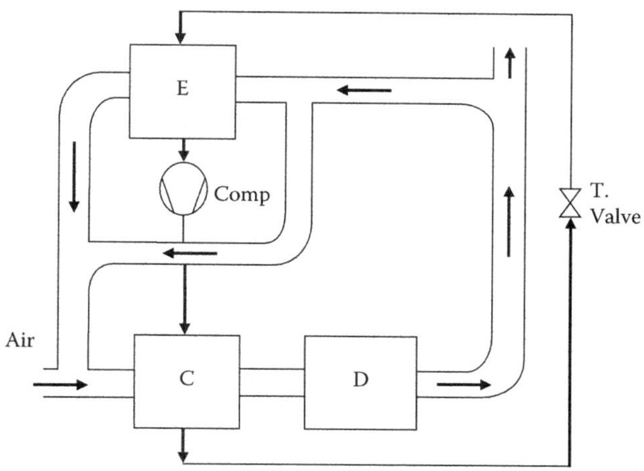

FIGURE 8.1 A simple schematic of a heat pump dryer (C, condenser; Comp, compressor; D, dryer; E, evaporator; T. Valve, throttling valve).

to retain aromatic compounds present in herbs and spices. A closed-loop HPD system can substantially reduce the operating cost as the same inert gas can be recycled with some minor modifications as needed.

Although mechanical compression is the most common heat pump system used, there are several other systems that are shown to provide much improved efficiency such as, multistage heat pumps, chemical heat pumps, or use of heat pipes (Jangam and Mujumdar, 2015). Other ways of improving the performance of HPD includes use of external heating modes such as MW, radio frequency (RF), infrared, and solar. These are useful during the final stages of drying in order to minimize possible damage to the physical structure of foods. Solar-assisted HPD is extremely useful for agricultural products. The concept of intermittent drying (to be discussed in Section 8.2.5) is another way of reducing the drying cost and improving the product quality in HPD. For more detailed discussion on advances in HPD (new designs and applications), readers may refer to Jangam and Mujumdar (2011), Mujumdar (2014), and Jangam and Mujumdar (2015). Table 8.3 provides a summary of the applications of HPD systems for herbs and spices.

8.2.5 Intermittent Drying

The idea of intermittent drying is used to enhance the energy efficiency and to get better product quality by achieving uniform drying. Intermittent drying may use the variations in drying air velocity, temperature, humidity, and/or other parameters such as system pressure and external parameters (MW, infrared, RF) during the drying cycle so that the required quality parameters are achieved (Kumar et al., 2014). Sometimes, cycles of different heating modes can be used. The rationale behind the time variation of the aforementioned operating parameters is to allow the internal moisture to migrate to the material surface during the non-active zone—known as tempering period. During the active period, the heat is supplied, whereas during tempering period the heat supply is completely stopped. As the moisture travels and accumulates at the surface during tempering period, the moisture evaporation rate is very high during the active period. The overall drying time may be longer as the drying rate is very small during the nonactive period; however, the overall energy consumption will be much lower (Mujumdar, 2014; Jangam and Mujumdar, 2015).

In the case of most of the food products, including herbs and spices, the drying rate is very slow in the final stages as the drying is controlled by the internal diffusion of moisture, and so the same drying conditions need not be used. To address this, intermittent drying by stepwise change in operating conditions can be used as appropriate. This approach of changing the operating conditions during drying cycle can save a lot of energy. Apart from this the product is exposed to high temperature for a shorter time, resulting in less damage to the dried product. The concept of intermittent drying has been applied to some herbs and spices using various dryer types.

There are few studies reported on intermittent drying of herbs and spices. Rosalizan et al. (2013) reported the use of intermittent cooling during drying of misai kucing (*Orthosiphon stamineus*). They showed that the use of intermittent cooling of 2 h during drying showed better retention of phytochemical contents in misai kucing as compared with continuous drying. The drying time and drying rate

TABLE 8.3
Application of Heat Pump-Assisted Drying for Herbs and Spices

Herbs/Spices	Research Outcome	References
Lavender	Closed-system drying of lavender using a heat pump system and a fixed bed dryer. Authors presented optimization of operating parameters for minimizing the loss of volatiles. The closed loop certainly helps retain the volatiles.	Krempski-Smejda et al. (2015)
Laurel clock vine (*Thunbergia laurifolia*)	Thermoelectric (TE) heat pump drying of laurel clock vine leaves was carried out as an alternative to hydrofluorocarbon-based heat pump system; Effect of drying-air temperature on the characteristics of the leaves was studied. Results showed that use of higher temperature of 50°C was better in terms of coefficient of performance (COP) and drying time compared to a temperature of 40°C. Quality parameters were not reported.	Wongsim et al. (2015)
Mint leaves	A condensation-type heat pump drying at lower temperature (between 35°C and 45°C) was successfully carried out. The dryer performance was evaluated in terms of coefficient of performance (between 3.81 and 2.29), and specific moisture extraction rate (between 0.034 and 0.044 kg/kWh).	Aktaş et al. (2014)
Thai sweet basil	Thin-layer drying characteristics and quality of sweet basil were studied using tray and heat pump-assisted dehumidified dryer. Drying rate and the quality evaluation by total phenolics, rehydration ratio, and color change showed that the best quality resulted from sweet basil leaves pretreated by blanching in boiling water for 1 min and dried at 40°C in a HPD dryer	Phoungchandang and Kongpim (2012)
Ginger	Effect of two-stage, tray and heat pump-assisted dehumidified drying on drying characteristics and qualities of dried ginger was studied. The heat pump dehumidified drying incorporated by the two-stage drying could reduce the drying time by 60% and increase 6-gingerol content by 6%. Other properties were better too for heat pump-dried ginger.	Phoungchandang and Saentaweesuk (2011)
Tom Yum herbs (chili, lemon grass, kaffir lime leaf, and galangal slice)	Thin-layer vacuum heat pump drying experiments were conducted at a constant pressure of 0.2 bars and temperatures ranging from 50°C to 65°C. The results showed that only falling rate appeared during drying of selected herbs, and drying time was reduced by increasing the temperature as the diffusion coefficient increased. No data on quality was provided.	Artnaseaw et al. (2010)
Jew's mallow, spearmint, and parsley	The effects of herb size, stem presence, surface load, and drying air temperature and air velocity on the drying characteristics of Jew's mallow have been predicted. The specific energy consumption for drying of various herbs was compared. It was found that parsley requires the lowest specific energy consumption (3684 kJ/kg), followed by spearmint (3982 kJ/kg) and Jew's mallow (4029 kJ/kg).The dryer productivity was correlated in terms of surface load, drying air velocity, and temperature.	Fatouh et al. (2006)

(Continued)

TABLE 8.3 (*Continued*)
Application of Heat Pump-Assisted Drying for Herbs and Spices

Herbs/Spices	Research Outcome	References
Ginger	A heat pump drying using air, nitrogen, and carbon dioxide was applied to dry sliced West Indian ginger. The inert gas heat pump drying showed an improved effective diffusivity. The inert gas also showed a better retention of flavor compared to other types of drying such as freeze and vacuum drying.	Hawlader et al. (2006)

were not significantly affected by intermittent cooling treatment. An et al. (2016) used the intermittent MD technique for Chinese ginger, which resulted in higher retention of gingerols. Lv et al. (2015) also showed similar results for ginger, with increased drying rate and quality of ginger slices. Balakrishnan et al. (2011) carried out intermittent spouted bed drying of cardamom. They found a considerable saving in thermal energy and better product quality in terms of color, flavor, and percentage yield of oleoresin extract. These studies show that intermittent drying can be a very useful drying option for herbs and spices to improve quality and reduce energy consumption.

8.2.6 Hybrid Drying Techniques

It is well known that the use of a single dryer may not allow much flexibility during the drying process, only HPD may intrinsically offer this flexibility in the operating parameters. Several research studies have shown that it is difficult to always find a drying technique that can offer both very good product qualities and very high energy efficiency (Mujumdar, 2014). Hence, it is a good idea to combine more than one drying technique together. These so-called hybrid drying systems can offer great flexibility of operating conditions. The concept of hybrid drying means the use of more than one drying technique for a particular product (multistage drying), use of more than one mode to transfer the required heat (multimode drying), or use of multiprocessing dryers (Jangam and Mujumdar, 2015). Hybrid dryers are found to enhance performance, improve the quality of dried products, and provide smaller footprint of equipment (Zhang et al., 2010).

A combination of various fluidized bed dryers (FBDs) with some other dryer can be used for particulate matter. For drying of liquid material, FBD is generally used after spray drying to reduce moisture content to an acceptable level, which may not be possible by spray dryer alone (Kudra and Mujumdar, 2009). In the case of many food products, infrared drying is useful to remove the final traces of moisture at a faster rate. The use of MW in combination with another drying technique has been used in several applications. These techniques are typically combined with other drying methods to overcome the limitations of uneven heating. However, these techniques have a considerably high startup cost (Xu et al., 2006). MW drying, as discussed earlier, can volumetrically heat the material, and a combination of MW and vacuum drying results in an improved product compared to air-dried products. The

MW-assisted FD is another recently used hybrid technique for various food products (Zhang et al., 2010). Mortezapour et al. (2012) used the heat pump-assisted hybrid photovoltaic-thermal solar dryer for saffron drying. It was found that applying a heat pump with the dryer led to a decrease in the drying time and energy consumption and an increase in electrical efficiency of the solar collector. The use of MW–FD combination resulted in the production of high-quality products in a much faster way (Huang and Zhang, 2012; Zhang and Jiang, 2014). This method has already been applied for numerous food products.

8.3 NEEDS AND OPPORTUNITIES IN DRYING OF HERBS AND SPICES

8.3.1 Sustainability in Industrial Drying of Foods

The amount of energy used is directly proportional to the emission of greenhouse gases. Drying is highly energy-intensive, and the required heat is mainly supplied by fossil fuels, which have the potential to cause high global warming of the system (Mujumdar and Piacentini, 2013; Wang et al., 2015b). Although the use of solar dryers is very common for herbs and spices, it may not be the best possible option, because the quality of dried products has to be compromised. The other drying techniques discussed so far consume a lot of energy, although the drying time is very short in some of the techniques such as MD. However, to achieve the best possible product quality, a lot of money is spent on the energy consumed. Recently, efforts have been made to make the industrial processes sustainable, which means reducing the pollutant-forming compounds and reducing the overall energy consumption (Moses et al., 2014; Hammond et al., 2015). In the previous sections, some ideas were presented such as HPD, intermittent drying, and hybrid drying that minimize the energy consumption. However, the renewable sources of energy have tremendous potential to make the industrial drying process pollutant-free and more sustainable. Hence, renewable energies, such as solar and wind, should be looked at seriously as current concerns over potential energy shortage and global climate change would likely result in legislative actions to minimize fossil fuel usage.

The use of solar dryers is very common for herbs and spices; however, many research opportunities are present in the field of improving quality of product using these drying techniques. This may include improving the air distribution in existing solar cabinet dryers, improving the quality of the air circulated in the dryers, and controlling the drying parameters so that the drying will be faster with minimal alteration of quality. In future, larger systems could be designed utilizing solar thermal, photovoltaic panels combined with wind power. Because solar and wind energy are necessarily intermittent, advances in thermal and electrical energy storage are needed to make use of renewable energy viable in drying. To minimize use of oil or gas, one could use biomass to provide backup heating in the absence of insolation and wind. As discussed by Jangam and Mujumdar (2015), use of thermal energy storage in water pools, pebble beds, and/or in phase change materials can be coupled with the use of intermittent energy sources such as solar and wind energy. Further R&D is needed at the systems level to make this concept commercially viable.

8.3.2 Smart Dryers

Mujumdar (2013) and Jangam and Mujumdar (2015) have mentioned the need to develop smart or intelligent dryers. Mujumdar (2013) proposed that, because drying is a highly energy-intensive operation that affects quality of product decisively and also has an adverse effect on the environment through greenhouse gas emissions, this has to be made a sustainable operation utilizing the latest developments in allied technologies, including advanced computer hardware and software. With the recent advances in mathematical modeling of dryers, advanced sensors for real-time measurement of variables of interest in automatic control of dryers, and robust control strategies, it is now feasible to design smart dryers. The key issue often is the cost involved.

The proposed criteria to define a smart or intelligent dryer/drying system is such that it provides actionable information regarding the performance of the drying system, accurately monitors and detects errors or deficiencies in dryer operation, and uses the tools, resources, and practices to contribute to energy conservation and environmental sustainability (Jangam and Mujumdar, 2015; Su et al., 2015). A smart dryer must be designed to sense local drying conditions and their effect on prespecified quality parameters, and to adjust the operating conditions such that the product quality is assured at minimal or optimal energy consumption and minimal environmental impact. This idea is extremely useful in the case of many herbs and spices, as the smart dryers can help to precisely control the product quality with optimum heat utilization. Martynenko and Kudra (2015) have proposed a similar nonisothermal drying system for medicinal plants, which makes use of an intelligent control system. In their study, a machine vision combined with neural network model provided real-time estimation of quality parameters. The operating parameters could be controlled based on the estimated quality parameters. They concluded that such a control system could be easily adapted for heat-sensitive batch processes such as drying. However, much work is needed and this topic needs to be explored in more detail.

8.4 CONCLUDING REMARKS

A brief overview of conventional and advanced drying technologies for herbs and spices has been presented. It is difficult to make a definitive recommendation of a specific dryer for a particular herb/spice. In general, solar cabinet drying, MD (convection or vacuum), FD, heat pump-assisted drying, and hybrid drying technologies are some of the more popular drying technologies for the products discussed in this chapter. As energy costs soar, energy efficiency will be a key criterion for marketing of dryers. However, the choice is made based on a specific product under study and the relative importance of quality and energy. Much R&D needs to be done to make some of the new concepts commercially attractive.

REFERENCES

Akpinar, E.K. Drying of parsley leaves in a solar dryer and under open sun: Modeling, energy and exergy aspects. *Journal of Food Process Engineering* 2011, 34 (1), 27–48.

Aktaş, M., Ceylan, I., Gürel, A.E. Testing of a condensation-type heat pump system for low-temperature drying applications. *International Journal of Food Engineering* 2014, 10 (3), 521–531.

An, K., Zhao, D., Wang, Z., Wu, J., Xu, Y., Xiao, G. Comparison of different drying methods on Chinese ginger (*Zingiber officinale* Roscoe): Changes in volatiles, chemical profile, antioxidant properties, and microstructure. *Food Chemistry* 2016, 197, 1292–1300.

Antal, T., Figiel, A., Kerekes, B., Sikolya, L. Effect of drying methods on the quality of the essential oil of spearmint leaves (*Mentha spicata* L.). *Drying Technology* 2011, 29 (15), 1836–1844.

Arslan, D., Özcan, M.M., Mengeş, H.O. Evaluation of drying methods with respect to drying parameters, some nutritional and colour characteristics of peppermint (*Mentha × piperita* L.). *Energy Conversion and Management* 2010, 51 (12), 2769–2775.

Artnaseaw, A., Theerakulpisut, S., Benjapiyaporn, C. Thin layer modeling of Tom Yum herbs in vacuum heat pump dryer. *Food Science and Technology International* 2010, 16 (2), 135–146.

Arulselvan, P., Ghofar, H.A.A., Karthivashan, G., Halim, M.F.A., Ghafar, M.S.A., Fakurazi, S. Antidiabetic therapeutics from natural source: A systematic review. *Biomedicine and Preventive Nutrition* 2014, 4 (4), 607–617.

Balakrishnan, M., Raghavan, G.S.V., Sreenarayanan, V.V., Viswanathan, R. Batch drying kinetics of cardamom in a two-dimensional spouted bed. *Drying Technology* 2011, 29 (11), 1283–1290.

Borah, A., Sethi, L.N., Sarkar, S., Hazarika, K. Effect of drying on texture and color characteristics of ginger and turmeric in a solar biomass integrated dryer. *Journal of Food Process Engineering* 2017, 40 (1), e12310.

Buchaillot, A., Caffin, N., Bhandari, B. Drying of lemon myrtle (*Backhousia citriodora*) leaves: Retention of volatiles and color. *Drying Technology* 2009, 27 (3), 445–450.

Calín-Sánchez, A., Figiel, A., Lech, K., Szumny, A., Martínez-Tomé, J., Carbonell-Barrachina, Á.A. Drying methods affect the aroma of *Origanum majorana* L. analyzed by GC–MS and descriptive sensory analysis. *Industrial Crops and Products* 2015, 74, 218–227.

Calín-Sánchez, A., Lech, K., Szumny, A., Figiel, A., Carbonell-Barrachina, Á.A. Volatile composition of sweet basil essential oil (*Ocimum basilicum* L.) as affected by drying method. *Food Research International* 2012, 48 (1), 217–225.

Calín-Sánchez, A., Szumny, A., Figiel, A., Jałoszyński, K., Adamski, M., Carbonell-Barrachina, Á.A. Effects of vacuum level and microwave power on rosemary volatile composition during vacuum-microwave drying. *Journal of Food Engineering* 2011, 103 (2), 219–227.

Chan, E.W.C., Eng, S.Y., Tan, Y.P., Wong, Z.C., Lye, P.Y., Tan, L.N. Antioxidant and sensory properties of Thai herbal teas with emphasis on *Thunbergia laurifolia* Lindl. *Chiang Mai Journal of Science* 2012, 39 (4), 599–609.

Danso-Boateng, E. Effect of drying methods on nutrient quality of basil (*Ocimum viride*) leaves cultivated in Ghana. *International Food Research Journal* 2013, 20 (4), 1569–1573.

Deepika, S., Sutar, P.P. Microwave assisted hybrid drying in food and agricultural materials. In *Drying Technologies for Foods: Fundamentals and Applications (Part I)*, Eds. Nema, P.K., Kaur, B.P., Mujumdar, A.S. New India Publishing Agency, New Delhi, 2015, 121–154.

Di Cesare, L.F., Forni, E., Viscardi, D., Nani, R.C. Changes in the chemical composition of basil caused by different drying procedures. *Journal of Agricultural and Food Chemistry* 2003, 51 (12), 3575–3581.

Eltawil, M.A., AbouZaher, S.E., El-Hadad, W.Z. Solar-wind ventilation to enhance the cabinet dryer performance for medicinal herbs and horticultural products. *Agricultural Engineering International: CIGR Journal* 2012, 14 (4), 56–74.

Fan, Y.-P. Problems and consideration on treatment of cerebrovascular diseases with Chinese medicine. *Chinese Journal of Clinical Rehabilitation* 2005, 9 (1), 254–256.

Fatouh, M., Metwally, M.N., Helali, A.B., Shedid, M.H. Herbs drying using a heat pump dryer. *Energy Conversion and Management* 2006, 47 (15–16), 2629–2643.

Hammond, S.T., Brown, J.H., Burger, J.R., Flanagan, T.P., Fristoe, T.S., Mercado-Silva, N., Nekola, J.C., Okie, J.G. Food spoilage, storage, and transport: Implications for a sustainable future. *BioScience* 2015, 65 (8), 758–768.

Hawlader, M.N.A., Perera, C.O., Tian, M. Comparison of the retention of 6-gingerol in drying of ginger under modified atmosphere heat pump drying and other drying methods. *Drying Technology* 2006, 24 (1), 51–56.

Huang, B., Wang, G., Chu, Z., Qin, L. Effect of oven drying, microwave drying, and silica gel drying methods on the volatile components of ginger (*Zingiber officinale* Roscoe) by HS-SPME-GC-MS. *Drying Technology* 2012, 30 (3), 248–255.

Huang, L.-I., Zhang, M., Trends in development of dried vegetable products as snacks. *Drying Technology* 2012, 30 (5), 448–461.

Jangam, S.V., Mujumdar, A.S. Heat pump assisted drying technology: Overview with focus on energy, environment and product quality. In Tsotsas, E., Mujumdar, A.S. (Eds.) *Modern Drying Technology*, Volume 4. Wiley-VCH, Weinheim, Germany, 2011, 121–162.

Jangam, S.V., Mujumdar, A.S. Advances in food drying. In *Drying Technologies for Foods: Fundamentals and Applications (Part I)*, Ed. Nema, P.K., Kaur, B.P., Mujumdar, A.S. New India Publishing Agency, New Delhi, 2015, 347–361.

Jarald, E., Joshi, S.B., Jain, D.C. Diabetes and herbal medicines. *Iranian Journal of Pharmacology and Therapeutics* 2008, 7 (1), 97–106.

Juhari, N.H., Lasekan, O., Kharidah, M., Ab Karim, S. Optimization of hot-air drying conditions on the physicochemical characteristics of torch ginger (*Etlingera elatior*). *Journal of Food, Agriculture and Environment* 2012, 10 (2), 64–72.

Krempski-Smejda, M., Stawczyk, J., Śmigielski, K., Prusinowska, R. Drying of herbal product in closed system. *Drying Technology* 2015, 33 (13), 1671–1677.

Kudra, T., Mujumdar, A.S. *Advanced Drying Technologies*, 2nd edition. CRC Press, Boca Raton, FL, 2009.

Kumar, C., Karim, M.A., Joardder, M.U.H. Intermittent drying of food products: A critical review. *Journal of Food Engineering* 2014, 121, 48–57.

Kumar, V.R., Kandagatla, K., Chillal, S. Review: Signature on medicinal plants used in hypertension. *International Journal of Pharmacy and Technology* 2011, 3 (3), 1333–1355.

Lv, W., Han, Q., Li, S., Zhang, X., Xu, T., Sun, L. Thermal dynamic and physical qualities of ginger (*Zingiber officinale*) slices in intermittent microwave fluidized drying. *International Agricultural Engineering Journal* 2015, 24 (1), 39–46.

Martynenko, A., Kudra, T. Non-isothermal drying of medicinal plants. *Drying Technology* 2015, 33 (13), 1550–1559.

Mortezapour, H., Ghobadian, B., Minaei, S., Khoshtaghaza, M.H. Saffron drying with a heat pump-assisted hybrid photovoltaic-thermal solar dryer. *Drying Technology* 2012, 30 (6), 560–566.

Moses, J.A., Norton, A., Alagusundaram, K., Tiwari, B.K. Novel drying techniques for the food industry. *Food Engineering Reviews* 2014, 6 (3), 43–55.

Mujumdar, A.S. An overview of innovation in industrial drying: Current status and R&D needs. *Transport in Porous Media* 2007, 66 (1–2), 3–18.

Mujumdar, A.S., Editorial: The extreme future. *Drying Technology* 2013, 31 (5), 489–490.

Mujumdar, A.S. *Handbook of Industrial Drying*, 4th edition. CRC Press, Boca Raton, FL, 2014.

Mujumdar, A.S., Law, C.L. Drying technology: Trends and applications in postharvest processing. *Food and Bioprocess Technology* 2010, 3 (6), 843–852.

Mujumdar, A.S., Piacentini, R.D. Guest editorial: Urgent need for reduction in greenhouse gas emissions in industrial processes: Are we past the tipping point for global warming? *Drying Technology* 2013, 31, 3–4.

Narayanaswamy, R., Ismail, I.S. Cosmetic potential of Southeast Asian herbs: An overview. *Phytochemistry Reviews* 2015, 14 (3), 419–428.

Nugboon, K., Intarapichet, K. Antioxidant and antibacterial activities of Thai culinary herb and spice extracts, and application in pork meatballs. *International Food Research Journal* 2015, 22 (5), 1788–1800.

Orphanides, A., Goulas, V., Gekas, V. Drying technologies: Vehicle to high-quality herbs. *Food Engineering Reviews* 2016, 8 (2), 164–180.

Phoungchandang, S., Kongpim, P. Modeling using a new thin-layer drying model and drying characteristics of sweet basil (*Ocimum basilicum* Linn.) using tray and heat pump-assisted dehumidified drying. *Journal of Food Process Engineering* 2012, 35 (6), 851–862.

Phoungchandang, S., Saentaweesuk, S. Effect of two stage, tray and heat pump assisted-dehumidified drying on drying characteristics and qualities of dried ginger. *Food and Bioproducts Processing* 2011, 89 (4), 429–437.

Rababah, T.M., Al-u'datt, M., Alhamad, M., Al-Mahasneh, M., Ereifej, K., Andrade, J., Altarifi, B., Almajwal, A., Yang, W. Effects of drying process on total phenolics, antioxidant activity and flavonoid contents of common Mediterranean herbs. *International Journal of Agricultural and Biological Engineering* 2015, 8 (2), 145–150.

Rahimmalek, M., Goli, S.A.H. Evaluation of six drying treatments with respect to essential oil yield, composition and color characteristics of *Thymus daenensis* subsp. daenensis. Celak leaves. *Industrial Crops and Products* 2013, 42 (1), 613–619.

Rosalizan, M.S., Ahmad Tarmizi, S., Osman, M., Rasali, M., Mohd. Shukri, M.A. The effect of intermittent cooling during drying on phytochemical content of misai kucing (*Orthosiphon stamineus*). *Acta Horticulturae* 2013, 1012, 1235–1240.

Shivanna, V.B., Subban, N. Carotenoids retention in processed curry leaves (*Murraya koenigii* L. Spreng). *International Journal of Food Sciences and Nutrition* 2013, 64 (1), 58–62.

Śledź, M., Nowacka, M., Wiktor, A., Witrowa-Rajchert, D. Selected chemical and physicochemical properties of microwave-convective dried herbs. *Food and Bioproducts Processing* 2013, 91 (4), 421–428.

Su, Y., Zhang, M., Mujumdar, A.S. Recent developments in smart drying technology. *Drying Technology* 2015, 33 (3), 260–276.

Tambunan, A.H., Yudistira, Kisdiyani, Hernani. Freeze drying characteristics of medicinal herbs. *Drying Technology* 2001, 19 (2), 325–331.

Tasirin, S.M., Puspasari, I., Lun, A.W., Chai, P.V., Lee, W.T. Drying of kaffir lime leaves in a fluidized bed dryer with inert particles: Kinetics and quality determination. *Industrial Crops and Products* 2014, 61, 193–201.

Tilburt, J.C., Kaptchuk, T.J. Herbal medicine research and global health: An ethical analysis. *Bulletin of the World Health Organization* 2008, 86, 577–656.

Torki-Harchegani, M., Ghanbarian, D., Ghasemi Pirbalouti, A., Sadeghi, M. Dehydration behaviour, mathematical modelling, energy efficiency and essential oil yield of peppermint leaves undergoing microwave and hot air treatments. *Renewable and Sustainable Energy Reviews* 2016, 58, 407–418.

Wang, W., Chen, H., Zhao, M. Technological parameters optimization for improving quality of heated-air dried alfalfa. *Transactions of the Chinese Society of Agricultural Engineering* 2015a, 31, 337–345. (In Chinese with English abstract).

Wang, D., Dai, J.-W., Ju, H.-Y., Xie, L., Xiao, H.-W., Liu, Y.-H., Gao, Z.-J. Drying kinetics of American ginseng slices in thin-layer air impingement dryer. *International Journal of Food Engineering* 2015b, 11 (5), 701–711.

WHO. General guidelines for methodologies on research and evaluation of traditional medicine, 2000. Retrieved from http://apps.who.int/iris/bitstream/10665/66783/1/WHO_EDM_TRM_2000.1.pdf. Accessed February 15, 2016.

WHO monographs on selected medicinal plants. 1999. Retrieved from http://apps.who.int/medicinedocs/pdf/s2200e/s2200e.pdf. Accessed February 15, 2016.

Wongsim, K., Jamradloedluk, J., Lertsatitthanakorn, C., Siriamornpun, S., Rungsiyopas, M., Soponronnarit, S. Experimental performance of a thermoelectric heat-pump drying system for drying herbs. *Journal of Electronic Materials* 2015, 44 (6), 2142–2145.

Woo, M.W., Mujumdar, A.S. Effects of electric and magnetic field on freezing and possible relevance in freeze drying. *Drying Technology* 2010, 28 (4), 433–443.

Xiao, H.-W. Guest editorial: Some mitigation strategies for climate change. *Drying Technology* 2015, 33 (14), 1679–1680.

Xie, Y.-J., Kong, W.-J., Yang, M.-H., Yang, S.-H. Research progress of Chinese herbal medicine raw materials in cosmetics. *Zhongguo Zhong Yao Za Zhi* 2015, 40 (20), 3925–3931. (In Chinese with English abstract).

Xie, L., Mujumdar, A.S., Xiao, H-W., Gao, Z-J. Recent technologies and trends in medicinal herb drying. In *Drying Technologies for Foods: Fundamentals and Applications (Part II)*, Eds. Nema, P.K., Kaur, B.P., Mujumdar, A.S. New India Publishing Agency, New Delhi, 2016, 69–90.

Xu, Y., Zhang, M., Mujumdar, A.S., Duan, X., Sun, J.C. A two-stage vacuum freeze and convective air drying method for strawberries. *Drying Technology* 2006, 24 (8), 1019–1023.

Yang, J., Di, Q., Zhao, J., Wang, L. Mechanism on mass transfer in micro-scale during the microwave drying of plant porous materials. In *Proceedings of the ASME 2009 Summer Heat Transfer Conference (HT2009)*, 2009, 3, 667–674.

Yarnell, E., Abascal, K. Botanical medicines for headache. *Alternative and Complementary Therapies* 2007, 13 (3), 148–152.

Yasurin, P. Antimicrobial properties of common herbs and spices used in Thai cooking. *Research Journal of Pharmaceutical, Biological and Chemical Sciences* 2015, 6 (1), 48–57.

Zhang, M., Jiang, H. Recent food drying R&D at Jiangnan University: An overview. *Drying Technology* 2014, 32 (15), 1743–1750.

Zhang, M., Jiang, H., Lim, R.X. Recent developments in microwave-assisted drying of vegetables, fruits, and aquatic products: Drying kinetics and quality considerations. *Drying Technology* 2010, 28 (11), 1307–1316.

Zhu, Z.-Z., Liu, C.-M., Feng, L. Chinese herbs in the treatment of senile dementia. *Chinese Journal of Clinical Rehabilitation* 2004, 8 (13), 2598–2600 (In Chinese with English abstract).

9 Drying of Vegetable Snacks

Hao Jiang and Shaojin Wang
Northwest A&F University

CONTENTS

9.1 INTRODUCTION

According to the Oxford dictionary, snacks or snack foods can be defined as a small meal or a small amount of food that are usually consumed in a hurry. Traditionally, snacks are prepared from commonly available home ingredients. After hundreds, or thousands, of years of development, modern snacks have become a huge business food category, and the ingredients for preparing snacks include almost all edible materials, such as fruit, vegetable, meat, egg, and grain. The form of snacks varies and includes dessert, jerky, ice-cream, preserved fruit, chips, dried fruit/vegetable, etc. Because convenience stores have become more and more popular nowadays, packaged snack foods are a significant business. Snack foods often contain substantial amounts of sweeteners, preservatives, and appealing ingredients, such as chocolate, peanuts, and specially formulated flavors (e.g., flavored potato chips).

It is difficult to define the accurate category of snacks, and it is also hard to determine how many kinds of snacks exist in the market. It is noted that the industry of snack-related domain is huge, and it still shows a strong growth power. It was reported by the International Trade Association that the snack food industry showed a steady growth pattern over the past decade (1990–1999). According to the sales statistics of potato chips from SFA (Snack Food Association) annual report in 2000, tortilla chips, corn chips, pretzels, meat snacks, nuts, pork rinds, and popcorn were the most popular snacks in the world (SFA data, 2010).

Based on the research report, in China 2005–2010 (Data from Research in China, 2014), the snack food industry of China had an annual growth of 16.9%, with a revenue of RMB 503.57 billion (about US$ 74.05 billion) in 2010. In 2010, the revenue of preserved fruit and vegetable snacks was RMB 16.8 billion (about US$ 2.47 billion), with a related manufacturing company number of 409. It was found that the growth rate of Chinese snack food industries between 2011 and 2013 was around 17%.

Consumers have high preferences for fruits and vegetables, which are important dietary sources of vitamins, phytochemicals, fibers, and minerals. The intake of fruits and vegetables has been associated with a wide range of beneficial health effects (Pomerleau et al., 2006). It is a predicament that youth fail to take the recommended amount of fruits and vegetables in many countries. The 2010 National Youth Physical Activity and Nutrition Studies data showed that 28.5% and 33.2% of high school students consumed fruits and vegetables less than once a day, respectively (Kim et al., 2011). In the case of investigation on health-related business, child-friendly fruit and vegetable snacks are most welcome by parents. Unfortunately, the majority of current fruit and vegetable snacks usually contain high fat, sugar, and/or sodium. It has to be admitted that current snacks are consumed by contemporary children/youth for its flavor and taste, but not for its nutrient values. However, it is a common view that snacks should meet the requirements not only with good taste but also with full nutrients.

Dehydration is one of the most important snack manufacturing technologies, because it can offer food the special taste, flavor, and shape. At the same time, most traditional drying methods, such as solar or hot air drying, are labor-intensive, but show less technical barriers in developing/undeveloped countries/areas. The disadvantage of most dehydrated fruits and vegetables is substantial degradation in quality, including appearance (shrinkage, drying up, darkening), nutrients and flavor, and low rate of rehydration (Huang and Zhang, 2012). This chapter focuses on the drying technologies for vegetable snack processing.

9.2 DRIED VEGETABLE SNACKS CLASSIFIED BY VEGETABLE VARIETIES

9.2.1 Root Vegetables (Potato, Carrot, Cassava, Sweet Potato, etc.)

Roots are the storage organs of vegetables, mainly in the form of carbohydrates, including sugar and starch. Apart from being important staple foods, starchy root vegetables are also good resources for snacks. For example, potato chips are a predominant snack food in Western country markets. The global potato chip market generated a total revenue of US$16.49 billion in 2005, accounted for 35.5% of the total savory snack food market ($46.1 billion) (China Snack Food Industry Report, 2014).

Some reports investigated the manufacturing method of potato chips. Ahmad Tarmizi and Ismail (2008) investigated the drainage methods after deep-frying. They studied the influence of atmospheric and vacuum drainage on free fatty acid (FFA) content, *p*-anisidine value (AnV), color, viscosity, fatty acid profile, and concentration of tocols. Meanwhile, the oil deterioration by oxidation and polymerization was also reduced by the use of vacuum drainage. The rate of FFA formation was found to be about half of that from atmospheric drainage. The AnV of the oil after vacuum drainage was lower by about 12%, the total color difference was improved by 14%, and the viscosity was slightly reduced after 5 days of frying, compared to the values for oil that had been drained at atmospheric pressure. There was a reduction in the loss of polyunsaturated fatty acids in the case of vacuum drainage after 5 days of frying, but differences in retention of tocols were only evident in the first 2 days of frying. Lioumbas and Karapantsios (2014) studied the effect of increased gravitational acceleration in potato deep fat frying. Scaling of gravitational acceleration allowed scaling of buoyancy forces, which controlled heat transfer from hot oil to potato surface. The data revealed the positive effect of increased gravitational acceleration on the evaporation front propagation and crust thickness evolution inside a deep-fried potato. The effect of altering the gravity level was appreciable for horizontal top surfaces of potatoes but was much less for vertical side and horizontal bottom surfaces. At high gravity levels, the crust eventually detached from the core, probably due to collapse of large pores of superheated steam and the ensuing significant tensions. Chouicha et al. (2013) studied the solar hybrid drying of sliced potatoes by forced convection in an indirect solar dryer using extra energy via a heater by joule effect generated by photovoltaic modules connected in parallel. The data indicated that with the use of one solar panel to reach the final water content of the potato, the drying time taken was 3 h. In the case of two panels, this time was 2 h 45 min. The results revealed that hybrid drying could reduce the drying time and energy consumption.

The Food and Agriculture Organization (FAO) of the United Nations reported that worldwide production of carrots and turnips in 2011 was almost 35.658 million tonnes, and almost half of the production was from China (FAO data, 2012). β-carotene, α-carotene, γ-carotene, lutein, and zeaxanthin were the main nutrients in carrot. To preserve these nutrients during drying was the hot research area in carrot snacks manufacturing. The objective of Dueik et al.'s research (2013) was to study the quality attributes of vacuum and atmospheric dried carrot snacks, to compare sensory maps with the atmospheric and vacuum dried samples, and to understand their potential as a healthy snack. Dried product quality was compared in terms of color, carotenoids content, Maillard reaction occurrence, and texture. The results indicated that vacuum drying did not offer any significant improvement in the quality parameters compared to a two-step atmospheric drying process. Carotenoids retention and color were similar, but texture was negatively affected. Overall, vacuum inclusion had a better effect in frying compared to drying in sensory perception. Lin et al. (1998) studied the characteristics of vacuum–microwave drying of carrot slices and compared them to air drying and freeze-drying based on rehydration potential, color, density, nutritional value, and textural properties. The results showed that the air-dried carrot slices were darker with less red and yellow hues. Compared with hot air drying, less color deterioration occurred when vacuum–microwave drying was applied. Although freeze-drying of carrot slices yielded a product with improved rehydration potential, appearance, and nutrient retention, the vacuum–microwave-dried carrot slices were rated as equal to or better than freeze-dried samples by a sensory panel for color, texture, flavor, and overall preference both in the dry and rehydrated states. Dueik et al. (2010) also studied how to reduce the oil uptake of carrot snacks under vacuum frying. They examined the most important quality parameters of vacuum and atmospherically fried carrot slices to identify the specific advantages of vacuum technology. The results showed that vacuum-fried crisps (driving force of 60°C) may reduce the oil content of carrot crisps by nearly 50% (d.b.) compared to atmospheric-fried chips produced using the same driving force. Furthermore, this method (vacuum frying) could preserve approximately 90% of trans α-carotene and 86% of trans β-carotene.

Sweet potato is a large, starchy, and sweet tasting root vegetable. Sweet potatoes are also rich in complex carbohydrates, dietary fiber, and β-carotene, with moderate contents of other micronutrients, including vitamin B_5, vitamin B_6, manganese, and potassium. According to the FAO statistics, in 2004, the worldwide production was 127 million tonnes (Singh and Pandey, 2012). Sweet potato was a good resource for snacks, as a raw material or as an ingredient. Guo et al. (2014) studied the interaction between sweet potato amylose/amylopectin and KCl during drying because starches treated with potassium ions showed lower rate of retrogradation than the respective native starches. They found that KCl was entrapped by starch in the crystallization process during drying. It showed positive functions on quality, acceptability, and shelf life of starch-containing foods. Singh and Pandey (2012) studied the convective air drying characteristics of sweet potato cubes, and the results indicated the effective diffusivity increased as flow rate and temperature increased. Qiao et al. (2012) used lychee and sweet potato mixture to manufacture new flavored snacks. The best ratio between sweet potato and lychee to produce desirable snacks was found to be 7:3 (W:W). In terms of the drying parameters, high-power density resulted in shorter drying time and chips with improved nutrients. Yang et al. (2012) studied the vacuum frying method for preparation of sweet potato snacks. Compared with traditional frying, they demonstrated that vacuum frying was the appropriate deep fat frying method for health-conscious people who prefer to take less fats and oils from food. Doymaz (2012) investigated the effect of infrared drying on sweet potato slices. The results showed that the drying rate and product quality were significantly influenced by infrared power, and the highest rehydration ratio values were obtained at an infrared power of 146 W.

Mitra et al. (2011) studied the color, flavor, and rehydration ratio of vacuum-dried onion compared with those prepared from convective air drying. The results indicated that all the aforementioned quality indexes from vacuum drying were better than that from convective air drying. Mota et al. (2010) investigated the kinetics and nutrition of convective drying of onion. It was verified that some chemical components (ash, fat, protein, and fiber) of the onions were not affected by drying, whereas some other characteristics (sugars, acidity, and vitamin C) were considerably influenced.

9.2.2 Stem Vegetables (Ginger, Lettuce, Celery, Bamboo Shoot, Mustard Tuber, etc.)

Stem vegetables, such as ginger, mustard tuber, celery, etc., could be good sources or ingredients for snack foods because of their high nutritional value and bright colors.

In China, ginger slice, candy, and tea is widely accepted. Gümüsay et al. (2015) investigated the effects of four different drying processes including sun drying, oven drying, vacuum oven drying, and freeze-drying of gingers and tomatoes. The results indicated that the freeze-dried samples showed the best antioxidant properties among the four drying methods (Figure 9.1). The quality of dried gingers by various drying methods, including sun drying, solar tunnel drying, and cabinet tray drying at different temperatures, was studied by Jayashree et al. (2014). They found a significant reduction in the volatile constituents of ginger essential oil including zingiberene, limonene, linalool, geraniol, and nerolidol when compared to fresh ginger, regardless of the drying method used. A study was undertaken to prepare ginger powder using various drying methods, and their nutritional evaluation was carried out by Sangwan et al. (2014). Ginger was dried using shade, solar, oven, and microwave drying methods. All the samples were ground into a fine powder. The results showed that ginger powder prepared from various drying methods had good sensory and nutritional profiles. It should be pointed out that ginger usually maintains good color after drying because of the high fiber content.

Some other stem vegetables (celery and lettuce, for instance) are also good for drying as snack foods. Białobrzewski (2007) researched the mass transfer coefficient during hot air drying of celery root. It was also observed that mass transfer coefficient determination based on the Chilton–Colburn analogy was burdened with gross errors. Wang et al. (2013) used a new drying technique called microwave-assisted pulse spouted bed freeze-drying to process stem lettuce snacks. Temperature distribution uniformity, microstructure, apparent density,

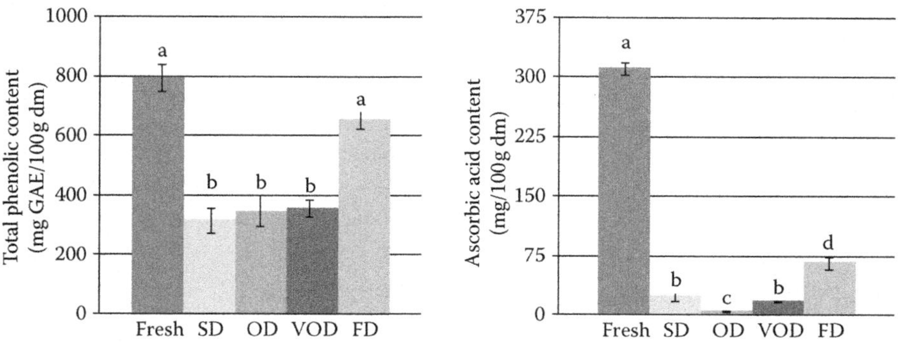

FIGURE 9.1 Total phenolic content and ascorbic acid content of fresh and dried tomato samples. SD, sun-dried; OD, oven-dried; VOD, vacuum-oven-dried; FD, freeze-dried; dm, dry matter. For each column, values followed by the same letter (a–c) are not statistically different at $p < .05$. (From Gümüsay, Ö.A. et al., *Food Chem.*, 173, 156–162, 2015. With permission.)

rehydration, and texture properties were measured to evaluate the quality of dried stem lettuce slices. Results showed that this drying technique led to better temperature distribution during drying as compared to those dried by microwave freeze-drying. Being an emerging drying technique, microwave-assisted pulse spouted bed freeze-drying can be an alternative method to microwave freeze-drying. It offered dried stem lettuce slices the competitive characteristics, such as lower discoloration, better drying uniformity, compact microstructure, higher rehydration capacity, and greater hardness. Wang et al. (2012) also studied the effect of blanching on microwave freeze-drying of stem lettuce cubes. The results showed that the electrical conductivity of samples blanched by microwave was two times higher than that of blanched with boiling water and five times higher than that of unblanched samples. Being a partly drying method, osmotic treatment can impart a special flavor and shape to food. Liu et al. (2009) aimed to study the changes in volatile compounds of pickled mustard tuber during the pickling process (involving osmotic treatment). The results indicated that the contents of sulfides, acids, aldehydes, alcohols, phenols, esters, nitriles, and heterocyclic compounds were increased with pickling time, but isothiocyanates increased at the first and second pickling stages and then started to decrease during the last pickling stage. It was also demonstrated that more than 60 days pickling time was necessary for the formation of the typical aroma notes.

9.2.3 Leafy Vegetables (Cabbage, Chinese Cabbage, etc.)

Leafy vegetables are plant leaves eaten as a vegetable, sometimes accompanied with tender petioles and shoots. They are typically low in calories and fat but high in dietary fiber, vitamin C, carotenoids, folate, manganese, and vitamin K.

Lekcharoenkul et al. (2014) studied the change of sulforaphane content after hybrid drying of cabbage at different drying temperatures. In their experiment, hybrid drying with low-pressure superheated steam drying at 60°C followed by vacuum drying at 45°C led to the highest retention of sulforaphane in the final dried cabbage powder. Tanongkankit et al. (2015) also studied the effects of blanching and drying methods on the antioxidant content, mainly on the content of phenolic compounds and vitamin C in the outer leaves of white cabbage during vacuum drying at 60°C, 70°C, and 80°C. The results showed that there were losses of antioxidants during steam blanching. The vacuum-dried and blanched leaves, nevertheless, had higher antioxidant contents and activity than dried, unblanched leaves. Losses of antioxidants during vacuum drying were also noted to be less than those during hot air drying.

9.2.4 Flower Vegetables (Broccoli, Cauliflower, etc.)

Vegetables using the bud, anthocaulus, or corolla as the edible part can be classified as the flower vegetables. With their powerful, unique flavors, textures, and colors, flower vegetables have gained popularity as a creative and innovative ingredient in the culinary world. Flower vegetables are popular in some compound vegetable/fruit snacks because of the bright color. In some places, such as Yunnan province in China, flowers are consumed as snacks directly.

Jin et al. (2014) used broccoli as the material and investigated the effect of energy-efficient drying strategies on retaining its nutritional components. Kinetics of the degradation of glucosinolates, vitamin C, and drying rate of broccoli were used to optimize the drying trajectories by controlling various airflow rates and temperatures. Jin et al. (2014) also used MRI (magnetic resonance imaging) to determine the moisture distribution of broccoli during hot air drying. Gupta et al. (2013) investigated the optimization of process parameters for cauliflower drying. In their experiment, different sizes (3, 4, and 5 cm) of cauliflowers were dehydrated in a thin layer at temperatures of 55°C, 60°C, and 65°C with air velocities of 40, 50, and 60 m/min, respectively. Statistical analysis indicated that drying time was dependent on the initial size of cauliflower, drying air temperature, and velocity, but the rehydration ratio was significantly affected by the combined effects of temperature and airflow velocity.

9.2.5 Fruit Vegetables (Tomato, Cucumber, Capsicum, Chilli, Green Bean, Peanuts, Pea, etc.)

Fruit vegetables including tomato, chili, cucumber, capsicum, pea, bean, etc., are widely consumed around the world. They can be eaten freshly or processed, or can be used as condiments in the form of powder, crisps, pickle or jam, and snacks. With high nutrition values, they are important daily food for human health. The research domain on fruit vegetable drying was mainly focused on nutrition protection during drying and accelerated drying rate.

As an important fruit vegetable, tomato was widely chosen as the material to study the drying techniques. The energy and some physical and nutritional quality properties of tomatoes dried with short-infrared radiation were studied by Kocabiyik et al. (2015). The results showed that drying time was prolonged with reducing air velocity, whereas it was shortened with increasing infrared radiation intensity. The lowest energy consumption occurred at an air velocity of 1.0 m/s and at infrared radiation intensity of 2640 W/m^2. In general, it was observed that short-infrared drying of tomato provided good nutrient retention and low cost of energy compared with traditional drying methods, such as hot air drying. Gümüsay et al. (2015) investigated the antioxidant properties of tomatoes during drying. Four drying methods—sun drying, oven drying, vacuum oven drying, and freeze-drying—were studied, and the results showed that freeze-dried tomato samples had the best antioxidant properties, but freeze-drying might not be an economical drying method. Zhao et al. (2013) used chili slices as the materials to investigate the microwave drying characteristics after osmosis pretreatment, because osmotic treatment can alter the dielectric properties of samples and alter the drying rate. The results showed that the samples osmotically treated in mixed solution (10% salt + 50% sucrose) had the best dehydration effect as compared with single salt or sugar solutions. When the moisture content was decreased, the activation energy was increased gradually. However, among all the processes, microwave drying at 60 W after osmotic drying presented the best vitamin C retention rate and the best color, compared with the fresh samples.

Leguminous plants are classified as fruit vegetables and also as good resources for snack food productions. The effect of multistage heat pump fluidized bed atmospheric freeze-drying and microwave vacuum drying were evaluated by Zielińska et al. (2013) on the drying kinetics, moisture diffusivities, microstructure, and physical parameters of green peas. The application of microwave vacuum drying accelerated the drying rate. Furthermore, microwave vacuum-dried samples exhibited extra crispness.

9.3 DRYING OF VEGETABLE SNACKS BY DIFFERENT METHODS

9.3.1 Convective Drying

Convective drying, including solar drying, is a well-established drying technology. It has been used since time immemorial to dry plants, seeds, fruits, meat, fish, wood, and other agricultural or forest products as a means of preservation. For large-scale production, the limitations of convective drying are well-known, such as high labor costs, large space requirement, lack of ability to control the drying process, possible degradation due to biochemical or microbiological reactions, and insect infestation, etc. (Mujumdar, 2007). However, convective drying shows the huge benefit on small equipment investment and simple operation. In some developing countries, the cost is always determined by technologies applied, but not labor. A typical convective dryer is shown in Figure 9.2.

Recently, a few studies reported the drying parameters or characteristics of convective drying. Most of these articles focused on the model, kinetics, hybrid drying, or improving the equipment to get better drying efficiency. Sallam et al. (2015)

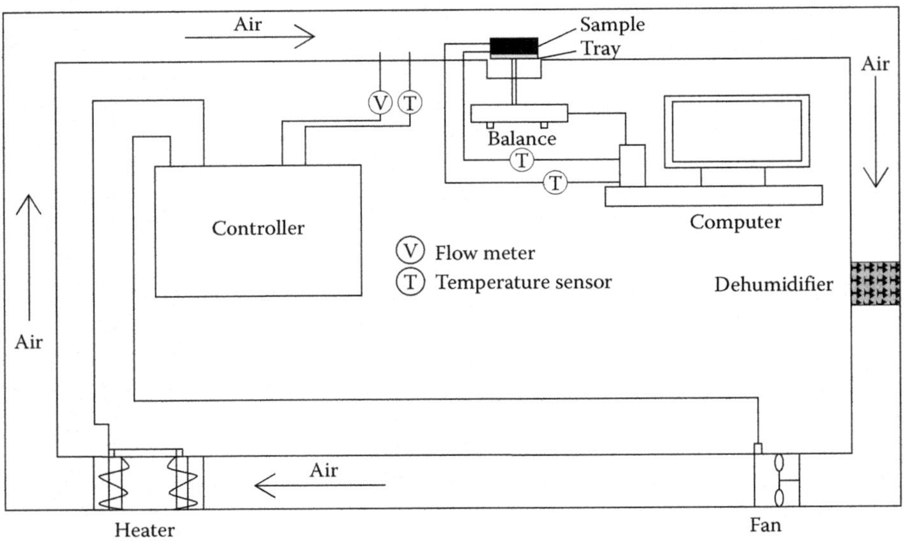

FIGURE 9.2 Convective dryer setup. (From Huang, Y.W. et al., *Appl. Therm. Eng.*, 96, 209–216, 2016. With permission.)

compared the natural and forced convection drying of whole mint plant heated by solar energy. The data revealed that the drying rate of mint under forced convection was higher than that under natural convection, especially during the first hour of drying because the mass transfer resistance was weak. Ayadi et al. (2014) and Hashim et al. (2014) studied the kinetics of convective drying on vegetables, and their data indicated that the main factor in controlling this rate was the drying air temperature. To optimize the shape of dried snack food, Curcio and Aversa (2014) studied the shrinkage model on convective drying of fresh vegetables. The effects of food shrinkage on drying performance were ascertained by analyzing the spatial distributions of temperature, moisture content, strain, and stress, as a function of operating conditions. It was proven that deformation strongly influenced the transport phenomena and could not be neglected when a comprehensive food-drying model had to be formulated. Calín-Sánchez et al. (2014) studied the drying kinetics, energy consumption, and product quality by combined convective drying/vacuum–microwave drying ginger slices, which can then be used as lozenges. The hybrid treatment significantly reduced the drying time from 630 min for convective drying to 49 min and contributed to a significant energy saving of about 54.4%–86.3%. In general, hybrid drying can be a good way to reduce high energy consumption during drying.

9.3.2 Conduction Drying

Conduction drying includes a group of different drying methods, such as rotary drying, fluidized bed drying, drum drying, freeze-drying, and vacuum drying (Figure 9.3). This drying technique allows materials to contact the heating plate directly to transfer the energy from the heating plate to materials. However, most foods, including vegetables, are bad conductors of heat, and this characteristic makes conduction drying classified as a high energy-consuming and labor-intensive drying method.

Pelegrina et al. (2002) optimized the drying condition of a vegetable rotary drier by the response surface method. They found that the temperature and relative humidity of the air were the key factors that determine the drying performance. Peressini et al. (2015) added edible fiber (inulin) into the dough and studied the dough rheological properties, texture, and sensory qualities of dried snacks. The production showed no negative impact on the quality of production, and 7% short-chain inulin lowered the extent of nonenzymatic browning. The vacuum drying effect of pretreated pumpkin and carrot slices was studied by Arévalo-Pinedo and Murr (2007), and freezing and blanching were applied as pretreatments. The results indicated that predrying/pretreatments affected moisture transport of the products. Kakade et al. (2011) evaluated the performance of a double drum dryer for potato flake production. The models developed in their study could be used to predict the values of responses at different values of the independent parameters including drum speed, steam pressure, and liquid level at the nip of the two drums. Besides, George and Datta (2002) established the heat and mass transfer models for freeze-drying of vegetable slices.

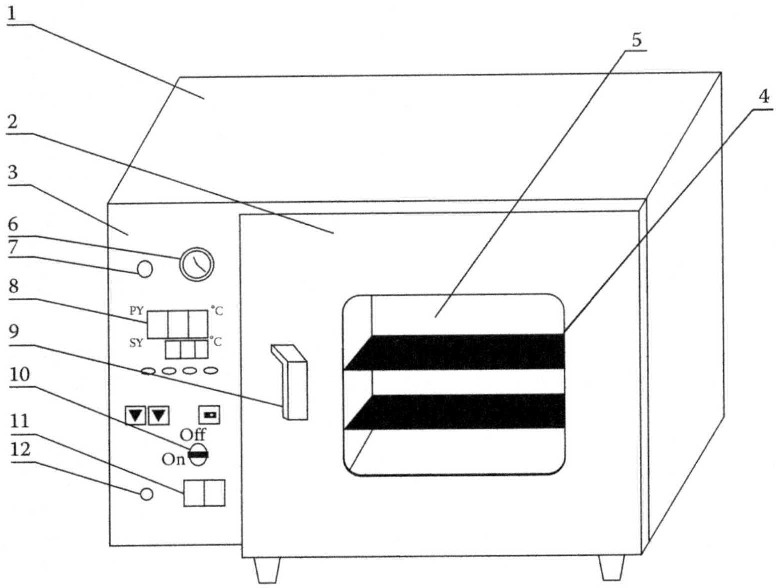

FIGURE 9.3 Schematic illustration of the vacuum dryer setup. 1, Case body; 2, case door; 3, panel; 4, shelf; 5, drying chamber; 6, vacuum gauge; 7, deflation valve; 8, temperature controller; 9, door handle; 10, vacuum valve; 11, power switch; and 12, power light. (From Zhang, Y. et al., *Energy Convers. Manag.*, 80, 266–275, 2014. With permission.)

9.3.3 Radiation Drying

In physics, radiation is the emission or transmission of energy in the form of waves or particles through space or through a material medium. It includes electromagnetic radiation, such as radiowaves, visible light, X-rays, particle radiation (e.g., α, β, and neutron radiation), and acoustic radiation (e.g., ultrasound, sound, and seismic waves) (Roberts, 2014). Radiation drying has immense advantages on drying rate. The most useful radiation drying techniques were dielectric and infrared drying.

Dielectric energy can penetrate materials and heat products without the aid of thermal gradients, which has a positive effect on dehydration. The basic physical phenomenon, which is responsible for the heating of food materials at microwave frequencies, is dipole rotation. The dipole rotation mechanism relies on the fact that water molecules are subject to a microwave field that rapidly change direction, and the dipoles try to align with the direction of electrical field, which generates heat (Zhang et al., 2006). Koné et al. (2013) combined hot air drying and microwave drying to improve the quality of dried tomatoes. The drying system could adjust the input power automatically. Hybrid drying systems can not only save energy, but also improve the quality of production significantly, such as color and flavor. It is one of the development trends in drying technologies. Some typical hybrid dryers are presented in Figure 9.4.

Microwave can be combined with other techniques, for instance, vacuum microwave drying and microwave freeze-drying. Relevant research on dried edamame (Hu et al., 2007), potato (Wang et al., 2010a), vegetable instant soup (Wang et al., 2010b), lettuce (Wang et al., 2012), etc. have proven that microwave-related hybrid treatments are fantastic drying methods for vegetable snack processing. Yan et al. (2010a,b) studied the characteristics of potato and carrot cube snacks dried by microwave-assisted spouted bed dryer, and the products showed good color and rehydration (Figure 9.5).

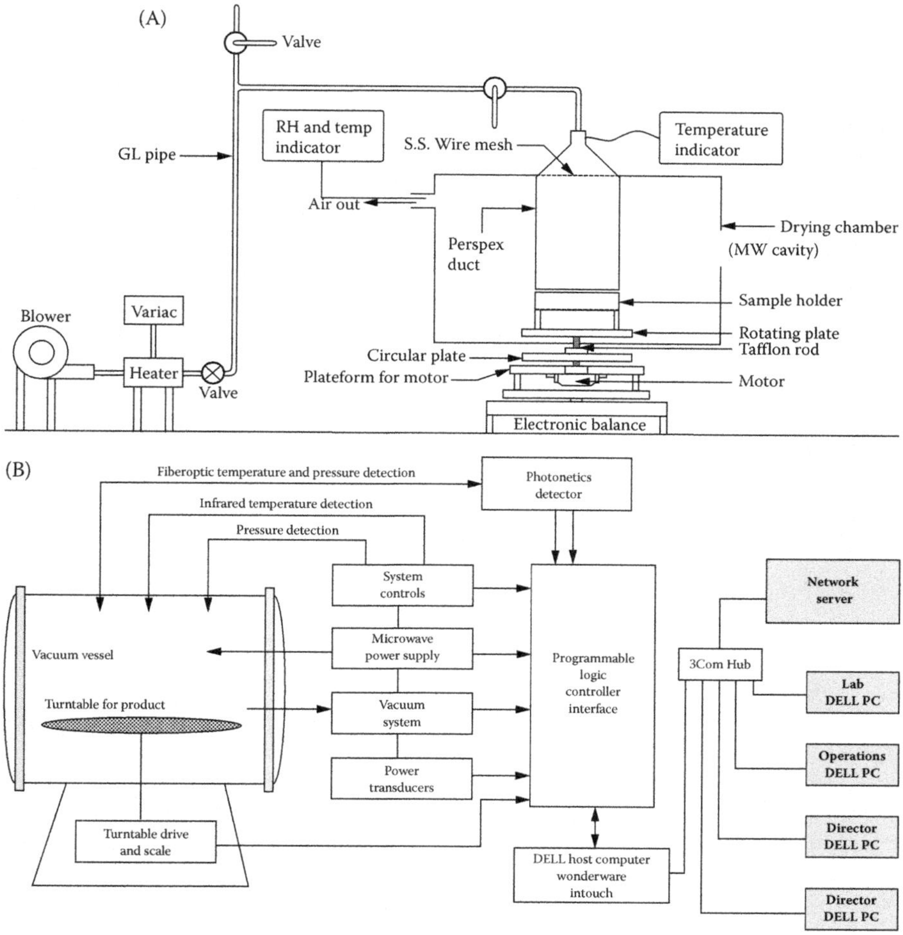

FIGURE 9.4 Typical hybrid dryer schematics: (A) Microwave-convective dryer; (B) laboratory microwave vacuum dehydration system; and (C) microwave and spouted bed drying system. (From Uprit, S., Mishra, H.N., *Food Bioprod. Process.*, 81(2), 89–96, 2003; Clary, C.D. et al., *J. Food Sci.*, 70(5), 344–349, 2005; Feng, H., Tang, J., *J. Food Sci.*, 63(4), 679–683, 1998.)

(*Continued*)

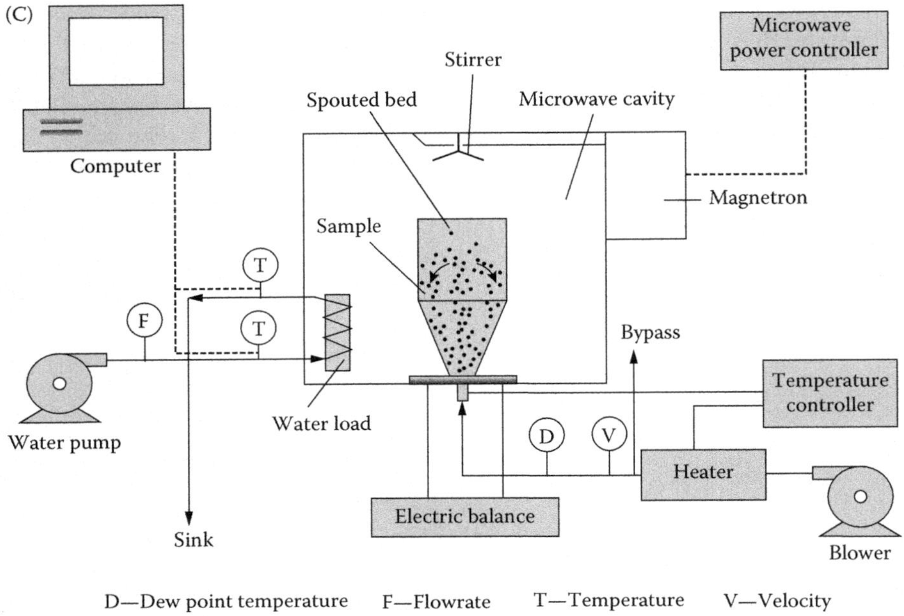

FIGURE 9.4 (CONTINUED) Typical hybrid dryer schematics: (A) Microwave-convective dryer; (B) laboratory microwave vacuum dehydration system; and (C) microwave and spouted bed drying system. (From Uprit, S., Mishra, H.N., *Food Bioprod. Process.*, 81(2), 89–96, 2003; Clary, C.D. et al., *J. Food Sci.*, 70(5), 344–349, 2005; Feng, H., Tang, J., *J. Food Sci.*, 63(4), 679–683, 1998.)

FIGURE 9.5 Potatoes (A) and carrots (B) dried by microwave-assisted spouted bed dryer. (From Huang, L., Zhang, M., *Drying Technol.*, 30, 448–461, 2012. With permission.)

Low-frequency dielectric drying, especially radio frequency drying, is a new drying technology (Huang et al., 2013). For food drying, the radio frequency drying is useful for further industrial applications. Wang et al. (2014a,b) found that radio frequency can not only be applied on nuts to dry the samples, but also can be used in sterilization and disinfection. Radio frequency offers better penetrability and controllability, which makes it a potential substitute for microwave drying (Wang et al., 2011). There are certain drawbacks (cost and size of radio frequency magnetic tubes of flux) that hamper the development of radio frequency. However, it would have a bright future if the technical problems are solved (Figure 9.6).

The principle of infrared heating is the transformation of radiation energy into molecules' movements in the absorbing materials. A typical infrared dryer can be found in Figure 9.7. In contrast to heat transfer by convection in a drying oven, the infrared thermal radiation supplies an efficient heating of the sample with the consequence that water is liberated faster. Infrared drying is considered as a mild drying method, which is well adapted to the requirement of vegetable snack drying. Kocabiyik et al. (2015) estimated the drying characteristic, energy, and physical and nutritional quality properties of tomato dried by short-infrared radiation. The results indicated that infrared drying of tomato provided good nutrient retention and low energy cost. Sadin et al. (2014) investigated the effect of temperature and slice thickness on the drying kinetics of tomato in an infrared dryer and found that drying rate was raised with increasing temperature and thinner slices. Similar to microwave drying, infrared drying could also be used in combination with other drying methods, such as vacuum drying (Hirun et al., 2015), heat pump drying (Deng et al., 2011), or vibratory bed dryer (Barzegar et al., 2015) (Figure 9.8) to increase the drying efficiency and to improve the product quality.

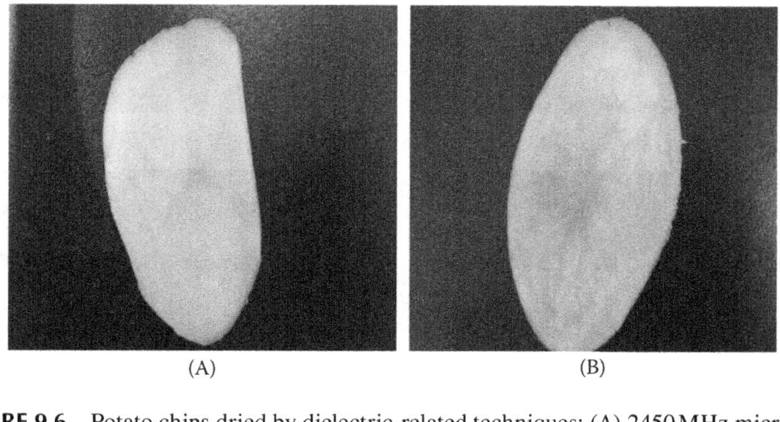

(A) (B)

FIGURE 9.6 Potato chips dried by dielectric-related techniques: (A) 2450 MHz microwave drying and (B) radio frequency (27 MHz) drying.

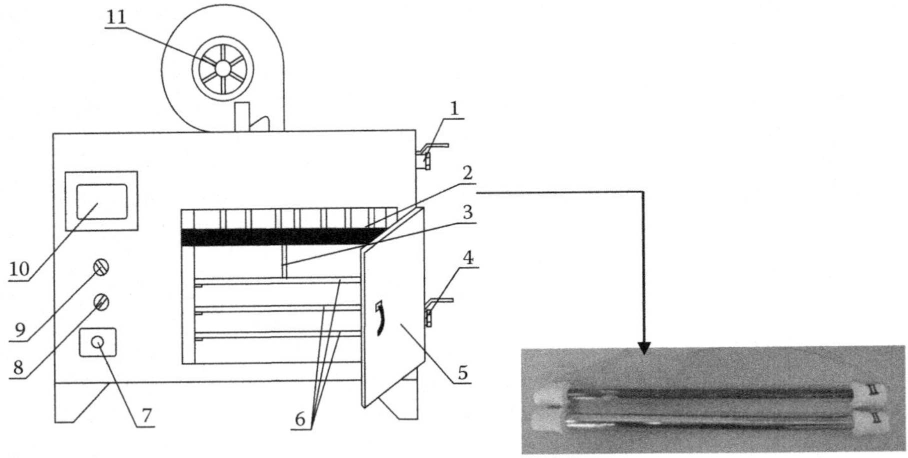

FIGURE 9.7 Schematic of mid-infrared-assisted convection dryer. 1, Air intake; 2, radiation tube; 3, temperature sensor; 4, air outlet; 5, door; 6, load material plate; 7, air regulator; 8, switch (control blower); 9, switch (control heating tube); 10, display screen; and 11, blower. (From Wang, H. et al., *Food Bioprod. Process.*, 94, 507–517, 2015. With permission.)

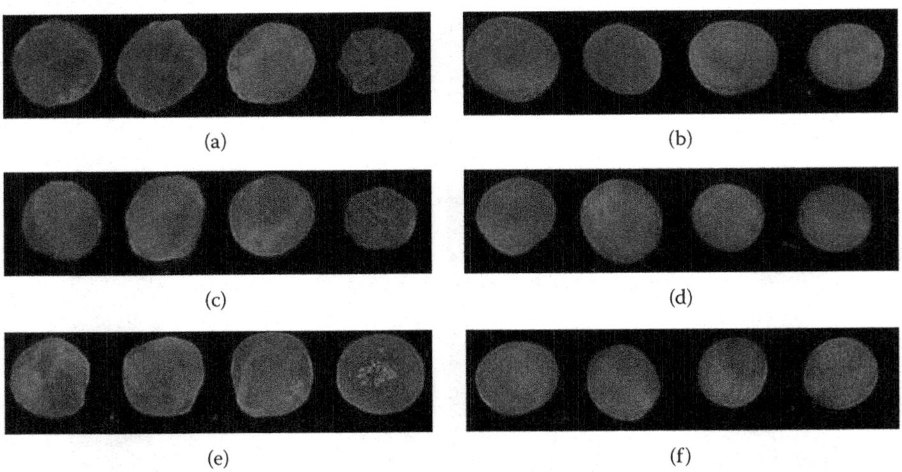

FIGURE 9.8 Color changes of dried green pea at different drying conditions for infrared radiation of 9000, 4000, and 2000 W/m^2 and no radiation (left to right, respectively): (a) fixed bed, 30°C, (b) vibratory bed, 30°C, (c) fixed bed, 40°C, (d) vibratory bed, 40°C, (e) fixed bed, 50°C, and (f) vibratory bed, 50°C. (From Barzegar, M. et al., *J. Food Eng.*, 166, 302–315, 2015. With permission.)

9.3.4 Deep-Frying

Snacks made by deep-frying are extensively employed in the domestic and industrial sectors because of their ability to create unique sensory characteristics in food (Ahmad Tarmizi and Ismail, 2008). Compared with other drying/manufacturing methods, deep-frying is a fast and easy way to prepare food because of high temperatures. The fact that deep-frying might not be a healthy cooking method should not be ignored because it has a high fat content. However, deep-frying is still popular as foods prepared by this method are always accomplished with unique crispness, color (golden brown), and most importantly, matchless taste (Figure 9.9).

The research hotspots of deep-frying were focused on food safety and health. Akil et al. (2015) aimed to improve the oxidative stability and investigated the changes in fatty acid and tocopherol composition of extra-virgin olive oil during short-term

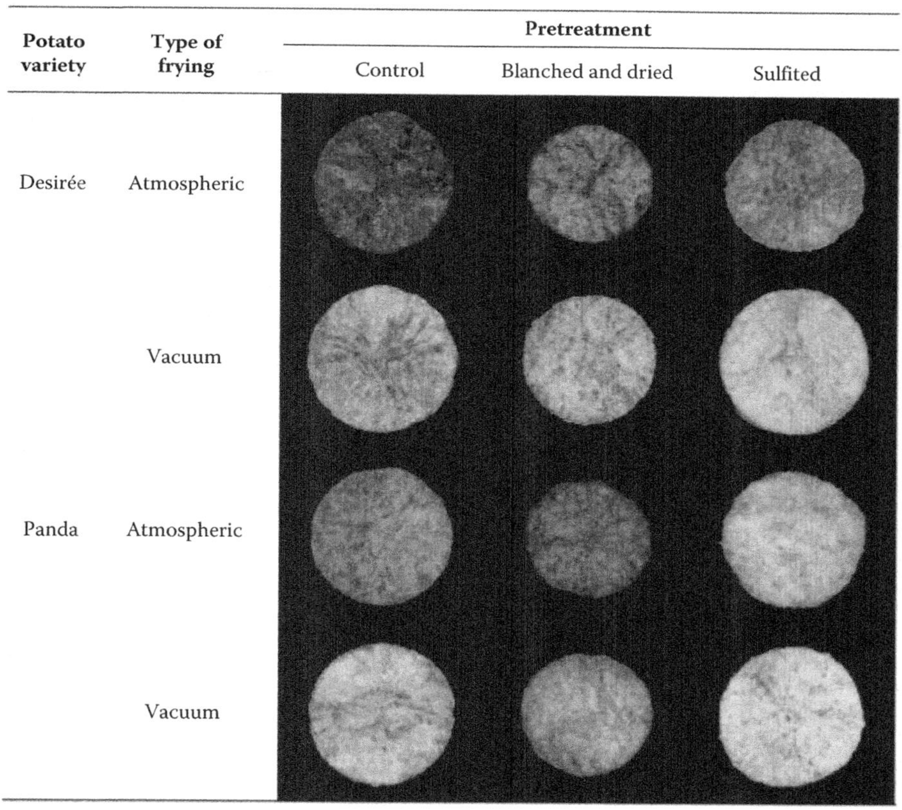

FIGURE 9.9 Image gallery of pretreated potato chips (Desirée and Panda varieties) fried at 120°C under vacuum and atmospheric frying. (From Troncoso, E. et al., *LWT-Food Sci. Technol.*, 42, 187–195, 2009. With permission.)

deep-frying of french fries. The results indicated that extra-virgin olive oil used as a frying medium showed high oxidative stability, preservation of unsaturated fatty acids, and low formation of FFAs and carbonyl compounds. Hassanien and Sharoba (2014) studied the rheological characteristics of vegetable oils after deep-frying of french fries. The fatty acid composition and rheological characteristics of sunflower oil, cottonseed oil, and palm olein during deep-frying for 4–16h were investigated. Fresh oils showed Newtonian behavior with correlation coefficients >0.99 at 25°C and slight nonNewtonian behavior after frying. Palm olein showed higher viscosity in comparison to cottonseed oil and sunflower oil. Rheological parameters of vegetables oils showed great changes, whereas the highest change in viscosity was recorded after 16h of frying.

9.3.5 Osmosis Drying

Osmosis drying is the method used to preserve vegetables or fruits in a vinegar sauce or salty water and is classified as osmosis treatment, which in turn is classified as partly drying. The main method of osmosis drying applied on food is pickling. The pickling procedure may typically affect the food's texture and flavor because of osmotic pressure. It can be used as a pretreatment combined with other drying methods. It should be noted that the content of nitroso compounds could be increased after pickling (Hou et al., 2013), which increased the potential risks on human health.

Weng et al. (2013) studied the growth rate of *Leuconostoc citreum* in pickled vegetables with low salinity. It was demonstrated that the temperature, initial pH, and the concentration of sodium chloride had a great influence on the growth of *L. citreum.* Therdthai and Visalrakkij (2012) found that the osmotic pretreatment showed influence on dielectric properties and also on the drying dynamics of dielectric drying, which was also confirmed by Wang et al. (2010b) and Al-Harahsheh et al. (2009). Corrêa et al. (2014) used pulsed vacuum osmotic dehydration to process blanched pumpkin. The results indicated that some factors (concentration of the osmotic solution, pressure, and time of vacuum pulse application) showed significant influences on blanched osmotic-dried pumpkin, while temperature was not a significant factor in the studied conditions.

9.4 CONCLUSIONS AND FUTURE TRENDS

As reported, the daily intake of vegetables by most children and adolescents is below the recommended level (Pomerleau et al., 2006). Vegetable snacks could be a nutritional supplement to increase the vegetable intake. It should be noted that snacks might contain high levels of sugars and solid fats, which could contribute to an intake of about 168 cal/day (Piernas and Popkin, 2010). To increase the daily intake of vegetable snacks, some important factors should be addressed as follows:

1. Lower production cost: Vegetable snack production is related to the thermal/dehydration processing, thus the energy consumption cannot be ignored. The balance between drying rate and production quality also motivated the progression of drying technologies. It is reported that world

TABLE 9.1
Energy Consumption of Freeze-Drying Banana Chips Processed with 400 W Heating Power and Microwave Freeze- Drying Banana Chips Processed with 1 W/g Microwave Power

Energy Consumption ($\times 10^6$ J)	1 h	2 h	3 h	4 h	5 h	6 h	7 h	8 h	9 h	10 h
FD Samples										
Total	4.932 (±0.33)	4.248 (±0.18)	4.104 (±0.24)	4.068 (±0.26)	3.780 (±0.13)	3.318 (±0.20)	2.844 (±0.23)	2.376 (±0.11)	2.160 (±0.13)	2.016 (±0.08)
Vacuum system	1.980 (±0.10)	1.484 (±0.15)	1.512 (±0.21)	1.476 (±0.12)	1.368 (±0.18)	1.052 (±0.09)	0.972 (±0.13)	0.936 (±0.11)	0.936 (±0.11)	0.900 (±0.07)
Cold trap	1.404 (±0.23)	1.252 (±0.05)	1.116 (±0.14)	1.044 (±0.20)	0.936 (±0.18)	0.796 (±0.16)	0.720 (±0.08)	0.756 (±0.12)	0.756 (±0.09)	0.684 (±0.11)
Heating system	1.476 (±0.18)	1.404 (±0.23)	1.440 (±0.15)	1.476 (±0.11)	1.440 (±0.03)	1.440 (±0.18)	1.116 (±0.13)	0.540 (±0.08)	0.432 (±0.04)	0.432 (±0.08)
MFD Samples										
Total	5.256 (±0.28)	4.932 (±0.14)	3.472 (±0.21)	3.240 (±0.26)	2.592 (±0.17)	2.268 (±0.20)				
Vacuum system	2.088 (±0.18)	1.944 (±0.23)	1.456 (±0.13)	1.116 (±0.28)	0.972 (±0.13)	0.900 (±0.14)				
Cold trap	1.548 (±0.23)	1.476 (±0.09)	1.044 (±0.16)	0.936 (±0.11)	0.756 (±0.18)	0.756 (±0.05)				
Heating system	1.440 (±0.16)	1.404 (±0.19)	1.440 (±0.06)	1.116 (±0.21)	0.828 (±0.13)	0.648 (±0.11)				

Source: Jiang, H. et al., *Food Bioprod. Process.*, 91(4), 464–472, 2013. With permission.
Note: Values are means ± SD of three determination.

energy consumption for industrial drying operations ranges from 10% to 15% (Mujumdar, 2007). Some methods, such as raw vegetable selection or appropriate pretreatment, can significantly reduce the energy consumption. Besides, application of new processing techniques is also a good way to reduce the cost. For instance, microwave freeze-drying can be a good substitute for freeze-drying, as more energy is saved because of the shortened drying time (Table 9.1).

2. More choices: It is reported that food shape and color could be crucial factors to determine the consumers' acceptance, especially for children (Olsen et al., 2012). It is recommended to manufacture different vegetable snacks specific for different situations, such as at home, school, or at leisure activities, or for specific consumers, such as diabetics, obese patients, or people with depression.
3. Healthier production: From a public health point of view, a high consumption of fruits and vegetables could reduce the risk of heart disease and some forms of tumors (Pomerleau et al., 2006). The fact that most snacks contain excessive fat and sugar could be a negative factor to human health. Preserving the nutritional value of fresh vegetables with the addition of minimum additives to the final products is an important topic in future scientific and industrial research for vegetable snacks.

REFERENCES

Ahmad Tarmizi, A. H., Ismail, I. 2008. Comparison of the frying stability of standard palm olein and special quality palm olein. *Journal of the American Oil Chemists' Society* 85(3): 245–251.

Akil, E., Castelo-Branco, V. N., Costa, A. M. M., Vendramini, A. L. A., Calado, V., Torres, A. G. 2015. Oxidative stability and changes in chemical composition of extra virgin olive oils after short-term deep-frying of French fries. *Journal of the American Oil Chemists' Society* 92: 409–421.

Al-Harahsheh, M., Al-Muhtaseb, A. H., Magee, T. R. A. 2009. Microwave drying kinetics of tomato pomace: Effect of osmotic dehydration. *Chemical Engineering and Processing* 48: 524–531.

Arévalo-Pinedo, A., Murr, F. E. X. 2007. Influence of pre-treatments on the drying kinetics during vacuum drying of carrot and pumpkin. *Journal of Food Engineering* 80: 152–156.

Ayadi, M., Mabrouk, S. B., Zouari, I., Bellagi, A. 2014. Kinetic study of the convective drying of spearmint. *Journal of the Saudi Society of Agricultural Sciences* 13: 1–7.

Barzegar, M., Zare, D., Stroshine, R. L. 2015. An integrated energy and quality approach to optimization of green peas drying in a hot air infrared-assisted vibratory bed dryer. *Journal of Food Engineering* 166: 302–315.

Białobrzewski, I. 2007. Determination of the mass transfer coefficient during hot-air-drying of celery root. *Journal of Food Engineering* 78(4): 1388–1396.

Calín-Sánchez, Á., Figiel, A., Wojdyło, A., Szarycz, M., Carbonell-Barrachina, Á. A. 2014. Drying of garlic slices using convective pre-drying and vacuum-microwave finishing drying: Kinetics, energy consumption, and quality studies. *Food and Bioprocess Technology* 7: 398–408.

China Snack Food Industry Report in 2010–2013, 2014. www.researchinchina.com/Htmls/Report/2011/6281.html.

Chouicha, S., Boubekri, A., Mennouche, D., Berrbeuh, M.H. 2013. Solar drying of sliced potatoes, an experimental investigation. *Energy Procedia* 36: 1276–1285.

Clary, C. D., Wang, S. J., Petrucci, V. E. 2005. Fixed and incremental levels of microwave power application on drying grapes under vacuum. *Journal of Food Science* 70(5): 344–349.

Corrêa, J. L. G., Ernesto, D. B., Alves, J. G. L. F., Andrade, R. S. 2014. Optimisation of vacuum pulse osmotic dehydration of blanched pumpkin. *International Journal of Food Science and Technology* 49: 2008–2014.

Curcio, S., Aversa, M. 2014. Influence of shrinkage on convective drying of fresh vegetables: A theoretical model. *Journal of Food Engineering* 123: 36–49.

Deng, Y., Liu, Y., Qian, B., Su, S., Wu, J., Song, X., Yang, H. 2011. Impact of far-infrared radiation-assisted heat pump drying on chemical compositions and physical properties of squid (*Illex illecebrosus*) fillets. *European Food Research and Technology* 232: 761–768.

Doymaz, İ. 2012. Infrared drying of sweet potato (*Ipomoea batatas* L.) slices. *Journal of Food Science and Technology* 49(6): 760–766.

Dueik, V., Marzullo, C., Bouchon, P. 2013. Effect of vacuum inclusion on the quality and the sensory attributes of carrot snacks. *LWT-Food Science and Technology* 50: 361–365.

Dueik, V., Robert, P., Bouchon, P. 2010. Vacuum frying reduces oil uptake and improves the quality parameters of carrot crisps. *Food Chemistry* 119: 1143–1149.

FAO data. 2012. Value of Agricultural Production http://www.fao.org/faostat/en/#data/QV.

Feng, H., Tang, J. 1998. Microwave finish drying of diced apples in spouted bed. *Journal of Food Science* 63(4): 679–683.

George, P. J., Datta, A. K. 2002. Development and validation of heat and mass transfer models for freeze-drying of vegetable slices. *Journal of Food Engineering* 52: 89–93.

Gümüsay, Ö. A., Borazan, A. A., Ercal, N., Demirkol, O. 2015. Drying effects on the antioxidant properties of tomatoes and ginger. *Food Chemistry* 173: 156–162.

Guo, J. J., Kang, H. Q., Feng, X., Lian, X. J., Li, L. 2014. The interaction of sweet potato amylose/amylopectin and KCl during drying. *Food Hydrocolloids* 41: 325–331.

Gupta, M. K., Sehgal, V. K., Arora, S. 2013. Optimization of drying process parameters for cauliflower drying. *Journal of Food Science and Technology* 50(1): 62–69.

Hashim, N., Daniel, O., Rahamana, E. 2014. A preliminary study: Kinetic model of drying process of pumpkins (*Cucurbita moschata*) in a convective hot air dryer. *Agriculture and Agricultural Science Procedia* 2: 345–352.

Hassanien, M. F. R., Sharoba, A. M. 2014. Rheological characteristics of vegetable oils as affected by deep frying of French fries. *Food Measurement* 8: 171–179.

Hirun, S., Choi, J., Ayarungsaritkul, J., Pawsaut, C., Sutthiwanjampa, C., Vuong, Q. V., Chockchaisawasdee, S., Heo, Y., Scarlett, C. J. 2015. Optimization of far-infrared vacuum drying conditions for Miang leaves (*Camellia sinensis* var. *assamica*) using response surface methodology. *Food Science and Biotechnology* 24(2): 461–469.

Hou, J., Jiang, C., Long, Z. 2013. Nitrite level of pickled vegetables in Northeast China. *Food Control* 29: 7–10.

Hu, Q. G., Zhang, M., Mujumdar, A. S., Xiao, G. N., Sun, J. C. 2007. Performance evaluation of vacuum microwave drying of edamame in deep-bed drying. *Drying Technology* 25(4): 731–736.

Huang, L., Zhang, M. 2012. Trends in development of dried vegetable products as snacks. *Drying Technology* 30: 448–461.

Huang, R., Wu, Y., Zhao, Y., Lu, J., Jiang, J., Chen, Z. 2013. Factors affecting the temperature increasing rate in wood during radio-frequency heating. *Drying Technology* 31(2): 246–252.

Huang, Y. W., Chen, M. Q., Jia, L. 2016. Assessment on thermal behavior of municipal sewage sludge thin-layer during hot air forced convective drying. *Applied Thermal Engineering* 96: 209–216.

Jayashree, E., Visvanathan, R., John Zachariah, T. 2014. Quality of dry ginger (*Zingiber officinale*) by different drying methods. *Food Science and Technology* 51(11): 3190–3198.

Jiang, H., Zhang, M., Liu, Y., Mujumdar, A. S., Liu, H. H. 2013. The energy consumption and color analysis of freeze/microwave freeze banana chips. *Food and Bioproducts Processing* 91(4): 464–472.

Jin, X., van der Sman, R. G. M., van Straten, G., Boom, R. M., van Boxtel, A. J. B. 2014. Energy efficient drying strategies to retain nutritional components in broccoli (*Brassica oleracea* var. *italica*). *Journal of Food Engineering* 123: 172–178.

Kakade, R. H., Das, H., Ali, S. 2011. Performance evaluation of a double drum dryer for potato flake production. *Journal of Food Science and Technology* 48(4): 432–439.

Kim, S. A., Grimm, K. A., Harris, D. M., Scanlon, K. S., Demissie, Z. 2011. Fruit and vegetable consumption among high school students. *Morbidity and Mortality Weekly Report* 60(46): 1583–1586.

Kocabiyik, H., Yilmaz, N., Tuncel, N. B., Sumer, S. K., Buyukcan, M. B. 2015. Drying, energy, and some physical and nutritional quality properties of tomatoes dried with short-infrared radiation. *Food and Bioprocess Technology* 8: 516–525.

Koné, K. Y., Druon, C., Gnimpieba, E. Z., Delmotte, M., Duquenoy, A., Laguerre, J. 2013. Power density control in microwave assisted air drying to improve quality of food. *Journal of Food Engineering* 119: 750–757.

Lekcharoenkul, P., Tanongkankit, Y., Chiewchan, N., Devahastin, S. 2014. Enhancement of sulforaphane content in cabbage outer leaves using hybrid drying technique and stepwise change of drying temperature. *Journal of Food Engineering* 122: 56–61.

Lin, T. M., Durance, T. D., Scaman, C. H. 1998. Characterization of vacuum microwave, air and freeze dried carrot slices. *Food Research International* 31(2): 111–117.

Lioumbas, J. S., Karapantsios, T. D. 2014. Effect of increased gravitational acceleration in potato deep-fat frying. *Food Research International* 55: 110–118.

Liu, M. C., Li, Z. G., Deng, W., Wang, G. M., Yang, Y. W. 2009. Changes in volatile compounds of pickled mustard tuber (*Brassica juncea* var. *tsatsai*) during the pickling process. *International Journal of Food Science and Technology* 44: 2278–2286.

Mitra, J., Shrivastava, S. L., Srinivasa Rao, P. 2011. Vacuum dehydration kinetics of onion slices. *Food and Bioproducts Processing* 89(1): 1–9.

Mota, C. L., Luciano, C., Dias, A., Barroca, M. J., Guinéa, R. P. F. 2010. Convective drying of onion: Kinetics and nutritional evaluation. *Food and Bioproducts Processing* 88: 115–123.

Mujumdar, A. S. 2007. *Handbook of Industrial Drying*. Taylor & Francis, Philadelphia, PA.

Olsen, A., Christian, R., Kramer, L., Møller, P. 2012. Serving styles of raw snack vegetables. What do children want? *Appetite* 59: 556–562.

Pelegrina, A., Elustondo, M., Urbicain, M. 2002. Setting the operating conditions of a vegetables rotary drier by the response surface method. *Journal of Food Engineering* 54: 59–62.

Peressini, D., Foschia, M., Tubaro, F., Sensidoni, A. 2015. Impact of soluble dietary fibre on the characteristics of extruded snacks. *Food Hydrocolloids* 43: 73–81.

Piernas, C., Popkin, B. M. 2010. Trends in snacking among U.S. children. *Health Affairs (Project Hope)* 29(3): 398–404.

Pomerleau, J., Lock, K., McKee, M. 2006. The burden of cardiovascular disease and cancer attributable to low fruit and vegetable intake in the European Union: Differences between old and new member states. *Public Health Nutrition* 9(5): 575–583.

Qiao, F., Huang, L. L., Xia, W. S. 2012. A study on microwave vacuum dried re-structured lychee (*Litchi chinensis* Sonn.) mixed with purple sweet potato (*Ipomoea batatas*) snacks. *Food and Bioproducts Processing* 90(4): 653–658.

Research in China, China Snack Food Industry Report, 2010–2013. 2014. www.researchinchina.com/Htmls/Report/2011/6281.html.

Roberts, P. B. 2014. Food irradiation is safe: Half a century of studies. *Radiation Physics and Chemistry* 105: 78–82.

Sadin, R., Chegini, G., Sadin, H. 2014. The effect of temperature and slice thickness on drying kinetics tomato in the infrared dryer. *Heat and Mass Transfer* 50: 501–507.

Sallam, Y. I., Aly, M. H., Nassar, A. F., Mohamed, E. A. 2015. Solar drying of whole mint plant under natural and forced convection. *Journal of Advanced Research* 6: 171–178.

Sangwan, A., Kawatra, A., Sehgal, S. 2014. Nutritional composition of ginger powder prepared using various drying methods. *Journal of Food Science and Technology* 51(9): 2260–2262.

Singh, N. J., Pandey, R. K. 2012. Convective air drying characteristics of sweet potato cube (*Ipomoea batatas* L.). *Food and Bioproducts Processing* 90: 317–322.

Snack Food Association annual report. 2000, http://www.supershareware.com/snack-food-association-2010-free/

Tanongkankit, Y., Chiewchan, N., Devahastin, S. 2015. Evolution of antioxidants in dietary fiber powder produced from white cabbage outer leaves: Effects of blanching and drying methods. *Journal of Food Science and Technology* 52(4): 2280–2287.

Therdthai, N., Visalrakkij, T. 2012. Effect of osmotic dehydration on dielectric properties, microwave vacuum drying kinetics and quality of mangosteen. *International Journal of Food Science and Technology* 47: 2606–2612.

Troncoso, E., Pedreschi, F., Zúniga, R. N. 2009. Comparative study of physical and sensory properties of pre-treated potato slices during vacuum and atmospheric frying. *LWT-Food Science and Technology* 42: 187–195.

Uprit, S., Mishra, H. N. 2003. Microwave convective drying and storage of soy-fortified paneer. *Food and Bioproducts Processing* 81(2): 89–96.

Wang, Y., Li, Y., Wang, S., Zhang, L., Gao, M., Tang, J. 2011. Review of dielectric drying of foods and agricultural products. *International Journal of Agricultural and Biological Engineering* 4(1): 1–19.

Wang, H., Zhang, M., Adhikari, B. 2015. Drying of shiitake mushroom by combining freeze-drying and mid-infrared radiation. *Food and Bioproducts Processing* 94: 507–517.

Wang, Y., Zhang, L., Gao, M., Tang, J., Wang, S. 2014a. Pilot-scale radio frequency drying of Macadamia nuts:Heating and drying uniformity. *Drying Technology* 32(9): 1052–1059.

Wang, Y., Zhang, L., Gao, M., Tang, J., Powers, J. R., Johnson, J. A., Wang, S. 2014b. Developing hot air assisted radio frequency drying process of in-shell Macadamia nuts. *Food and Bioprocess Technology* 7(1): 278–288.

Wang, R., Zhang, M., Mujumdar, A. S. 2010a. Effects of vacuum and microwave freeze drying on microstructure and quality of potato slices. *Journal of Food Engineering* 101(2): 131–139.

Wang, R., Zhang, M., Mujumdar, A. S. 2010b. Effect of food ingredient on microwave freeze drying of instant vegetable soup. *LWT-Food Science and Technology* 43(7): 1144–1150.

Wang, Y. C, Zhang, M., Mujumdar, A. S., Mothibe, K. J. 2013. Microwave-assisted pulse-spouted bed freeze drying of stem lettuce slices—effect on product quality. *Food and Bioprocess Technologies* 6: 3530–3543.

Wang, Y. C., Zhang, M., Mujumdar, A. S., Mothibe, K. J., Roknul Azam, S. M. 2012. Effect of blanching on microwave freeze drying of stem lettuce cubes in a circular conduit drying chamber. *Journal of Food Engineering* 113: 177–185.

Weng, P. F., Wu, Z. F., Lei, L. L. 2013. Predictive models for growth of *Leuconostoc citreum* and its dynamics in pickled vegetables with low salinity. *Journal of Food Process Engineering* 36: 284–291.

Yan, W. Q., Zhang, M., Huang, L., Tang, J., Mujumdar, A. S., Sun, J. 2010a. Study of the optimization of puffing characteristics of potato cubes by spouted bed drying enhanced with microwave. *Journal of the Science of Food and Agriculture* 90(8): 1300–1307.

Yan, W. Q., Zhang, M., Huang, L., Tang, J., Mujumdar, A. S., Sun, J. 2010b. Studies on different combined microwave-drying of carrot pieces. *International Journal of Food Science and Technology* 45: 2141–2148.

Yang, J. H., Park, H. Y., Kim, Y. S., Choi, I. W., Kim, S. S., Choi, H. D. 2012. Quality characteristics of vacuum-fried snacks prepared from various sweet potato cultivars. *Food Science and Biotechnology* 21(2): 525–530.

Zhang, Y., Chen, H., Chen, T. 2014. Drying kinetics of RDX under atmospheric pressure and vacuum conditions. *Energy Conversion and Management* 80: 266–275.

Zhang, M., Tang, J., Mujumdar, A. S., Wang, S. 2006. Trends in microwave-related drying of fruits and vegetables. *Trends in Food Science and Technology* 17(10): 524–534.

Zhao, D., Zhao, C., Tao, H., An, K., Ding, S., Wang, Z. 2013. The effect of osmosis pretreatment on hot-air drying and microwave drying characteristics of chili (*Capsicum annuum* L.) flesh. *International Journal of Food Science and Technology* 48: 1589–1595.

Zielińska, M., Zapotoczny, P., Alves-Filho, O., Eikevik, T. M., Błaszczak, W. 2013. A multistage combined heat pump and microwave vacuum drying of green peas. *Journal of Food Engineering* 115: 347–356.

10 Drying of Edible Flowers

Serkan Boyar, Erkan Dikmen, and Sabri Erbaş
Süleyman Demirel University

CONTENTS

10.1 INTRODUCTION

The part of a plant that can be used for the purposes of nutrition, health, and beauty are the flowers. Flowers are used to enhance the taste, color, aroma, and appearance of food (Shylaja and Peter 2004; Şahin and Kılıç 2014; Toker et al. 2015).

Studies have shown that there are approximately 450,000 kinds of flowering plants in the world (Pimm and Joppa 2015). Another source states that out of 422,000 (Govaerts 2001) species of flowering plants, an estimated 17.1% (72,000) (Schippmann et al. 2006) are medicinal and aromatic plants.

Various parts of plants have medicinal or aromatic value. These are, in order of importance, the root, the whole plant, the bark, and the fruit. Flowers have only a 5% share (Figure 10.1).

Edible flowers are either gathered seasonally from the wild, or produced in fields or gardens (Martinov and Konstantinovic 2007; Pirnă et al. 2011; Javani et al. 2015). They are grown at home in gardens or pots, as a hobby, while those with commercial value may be collected from the wild or grown in fields as agricultural crops (Kathe et al. 2003; Baydar et al. 2013; Kırıcı 2015). By examining the sensitivity of plants to overcollection of their usable parts (roots, leaves, flowers, fruit, seeds, etc.) from the wild, it is seen that annual and biennial plants are affected to a medium extent, whereas perennial plants, shrubs, and trees are affected to a lesser extent (Schippmann et al. 2002). This clearly shows the necessity of bringing plants with edible flowers into cultivation.

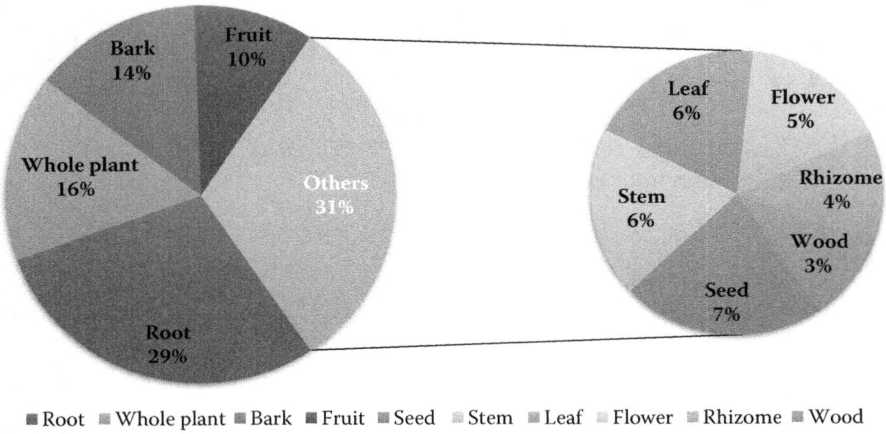

FIGURE 10.1 Distribution of medicinal and aromatic plants according to their useful parts. (From Jakhar, M.L. et al., *Medicinal Plants Utilisation and Conservation*, 2nd Revised and Enlarged Edition, Aavishkar Publishers, Jaipur, India, 2009.)

Although investment is needed to enlarge the product area and to ensure controlled growing conditions, it is seen that the market value and added value of the product is increasing, a standard is being achieved in raw materials and the processed product, new products are being introduced from the local to the global market, and product quality and value are increasing (Schippmann et al. 2002; Kathe et al. 2003; Manzoor Rashid 2009; Faydaoğlu and Sürücüoğlu 2011). The flowers of edible plants, many of which have medicinal and aromatic qualities, are the most delicate parts of the plant, and great care must be taken in processing and marketing them. The characteristics of the producer organization have a great effect on the characteristics of edible flowers such as material quality, purity, or traces of agricultural chemicals (Figure 10.2).

Drying is the basic method for preserving and extending the shelf life of plants with edible flowers after harvesting. Traditional drying methods are used by local collectors or producers (Shahidullah 2007; Javani et al. 2015), but modern methods are preferred for high-value products by industrial producers relying more on agricultural production. It can be said that methods used to dry the flowers vary according to the region where the flowers are produced, the climatic characteristics of that region, the way they are used, the characteristics of the flowers, the market value of the product, and the demands and preferences of the consumers.

10.2 PHYSICAL AND CHEMICAL PROPERTIES OF EDIBLE FLOWERS

Edible flowers are mostly from plants with medicinal or aromatic uses and are used as a raw material either in fresh or dried form in various sectors (Shahidullah 2007; Gül 2014; Mahboubi 2015).

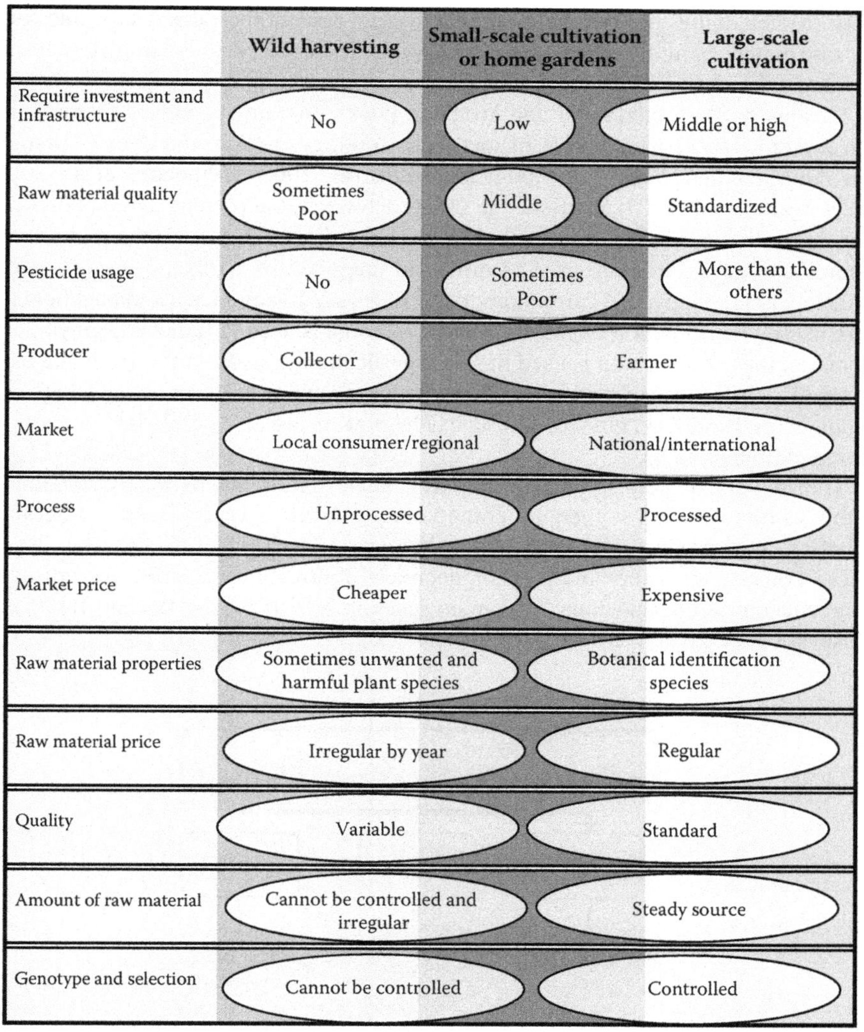

FIGURE 10.2 Methods of producing, processing, and marketing medicinal and aromatic plants and assessment.

Edible flowers are often dried and used as herbal teas, herbal medicines, food additives, spices, etc., or they reach the customer in various processed forms (Cai et al. 2004; Bayhan et al. 2011; Faydaoğlu and Sürücüoğlu 2011; Zeng et al. 2014; Mahboubi 2015; Toker et al. 2015). On the other hand, the consumption of inedible or unidentified flowers, especially those gathered from the wild, presents a considerable risk to human life and health. For this reason, it is recommended that edible flowers should be obtained from reliable sources (Surveswaran et al. 2007; Toker et al. 2015). It is seen that edible flowers, which have an important place in human nutrition when fresh (Mlcek and Rop 2011), have become widely used in their dried state and form the raw materials in the manufacture of new ready-to-use products in different sectors and using different processing techniques (Figure 10.3).

Various parts of medicinal and aromatic plants contain secondary compounds, and therefore serve as a source of medicine, food, cosmetics, and dyes (Al-Ismail and Aburjai 2004; Cai et al. 2004; Ivanova et al. 2005; Hinneburg et al. 2006; Surveswaran et al. 2007; Youwei et al. 2008). However, the preferred plant parts can vary according to the field in which they are used. For example, all the parts of the plant above the ground are used in mint and oregano; the fruits are used in anise, fennel, pepper, poppy, and cardamom; the root is used in echinacea and angelica; the bark is used as cinnamon or chinchona; the rhizome is used in liquorice, ginger, and turmeric; the tuber or bulb is used in salep, garlic, and snowdrop; the seeds are used in nigella, coffee, cacao, and cumin; and the leaves are used in rosemary, aloe vera, lemon balm, basil, bay, gingko, and sage. The flowers are also edible for some of the aforementioned plants.

The flowers of plants, like other parts, also contain flavonoids, carotenoids, anthocyanins, and other phenolic compounds, as well as waxes, resins, fragrance volatiles, and essential oil components (Youwei et al. 2008; Kaisoon et al. 2012). These compounds either increase or decrease antioxidant capacity according to their concentration, and thus have many uses in health protection and treatment (Tusevski et al. 2014).

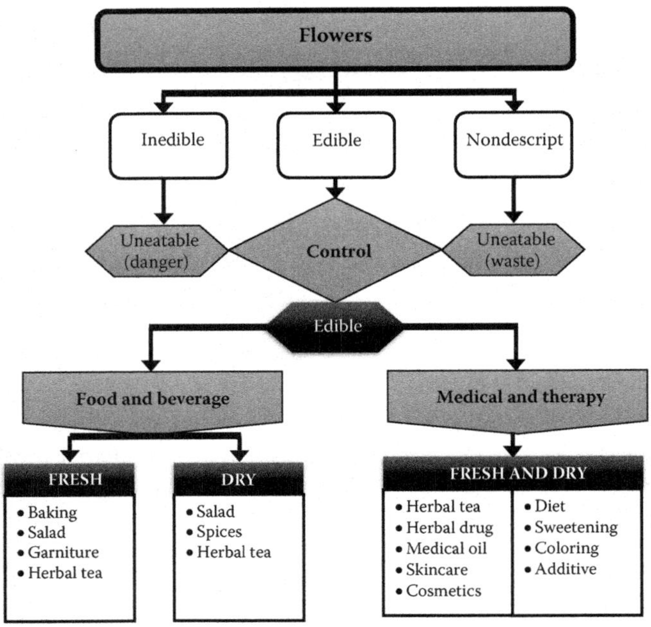

FIGURE 10.3 The uses of edible flowers.

The details of the local and traditional knowledge of preparation and application of dried edible flowers according to their use is described in many sources (Küçük et al. 2000; Asımgil 2009; Yalçın 2010; Tezer et al. 2010; Özer 2013; Pamuk 2013; Baydar 2016). Examining these sources, it is seen that along with rose, chamomile, lavender, and rosemary flowers, the flowers of gentian, primula, poppy, marigold, and echinacea are also widely consumed (Figures 10.4 and 10.5). These flowers are principally used as herbal teas and in various combinations in salads, garnish, jams, and syrups or to provide color, flavor, or aroma to food. The importance and interest accorded to medicinal and aromatic plants in the world, and their consumption and use in the light of local knowledge are steadily increasing as a result of an increasing number of scientific studies and information spread through the mass media. The creation of standards in the use of edible flowers, particularly in medical treatment, and recommendation and control by qualified experts are gaining importance.

The physical and chemical properties of edible flowers play an important role in preserving their quality after they have been dried. Thus, physical properties such as the flower's stage of development, the number of petals, color, density, the size of the hypanthium and ovary (Figures 10.5, 10.6, and 10.13), the presence and characteristics of volatile oils in the flower's tissues (Figure 10.7), and chemical properties such as the type and structure of secondary compounds are important parameters that must be taken into consideration. A good analysis of these properties will help in a successful choice of the best drying methods.

A good example of a flower with these properties is the Damask rose (*Rosa damascena* Mill.), known in Turkey as the Isparta Rose, the region where it is grown. It is pink in color, semi-double, has volatile compounds and a strong odor (Erbaş et al. 2015) (Figure 10.4). Apart from being a basic raw material in the cosmetic industry across the world, it is also widely favored in different types of

FIGURE 10.4 Isparta Rose (*Rosa damascena* Mill.). (Illustration by Korkmaz, Ö., Isparta Gülü, Isparta Rose (*Rosa damascena* Mill.), 2015a).

German Chamomilla (*Matricaria chamomilla* L.)
(Illustration by Ö. Korkmaz, 2016b)

Lavandin (*Lavandula x intermedia* Emeric ex Loisel.)
(Illustration by Ö. Korkmaz, 2016c)

Marigold (*Tagates patula* L.)
(Illustration by Ö. Korkmaz, 2014a)

St. John's Wort (*Hypericum perforatum* L.)
(Illustration by H. Korkmaz, 2015d)

Cowslip (*Primula officinalis* L.)
(Illustration by Ö. Korkmaz, 2014b)

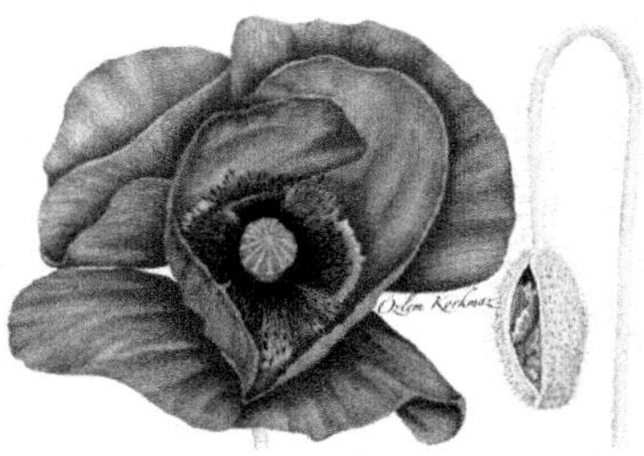

Common Poppy (*Papaver rhoeas* L.)
(Illustration by Ö. Korkmaz, 2014c)

Rosemary (*Rosmarinus officinalis*)
(Illustration by Ö. Korkmaz, 2015b)

Coneflower (*Echinacea purpurea* L.)
(Illustration by Ö. Korkmaz, 2015c)

FIGURE 10.5 Various widely consumed edible flowers.

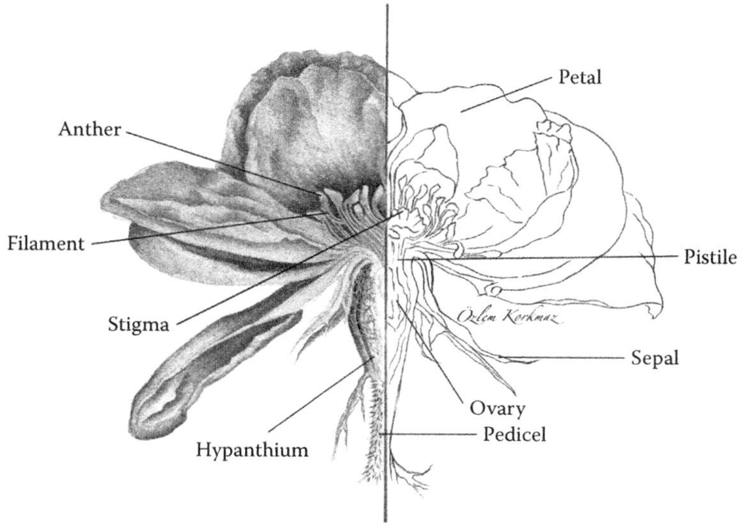

FIGURE 10.6 General view and structure of the Damask (Isparta) Rose. (Illustration by Korkmaz, Ö., General view and structure of the Damask (Isparta) Rose, 2016a.)

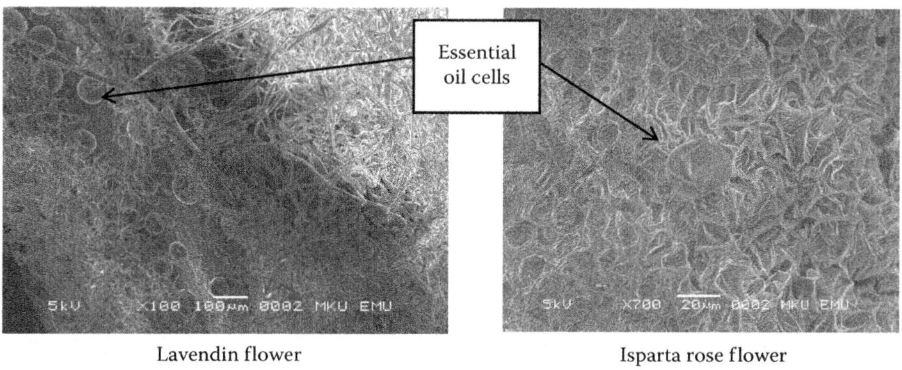

FIGURE 10.7 SEM images of dried lavandin and Isparta rose flower tissue, showing essential oil cells. (From Pehlivan, M., Doğaner, A., Isparta yağ gülü (*Rosa damascena* Mill.) ayırma-sınıflandırma ve kurutma sistemlerinin geliştirilmesi Project Report. TÜBİTAK-TEYDEB-7100075 (in Turkish), 2012; Photos by Kemal Sangun, 2010).

food. In the world market, Turkey and Bulgaria dominate the production of rose oil, rose concrete, and rose absolute, and in these countries, the production of rose syrup and rose jam from fresh roses is also widespread. In addition, Iran and India are major producers of rose water, as is Morocco in producing dried roses (Baydar and Kazaz 2013).

The amount of volatile oil that the edible flowering plants have and its location in their tissues plays an important part in whether or not they are lost in the course of drying. For example, while the volatile oils in the lavender plant are contained in the secretory hairs (Harborne and Williams 2002), in the Damask rose they are found particularly in the epidermal cells of the petal (Baydar et al. 2013). This oil, known as Turkish rose oil, is made up of important aromatic molecules such as the monoterpenic alcohols linalool, citronellol, nerol and geraniol; long-chain hydrocarbons such as nonadecane, nonadecene, eicosane, heneicosane and tricosane; oxides and ethers such as methyl eugenol; esters and aldehydes such as geranyl acetate and geranial; and phenols such as eugenol. In general, rose flowers contain 0.035%–0.055% of volatile oils. The composition of this oil is 6%–8% geranyl acetate, 8%–13% nerol, 20%–25% citronellol, 30%–35% geraniol, 8%–12% compounds of hydrocarbons, and 0.5%–2.5% eugenol and methyl eugenol (Anaç 1984; Kürkçüoglu 1988; Başer 1992; Bayrak and Akgül 1994). Citronellol, nerol, and geraniol determine the quality of the rose and are used worldwide in the perfume industry as fixators. For this reason, the hydrocarbon compounds affect the quality negatively. Moreover, methyl eugenol is not included in rose oil because of alleged allergic and mutagenic reactions (Harris 2002; Rusanov et al. 2012). In addition, trace amounts of compounds such as β-damascenone, β-damascone, and β-ionone that impart the characteristic scent to rose oil are also present (David et al. 2006). The Isparta rose also contains large amounts of phenolic and flavonoid compounds, which increase its potential for use as a natural antioxidant (Table 10.1). Lavender flowers also contain a large amount of volatile oil, which consists of linalool, linalyl acetate, 1,8-cineole, β-caryophyllene, and camphor. However, when flowers such as lavender, in which the level of volatile oils is high, are dried under natural conditions in the shade, the oil cells remain intact, but in roses, where the proportion of oil is low, the oil is easily separated from the tissues during drying because of the location of the oil cells (Figures 10.7 and 10.19). Apart from these two flowers, chamomile, rosemary, gentian, primula, common poppy, marigold, and echinacea flowers contain secondary compounds that make them highly effective as antioxidants (Table 10.1).

On examining the chemical contents of edible flowers, it is seen that they contain significant amounts of essential oils, phenolics, and flavonoids (Table 10.1).

There are many edible flowers in the world (Lim 2013) that are widely collected or produced and used for different purposes; these are grouped into three categories as fruit flowers (Table 10.2), vegetable flowers (Table 10.3), and medicinal aromatic flowers (Table 10.4). It is seen that most of the edible flowers fall under the third category (Buchbauer et al. 1992; Kathe et al. 2003; Al-Ismail and Aburjai 2004; Ivanova et al. 2005; Heindl and Müller 1997 from Müller and Heindl 2006; Minaei et al. 2014).

10.3 PROCESSING OF EDIBLE FLOWERS

Edible flowers are harvested when the plant's flowers are in full bloom, early in the morning (Kazaz 1997; Martinov and Konstantinovic 2007; Baydar 2016) when the active compounds are at their highest levels (Pirnă et al. 2011). Relative to the other parts of the plant, the flowers can be easily gathered without much effort

TABLE 10.1
Secondary Compounds in Some Widely Consumed Edible Flowers

Species	Essential Oil Content	Essential Oil Composition	Total Phenolic Content	Polyphenols	Total Flavonoid Content	Flavonoids
Rosa damascena	0.035%–0.055% (Baydar et al. 2013)	Citronellol (20%–25%), geraniol (30%–35%), nerol (8%–13%), hydrocarbons compounds (8%–12%) (Baydar et al. 2013)	344.5 mg gallic/g extract (MeOH) (Göktürk Baydar and Baydar 2013)	Gallic acid, syringic acid, quercetin (Göktürk Baydar and Baydar 2013)	29.8 mg quercetin/100 g extract (MeOH) (Göktürk Baydar and Baydar 2013)	Cyanidin, quercetin, kaempferol (Velioglu and Mazza 1991)
Matricaria chamomilla	0.3%–1.5% (Schilcher et al. 2005)	(E)-β-farnesene (4.9%–8.1%), α-bisabolol oxide B (12.2%–30.9%), α-bisabolol (4.8%–11.3%), chamazulene (2.3%–10.9%), α-bisabolol oxide A (25.5%–28.7%), (Schilcher et al. 2005)	36.7% (w/w, MeOH) (Singh et al. 2011)	Chlorogenic acid, caffeic acid, herniarin, umbelliferone (Schilcher et al. 2005)	50.7% (w/w, MeOH) (Singh et al. 2011)	Luteolin, Luteolin-7-glucoside, Apigenin, Apigenin-7-glucoside, Quercetin, Rutin, Naringenin (Singh et al. 2011)
Lavandula x intermedia	0.5%–3.0% (Harborne and Williams 2002)	Linalool (19%–26%), Linalyl acetate (20%–23%), 1,8-cineole (8%–12%), β-caryophyllene (2.7%–6.0%), camphor (10%–14%) (Harborne and Williams 2002)	8.5 mg gallic/kg extract (MeOH) (Harborne and Williams 2002)	Ferulic acid, rosmarinic acid, chlorogenic acid, caffeic acid, p-coumaric (Ferreres et al. 1986; Upson et al. 2000)	0.26 mg gallic/100 g extract (MeOH) (Harborne and Williams 2002)	Apigenin, Genkwanin, Luteolin, Apigenin 7-glikozit, Luteolin 7-glikozit, 7-rutinosit, Flavon *C*-glikozitler, Viteksin, Visenin-2 (Ferreres et al. 1986; Upson et al. 2000)

(Continued)

TABLE 10.1 (*Continued*)
Secondary Compounds in Some Widely Consumed Edible Flowers

Species	Essential Oil Content	Essential Oil Composition	Total Phenolic Content	Polyphenols	Total Flavonoid Content	Flavonoids
Tagatespatula	0.1%–0.4% (Sagar et al. 2005)	Limonene (13.6%), (Z)-β-ocimene (8.3%), terpiolene (11.2%), p-cymen-8-ol (11.0%), piperitone (10.1%) (Sagar et al. 2005)	4.6 g gallic/kg fresh matter (MeOH) (Lim 2013)	Patuletin, patulitrin (patuletin 7-*O*-glucoside), methyl protocatechuate (Lim 2013)	1.9 g rutin/kg fresh matter (MeOH) (Lim 2013)	Quercetagetin, quercetagetin 7-*O*-glucoside, luteolin (Lim 2013)
Hypericum perforatum	0.08% (Đorđević 2015)	β-caryophyllene (4.1%–5.9%), γ-muurolene (5.0%–9.6%), β-selinene (5.1%–19.6%), α-selinene (4.1%–10.4%), caryophyllene oxide (6.0%–12.2%) (Çırak et al. 2010)	319.3 mg gallic/g (MeOH) (Öztürk et al. 2009)	Chlorogenic, caffeic acids, protocatechuic acid (Öztürk et al. 2009)	10.3 mg gallic/g (MeOH) (Öztürk et al. 2009)	Rutin, hyperoside, isoquercitrin, miquelianin, astilbin, guaijaverin, quercitrin, quercetin, isoorientin, cyanidin (Ernst 2003)
Primula vulgaris	Trace (Colombo et al. 2014)	β-caryophyllene (28.1%), germacrene D (22.3%), β-elemene (14.3%), α-cubebene (10.6%) (Colombo et al. 2014)	7.6 mg gallic/g extract (MeOH) (Orhan et al. 2012)	Rosmarinic acid, kaempferol, quercetin (Colombo et al. 2014)	-	Malvin, malvidin 3-glucoside, delphinidin, primuletin, chrysin, primetin, hydroxyflavone (Colombo et al. 2014)

(Continued)

TABLE 10.1 (*Continued*)
Secondary Compounds in Some Widely Consumed Edible Flowers

Species	Essential Oil Content	Essential Oil Composition	Total Phenolic Content	Polyphenols	Total Flavonoid Content	Flavonoids
Papaver rhoaes	0.2% (Doğan and Bağcı 2014)	2-fitol (52.8%), trikosan (7.8%), pentadekanon (6%), heneikosan (5.3%) (Doğan and Bağcı 2014)	4.9 mg quercetin/g (Zeng et al. 2014)	Cyanidin-3-O-Glikoside (Zeng et al. 2014)	47.7 mg quercetin/g (MeOH) (Zeng et al. 2014)	Chrysin, apigenin, luteolin-glycosyde, kaempferol, quercetin (Hillenbrand et al. 2004)
Rosmarinus officinalis	0.5%–2.5% (Cosmetic Ingredient Review 2013)	1,8-cineole (15%–30%), camphor (5%–25%), borneol (10%–20%), isorosmanol (17%-detected in only flower) (Cosmetic Ingredient Review 2013)	214.2 mg gallic/kg extract (MeOH) (Cosmetic Ingredient Review 2013)	Rosmarinic acid, carnosic acid, carnosol (Cosmetic Ingredient Review 2013)	3367.3 mg gallic/100 g extract (MeOH) (Cosmetic Ingredient Review 2013)	Luteolin, genkwanin, diosmetin, diosmin, 6-hispidulin, apigenin (Cosmetic Ingredient Review 2013)
Echinacea purpurea	1.5% (Diraz et al. 2012)	Germacrene D (11.3%), caryophyllene oxide (8.7%), β-caryophyllene (7.2%), α-cadinol (6.2%) (Lim 2013)	7.9%–10.9% (Lim 2013)	Chlorogenic acid, caftaric acid, cynarin, cichoric acid (Lim 2013)	0.80% (Lim 2013)	Rutin, nicotiflorin (3-*O*-rutinoside campherol) (Seemannová et al. 2006)

TABLE 10.2
Various Edible Fruit Flowers

Fruit Flowers					
Species					
Scientific Name	Common Name	Collected/ Cultivated	Consumption (Fresh/Dry)	Uses	References
Citrus lemon	Lemon				
Citrus sinensis	Orange	Cultivated	Fresh	Fruit salads, soup, garnish	Creasey (1999)
Malus communis	Apple	Cultivated	Fresh	Syrup, sauce, tarts, fruit soups	Creasey (1999)
Pistacia terebinthus	Terebinth tree	Cultivated	Fresh	Raw	Deane (2007–2012)
Prunus persica	Peach	Cultivated	Fresh	Sweetener	Creasey (1999)
Punica granatum	Pomegranate	Cultivated	Fresh	Fruit salads, soup, garnish	Creasey (1999)
Viburnum edula Raf.	Moosebery	Collected	Dry	Garnish	Schofield (2003)

TABLE 10.3
Various Edible Vegetable Flowers

Vegetable Flowers					
Species					
Scientific Name	Common Name	Collected/ Cultivated	Consumption (Fresh/Dry)	Uses	References
Allium cepa L.	Onion	Cultivated	Fresh/dry	Raw, salad	Schofield (2003)
Allium sativum L.	Garlic	Cultivated	Fresh	Raw, salad	Woodward (2000)
Anethum gravelens L.	Dill	Cultivated	Dry	Spice, salad, garnish	Newman and O'Connor (2009)
Borago officinalis L.	Borage	Collected	Dry	Salad, garnish, dessert, coloring, and flavoring in drinks	
Cucurbita spp.	Pumpkin, squash	Cultivated	Fresh/dry	Salad, garnish	
Luffa aegyptiaca L.	Loafah	Cultivated	Fresh/dry	Salad, garnish	Lim (2012)
Petroselinum cripum L.	Parsley	Cultivated	Fresh/dry	Salad, eat	
Raphanus sativus L.	Radish	Collected	Fresh	Cooked vegetable	Brown (2011)

TABLE 10.4
Various Edible Medicinal and Aromatic Flowers

Medicinal and Aromatic Flowers					
Species					
Scientific Name	Common Name	Collected/Cultivated	Consumption (Fresh/Dry)	Uses	References
Achilla millefolium L.	Yarrow	Collected	Dry	Herbal tea, add to alcoholic drinks	Schofield (2003)
Agastache foeniculum L.	Anise hyssop	Cultivated	Fresh/dry	Drinks, custard, ice cream, sorbets	Tenenbaum (1999)
Althea rosea L.	Hollyhock	Collected	Fresh/dry	Herbal tea, raw	Tenenbaum (1999)
Anchusa officinalis L.	Alkanet	Collected	Dry	Raw, salad, garnish	Facciola (1990)
Anthemis nobilis L.	Roman chamomile	Collected	Dry	Flavoring and herbal tea	Newman and O'Connor (2009)
Artemisia vulgaris L.	Mugwort	Collected	Dry	Added beer or tea	Facciola (1990)
Calendula officinalis L.	Calendula	Collected/cultivated	Fresh/dry	Salad, soup, butter, sauce, drink	Creasey (1999)
Carthamus tinctorius L.	Safflower	Cultivated	Fresh/dry	Coloring of food, herbal tea, salad, garnish	Newman and O'Connor (2009)
Chamomilla recuita L.	German chamomile	Collected	Dry	Herbal tea	Garland (1993)
Chrysanthemum morifolium Ramat.	Chrysanthemum	Cultivated	Dry	Salad, garniture, herbal tea, spice	Facciola (1990)
Coriandrum sativum L.	Coriander	Cultivated	Fresh/dry	Spice, herbal tea, salad	Tenenbaum (1999)
Crocus sativus L.	Saffron	Cultivated	Dry	Coloring of food	Creasey (1999)
Dianthus carryophylus L.	Carnation Dianthus	Collected/cultivated	Dry	Syrup, flavoring of food and drink, herbal tea	Rop et al. (2012)

(Continued)

TABLE 10.4 (*Continued*)
Various Edible Medicinal and Aromatic Flowers

Medicinal and Aromatic Flowers					
Species					
Scientific Name	Common Name	Collected/Cultivated	Consumption (Fresh/Dry)	Uses	References
Echinaceae purpurea L. Moench	Coneflower	Cultivated	Dry	Salad, herbal tea	Roberts (2000)
Foeniculum vulgare L.	Fennel	Collected/cultivated	Fresh/dry	Salad, garnish, spice	Creasey (1999)
Helichrysum arenarium L.	Everlasting Flower	Collected	Dry	Herbal tea, essential oil	Facciola (1990)
Humulus lupuus L.	Hops	Collected/cultivated	Dry	Beer, liquori herbal tea	Hu (2005)
Hyssopus officinalis L.	Hyssop	Collected/cultivated	Fresh/dry	Flavoring liqueur, herbal tea	Newman and O'Connor (2009)
Jasminum sambac L.	Jasmine	Cultivated	Dry	Sorbet, ice creams	Creasey (1999)
Lavandula officinalis	Lavender	Cultivated	Dry	Herbal tea, syrup, jellies, sorbets, caramel, custard, ice cream	Creasey (1999)
Lilium longiflorum Thumb.	Easter lily	Collected	Dry	Herbal tea	Zeng et al. (2014)
Lonicera japonica L.	Honeysuckle	Cultivated	Fresh/dry	Herbal tea, salad, garnish	Creasey (1999)
Matricaria chamomilla L.	Chamomilla	Collected	Dry	Raw, salad, vegetable, herbal tea	Schofield (2003)
Muscari neglectum	Hyacinth	Collected	Fresh/dry	Sweetener	Facciola (1990)
Narcissus jonquilla L.	Jonquil	Collected	Dry	Raw, sweetener, dessert	Facciola (1990)
Origanum onites L.	Oregano	Collected/cultivated	Fresh/dry	Spice, dressings, sauces, herbal tea	Kintzios (2002)
Osmanthus fragrans (Thunb.) Lour.	Tea olive	Collected	Dry	Herbal tea	Zeng et al. (2014)

(*Continued*)

TABLE 10.4 (*Continued*)
Various Edible Medicinal and Aromatic Flowers

Medicinal and Aromatic Flowers					
Species					
Scientific Name	Common Name	Collected/Cultivated	Consumption (Fresh/Dry)	Uses	References
Paeonia lactiflora Pall.	Garden peony	Collected	Dry	Herbal tea	Zeng et al. (2014)
Primula officinalis L.	Cowslip	Collected	Fresh	Salad	Tenenbaum (1999)
Rosa damascena Mill.	Oil-bearing rose	Cultivated	Fresh/dry	Herbal tea, jam, vinegar, sorbet	Baydar (2016)
Rosmarinus officinalis L.	Rosemary	Collected/cultivated	Fresh/dry	Garnish	Creasey (1999)
Sambucus canadensis L.	Elderberry	Cultivated	Dry	Syrup, marmalade, yogurt, desserts	McCullough (2007)
Satureja hortensis L.	Savory	Collected/cultivated	Fresh/dry	Flavoring, herbal tea, vinegars	Creasey (1999)
Solidago canadensis L.	Rock goldenrod	Collected	Fresh/dry	Herbal tea, bread, biscuits	Schofield (2003)
Tagates patula L.	Marigold	Cultivated	Fresh/dry	Herbal tea, coloring of foods	Creasey (1999)
Tanacetum vulgare L.	Fever few	Collected	Dry	Garnish, herbal tea	Facciola (1990)
Taraxacum officinale Webb.	Dandelion	Collected	Fresh/dry	Pancakes, omelette, fritter, salad	Newman and O'Connor (2009)
Thyme vulgaris L.	Thyme	Collected/cultivated	Fresh/dry	Garnish, spice, herbal tea	Stahl-Biskup and Saez (2002)
Tilia platyphyllos Scop.	Common lime	Collected	Dry	Herbal tea	Ivanova et al. (2005)

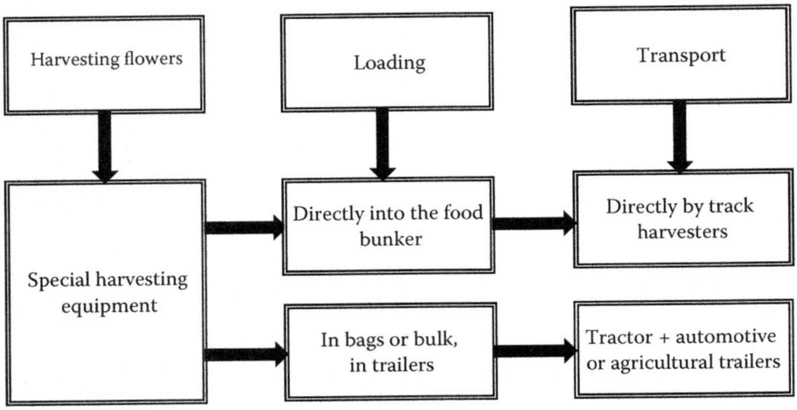

FIGURE 10.8 Specific work for flower-harvesting technologies. (From Pirnă, I. et al., Course harvest technologies for aromatic and medicinal plants in Calarasi-Silistra area, MedPlaNet Project, http://medplanet.dbioro.eu/doc/Courses_EN.pdf, 2011.)

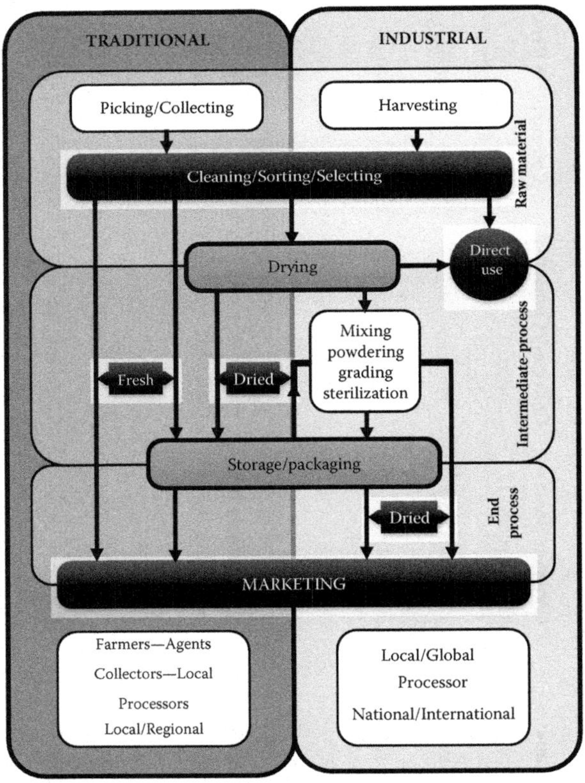

FIGURE 10.9 Processing steps of edible flowers.

(WHO 1993). The harvesting of edible flowers can mostly be carried out by hand without the use of any special equipment, or, as in the case of chamomile flowers, a simple handheld device is used. Lavender is harvested in certain places by mechanical means. Various methods may be used for loading and carrying (Figure 10.8) (Martinov and Konstantinovic 2007; Öztekin and Martinov 2007; Pirnă et al. 2011).

There are two basic types of processing methods for edible flowers—traditional and industrial. Traditional processing covers raw materials mostly gathered in the wild, dried in the same area, and then directly consumed or sold in a nearby market. Industrial processing may be characterized as raw materials either collected in the wild or cultured processed in a standardized way by using technological equipment and sold on the global market. The traditional process starts with collection from the spontaneous flora, while the industrial process starts with harvesting from the cultivated flora (Pirnă et al. 2011). After drying, these kind of products can be marketed as raw material, semifinished, or as finished goods (Shahidullah 2007; Toker et al. 2015); the products can also be consumed fresh or dried directly after harvesting. This is more relevant to local consumption. Studies on the steps of processing show that there can be differences between the traditional and industrial processes. In the traditional process, the products mostly go directly from the producer to the market while still in the raw material form. In the industrial process, however, the product reaches a semifinished or finished form usually after drying and is then put on the market after it is milled, converted to a powder, or mixed with other products (WHO 1993; Schippmann et al. 2002; Kathe et al. 2003; Lange 2006; Shahidullah 2007). Edible flowers gain added value and are put on the market after processing through various sales chains at national or international levels (Figure 10.9).

As an example of the processing of edible flowers, Isparta Rose flowers are harvested carefully by hand, sorted mechanically, predried, and then dried naturally or artificially to bring their moisture content to a suitable level (Figure 10.10).

10.4 PRINCIPLES OF EDIBLE FLOWER DRYING

After harvesting, one of the first steps in processing is drying. The moisture content of agricultural products has an effect on the physical and chemical properties of the material. The moisture content of a dried product after harvest is directly related to its shelf life, especially when it is presented for consumption. The importance of the process of drying edible flowers is increased even more by the necessary active compounds, which are the source of flavor and aroma in the kitchen, color, antioxidant capacity, and volatile compounds used in treatment and the preservation of health.

It is recommended that the maximum final moisture content of edible flowers should be 10%–12% (MCf, w.b.) (Müller and Heindl 2006; Rocha et al. 2011). Flowers are more sensitive to heat than other parts of the plant; therefore, there are many benefits in drying them at a low temperature.

FIGURE 10.10 Processing steps of drying for Isparta Rose Flowers. (Photos by S. Boyar and S. Erbaş).

In the drying process, three requirements must generally be met (Tanko et al. 2005; Martinov et al. 2007). They are

1. Reducing moisture content to equilibrium level
2. Preserving qualities such as active ingredients, color, flavor and aroma and reducing losses to a minimum
3. Keeping microbial development within limits

The physicomechanical and chemical properties of flowers and their growing conditions directly or indirectly affect the processes of harvesting and drying (Figure 10.11). Controlled drying of flowers that have been harvested at the right time and at the flowering stage will make it possible to achieve success. Factors such

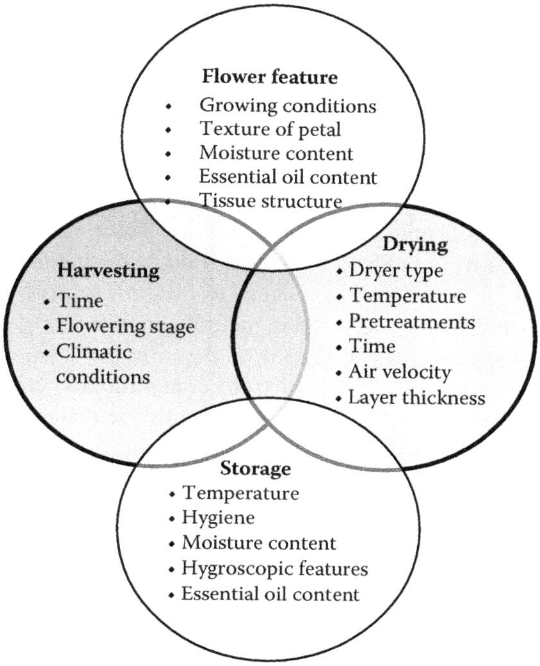

FIGURE 10.11 Parameters affecting the quality of dried edible flowers.

as the flower's moisture content, the climatic conditions on the day of harvest (rain or dew), the harvesting method, the uniformity of the flowers, and the depth of heaping (Cemeroğlu 2011) affect the steps in harvesting and drying processes, (Boyar and Bayhan 2012; Ravichandra and Pedapati 2014). The delay between harvesting and drying, drying temperature, environmental and product moisture, and storage techniques have an important effect on product quality.

The ideal drying temperature recommended to preserve the quality and amount of active compounds in many medicinal and aromatic plants is 50°C–60°C, and 45°C–65°C for flowers (Rani and Reddy 2015); therefore, the drier must be selected taking into account the drying method and the physical and chemical properties of the flower.

10.5 EDIBLE FLOWER DRYING METHODS AND SELECTION

Various methods for drying raw materials are employed by the industry in different parts of the world, such as drying in the sun, drying in the shade, solar energy-supported drying, or hot air drying (Müller et al. 1993 from Shylaja and Peter 2004; Dilta et al. 2011; Pirnă et al. 2011). Apart from this, flowers are dried by methods such as microwave drying, desiccant drying (sands, silica gel, borax powder, alum powder, corn powder, sawdust, etc.), pressing drying, and water drying (Rani and Reddy 2015), although flowers dried by these methods are more often used for decoration.

Sun and shade drying are traditional methods, and because they are cheap and simple they are in widespread use (Shylaja and Peter 2004; Janjai and Tung 2005; Tanko et al. 2005). Drying of flowers can be completed in 1–3 weeks in places where there is good ventilation (Rani and Reddy 2015).

In two separate studies using two different hot air dryers fitted with solar collectors on the roof, Rosella flowers (*Hibiscus sabdariffa*) with a starting moisture content of 92% and 90% were dried with a drying air speed of 0.1 m/s (Janjai and Tung 2005) and 0.2 m/s (Janjai et al. 2008) (Figure 10.12), respectively. In the first study, moisture levels were reduced to 16% in 4 days (drying time 27 solar hours), and in the second study, moisture was reduced to 18% in 3 days (drying time 27 solar hours). When the same flower was dried in natural sunlight for the purposes of comparison, moisture was only reduced to 28% on the third day (Janjai et al. 2008). Sun or shade drying areas are not protected from contamination by birds, insects, or rodents. However, with dryers fitted with integrated solar collectors on the roof and keeping the product in an area protected from rain, protection from insects and other damage can also be achieved (Janjai and Tung 2005).

In a study in which the stigmas were collected from half-open and fully open saffron flowers before sunrise, after sunrise, and at 10 o'clock and dried at temperatures of 40°C, 50°C, and 60°C and at an air speed of 0.5 and 1 m/s, it was found that the best color, aroma, and flavor were obtained at a temperature of 60°C and at an air speed of 0.5 m/s (Saeidirad et al. 2014).

To preserve the product with minimum losses of active ingredients, color, flavor, and aroma in the drying process, the choice of drying method is important. Thus, the oil content of Roman chamomile flowers dried in shade (1.9% w/w) is higher than that in flowers dried in the sun (0.4%) or at 40°C in an oven (0.9%). The drying method can have an important effect on the proportions of different components (Omidbaigi et al. 2004).

Hot air drying methods reduce product flavor and quality losses and allow drying in a short time, but they can also cause losses of volatile components (Tanko et al. 2005). Freeze-drying has been proposed as an ideal way of preserving the quality of the product, but investment and operating costs mean that it can only be used in products with a high added value (Shylaja and Peter 2004; Tanko et al. 2005).

With flowers such as saffron, which take up a small volume, homogeneous drying can be achieved by placing the material in the dryer inside a sterile silk net. Each net contains 150–200 g of product, and they are arranged in layers 2–3 cm deep (Dadkhah et al. 2003 from Gohari et al. 2013).

Feverfew flowers are dried in the shade, in the sun, and in ovens at 40°C. The highest content of volatile oils (0.48% w/w) of these flowers has been determined dried in the shade method which is greater than the other two methods. The volatile oil contents of flowers dried in the sun and at 40°C in a dryer were 0.27% and 0.42%, respectively. The proportion of camphor which has as an antiseptic, antipruritic, rubifacient, abortifacient, aphrodisiac, contraceptive, and lactation suppressant effect (Zuccarini 2009) of dried flowers in the shade was greater than the other two methods (Omidbaigi et al. 2004).

The market value of a product has a great effect on the choice of drying system. Thus, industrial drying methods are preferred in the drying of plants that have a high

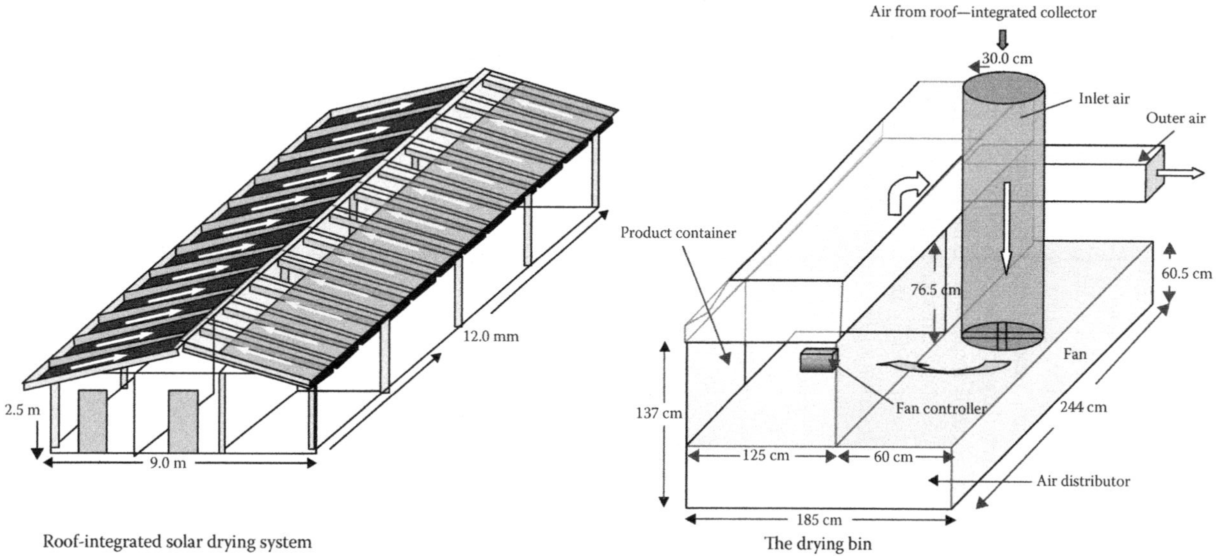

FIGURE 10.12 Schematic diagram of a solar dryer using hot air from roof-integrated solar collectors and dryer. (From Janjai, S. et al., *Energy*, 33(1), 91–103, 2008.)

added value (HAV), high market value, volatile oils and color qualities that need to be preserved, a high risk of mold and other microorganism development during drying, high sensitivity to heat, and in which only the flowers are used. In the case of flowers of low added value (LAV), which have a large volume, which are used together with the other parts of the plant, where the appearance and attractiveness are not important, and which have low moisture content, drying in shady conditions can also be an acceptable method. Assessing the flowering parts, drying flowers with HAV in the sun is seen to have a negative effect on product value (Table 10.5).

Differences can be seen in the drying process between traditional and industrial methods. Traditional methods are more subject to environmental factors, whereas industrial methods are dominated by controlled conditions. With the technology used in industrial drying systems, heat is kept under control and continuity is ensured, so that the product can be dried in a short time and quality can be preserved.

The drying method that is most widely used and preferred today is hot air drying, which is more expensive in terms of operating costs than natural methods. Heat pump systems are shown to provide a good alternative for energy efficiency, because the product is dried quickly and is protected from external factors such as rain, dust, and insects. Solar energy dryers with heat pump support shorten drying time and reduce specific energy consumption (Minaei et al. 2014). In addition, because products dried with heat pumps are dried at a suitable temperature, they maintain their color, appearance, aroma, and other qualities for a longer time than naturally dried products (Karabacak et al. 2011).

This method is especially good for preserving morphological qualities such as color, shape, and chemical qualities such as phenolic and essential oil compounds. Desiccant materials that have been found to be suitable for plants and flowers include silica gel, borax, coarse cornmeal, and alum (aluminum sulfate). Examples of flowers on which desiccants can be used are roses, lavender, and chamomile. With this method, the whole plant or just the petals can be dried (Chua and Chou 2003).

TABLE 10.5
Choices and Preferences in Drying Method according to Flower Parts and Product Value

	Flower Parts					
	Petal		Bud		Opened	
Drying Method	HAV	LAV	HAV	LAV	HAV	LAV
Sun	X	X	X	[a]	X	[a]
Shade	[a]	[b]	[a]	[a]	[a]	[b]
Hybrid systems	[c]	[b]	[c]	[b]	[c]	[b]

[a] Low market value and quality.
[b] Can meet expectations.
[c] Market value and product quality high.

In a study in which saffron flowers were dried at three different temperatures with and without heat pump support, coloring and aromatic strength values of the dried flowers were found to be higher when the heat pump was used than when it was not (Mortezapour et al. 2014).

To determine the effect of drying on antioxidant activity and volatile oil content, chrysanthemi flos flowers were dried in a hot air dryer and in a far-infrared dryer at temperatures of 40°C, 50°C, and 60°C, to a moisture content of 22% ± 1%. It was found that the antioxidant activities of the extracts increased with an increase in temperature when hot air drying was used, but decreased when the far-infrared dryer was used. The highest content of luteonin, the most important flavonoid in chrysanthemi flos flowers, was obtained with the far-infrared dryer at a temperature of 60°C. In the same way, the highest volatile oil contents were found at 50°C in the hot air dryer and at 40°C in the infrared dryer (Bae et al. 2009).

Preservation of the antioxidant content and the volatile oil components of the flowers to be dried (Bae et al. 2009), choice of the preferred drying method, and planning and putting into practice the drying conditions must be considered. Therefore, methods that ensure controlled conditions for drying are to be preferred.

Because the method of drying can greatly influence many things, from investment costs to final product characteristics, various criteria must be considered when choosing the method (Table 10.6).

TABLE 10.6
Basic Comparison Criteria for Choosing Drying Methods and the Suitability for Flowering Plants

	Drying Methods			
Criteria	**Sun**	**Shade**	**Hot Air**	**Hybrid Systems**[a]
Investment cost	Low	Medium/high	High	High
Operating cost	Low	Medium	Medium/high	Medium/high
Energy consumption	—	Low/medium	Medium/high	High
Suitability for flowering plants	Not good	Good	Good/very good	Very good
Material loading method	Batch	Batch	Batch/continuous	Batch/continuous
Drying time	Medium	High	Low	Low
Drying speed	Medium	Low	High	High
Quality properties (color, smell and flavor, aroma, volatile components, appearance, etc.)	Bad	Good	Medium/high	Medium/high
Effect of environmental factors	High	Medium/high	Low	Low

[a] Innovative systems using more than one drying method.

10.6 A CASE STUDY: DRYING OF ISPARTA ROSE (*ROSA DAMASCENA* MILL.)

The rose is a flower whose scent, beauty, and mystical qualities have appealed to the human soul, and it is one of the main symbols of beauty and sanctity in works of literature. In particular, the oil rose is regarded as a cultural heritage. It is produced in just a few places in the world and is known as a strategic product with historical and traditional qualities (Altıntaş 2010; Boyar 2013; Baydar 2016).

In the areas where it is widely used, it is appreciated for its physical qualities and also has a place in human history as "glorious" and is accorded a value in literature as a symbol of love and of the Prophet in Islamic mysticism (Altıntaş 2010).

The oil rose, when fresh, has a very wide range of applications such as its use as basic products of volatile oil, rosewater, rose concrete, and rose absolute, and when dried, it has its uses in medicine, food, cosmetics, skin care, aromatherapy, and phytotherapy (Boyar and Bayhan 2012).

10.6.1 Characteristics of Isparta Rose (*Rosa damascena* Mill.)

Known around the world by such names as Rose of Damascus, gol-e Muhammedi, and Rose of Castile, the Damask rose is known in Turkey as the Isparta rose. It is grown in Bulgaria in the regions of Kazanlık, Stara Zagora, Plovdiv and Pazarjic, in Iran in Kerman, Isfahan, Shiraz, and Tabriz, and in Turkey 80% of roses are grown in Isparta, Burdur, Denizli, and Afyonkarahisar.

Various physical properties of the flowers of Isparta roses are given in Table 10.7. In terms of weight, 65% of the flower is made up of the petals, 10% the sepals (the epicalyx and hypanthium), 5% the stamens (the male organs), and 20% the pistil (female organ: stigma and ovary) (Baydar 2016).

The areas where the rose is grown in Turkey are at an altitude of 800–1400 m, and the flowering season lasts for about a month in the period from the first week in May to the end of June. Harvesting time for the Isparta rose is determined by monitoring the stage of flowering (Figure 10.13).

The scent molecules in the petals of the rose are manufactured in the cells of the epidermis. Volatile oil synthesis and scent molecules show great differences as the flower develops from a bud (a) to a fully open flower (e). In this way, it has been established that the best stage to harvest the flowers is in the form (d), when the volatile oils and their compounds are at their highest level (Table 10.8; Figure 10.13).

Flowers at this stage are picked by hand very early in the morning between 04.00 am and 10.00 am by breaking them below the ovary (Figure 10.10). Generally, approximately 3.5 tons of flowers can be gathered in a season from a 5000 m^2 garden, and these flowers, when fresh, will yield 1 kg of rose oil (Baydar and Kazaz 2013). However, the content and composition of the volatile oil in the Isparta rose can easily be affected by many factors. The average yield of rose oil from one rose garden over the 20-day flowering season was found to be 0.030% at the beginning of the period, 0.045% in the middle, and 0.020% at the end. From the beginning of the flowering period toward the end, the proportion of citronellol increases from 19.3% to 37.7%, while the proportion of gerianol declines from 42.4% to 13.0%. At the same time, the

TABLE 10.7
Various Physical Properties of the Isparta Rose

Organs	No[a]	Organs	Diam. (mm)	Organs	Length (mm)
Petal	30	Flower	50	Filament	5–5.5
Sepal	5	Anther	2.4	Sepal	4–5
Anther	90	Hypanthium	5	Pistil	6–8
Stigma	40	Anther weight	0.10 g		
Pollen (in the anther)	50,000	Average flower weight	2 g		

Source: Erbaş, S. et al., *Süleyman Demirel Univ. J. Fac. Agric.*, 10(2), 40–50, 2015.
[a] The number in one flower.

TABLE 10.8
Volatile Compound Properties of Isparta Roses according to Flower Form

	Stages of Flower Development				
Compounds (%)	a	b	c	d	e
Nerol	0.6	4.2	8.9	10.3	14.3
Citronellol	2.3	33.6	41.8	43.5	23.3
Geraniol	3.4	20.3	21.9	21.4	33.6
Hydrocarbon compounds	85.9	36.6	22.9	21.0	17.5
Essential oil content (%)	0.008	0.015	0.032	0.045	0.043

Source: Baydar, H. et al., *Süleyman Demirel Univ. J. Fac. Agric.*, 8(1), 1–11, 2013.

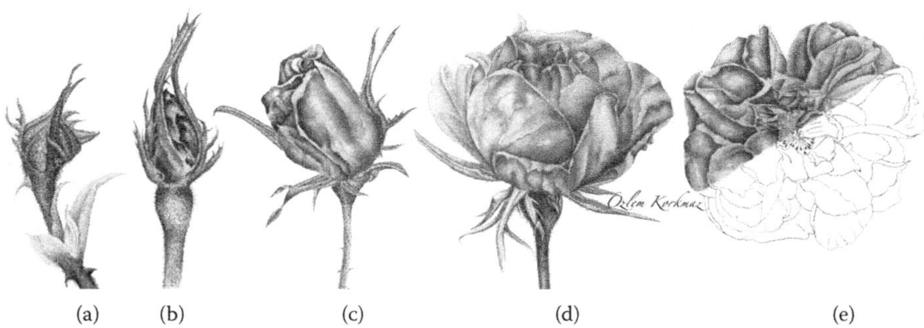

FIGURE 10.13 (a) budding flower, (b) early opening flower, (c) half opening flower, (d) opening flower, (e) full opening flower. The various stages of flowering of the Isparta rose (*Rosa damascena* Mill.). (Illustration by Korkmaz, Ö., The various stages of flowering of the Isparta rose (*Rosa damascena* Mill.), 2016d.)

proportion of nerol decreases, and the proportions of hydrocarbon compounds and methyl eugenol increase (Baydar et al. 2013).

The content of volatile oil and its compounds in the oil rose vary widely according to the flowering period, the state of opening of the flower, the various organs of the flower, the time of the day when it was picked, the length of time it is kept after harvest, how it is stored, and whether the flowers are fresh or dried. In one study, 0.035% of volatile oil was found in the whole flowers, varying between 0.057% in the petals and 0.013% in the sepals. In the petals, the proportions of gerianol, citronellol, and nerol in the volatile oil were higher and those of hydrocarbons and methyl eugenol were lower than that in the whole plant. In the sepals, hydrocarbons and especially methyl eugenol proportions were 8–12 times higher (Table 10.9).

From early the morning to the evening during the picking time, the essential oil content of the flowers steadily decreases from 0,055% to 0,017%, and rose oil quality is negatively affected. During day from 08.00 AM to 02.00 PM, the rate of citronellol of rose oil increases from 19.4% to 35.8%, and the rates of geraniol and nerol decrease from 30.55% to 3.3% and from 8.5% to 1.1%, respectively (Kazaz 1997).

The length of time between harvesting and distillation is especially important. This is because the amount of volatile oil present in the fresh flowers processed immediately after harvesting is in a proportion of 0.055%, but this falls to 0.035% after 18 h and to 0.025% after 36 h. In the fermentation, it is seen when the flowers are kept unprocessed, the volatile oils are broken down enzymatically, and the compounds forming them react with each other. In this way, the proportions of nerol and gerianol in the fresh flowers decrease significantly, and the proportions of citronellol, methyl eugenol, and hydrocarbon compounds increase (Baydar et al. 2007; Baydar et al. 2008a). Attempts have been made to solve this problem partially by not keeping them in sacks but spreading them in the shade in heaps with a depth of 20–25 cm. Even then, there is serious loss of volatile oils and deterioration of quality. Studies have shown that

TABLE 10.9
Essential Oil Content and Its Composition in Different Flower Parts of Oil Rose (%)

Essential Oil Content and Composition (%)	Parts of Flower		
	Full Flower	Petal	Sepal
Essential oil content (%)	0.035	0.057	0.013
Sitronellol	25.6	26.7	4.7
Nerol	12.2	14.9	5.1
Geraniol	31.4	36.9	15.5
Total hydrocarbon content	11.6	10.8	27.4
Öjenol	1.8	0.4	8.1
Metil öjenol	1.0	0.4	6.9

Source: Baydar, H. et al., *Süleyman Demirel Univ. J. Fac. Agric.*, 8(1), 1–11, 2013.

if flowers are kept in cold storage at temperatures between 0°C and 3°C, there is no change in the proportion of volatile oils or quality for approximately 7 days (Kazaz et al. 2009).

10.6.2 Drying Process

One of the best ways of making use of the flowers of the Isparta rose is drying. In a study in which rose flowers were dried naturally in the shade, an average volatile oil proportion of 0.060% was reported. However, the quality of the rose oil after drying was much lower than that of oil produced from fresh roses. The proportion of nerol, citronellol, and gerianol from the distillation of dried roses does not exceed 10%, whereas the proportion of hydrocarbon compounds is above 70% (Baydar et al. 2008b).

Rose flowers are picked at all stages of development, from bud to fully open flower (Figure 10.13). After harvest, and during transport after harvest, the sepals and petals can become separated from one another. The various development stages of the flowers react differently to drying. Petals dry very quickly, whereas new buds take a long time to dry. If the flowers are sorted by stage of development before the drying process, the drying period will be shortened. Uniformity of the product to be dried makes it easier to make correct decisions and estimates on the completion of drying (Boyar et al. 2013).

In four separate experiments to determine the behavior of Isparta roses dried in the shade in natural conditions (EXP I, EXP II, EXP III, and EXP IV), it was observed that samples dried between 72 and 162 h (3–7 days) and that both the fresh flowers before drying and the product during the drying process were greatly affected by the climatic conditions of the area in May and June. Based on this it was determined that the speed of drying caused differences between the experiments and that this affected the quality (Figure 10.14) (Boyar et al. 2013).

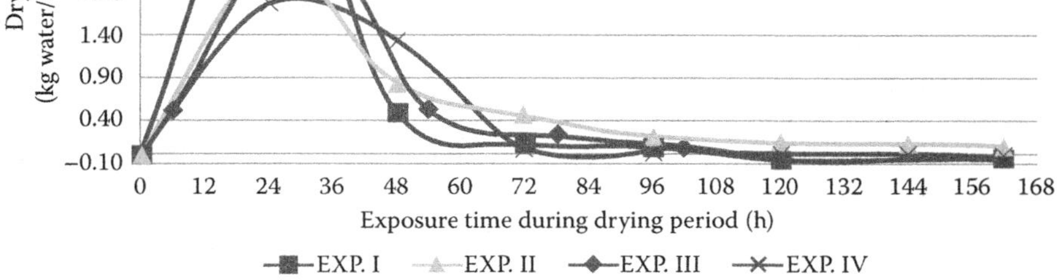

FIGURE 10.14 Changing drying rate values of Isparta Rose flowers in experiments. (From Boyar, S. et al., *Bulgarian J. Agric. Sci.*, 19(2), 361–74, 2013.)

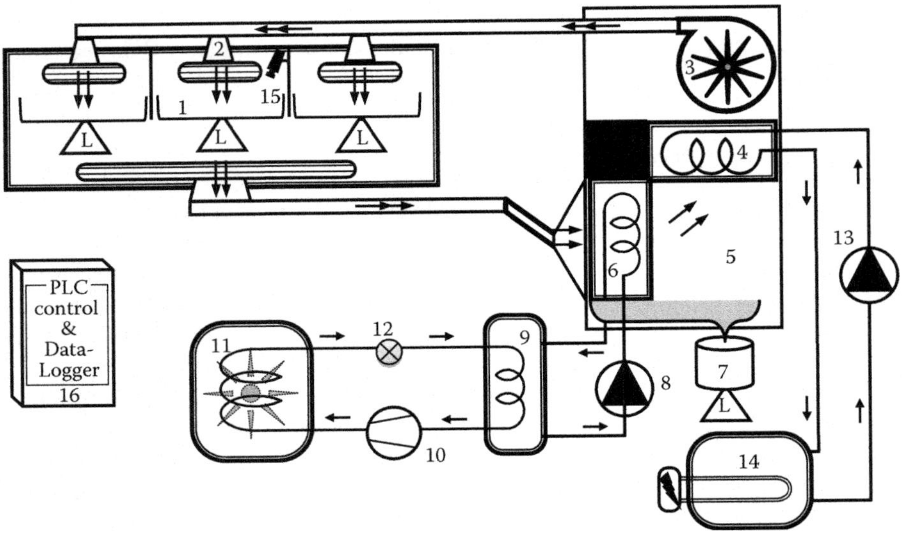

FIGURE 10.15 Condensation drying system with image processing control (This drying system was developed under project number TÜBİTAK-112O092). 1, Product tray; 2, hot dry air; 3, fan; 4, dry air heater; 5, dry air-conditioning (heating + dehumidification); 6, cooling the dry air; 7, water condenser; 8, cool water circulation pump; 9, evaporator of heat pump; 10, R407c compressor; 11, condenser of heat pump; 12, expansion valve; 13, hot water circulation pump; 14, heater; 15, camera; 16, PLC control-logger system. (From Boyar, S. et al., Görüntü işleme ile kontrollü kondenzasyonlu Isparta gülü kurutma sisteminin geliştirilmesi TÜBİTAK 1020092 Proje Sonuç Raporu, Isparta (in Turkish), 2015.)

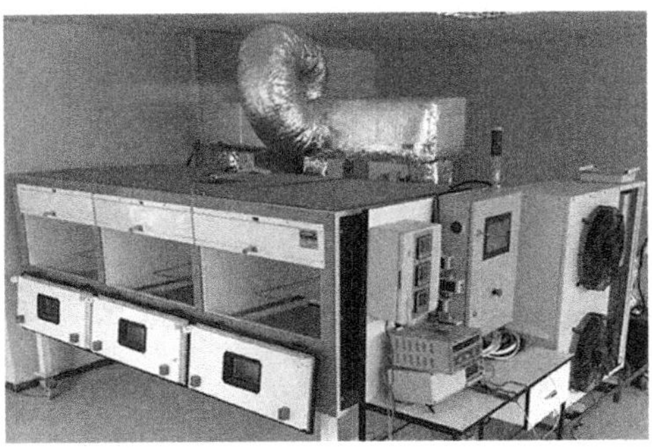

FIGURE 10.16 Condensation drying system with image processing control. (From Boyar, S. et al., Görüntü işleme ile kontrollü kondenzasyonlu Isparta gülü kurutma sisteminin geliştirilmesi TÜBİTAK 1020092 Proje Sonuç Raporu, Isparta (in Turkish), 2015.)

A condensation drying system designed as a laboratory prototype to develop an industrial heat pump system for drying the Isparta rose, an edible flower (Figures 10.15 and 10.16), consists of five parts: a dryer, an air-conditioning unit, preparation system for cooled water, preparation system for heated water, automatic control and datalogger system with Programming Logic Control (PLC):

- *Dryer*: Drying is performed on three trays with a total area of $3\,m^2$.
- *Air-conditioning unit*: The air-conditioning unit consists of a dehumidifier, a heater, and a fan. Warm humid drying air is cooled and dehumidified, and the moisture condenses into water. Cool humid drying air is heated to the required temperature. Heated drying air is directed to the dryer by a fan.
- *Preparation system for cooled water*: The cooled water preparation system consists of a water-to-air type heat pump using R407c refrigerant. Cooled water at the required temperature is prepared in an evaporator and circulated between the evaporator and a dehumidifier by a pump.
- *Preparation system for heated water*: The heated water preparation system consists of an electric water heater, where water is heated to the required temperature.
- *Automatic control and datalogger system with PLC*: The PLC unit continually monitors and logs temperature, humidity, air velocity, rates of airflow and water flow, weight change of product, and energy consumption.

In the heat pump drying system, images of the drying process were taken by camera, and changes during drying were monitored instantaneously (Figures 10.15 through 10.18). To determine the color scale of the product in the images taken during the drying process of the roses, values of brightness (L), redness (a), and yellowness (b) were obtained (McGuire 1992). It was observed that at temperatures of 35°C, 40°C, and 45°C, the product shrank with the loss of moisture, and toward the end of the drying process its color darkened and took on a matt quality, and the redness value increased (Table 10.10; Figures 10.17 and 10.20).

In another study in which Isparta roses were dried under controlled conditions in a drying system with heat pump-supported hot air at temperatures of 35°C, 40°C,

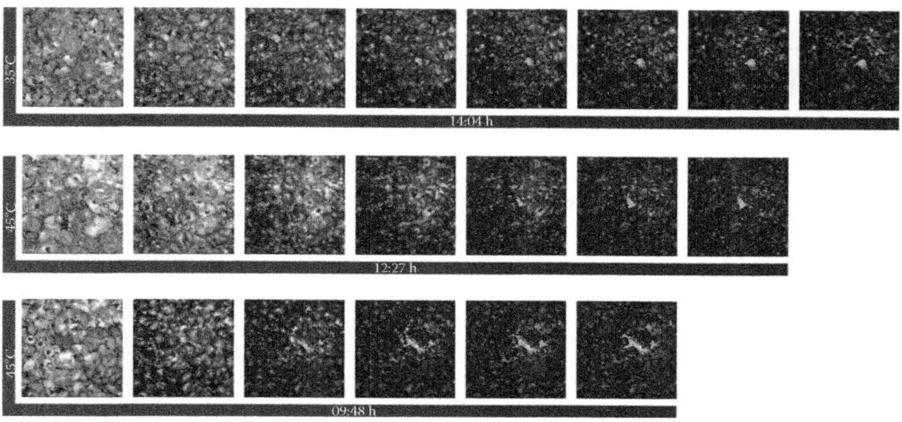

FIGURE 10.17 Isparta roses (*Rosa damascena* Mill.) in an image processing controlled heat pump-supported drying system during the drying process. (From Boyar, S. et al., Görüntü işleme ile kontrollü kondenzasyonlu Isparta gülü kurutma sisteminin geliştirilmesi TÜBİTAK 1020092 Proje Sonuç Raporu, Isparta (in Turkish), 2015.)

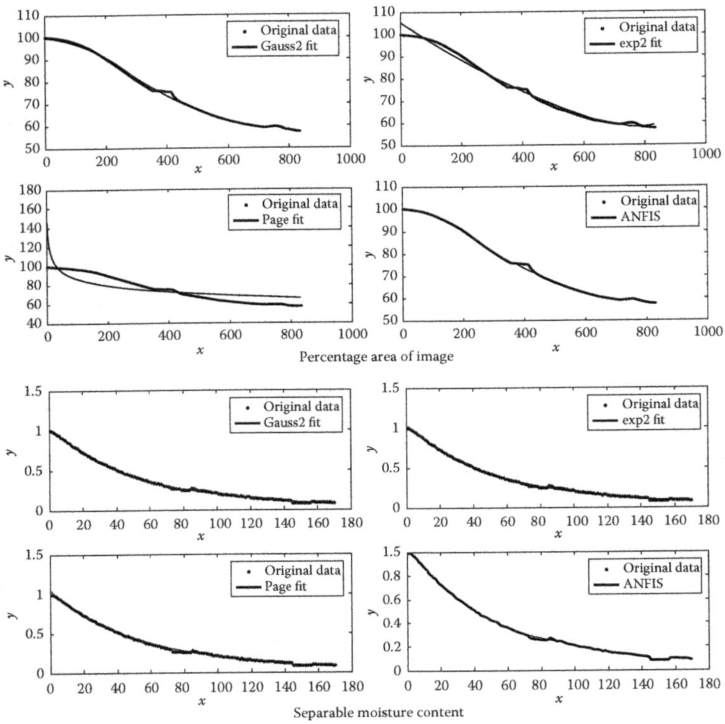

FIGURE 10.18 Convergence of curves of percentage area of image of roses during the drying process (%) and variations in separable moisture content when dried at 35°C. (From Boyar, S. et al., Görüntü işleme ile kontrollü kondenzasyonlu Isparta gülü kurutma sisteminin geliştirilmesi TÜBİTAK 1020092 Proje Sonuç Raporu, Isparta (in Turkish), 2015.)

TABLE 10.10
Color Changes in Samples of Fresh and Dry Isparta Roses

Temperature	35		40		45		Mean	
(°C)	Fresh	Dry	Fresh	Dry	Taze	Fresh	Dry	Fresh
Brightness[a] (L)	60.74	40.77	61.94	39.06	59.96	40.48	60.88 ± 0.997	40.10 ± 0.991
Redness[b] (A)	33.79	28.39	31.98	28.54	29.70	25.02	31.82 ± 2.049	27.32 ± 1.274
Yellowness[c] (B)	−7.12	−5.54	−7.12	−5.88	−6.45	−4.88	−6.90 ± 0.387	−5.43 ± 0.341

Source: Boyar, S. et al., Görüntü işleme ile kontrollü kondenzasyonlu Isparta gülü kurutma sisteminin geliştirilmesi TÜBİTAK 1020092 Proje Sonuç Raporu, Isparta (in Turkish), 2015.

[a] L is between 0 and 100. 0 indicates no reflection and therefore black color; 100 indicates total reflection and therefore white.

[b] When A is positive, it represents redness, and when negative, it represents greenness.

[c] When B is positive, it represents yellowness, and when negative, it represents blueness.

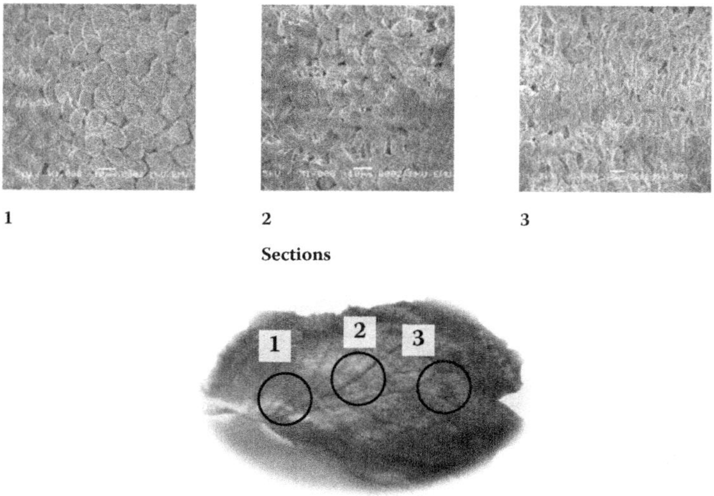

FIGURE 10.19 SEM images (×1000) of different parts of the petal of dried Isparta rose. (From Pehlivan, M., Doğaner, A., Isparta yağ gülü (*Rosa damascena* Mill.) ayırma-sınıflandırma ve kurutma sistemlerinin geliştirilmesi Project Report. TÜBİTAK-TEYDEB-7100075 (in Turkish), 2012; Photos by Kemal Sangun.)

FIGURE 10.20 Opened flowers and petals before and after drying. (Photos by S. Boyar.)

TABLE 10.11
Variation in Oil Proportion in Dried Roses according to Drying Temperature (%)

Temperature (°C)	35	40	45
Mean	40.1	37.8	36.7
Standard deviation	16.21	9.69	12.25
Min.	23.2	28.4	21.8
Max.	68.1	49.3	53.7

and 45°C, drying was achieved in 14.04, 12.27, and 9.48 h, respectively (Figure 10.17) (Boyar et al. 2015).

Comparing fresh roses with the product of the drying process, it is seen that the dried roses are less bright and darker, redness and blueness values are somewhat reduced, and they are duller (Table 10.10).

It was determined that there was a close relationship between the percentage image area changes obtained by processing the images taken during the drying process and the proportion of separable moisture, and that the mathematical model and the derived curves showed great conformity (Boyar et al. 2015) (Figure 10.18).

Examining the proportions of oil in relation to the calculated amounts of oil from dry matter according to the temperature values selected in the drying experiments, it was determined that the oil retained by dried roses was 40.1% at 35°C, 37.8% at 40°C, and 36.7% at 45°C (Table 10.11).

Examining scanning electron microscope (SEM) tissue photographs of the root, middle, and tip of rose petals dried at 43°C in a hot air shelf-type dryer, it was seen that after drying, the oil cells were easily separated from the tissue by heat and the volatile compounds were lost (Figure 10.19) (Pehlivan and Doğaner 2012).

Sorting the roses into buds, opened flowers, and petals, according to the purpose for which they will be used, shortens the drying period and ensures homogeneous drying. The sorting process also prevents excessive drying especially of petals when they are mixed, and the rapid fermentation of the middle part of the buds when they are dried at a temperature of over 40°C (Öztekin and Soysal 2000; Boyar et al. 2013).

10.7 CONCLUSION

The technical and functional characteristics of the drying process, which is one of the most effective ways of extending the shelf life of edible flowers, vary according to the type of flower and the qualities expected from the finished product. The quality characteristics that define market value of dried edible flowers such as color, aroma, appearance, and appeal are affected not only by drying but also by harvesting and storage conditions.

Monitoring the drying of edible flowers by constantly following the stages in the process from harvest to storage and setting technical values for each type of flower have been seen to be the principal rule in achieving the desired quality without product loss.

Because edible flowers are more sensitive to hot air than other products, it is seen that characteristics such as the species of flower, the petal structure, its characteristics, the amount of volatile oil, compounds, and color must be considered. First, laboratory studies must be carried out on each kind of edible flower and its physical and chemical characteristics, and then the optimum time and conditions for harvest must be determined and the best drying methods and system must be determined. In this way, product quality criteria of edible flowers such as color, appearance, aroma, and volatile oils can be maintained at a high level.

While drying edible flowers, not leaving them in direct sunlight but using shade and moving air or controlled conditions and applying low temperature that is suitable for every type of flower is a necessity. When flowers are dried in the sun, there are significant losses of quality, and while the shade conditions vary according to the species of flower, expectations can be met. However, considering drying speed and length of drying time, it takes longer than that in industrial drying (Figure 10.21).

For edible flowers to possess the product quality sought on the international market, they must have the desired values, especially of color and volatile oil content. Because they have a clear effect on the final product, it is necessary to develop drying technologies that can be applied in industry; to conduct studies of desiccant technologies that are applicable to foodstuff, such as microwave, infrared, dehydration, and lyophilization techniques, and to develop intelligent control systems and drying processes and systems that can be dynamically monitored.

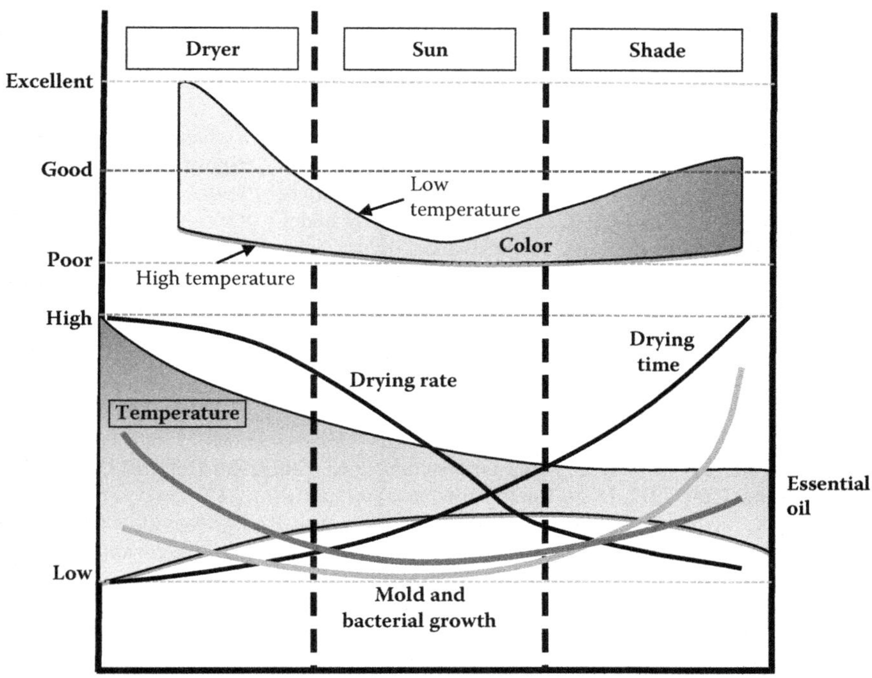

FIGURE 10.21 The effectiveness of different methods of drying.

REFERENCES

Al-Ismail, K.M. and T. Aburjai. 2004. Antioxidant activity of water and alcohol extracts of chamomile flowers, anise seeds and dill seeds. *Journal of the Sciences of Food and Agriculture* 84:173–8.

Altıntaş, A. 2010. *Rose, Rose Water: Historical, Therapeutic and Cultural Perspectives. Pharmacy and the History of Medicine Series*, Volume 1. Maestro Publishing, Istanbul, Turkey.

Anac, O. 1984. Gas chromatographic analysis on Turkish rose oil, absolute and concrete. *Perfumer and Flavorist* 9:1–14.

Asımgil, A. 2009. *Şifalı Bitkiler. Sağlık Dizisi 2*, 19. Baskı, Timaş Yayınları (in Turkish).

Bae, S.M., A.S. Na, H.K. Seo, and S.C. Lee. 2009. Effects of drying conditions on the antioxidant activities and volatile compounds of chrysanthemi flos flowers. *Preventive Nutrition and Food Science* 14(4):329–34.

Başer, K.H.C. 1992. Turkish rose oil. *Perfumer and Flavorist* 17:45–52.

Baydar, H. 2016. *Medicinal and Aromatic Plant Science and Technology*, 5th Edition. Süleyman Demirel University Press, Isparta, Turkey (in Turkish).

Baydar H., S. Erbaş, S. Kıneci, and S. Kazaz. 2007. Effect of Tween-20 adding to distillation water on rose oil yield and quality in fresh and fermented flowers of oil-bearing rose (*Rosa damascena* Mill.). *Süleyman Demirel University Journal of Faculty Agriculture* 2(1):15–20.

Baydar, H. and S. Kazaz. 2013. *Oil Bearing Rose and Isparta Rose Production*. Gülbirlik Press, Isparta, Turkey.

Baydar, H., S. Kazaz, and S. Erbaş. 2013. Morphogenetic, ontogenetic and diurnal variabilities of turkish oil-bearing rose (*Rosa damascena* Mill.). *Süleyman Demirel University Journal of Faculty Agriculture* 8(1):1–11 (in Turkish).

Baydar H., S. Kazaz, S. Erbaş, and Ö.K. Örücü. 2008b. The effects of cold storage and drying on essential oil content and composition in oil-bearing rose (*Rosa damascena* Mill.) flowers. *Süleyman Demirel University Journal of Faculty Agriculture* 3(1):42–8 (in Turkish).

Baydar H., H. Schulz, H. Krüger, S. Erbaş, and S. Kıneci. 2008a. Influence of fermentation time, hydro-distillation time and fractions on essential oil composition of Damask rose (*Rosa damascena* Mill.). *Journal of Essential Oil Bearing Plants* 11(3):224–32.

Bayhan, A.K., S. Boyar, M.F. Caglar, R.C. Akdeniz, and O. Kayaalp. 2011. Design of the experimental dryer for medical and aromatic plants. "Hungarian Agricultural Engineering" No: 23, December, 2011, 5–8 Szent Istvan University, Faculty of Mechanical Engineering, Gödöllő, Hungary.

Bayrak, A. and A. Akgül. 1994. Volatile oil composition of Turkish rose (*Rosa damascena* Mill.). *Journal of the Science of Food and Agriculture* 64:441–8.

Boyar, S. 2013. *Kültürel Miras: Isparta Gülü*, T.C. Isparta Valiliği, Isparta Avrupa Dergisi, 3:30–1, Isparta (in Turkish).

Boyar, S. and A.K. Bayhan. 2012. Isparta Gülü (*Rosa damascena* Mill.) 'nün kurutulması, "Her Yönüyle Gül Sempozyumu Özet Kitabı" s:63. Süleyman Demirel Üniversitesi, Haziran 07–09, 2012. Doğu Kampusu/Isparta (in Turkish).

Boyar, S., A.K. Bayhan, and E. Dikmen. 2013. Investigation on drying behavior of Isparta rose flowers (*Rosa damascena* Mill.) under natural shade conditions. *Bulgarian Journal of Agricultural Science* 19(2):361–74.

Boyar, S., A.K. Bayhan, E. Dikmen, B. Çetişli, and M.F. Çağlar. 2015. Görüntü işleme ile kontrollü kondenzasyonlu Isparta gülü kurutma sisteminin geliştirilmesi TÜBİTAK 1020092 Proje Sonuç Raporu, Isparta (in Turkish).

Brown, K. 2011. *Edible Flowers: From Garden to Plate: 25 Recipes and an A-Z Pictorial Directory of Culinary Flora*. Aquamarine Publishers, London, 186 pp.

Buchbauer, G., L. Jirovetz, B. Remberg, G. Remberg, and A. Nikiforov. 1992. Headspace analysis of the dried herb of passion flower (*Herba Passiflorae*) and dried flowers of lime tree (*Flores Tiliae*). *Flavour and Fragrance Journal* 7(6):329–32.

Cai, Y., Q. Luob, M. Sunc, and H. Corke. 2004. Antioxidant activity and phenolic compounds of 112 traditional Chinese medicinal plants associated with anticancer. *Life Sciences* 74:2157–84.

Cemeroğlu, B. 2011. *Meyve ve Sebze İşleme Teknolojisi*. Nobel Yayınevi Yayın no: 192.

Chua, K.J. and S.K. Chou. 2003. Low-cost drying methods for developing countries. *Trends in Food Science and Technology* 14(12):519–28.

Çırak, C., A. Bertoli, L. Pistelli, and F. Seyis. 2010. Essential oil composition and variability of hypericum perforatum from wild populations of northern Turkey. *Pharmacutical Biology* 48(8):906–14.

Colombo, P.S., G. Flamini, M.S. Christodoulou, G. Rodondi, S. Vitalini, D. Passarella, and G. Fico. 2014. Farinose alpine Primula species: Phytochemical and morphological investigations. *Phytochemistry* 98:151–9.

Cosmetic Ingredient Review. 2013. Safety assessment of *Rosmarinus officinalis* (Rosemary)-derived ingredients as used in cosmetics. Washington, DC, 91 pp.

Creasey, R. 1999. *The Edible Flower Garden*, 1st Edition. 106 pp. Singapore (ebook).

Dadkhah, M.R., M. Ehtesham, and H. Fekrat. 2003. *Tehran: Shahr Ashoob Publication*; Iranian saffron an unknown jewel, 1–20.

David, F., C. De Clercq, and P. Sandra. 2006. GC/MS/MS analysis of β-damascenone in rose oil. *Varian GC/MS App*. Note 52.

Deane G. 2007–2012. Edible flowers: Part two. www.eattheweeds.com/edible-flowers-part-two (Accessed December 1st, 2015).

Diraz, E., S. Karaman and N. Koca. 2012. Fatty acid and essential oil composition of *Echinacea purpurea* (L.) Moench, growing in Kahramanmaras-Turkey. In: International conference on environmental and biological sciences (ICEBS'2012), Bangkok, 35–7.

Dilta, B.S., B.P. Sharma, H.S. Baweja, and B. Kashyap. 2011. Flower drying techniques: A review. *International Journal of Farm Sciences* 1(2):1–16.

Doğan, G. and E. Bağcı. 2014. Essential oil composition of *Papaver rhoeas* L. (Corn poppy) (Papaveraceae) from Turkey. *Hacettepe Journal of Biology and Chemistry* 42(4):545–9.

Đorđević, A.S. 2015. Chemical composition of *Hypericum perforatum* L. essential oil. *Advanced Technologies* 4(1):64–8.

Erbaş, S., M. Alagöz, and H. Baydar. 2015. A research on flower morphology and pollen viabilty of oil-bearing rose (*Rosa damascena* Mill.) *Süleyman Demirel University Journal of Faculty Agriculture* 10(2):40–50 (in Turkish).

Ernst, E. 2003. *Hypericum: The Genus Hypericum*. Taylor & Francis, New York, 233.

Facciola, S. 1990. *Cornucopia: A Source Book of Edible Plants*. Kampong Publication, Vista, CA, 677.

Faydaoğlu, E. and M.S. Sürücüoğlu. 2011. History of the use of medical and aromatic plants and their economic importance. *Kastamonu University Journal of Forest Faculty* 11(1):52–67 (in Turkish).

Ferreres, F., F.A.T. Barberan, and F. Tomas. 1986. Flavonoids from *Lavandula dentata*. *Fitoterapia* 57:199–200.

Garland, S. 1993. *The Complete Book of Herbs and Spices*. Holder & Stoughton, Sydney, Australia, 288.

Gohari, A.R., S. Saeidnia, and M.K. Mahmoodabadi. 2013. An overview on saffron, phytochemicals, and medicinal properties. *Pharmacognosy Reviews* 7(13):61–6.

Göktürk Baydar, N. and H. Baydar. 2013. Phenolic compounds, antiradical activity and antioxidant capacity of oil-bearing rose (*Rosa damascena* Mill.) extracts. *Industrial Crops and Products* 41:375–80.

Govaerts, R. 2001. How many species of seed plants are there? *Taxon* 50(4):1085–90.

Gül, V. 2014. A general view to medicinal and aromatical plants belonging to Rize province. *Iğdır University Journal of the Institute of Science and Technology* 4(4):97–107.

Harborne, J.B. and C. Williams. 2002. Phytochemistry of the genus *Lavandula*. In: Lis-Balchin, M. (ed.) *Lavender: The Genus Lavandula*. Taylor & Francis, New York.

Harris, B. 2002. Methyl eugenol: The current bête noir of aromatherapy. *International Journal of Aromatherapy* 12(4):193–201.

Heindl, A. and Müller, J., 1997. Trocknung von Arznei-und Gewürzpflanzen. *Zeitschrift für Arznei-und Gewürzpflanzen* 2(2):90–7.

Hillenbrand, M., J. Zapp, and H. Becker. 2004. Depsides from the petals of *Papaver rhoeas*. *Plant Medica* 70(4):380–2.

Hinneburg, I., H.J.D. Dorman, and R. Hiltunen. 2006. Antioxidant activities of extracts from selected culinary herbs and spices. *Food Chemistry* 97:122–9.

Hu, S.Y. 2005. *Food Plants of China*. The Chinese University Press, Hong Kong, 844.

Ivanova, D., D. Gerova, T. Chervenkov, and T. Yankova. 2005. Polyphenols and antioxidant capacity of Bulgarian medicinal plants. *Journal of Ethnopharmacology* 96:145–50.

Jakhar, M.L., B.L. Kakralya, S.J. Singh, and K. Singh. 2009. Enhancing the export potential of medicinal plants through biodiversity conservation and development under multi-adversity environment, 3rd Edition. In: Trivedi, P.C. (ed.) *Medicinal Plants Utilisation and Conservation*, 2nd Revised and Enlarged Edition. Aavishkar Publishers, Jaipur, India.

Janjai, S., N. Srisittipokakun, and B.K. Bala. 2008. Experimental and modelling performances of a roof-integrated solar drying system for drying herbs and spices. *Energy* 33(1):91–103.

Janjai, S. and P. Tung. 2005. Performance of a solar dryer using hot air from roof-integrated solar collectors for drying herbs and spices. *Renewable Energy* 30(14):2085–95.

Javani, M., N. Arslan, and M. Taher. 2015. Tıbbi aromatik bitkilerin doğadan toplanmasında iyi tarım uygulamaları. *TÜRKTOB Dergisi*. 4(15):28–33 (in Turkish).

Kaisoon, O., I. Konczak, and S. Siriamornpun. 2012. Potential health enhancing properties of edible flowers from Thailand. *Food Research International* 46(2):563–71.

Karabacak, R., Ş.G., Acar, and Ö. Atalay. 2011. Toprak kaynaklı ısı pompası ile güneş enerjisi destekli bir pilot tesiste kurutma ve iklimlendirme uygulamaları. Ulusal İklimlendirme Kongresi (İKLİM 2011), 18–20 Kasım 2011, Antalya (In Turkish).

Kathe, W., S. Honnef, and A. Heym. 2003. Medicinal and aromatic plants in Albania, Bosnia-Herzegovina, Bulgaria, Croatia and Romania, A study of the collection of and trade in medicinal and aromatic plants (MAPs), relevant legislation and the potential of MAP use for financing nature conservation and protected areas. Federal Agency for Nature Conservation (BfN) Bonn-Germany, 201. www.bfn.de/fileadmin/MDB/documents/skript91.pdf (Accessed September 15, 2015).

Kazaz, S. 1997. Hasat zamanı ve hasat sonrası bekleme süresinin yağ gülünde (*Rosa damascena* Mill.) yağ miktarı ve kalitesine etkileri üzerine bir araştırma. SDÜ Fen Bilimleri Enstitüsü, Bahçe Bitkileri Anabilim Dalı, Yüksek Lisans Tezi, Isparta (In Turkish).

Kazaz S., S. Erbaş, and H. Baydar. 2009. The effects of storage temperature and duration on essential oil content and composition oil rose (*Rosa damascena* Mill.). *Journal of Field Crops* 14(2):89–96.

Kintzios, S.E. 2002. *Oregano: The Genera Origanum and Lippia*. CRC Press, Boca Raton, FL, 267.

Kırıcı, S. 2015. Türkiye'de tibbi ve aromatik bitkilerin genel durumu. *TÜRKTOB-Türkiye Tohumcular Birliği Dergisi* 4(15):4–11.

Korkmaz, Ö. 2014a. Kadife Çiçeği, Marigold (*Tagates patula* L.).

Korkmaz, Ö. 2014b. Çuha Çiçeği, Cowslip (*Primula officinalis* L.).

Korkmaz, Ö. 2014c. Gelincik Çiçeği, Common Poppy (*Papaver rhoeas* L.).

Korkmaz, Ö. 2015a. Isparta Gülü, Isparta Rose (*Rosa damascena* Mill.).
Korkmaz, Ö. 2015b. Biberiye, Rosemary (*Rosmarinus officinalis*).
Korkmaz, Ö. 2015c. Ekinezya, Coneflower (*Echinacea purpurea* L.).
Korkmaz, H. 2015d. Kantaron, St. John's Wort (*Hypericum perforatum* L.).
Korkmaz, Ö. 2016a. General view and structure of the Damask (Isparta) Rose.
Korkmaz, Ö. 2016b. Papatya, German Chamomilla (*Matricaria chamomilla* L.).
Korkmaz, Ö. 2016c. Lavanta, Lavandin (*Lavandula x intermedia* Emeric ex Loisel.).
Korkmaz, Ö. 2016d. The various stages of flowering of the Isparta rose (*Rosa damascena* Mill.).
Küçük, Ş., F. Çetiner, and T. Ulu. 2000. Medicinal and aromatic commercial native plants in The Eastern Black Sea Region of Turkey. In: *Seminar Proceedings Harvesting of Non-wood Forest Products*, Izmir-Turkey, 33–44. www.fao.org/docrep/019/y4496e/y4496e.pdf (Accessed December 15, 2015).
Kürkçüoglu, M. 1988. Analysis and production of Turkish rose oil. Masters dissertation, Anadolu University.
Lange, D. 2006, International trade in medicinal and aromatic plants actors, volumes and commodities, Chapter 11. In: Bogers, R.J., L.E. Craker and D. Lange. (eds) *Medicinal and Aromatic Plants: Agricultural, Commercial, Ecological, Legal, Pharmacological and Social Aspects*, Volume 17. Springer, the Netherlands, 155–170.
Lim, T.K. 2012. *Edible Medicinal and Non-medical Plants: Fruits*. Springer, Dordrecht, the Netherlands, 1110.
Lim, T.K. 2013. *Edible Medicinal and Non-medicinal Plants: Flowers*, Springer, New York (eBook).
Mahboubi, M. 2015. *Rosa damascena* as holy ancient herb with novel applications. *Journal of Traditional and Complementary Medicine* 6(1):10–16.
Manzoor Rashid, A.Z.M. 2009. Management of medicinal plants in Bangladesh: Issues and challenges of sustainability. Second cycle, A2E. Masters dissertation, SLU, Swedish Biodiversity Centre. stud.epsilon.slu.se/657/1/rashid_m_091125.pdf (Accessed September 13, 2015).
Martinov, M. and M. Konstantinovic. 2007. Harvesting, Chapter 2. In: Öztekin, S. and M. Martinov. (eds) *Medicinal and Aromatic Crops: Harvesting, Drying and Processing*. Haworth Food Agricultural Products Press, New York.
Martinov, M., S. Öztekin, and J. Müller. 2007. Drying, Chapter 3. In: Öztekin, S. and M. Martinov. (eds) *Medicinal and Aromatic Crops: Harvesting, Drying and Processing*. Haworth Food Agricultural Products Press, New York.
McCullough, J. 2007. *The Ultimate Guide to U.S. Army Survival Skills, Tactics, and Techniques*. Department of the Army, Skyhorse Publishing, New York.
McGuire, R.G. 1992. Reporting of objective color measurements. *HortScience* 27:1254–55.
Minaei, S., A. Motevali, B. Ghobadian, A. Banakar, and S.H. Samadi. 2014. An investigation of energy consumption, solar fraction and hybrid photovoltaic–thermal solar dryer parameters in drying of chamomile flower. *International Journal of Food Engineering* 10(4):697–711.
Mlcek, J. and O. Rop. 2011. Fresh edible flowers of ornamental plants–A new source of nutraceutical foods. *Trends in Food Science and Technology* 22(10):561–9.
Mortezapour, H., B. Ghobadian, M.H. Khoshtaghaza, and S. Minaei. 2014. Drying kinetics and quality characteristics of saffron dried with a heat pump assisted hybrid photovoltaic-thermal solar dryer. *Journal of Agricultural Science and Technology* 16(1):33–45.
Müller J., T. Conrad, M. Tešıc, and J. Sabo. 1993. Drying of medicinal plants in a plastic-house type solar dryer. *Acta Horticulture* 344, 79–85.
Müller J. and A. Heindl. 2006. Drying of medicinal plants, Chapter 17. In: Bogers, R.J., L.E. Craker, and D. Lange (eds) *Medicinal and Aromatic Plants: Agricultural, Commercial, Ecological, Legal, Pharmacological and Social Aspects*. Medicinal and Aromatic Plants, Springer, Dordrecht, the Netherlands, 237–252.

Newman S.E. and A.S. O'Connor. 2009. Edible flowers. Colorado State University Extension. Fact sheet no. 7.237. www.ext.colostate.edu/pubs/garden/07237.pdf (Accessed December 1st, 2015).

Omidbaigi, R., F. Sefidkon, and F. Kazemi. 2004. Influence of drying methods on the essential oil content and composition of Roman chamomile. *Flavour and Fragrance Journal* 19(3):196–8.

Orhan, D.D., B. Özçelik, S. Hoşbaşı, and M. Vural. 2012. Assessment of antioxidant, antibacterial, antimycobacterial, and antifungal activities of some plants used as folk remedies in Turkey against dermatophytes and yeast-like fungi. *Turkish Journal of Biology* 36:672–86.

Özer, M. 2013. Tabiat Eczanesi Şifalı Bitkiler Ansiklopedisi. Ipek Yayın (in Turkish).

Öztekin, S. and M. Martinov. 2007. Introduction, Chapter 1. In: Öztekin, S. and M. Martinov (eds) *Medicinal and Aromatic Crops: Harvesting, Drying and Processing.* Haworth Food Agricultural Products Press, New York.

Öztekin, S. and Y. Soysal, 2000. Çiçeğinden yararlanılan aromatik bitkiler için tasarlanan bir raflı kurutucu: Çukurova II," 19. Ulusal Tarımsal Mekanizasyon Kongresi, Haziran 1–2, 2000, Bildiriler, 357–362, Erzurum, Turkey (in Turkish).

Öztürk, N., M. Tunçel, and İ.P. Erkara. 2009. Phenolic compounds and antioxidant activities of some *Hypericum* species: A comparative study with *H. perforatum*. *Pharmaceutical Biology* 47(2):120–7.

Pamuk, A. 2013. *Şifali Bitkiler Ansiklopedisi.* Pamuk Yayıncılık, Istanbul, Turkey (in Turkish).

Pehlivan, M. and A. Doğaner. 2012. Isparta yağ gülü (*Rosa damascena* Mill.) ayırma-sınıflandırma ve kurutma sistemlerinin geliştirilmesi Project Report. TÜBİTAK-TEYDEB-7100075 (in Turkish).

Pimm, S.L. and L.N. Joppa. 2015. How many plant species are there, where are they, and at what rate are they going extinct? *Annals of the Missouri Botanical Garden* 100(3):170–6.

Pirnă I., V. Emil, M. Radu, C. Radu, and M. Adriana. 2011. Course harvest technologies for aromatic and medicinal plants in Calarasi-Silistra area, MedPlaNet Project. http://medplanet.dbioro.eu/doc/Courses_EN.pdf (Accessed December 15, 2015).

Rani, P.R. and M.V. Reddy. 2015. Dehydration techniques for flowers. *International Journal of Applied Research* 1(10):306–11.

Ravichandra, S and A. Pedapati. 2014. Effect of pre-drying treatment on carnation dried flower quality. *Journal of AgriSearch* 1(1):35–9.

Roberts, M.J. 2000. *Edible and Medicinal Flowers.* New Africa Publishers, Cape Town, South Africa, 160 pp.

Rocha, R.P., E.C. Melo, and L.L. Radunz. 2011. Influence of drying process on the quality of medicinal plants: A review. *Journal of Medicinal Plants Research* 5(33):7076–84.

Rop, O., J. Mlcek, T. Jurikova, J. Neugebauerova, and J. Vabkova. 2012. Edible flowers—a new promising source of mineral elements in human nutrition. *Molecules* 17:6672–83.

Rusanov, K., N. Kovacheva, M. Rusanova, and I. Atanassov. 2012. Reducing methyl eugenol content in *Rosa damascena* Mill., rose oil by changing the traditional rose flower harvesting practices. *Journal of European Food Research and Technology* 234:921–6.

Saeidirad, M.H., P. Sharayei, and S. Zarifneshat. 2014. Effect of drying temperature, air velocity and flower types on dried saffron flower quality. *Agricultural Engineering International: The CIGR Journal* 16(4):251–4.

Sagar, D.V., S.N. Naik, P.K. Rout, and Y.R. Rao. 2005. Composition of essential oils of *Tagetes patula* L. growing in Northern India. *Journal of Essential Oil Research* 17:446–8.

Şahin, Ö. and B. Kılıç. 2014. In: Özdoğan, O.N. (ed.) Yenilebilir çiçekler. *Yiyecek içecek endüstrisinde trendler.* Detay Yayıncılık, 189–212, Ankara, Turkey (in Turkish).

Schilcher, H., P. Imming, and S. Goeters. 2005. Active chemical constituents of *Matricaria chamomilla* L. syn. *Chamomilla recutita* (L.) Rauschert. In: Franke, R., and Schilcher, H. (eds) *Chamomile-Industrial Profiles*. CRC Press, Boca Raton, FL, 55–76.

Schippmann, U., D.J. Leaman, and A.B. Cunningham. 2002. Impact of cultivation and gathering of medicinal plants on biodiversity: Global trends and issues. *Biodiversity and the Ecosystem Approach in Agriculture, Forestry and Fisheries*. Satellite event on the occasion of the Ninth Regular Session of the Commission on Genetic Resources for Food and Agriculture. October 12–13, 2002. Inter-Departmental Working Group on Biological Diversity for Food and Agriculture. Rome. (ftp://ftp.fao.org/docrep/fao/005/aa010e/AA010E00.pdf) (Accessed April 10, 2015).

Schippmann, U., D.J. Leaman, and A.B. Cunningham. 2006. A comparison of cultivation and wild collection of medicinal and aromatic plants under sustainability aspects, Chapter 6. In: Bogers, R.J., L.E. Craker, and D. Lange (eds) *Medicinal and Aromatic Plants: Agricultural, Commercial, Ecological, Legal, Pharmacological and Social Aspects*, Volume 17. Medicinal and Aromatic Plants, Springer, the Netherlands, 75–95.

Schofield, J. 2003. *Discovering Wild Plants: Alaska, Western Canada, The Northwest*, 1st edition. Alaska Northwest Books, Anchorage, AK, 353 pp.

Seemannová, Z., I. Mistríková, and S. Vaverková. 2006. Effects of growing methods and plant age on the yield, and on the content of flavonoids and phenolic acids in *Echinacea purpurea* (L.) Moench. *Plant Soil and Environment* 52:449–53.

Shahidullah, A.K.M. 2007. The role of medicinal plants in livelihood improvement and ecological sustainability in Bangladesh: An application of a participatory approach to management and marketing. Faculty of Graduate Studies of the University of Manitoba, MSc Thesis, Natural Resources Institute.

Shylaja, M.R. and K.V. Peter. 2004. The functional role of herbal spices. In: Peter, K.V. (ed.) *Handbook of Herbs and Spices*, Volume 2. Woodhead Publishing, Cambrigde, 11–21.

Singh, O., Z. Khanam, N. Misra, and M.K. Srivastava. 2011. Chamomile (*Matricaria chamomilla* L.): An overview. *Pharmacognosy Reviews* 5(9):82–95.

Stahl-Biskup, E. and F. Saez. 2002. *Thyme: The Genus Thymus*. CRC Press, Boca Raton, FL, 330.

Surveswaran, S., Y. Cai, H. Corke, and M. Sun. 2007. Systematic evaluation of natural phenolic antioxidants from 133 Indian medicinal plants. *Food Chemistry* 102:938–53.

Tanko, H., D.J. Carrier, L. Duan, and E. Clausen. 2005. Pre- and post-harvest processing of medicinal plants. *Plant Genetic Resources: Characterization and Utilization* 3(2):304–13.

Tenenbaum, F. 1999. *Herbs and Edible Flowers: Easy Plant for More Beautiful Gardens*. Houghton Mifflin Company, New York.

Tezer, M.N., İ. Gökler, and T. Kesercioğlu. 2010. *Çiçekli Bitkiler*. Yayın no: 1495. 1. Baskı. Nobel Yayın, Ankara., Turkey.

Toker, R., M. Görlükcü, and H. Tokgöz. (2015). Tıbbi ve aromatik bitkilerin gıda sanayisinde kullanim alanları, *TÜRKTOB-Türkiye Tohumcular Birliği Dergisi* 4(15):54–9 (in Turkish).

Tusevski, O., A. Kostovska, A. Iloska, L. Trajkovska, and S. Simic. 2014. Phenolic production and antioxidant properties of some Macedonian medicinal plants. *Open Life Sciences* 9(9):888–900.

Upson, T.M., R.J. Grayer, J.R. Greenham, C.A. Williams, F. Al-Ghamdi, and F.H. Chen. 2000. Leaf flavonoids as systematic characters in the genera *Lavandula* and *Sabaudia*. *Biochemical Systematics and Ecology* 28:991–1007.

Velioglu, Y.S. and G. Mazza. 1991. Characterization of flavonoids in petal of *Rose damascena* by HPLC and spectral analysis. *Journal of Agricultural and Food Chemistry* 39:463–67.

WHO. 1993. Guidelines on the conservation of medicinal plants. The World Health Organization (WHO), IUCN-The World Conservation Union, WWF-World Wide Fund for Nature. Gland & Geneva, Switzerland. http://apps.who.int/medicinedocs/documents/s7150e/ s7150e.pdf (Accessed September 15, 2015).

Woodward, P. 2000. *Asian Herbs and Vegetables.* Hylan House Publishing, Flemington, VIC, 146 pp.

Yalçın, A. 2010. *A'dan Z'ye Şifalı Bitkiler, Ev Ilaçları, Şifalı Sular Ansiklopedisi.* 16. Baskı, Türkmen Kitabevi (In Turkish).

Youwei, Z., Z. Jinlian, and P. Yonghong. 2008. A comparative study on the free radical scavenging activities of some fresh flowers in southern China. *LWT-Food Science and Technology* 41(9), 1586–91.

Zeng, Y., M. Deng, L. Zhencheng, and Y. Peng. 2014. Evaluation of antioxidant activities of extracts from 19 Chinese edible flowers. *SpringerPlus* 3:315.

Zuccarini, P. 2009. Camphor: Risks and benefits of a widely used natural product. *Journal of Applied Sciences and Environmental Management* 13(2): 69–74.

11 Drying of Mushrooms

Xin Jin
Soochow University

CONTENTS

11.1 INTRODUCTION

Mushroom is a macrofungus with a distinctive fruiting body, which can be either epigenous or hypogenous, and is large enough to be seen with the naked eye and to be picked by hand (Chang and Miles, 1992). History of mushrooms can be traced back to 300 million years ago (Editorial, 1997). In ancient times, Greeks, Egyptians, Romans, Chinese, and Mexicans treated mushrooms as a delicacy, prized them for their purported therapeutic values, and in some cases, prized them as treasures in religious rites (Chang, 2006). The Egyptians believed that mushroom was a gift from God, while the ancient Romans called them a "divine food," and the Chinese considered them "the elixir of life" (Smith et al., 2002). There are more than 2000 species of mushroom existing in nature that are considered as edible mushrooms. Among the edible mushrooms, about 25 species are widely accepted as food, and 10 species of these are reaching the industrial scale of production. A few species have a long history of application in the field of medicine (mainly in Asian countries, especially in China) (Smith et al., 2002; Chang, 2006; Barros et al., 2007). The term "mushroom" is used differently across the world. In western countries, *Agaricus bisporus* (white button mushroom) is referred to as mushroom, whereas in Asian countries, the term "mushroom" includes many other types of mushrooms that are either edible or can be used as medicine. Among the available edible mushrooms in Asia, the most common ones are shiitake mushroom (*Lentinula edodes*), oyster mushroom (*Pleurotus flavus*), enokitake (*Flammulina velutipes*), white jelly fungus (*Tremella fuciformis*), black fungus (*Auricularia auricula*, Jew's ear, wood ear), *Boletus edulis* mushroom, and the white button mushroom (*A. bisporus*). There are more than 270 species of mushrooms that are known to have various therapeutic properties, and the term "medical mushroom" is now increasingly gaining worldwide recognition (Smith et al., 2002).

Mushrooms as daily foods for human beings, either in their natural form or in processed form, have been characterized as tasty and healthy foods since a long time ago. Although their nutritional values have been investigated only in recent years, the health benefits of mushrooms to human beings have gained a lot of attention both from the scientific point of view and from the customer's point of view in the industrial world. Mushrooms are low in calories and high in vegetable proteins and can be used as good sources for protein supplement. Mushrooms also contain other health-promoting compounds, such as chitin, iron, zinc, fiber, essential amino acids, vitamins, and minerals (shown in Table 11.1). Therefore, mushrooms can be used as therapeutic foods to reduce the incidence of cardiovascular diseases, diabetic diseases, hypertension, hypercholesterolemia, and certain cancers (Mattila et al., 2001; Jeong et al., 2010).

As shown in Table 11.1, the average moisture content of mushrooms is $>90\%$; therefore, the shelf life of fresh cultivated mushroom is very short, <24 h at ambient conditions (Giri and Prasad, 2007). Despite storage under controlled conditions, the shelf life for mushrooms is very short. Various physiological and morphological changes occur after harvest. Because of higher rate of respiration, extensive moisture loss can occur. Quality deterioration is also significant, such as loss of color, flavor, nutritional values, and potential of microbial spoilage. Therefore, fresh cultivated mushroom should be consumed immediately or should be processed promptly after harvest. For centuries, various processing technologies have been used to transform fresh mushrooms to different forms of products, to extend the shelf life, and to serve as ingredients for further processing. After food processing, the shelf life of the vegetables can be extended up to more than 1 year, which can be consumed all year round and can be easily transported anywhere in the world (Walde et al., 2006). The most common technologies for food preservations are freezing, canning, and drying. Among them, drying is known to be the simplest and a more economical process (Walde et al., 2006; Zhang et al., 2013). After drying, water activity is lowered, and as a result, the microbial activity will also be low. Therefore, it allows the preservation of foods over an extended period. Drying is an old technology that can be traced back to ancient times when our ancestors used sun and wind to dry food naturally. With thousands of years of experience and modern research development, from ancient solar drying to modern freeze-drying, many drying methods have been developed, such as hot air drying, microwave drying, solar drying, spray drying, freeze-drying, vacuum drying, combination drying, etc. All the aforementioned drying methods

TABLE 11.1
Nutritional Values of Different Types of Mushrooms

Nutrient	Unit (per 100 g)	White Button Mushroom	Oyster Mushroom	Shiitake	Jew's Ear
Water	g	92.45	89.18	89.74	92.59
Energy	kcal	22	33	34	25
Protein	g	3.09	3.31	2.24	0.48
Total lipid (fat)	g	0.34	0.41	0.49	0.04
Carbohydrate, by difference	g	3.26	6.09	6.79	6.75
Fiber, total dietary	g	1	2.3	2.5	N.A.
Sugars, total	g	1.98	1.11	2.38	N.A.
Minerals					
Calcium, Ca	mg	3	3	2	16
Iron, Fe	mg	0.5	1.33	0.41	0.56
Magnesium, Mg	mg	9	18	20	25
Phosphorus, P	mg	86	120	112	14
Potassium, K	mg	318	420	304	43
Sodium, Na	mg	5	18	9	9
Zinc, Zn	mg	0.52	0.77	1.03	0.66
Vitamins					
Thiamin	mg	0.081	0.125	0.015	0.081
Riboflavin	mg	0.402	0.349	0.217	0.204
Niacin	mg	3.607	4.956	3.877	0.070
Vitamin B6	mg	0.104	0.11	0.293	0.088
Folate	μg	17	38	13	19
Vitamin D (D2 + D3)	μg	0.2	0.7	0.4	0
Vitamin D	IU	7	29	18	0

Source: Data taken from USDA nutrition database.

have been reported in recent years for different kinds of mushrooms (Loch-Bonazzi et al., 1992; Pappas et al., 1999; Torringa et al., 2001).

11.2 QUALITY CHANGES DURING DRYING

Originally, drying was applied to extend the shelf life of fresh food materials for safe products. Gradually, with the development of modern technologies, processors and consumers in the industrialized world have started to pursue high-quality food products. The overall quality attributes for any food product that are important for both food industry and customers are optical properties (color, appearance, etc.), sensory properties (odor, taste, flavor, etc.), structural properties (density, porosity, volume, etc.), textural properties (hardness, chewiness, etc.), rehydration properties (rehydration rate, rehydration capacity), and nutritional characteristics (vitamins, enzymes, antioxidants, etc.) (Krokida et al., 2000; Vadivambal and Jayas, 2007). Almost all the drying methods have some adverse effect on one or more of the quality attributes. During hot air drying, with the increased heat load during drying, significant quality deterioration and nutrient degradation occur such as darkening in color, loss of flavor, decrease in rehydration capacity, structure deformation, and loss of nutritional values (Lewicki, 2006). Kotwaliwale et al. (2007) reported significant changes in textural and optical properties during hot air drying of oyster mushroom due to the soft texture of mushrooms. Hardness and chewiness increased, whereas cohesiveness and springiness increased initially and decreased in the end. For color, white index decreased but yellow index increased, which is not acceptable for customers. Microwave drying can result in uneven heating, texture damage, and charring of edges, etc. Although freeze-dried product has the highest quality compared to others in terms of optical properties and sensory properties, the highly porous structure makes it highly fragile.

11.3 PRETREATMENTS PRIOR TO DRYING

Mushrooms are extremely sensitive to drying conditions, and various physiological and morphological changes occur during the processing. Pretreatments such as blanching with either water or steam and soaking in sulfite, potassium metabisulfite (KMS), citric acid, fermented whey, and fermented milk i.e., curd, or other solutions are quite often applied prior to drying for better quality (color, texture, etc.) preceding the drying step (Gothandapani et al., 1997; Prajapati et al., 2011; Severini et al., 2005; Walde et al., 2006; Tunde-Akintunde, 2010). Although the mushrooms are different in types and in some aspects of physical properties, the pretreatments are quite similar with the main purpose of retaining the color, which easily degrades during processing. Color has been considered to be of great importance in determining the quality of food product, because this is the most transparent criterion for consumers to decide whether to purchase the food product or not. Among all the chemical treatments, KMS and citric acid are often used on an industrial scale, for preventing browning reactions and color degradation during drying and storage (Brennan et al., 1999, 2000; Argyropoulos et al., 2011b). For example, the Irish mushroom industry has been using 1 g/L sodium metabisulfite to treat mushrooms prior to slicing to whiten the slices and to keep them whiter for several days (Brennan et al., 2000). Looking into literature, various combinations of treatments can be found, with different ratio of chemicals and different processing times. Kumar et al. (2013) evaluated three batches of mushrooms pretreated with different concentrations of KMS and citric acid solutions for 15 min: 1.0% KMS, 0.3% citric acid, 0.5% KMS with 0.2% citric acid. Their results showed that the mushrooms pretreated with 1.0% KMS gave the maximum values of whiteness and better rehydration properties. Walde et al. (2006) evaluated different pretreatment influences on drying rate for white button mushroom and oyster mushroom, which includes blanching; soaking in fermented milk i.e., curd; blanching followed by soaking in curd; soaking in fermented whey; and blanching followed by soaking in fermented whey. The blanching treatment was conducted with 2% salt solution at 90°C, followed by 1000 ppm KMS treatment. For oyster mushrooms, the blanching time was 3 min, whereas for button mushrooms, the blanching time was for 4 min. They reported that for both oyster and button mushrooms, drying was significantly enhanced by soaking in curds or fermented whey compared to other treatments in all types of dryers. In the work of Argyropoulos et al. (2011b), the samples were immersed in a solution of 0.25% KMS and 0.1% citric acid for 5 min at room temperature. In another study for *B. edulis* mushroom, the same author conducted two thermal treatments (water or steam blanching at 98°C for 3 min) and two chemical treatments (0.25 citric acid, 0.25% KMS, 10 min) at room temperature. The results indicated that both thermal treatments are not recommend for *B. edulis* mushrooms as they have adverse effects on both color and texture. Besides, they found that chemical treatment did not influence the color of mushrooms positively, which is in conflict with other studies. From these examples, we are convinced that the guidance of pretreatment is still based on experience. Systematic study is still lacking in this area. For example, blanching or steaming are often applied to inactivate the enzymes, which are responsible for the enzymatic browning reaction, and the common indicators are peroxidase or lipoxygenase. The processing time should be based on two criteria: (1) time for heat transfer, which is linked with the diffusion path and (2) time for inactivation. Although blanching can enhance drying rate significantly, it has adverse effects on hardness compared to other treatments (Figure 11.1). The chemical treatment is normally applied at room temperature, and the processing time varies between 3 and 20 min. The chemicals absorbed on the surface of the sample can protect the sample against oxidation. The drawback of this aqueous treatment is that it reduces the nutrient content as a result of removal of water-soluble nutrients (Gothandapani et al., 1997). Therefore, the dosage of chemicals and the processing time should be optimized to maximize the effects of those chemicals as well as to lower the leakage of soluble nutrients.

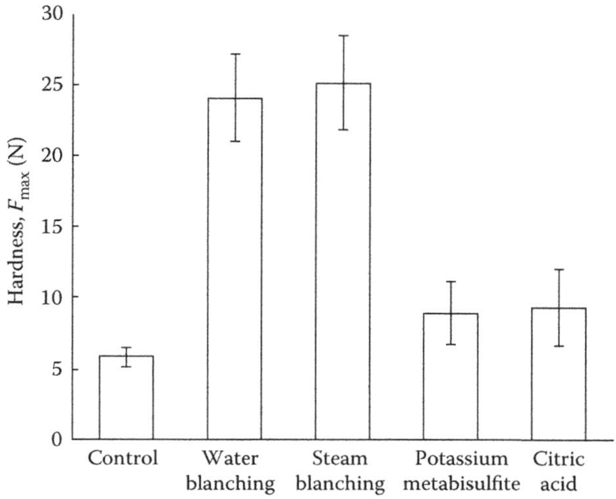

FIGURE 11.1 Effect of pretreatments on hardness of *B. edulis* mushrooms dried at 60°C. (From Argyropoulos et al., 2011b.)

11.4 DRYING OF WHITE BUTTON MUSHROOMS (*AGARICUS BISPORUS*)

Because the white button mushroom (*A. bisporus*) is the most widely cultivated and consumed mushroom, which contributes to about 40% of the total world mushroom production (Giri and Prasad, 2007), the majority of the literature that concerns drying of mushrooms is for button mushroom. This section focuses on every aspect of drying of button mushrooms, such as the characteristics of each drying method, the impact on quality attributes, and the pretreatments that can be applied before drying to retain food quality.

11.4.1 Drying Methods and Their Impact on Quality Aspects

For drying of fruits and vegetables, there are various drying methods available, such as simple use of solar energy to dry food product naturally, convective drying with heated air, and more advanced and complex methods, for instance, freeze-drying, osmotic dehydration, microwave drying, and combination drying, which combines two or more technologies together. Because of the special natural status, mushrooms are very sensitive to temperatures and drying medium; therefore, choosing the right drying method and proper drying conditions will be the key factor for a successful drying operation.

11.4.1.1 Convective Drying

Among the available drying methods, convective drying is one of the most applied and the most economical drying methods. Because convective drying is a heat-intensive process, during drying, with the increased heat load, significant quality deterioration may happen such as degradation of color, texture, and nutrients. Normally, the drying temperature is maintained between 50°C and 70°C (Giri and Prasad, 2007). Temperature has significant influence on texture and color. Higher temperature caused intensive color deterioration as well as increased hardness. Kotwaliwale et al. (2007) found that during hot air drying, textural and optical properties such as hardness, cohesiveness, and color changed significantly. Temperature is the key factor that had an inverse effect on whiteness of mushrooms. Nijhuis et al. (1998) reported significant aroma loss in hot air dried button mushrooms. As these quality attributes are very sensitive to temperature, therefore, choosing a moderate temperature and designing optimal temperature trajectories that the product undergoes, which can minimize the quality deterioration, is crucial. Because of the special nature status and the requirement for high quality food product, application of convective drying in mushroom dried products is limited. Model-based dynamic optimization can be an option to find the optimal drying operation conditions that can retain product quality in all the aspects. Dynamic optimization has a long history of improving the quality of food during food processing (Mishkin et al., 1984; Banga et al., 1991; Banga and Singh, 1994; Madamba, 1997; Kiranoudis and Markatos, 2000; Jin et al., 2014). However, the application of dynamic optimization in drying of mushrooms is still limited, which can be a topic for future research.

11.4.1.2 Microwave Drying and Microwave-Related Combined Drying Methods

Microwave heating can transfer the electromagnetic field energy into thermal energy directly, which means that materials can absorb microwave energy internally and convert it into heat (Vadivambal and Jayas, 2007). The most significant characteristic of microwave heating compared with conventional heating is volumetric heating (Mullin, 1995). In microwave heating, heat can be generated throughout the whole material, as a result microwave drying has the advantage of more rapid and uniform heating. It is a more energy-efficient drying process, compared to the conventional convective drying method. Among the quality aspect, in the case of mushroom drying, microwaves can inactivate the enzymes that are responsible for enzymatic browning (Devece et al., 1999). Lombraña et al. (2010) studied drying of sliced mushroom with microwave drying at different operational conditions. They pointed out the importance of position effect of temperature control and pressure control, which would affect the product quality. With the possibility of combined drying and enzyme inactivation, these results confirm that microwave drying has a great potential for industrial application. However, microwave drying alone can have several drawbacks, which include uneven heating, texture damage, charring of edges, and limited microwave radiation penetration into the product (Nijhuis et al., 1998; Zhang et al., 2006; Çelen et al., 2008). Therefore, microwave drying alone is not recommended for quality considerations (Bondaruk et al., 2007; Lombraña et al., 2010).

To overcome the limitations of microwave drying alone and other conventional drying methods, many investigations have been carried out for combining microwave heating with a variety of other effective drying methods such as microwave-assisted hot air drying, microwave–vacuum drying, and microwave-assisted freeze-drying. In all the cases with the combined drying methods, drying rate is enhanced and the quality is improved or maintained as microwave alone dried or other conventionally dried (Zhang and Xu, 2003).

Microwave energy can be supplied either by transverse magnetic modes or multimode cavity microwave oven. Funebo and Ohlsson (1998) evaluated these two methods. They found that for microwave-assisted hot air drying of mushroom, with the supply of microwave energy, the drying time was reduced by a factor of 4 compared to hot air drying only. Furthermore, combined microwave hot air drying can have less shrinkage and high open pore porosity, thus, improving rehydration properties (Funebo and Ohlsson, 1998; Torringa et al., 2001).

Microwave–vacuum drying combines the advantages of both microwave and vacuum drying, which can achieve fast mass transfer at lower temperature, and rapid energy transfer by vacuum drying and microwave heating, respectively (Zhang et al., 2006; Vadivambal and Jayas, 2007). Thus, this combined drying method has the potential to increase drying efficiency and product quality. Giri and Prasad (2009) reported that microwave–vacuum dried mushrooms had significantly higher rehydration ratio, lower density, better color, and softer texture than those dried with hot air. Furthermore, a sensory panel reported that microwave–vacuum dried mushrooms had comparable quality with freeze-dried mushrooms, in terms of appearance, color, and overall acceptability.

11.4.1.3 Freeze-Drying and Vacuum Drying

Freeze-drying has two steps: (1) freezing the product; (2) lowering the vacuum to sublimate the frozen water directly from the solid phase to gas phase at a lower temperature; as a result, the structure of food will not be damaged severely (Nijhuis et al., 1998; Vega-Mercado et al., 2001). Vacuum drying has a similar characteristic, which can dry the product under low vacuum, which in turn lowers the drying temperature. Both the technologies have the advantage of drying under lower temperatures, which is beneficial for heat-sensitive quality attributes. Low processing temperature and absence of air can minimize the product quality deterioration. Therefore, among the available drying methods, these two are the best ones to obtain high-quality dried food product, but far more expensive than convective drying. Giri and Prasad (2009) compared freeze-dried, microwave–vacuum dried, and conventional air-dried button mushrooms. They found that freeze-dried product had the best quality, which involves maximum rehydration ratio, best color preservation, and lowest hardness. Nijhuis et al. (1998) and Argyropoulos et al. (2011a) evaluated several drying methods in different aspects of characteristics including several quality attributes. Freeze-drying gained the highest score in maintenance of color, nutritional value, rehydration capacity, crispiness, and fresh appearance. Owing to the high investment cost, operational cost, high energy consumption, and maintenance costs, industrial application of these technologies is very limited (Giri and Prasad, 2009).

11.4.1.4 Desiccant Drying

Drying is an energy-intensive process, which contributes to about 15% of industrial energy consumption (Kemp, 2005) and in which 10% of the energy is used in the food industry (Mujumdar, 1997). For conventional convective drying processes, the energy efficiency is always low. Therefore, in recent years, some studies have been done to investigate the potential to improve energy efficiency of the dryer by using desiccant dehumidification as an add-on to the convective dryer process (Atuonwu et al., 2011). Since the pretreated air has lowered humidity, it can increase the drying potential of the air (Lewicki, 2006) and enhance the drying rate, which in turn has an impact on the residence time. As a result, drying can be achieved at a lower temperature. Quality degradation during drying depends on temperature, moisture content, and residence time in the dryer; thus, desiccant drying is beneficial to retain food quality during drying (Tutova and Fel'dman, 1976; Strumillo et al., 1995; Tadayyon et al., 1997; Gurtas Seyhan and Evranuz, 2000; Nagaya et al., 2006). This is also applicable for drying of mushrooms, which are quite heat-sensitive. Gurtas Seyhan and Evranuz (2000) investigated two types of desiccants—zeolite and silica gel. With the dehumidified air, drying can be performed at lower temperatures ranging from 20°C to 30°C, which is beneficial for obtaining the optimum quality in terms of appearance and rehydration capacity.

11.5 DRYING OF SHIITAKE MUSHROOMS (*LENTINUS EDODES*)

Shiitake mushrooms (*Lentinus edodes*) are one of the most common edible mushrooms, and it is the second most cultivated edible mushroom in the world (Chandra et al., 2011). The cultivation and consumption of shiitake mushrooms as both food and medicine dates back to thousands of years ago. It is more popular in Asian countries such as China, Japan, Korea, Thailand, etc. China accounts for more than 70% of the world's shiitake production (Zhang et al., 2007). Shiitake mushroom has high nutritional contents, including 18 types of amino acids, which can provide nearly the ideal ratios of all essential amino acids that are needed for human nutrition (Turło et al, 2008). Lentinan and vitamin B12 can support the human immune system (Zhang et al., 2011). Besides, it also has high content of other vitamins (B1, B2, C) and minerals. Various studies have shown that shiitake mushrooms have antitumor, antimicrobial, liver-function-improving, and cholesterol-lowering effects (Mizuno et al., 1995). In Japan and China, several chemical constituents from shiitake mushroom have been analyzed and extracted to be used in treating cancer and other diseases (Mizuno, 1995; Zhang et al., 2013). Similar to other mushrooms, fresh shiitake mushroom is highly perishable, and the quality starts to deteriorate immediately after harvest. As a result, the shelf life is normally less than 24 h. Therefore, it has to be processed to prolong the shelf life. There are several methods that can be used, such as drying and canning; of these, drying was found to be a comparatively cheaper method. Although shiitake mushroom is the second most cultivated mushroom in the world, there are comparatively less studies on it compared with white button mushroom. Only in recent years, more research is being carried out on shiitake mushroom, however, mainly focusing on the functional values of the shiitake mushroom. In the 1970s, β-D-glucan was identified as the active component that was effective for the treatment of cancer. Since then, studies on pharmacologically active components have been accelerated (Mizuno, 1995). For example, Minato et al. (1999) evaluated the degradation of an antitumor polysaccharide lentinan during storage at different conditions. Choi et al. (2006) reported that heat treatment could significantly enhance the overall antioxidant activities of shiitake mushroom.

11.5.1 Drying Methods and Their Impacts on Quality Aspects

In spite of the substantial increase in the consumption of shiitake mushroom, the number of researches that focused on drying of shiitake is still limited, and most of them were done by researchers from China, Japan, Korea, or Thailand. Most of them focused on the comparison of different drying methods, such as hot air drying, solar drying, freeze-drying, vacuum drying, microwave drying, and combination drying methods. The quality attributes that researchers have focused on are color, active amino acids, texture, flavor, sensory properties, and rehydration ratio. Moreover, exceptional attention has been paid to the antioxidant properties of the dried mushrooms.

11.5.1.1 Hot Air Drying

In the mushroom drying industry, hot air drying is still the dominant drying method because of its simplicity and cost-effective characteristics. Although it is the most widely used technology, research on hot air drying is limited, because of its low-quality food product, time consumption, reduced hydrophilic properties, poor rehydration properties, dark color, collapsed structure, and hard texture of the products (García-Segovia et al., 2011; Kantrong et al., 2014; Qi et al., 2014). Nevertheless, some works showed positive results. Zhang et al. (2012) have shown that hot air drying at 50°C resulted in high total phenolic, amino acid, uronic acid, and neutral sugar content and antioxidant activity. It was very interesting to see that the stipes were more nutritional than caps in some quality aspects in terms of total phenolic, amino acid, and neutral sugar contents, which was unexpected. Similarly, Tian et al. (2016) discovered that hot air drying resulted in an increase in the vitamin B12 content as well as total free amino acids and total amount of volatile compounds. In term of drying kinetics, the literature is limited. Guo et al. (2014) found that nonuniform intermittent drying reached a high chemical composition, better color retention, and improved rehydration properties. Rhim and Lee (2011) did a comprehensive study on the drying kinetics of whole and sliced shiitake mushrooms at different drying temperatures of 40°C, 50°C, 60°C, and 70°C. Obviously, increased temperature resulted in increased drying rate, and sliced mushrooms showed increased drying rate compared to the whole mushrooms.

11.5.1.2 Other Drying Methods

In recent years, besides hot air drying, new drying methods and dryers, especially hybrid drying methods, are being developed to overcome the drawbacks of hot air drying method. For instance, the combination of freeze-drying followed by mid-infrared drying can save 48% of drying time compared to freeze-drying alone while maintaining the product quality, which takes advantage of infrared heating to shorten drying time, improve energy efficiency, and improve product quality (Wang et al., 2015). Besides, the same group conducted research on combined mid-infrared-assisted convection drying of shiitake mushrooms (Wang et al., 2014). Their results showed that this hybrid drying method yielded minimal shrinkage, reduced hardness, better color retention, and improved nutrient retention (Figure 11.2).

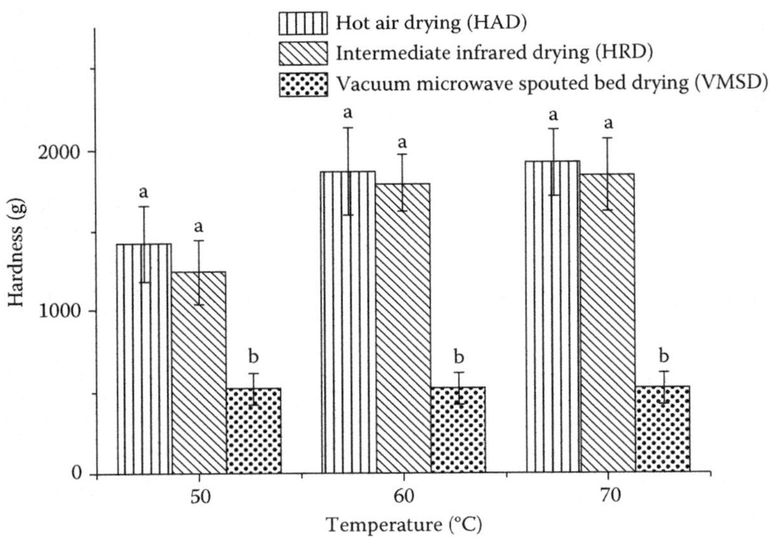

FIGURE 11.2 Effect of different drying methods on the hardness of shiitake mushroom cubes. (From Qi, L.-L. et al., *Drying Technol.*, 32(15), 1751–1761, 2014.)

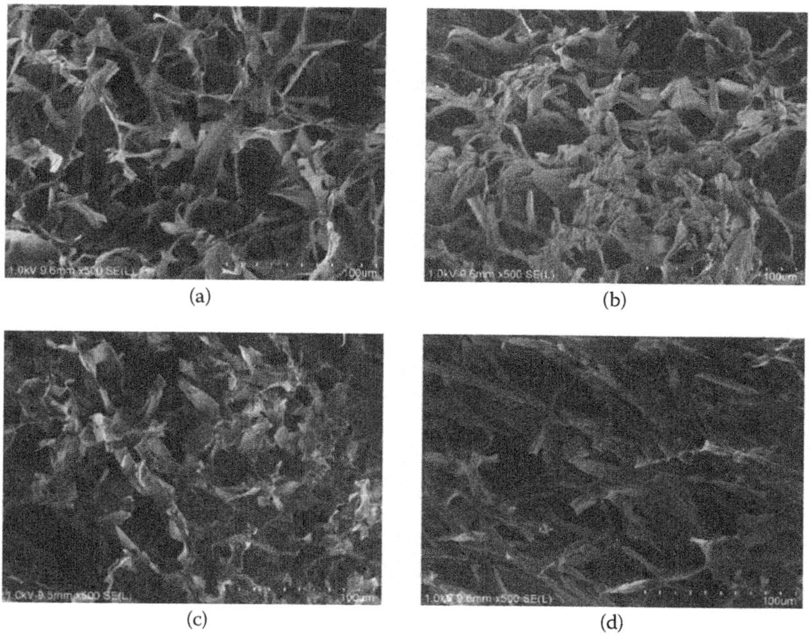

FIGURE 11.3 Scanning electron micrographs of shiitake mushroom dried with different methods. The magnification was set as ×500: (a) Mid-infrared-assisted convection-dried, (b) hot air coupled with radio frequency-dried, (c) hot air coupled with microwave dried, and (d) hot air-dried. (From Wang et al., 2014.)

Recent studies have shown that microwave–vacuum drying is a potential drying method for high-quality dried biological products. Under vacuum, the drying temperature can be lower than ordinary microwave drying conditions due to the lower boiling point (Kantrong et al., 2014). With the benefit of lower operational temperature, microwave–vacuum drying can maintain larger amounts of taste-active amino acids and other nutritional values; the color is very similar to that of fresh shiitake mushroom. Furthermore, the microstructure of shiitake mushrooms after being dried with different drying methods were significantly different, which can be seen from the scanning electron micrograph photos of shiitake mushrooms (Figure 11.3). The microwave-assisted dried mushroom has the unique honeycomb-like network and less-collapsed structure, which results in high rehydration ratio (Qi et al., 2014; Wang et al., 2014; Tian et al., 2016). In addition, Kantrong et al. (2014) also indicated that drying that involves microwave–vacuum combined with infrared drying could provide better qualities in terms of color, rehydration ratio, and texture.

11.6 DRYING OF OYSTER MUSHROOM

Oyster mushroom (*Pleurotus* spp) is the third largest commercially cultivated mushroom in the world (Obodai et al., 2003). The species of the genus *Pleurotus* have the potential to increase ruminate digestibility and nutritional content of wheat and rice straw (Rajarathnam et al., 1989). Among all the available *Pleurotus* species, *Pleurotus ostreatus* is the most cultivated species (Obodai et al., 2003). Because of its high moisture content and open gills, it is highly perishable as compared to button mushroom, which has closed gills and a membrane-type outer layer over the fruit body (Arumuganathan et al., 2010). Therefore, it has to be processed to prolong its shelf life and for off-season sale as well. Various drying methods can be applied to dry oyster mushroom, such as hot air drying, freeze-drying, solar drying, vacuum drying, etc. Prior to drying, pretreatments of mushroom in one form or the other, by soaking in KMS, citric acid, curd, and fermented whey, either alone or in combination, can improve the product quality or enhance the drying process. Blanching has a negative effect on color, texture, and rehydration capacity (Martínez-Soto et al., 2001; Kotwaliwale et al., 2007). Dipping in KMS or citric acid can improve the color as well as the rehydration capacity (Pal and Chakraverty, 1997; Martínez-Soto et al., 2001; Kotwaliwale et al., 2007). Gothandapani et al. (1997) found that storage of mushrooms after treatment with KMS at a higher concentration (1.5%) reduces the microbial spoilage. In contrast to others, Gothandapani et al. (1997) found that KMS concentration has no effect on rehydration capacity, but it has a negative effect on nutritional values. Water blanching or dipping the samples in fermented whey or curds are found to significantly enhance the drying process (Pal and Chakraverty, 1997; Walde et al., 2006; Srivastava et al., 2009).

The quality of dried oyster mushrooms depends highly on the drying methods. Hot air drying by convection in cabinet or tunnel dryers is still the dominant method, although it has a negative effect on color and texture (Gothandapani et al., 1997; Kotwaliwale et al., 2007). Drying temperature normally varies between 40°C and 80°C, and it has adverse effect on the whiteness of mushroom (Walde et al., 2006). Vacuum drying is a slow drying process due to the limited heat and mass transfer rate, although the product quality is improved compared to that in hot air drying (Walde et al., 2006). Freeze-drying gives better product quality than hot air drying or vacuum drying with respect to color and rehydration capacity (Martínez-Soto et al., 2001), whereas the structure of the freeze-dried product was fragile (Martínez-Soto et al., 2001; Arumuganathan et al., 2010). Nevertheless, the slow drying rates and high energy cost make freeze-drying a relatively expensive process. Regarding the drying rate, microwave drying alone can enhance drying progress significantly; however, the product case hardening is still a serious problem. Due to the unique physical properties of oyster mushroom, new and advanced drying methods should be developed to overcome the drawbacks of the available drying methods. Hybrid drying methods that take advantages of single drying methods have been applied successfully in drying white button mushrooms.

11.7 DRYING OF *BOLETUS EDULIS* MUSHROOM

Following white button mushroom, shiitake mushroom, and oyster mushroom, *B. edulis* is another type of mushroom that is the most frequently acquired wild edible mushroom worldwide. They occur naturally between summer and autumn in temperate forests of the Northern hemisphere across Europe, Asia, and North America, and they have also been introduced to South Africa and New Zealand

(Argyropoulos et al., 2011b). Because it is only seasonal, they are served worldwide as frozen, brined, and dried products (Jaworska and Bernas, 2010; Argyropoulos et al., 2011b). Although there are various drying technologies available for drying of mushrooms, hot air drying is still the most commonly applied method for drying of several wild mushrooms. Similar to other mushrooms, pretreatment is necessary prior to drying to retain the quality attributes. The drying temperature ranges between 50°C and 70°C (García-Pascual et al., 2005; Hernando et al., 2008; Argyropoulos et al., 2011b). According to Argyropoulos et al. (2011b), chemical pretreatments such as KMS or citric acid did not have a positive influence on the color, and during drying, lightness decreased slightly while yellowness and redness increased. Blanching either by water or by steam is not recommended because it results in harder texture and darker color. They also suggested that 60°C is the temperature limit for drying of *B. edulis* mushrooms. Besides, freeze-drying is another drying technique that was mentioned in literature. Hernando et al. (2008) reported that the rehydration capacity of freeze-dried *B. edulis* mushrooms was higher than that of air-dried ones. This is probably because the porous structure generated during freeze-drying enables the rehydration to take place mainly at the extracellular level. Therefore, regardless of the rehydration temperature, the product can absorb water more quickly.

Moreover, there are many edible wild mushrooms in the world. For example, in China, Yunnan province is a specific region abundant in wild-grown mushrooms, and more than 880 species are identified as edible. Among them, *Tricholoma matsutake* is the most valuable wild-grown mushroom. However, research on this specific mushroom are very limited. In recent years, an increasing number of works focused on the active components of matsutake mushroom.

11.8 DRYING OF JEW'S EAR AND WHITE JELLY MUSHROOM

Among the available edible mushrooms, *A. auricula*, commonly known as Jew's ear or tree ear, and *T. fuciformis*, known as white jelly mushroom or white jelly leaf, are most popular in Asian countries, mostly in China. Both mushrooms are edible mushrooms of Heterobasidiae, a subclass of Basidiomycetes.

Since ancient times, these mushrooms have been widely used in Chinese cuisine as well as for medicinal purposes. A significant number of works have been focused on the functionalities of these mushrooms and the active compounds within these mushrooms. For example, white jelly mushroom can improve immunodeficiency and prevent senile degradation of microvessels (Cheung, 1996; Cha et al., 2014). Because of the high dietary fiber content, both mushrooms have the potential of hypocholesterolemic effect. Besides, both mushrooms also have the potential to lower total plasma cholesterol levels (Cheung, 1996). Furthermore, the significant antitumor effect is more attractive, which have been used to prevent and cure tumors for centuries (Yu et al., 2009; Cha et al., 2014).

Because of the extremely high moisture content, fresh Jew's ear mushroom and fresh white jelly mushroom are highly perishable compared to other mushrooms. Therefore, in contrast to other edible mushrooms, they are often served in the dried form. Because of the geographical restrictions, most of the research is conducted in Asian countries, such as China, Japan, and Korea. In the industry, most white jelly mushrooms or Jew's ear mushrooms are dried with conventional drying methods such as sun drying or hot air drying. The air temperature is set between 40°C and 50°C, and the drying process lasts about 8 h. Temperature and airflow control are the key factors influencing the quality of dried product. In recent years, an increasing amount of research has focused on the investigation of new drying methods for better-improved quality of dried product, for instance, microwave drying, vacuum drying, freeze-drying, or combination drying method such as microwave–vacuum drying. Due to the geographical limitations, most of these works are published within the national journals in the national languages. Nevertheless, all these mushrooms have similarities in characteristics in one way or another. We may learn from the accumulated experience on drying of other well-researched mushrooms.

11.9 CONCLUSIONS AND FUTURE PERSPECTIVES

The global market for the mushroom industry, including edible mushroom, medical mushroom, and wild mushroom, was estimated at over $45 billion for 2005 (Chang, 2006). Since mushrooms are treated as delicacies, they become more and more popular because of their high nutritional values and potential positive health effects to the human body. Thus, we believe that the word market for mushrooms will increase continuously in the coming years. With the highly perishable properties, drying will still be the dominant preservation technology.

Drying of mushrooms has a long history since ancient times, when our ancestors preserved the natural foods for off-season use. With the centuries' research and development, various drying methods and drying technologies have been developed, such as hot air drying, microwave drying, vacuum drying, freeze-drying, and combination drying methods. All these drying methods have their advantages and disadvantages in different aspects. For the food industry, quality, especially nutritional values, and energy serve as a driving force for research and development. Because drying is an energy-intensive process, considering the global energy shortage and global warming, there is an urgent need for energy-efficient processes.

Furthermore, considering the quality aspect, in spite of the different physical properties of different mushrooms, regardless of the drying method used, pretreatment is necessary, especially for the retention of the color of the mushrooms. Because of the unique characteristics, blanching either with hot water or with hot steam is not recommended. Pretreatment with aqueous solution is often used, even on industrial scale. However, from the author's point of view, almost all the treatment with aqueous solutions are based on experience. Systematic study is required on both the dosage of chemical and the treatment time, as both factors influence nutrient retention, especially the water-soluble nutrients.

Among all the available mushrooms, white button mushrooms are well studied; there are other species that are treated as nutritional food sources or good medical sources. However, most of them are treated with conventional drying methods, such as conventional solar drying or hot air drying. Further studies on the retention of these functional compounds are required. Considering the traditional convective drying method, optimization of drying trajectories can be an option to increase both energy efficiency and product quality. Otherwise, the modern hybrid drying technology can be applied for these purposes, which can take advantage of each of the single drying methods.

REFERENCES

Argyropoulos, D., Heindl, A., and Müller, J. (2011a). Assessment of convection, hot-air combined with microwave-vacuum and freeze-drying methods for mushrooms with regard to product quality. *International Journal of Food Science and Technology*, *46*(2), 333–342.

Argyropoulos, D., Khan, M. T., and Müller, J. (2011b). Effect of air temperature and pretreatment on color changes and texture of dried *Boletus edulis* mushroom. *Drying Technology*, *29*(16), 1890–1900.

Arumuganathan, T., Manikantan, M. R., Indurani, C., Rai, R. D., and Kamal, S. (2010). Texture and quality parameters of oyster mushroom as influenced by drying methods. *International Agrophysics*, *24*, 339–342.

Atuonwu, J. C., van Straten, G., van Deventer, H. C., and van Boxtel, A. J. B. (2011). Model-based energy efficiency optimization of a low-temperature adsorption dryer. *Chemical Engineering and Technology*, *34*(10), 1723–1732.

Banga, J.R., and Singh, R.P. (1994). Optimization of air drying of foods. *Journal of Food Engineering*, *23*, 189–211.

Banga, J. R., Perez-Martin, R. I., Gallardo, J. M., and Casares, J. J. (1991). Optimization of the thermal processing of conduction-heated canned foods: Study of several objective functions. *Journal of Food Engineering*, *14*(1), 25–51.

Barros, L., Baptista, P., Correia, D. M., Casal, S., Oliveira, B., and Ferreira, I. C. F. R. (2007). Fatty acid and sugar compositions, and nutritional value of five wild edible mushrooms from Northeast Portugal. *Food Chemistry*, *105*(1), 140–145.

Bondaruk, J., Markowski, M., and Błaszczak, W. (2007). Effect of drying conditions on the quality of vacuum-microwave dried potato cubes. *Journal of Food Engineering*, *81*(2), 306–312.

Brennan, M., Le Port, G., and Gormley, R. (2000). Post-harvest treatment with citric acid or hydrogen peroxide to extend the shelf life of fresh sliced mushrooms. *LWT-Food Science and Technology*, *33*(4), 285–289.

Brennan, M., Le Port, G., Pulvirenti, A., and Gormley, R. (1999). The effect of sodium metabisulphite on the whiteness and keeping quality of sliced mushrooms. *LWT-Food Science and Technology*, *32*(7), 460–463.

Çelen, S., Kahveci, K., Akyol, U., and Haksever, A. (2010). Drying behavior of cultured mushrooms. *Journal of Food Processing and Preservation*, *34*(1), 27–42.

Cha, M.-H., Heo, J.-Y., Lee, C., Lo, Y. M., and Moon, B. (2014). Retracted: Quality and sensory characterization of white jelly mushroom (*Tremella fuciformis*) as a meat substitute in pork patty formulation. *Journal of Food Processing and Preservation*, *38*(3), 1018–1023.

Chandra, L. C., Smith, B. J., Clarke, S. L., Marlow, D., D'Offay, J. M., and Kuvibidila, S. R. (2011). Differential effects of shiitake- and white button mushroom-supplemented diets on hepatic steatosis in C57BL/6 mice. *Food and Chemical Toxicology*, *49*(12), 3074–3080.

Chang, S.-T. (2006). The world mushroom industry: Trends and technological development. *International Journal of Medicinal Mushrooms*, *8*(4), 297–314.

Chang, S.-T., and Miles, P. G. (1992). Mushroom biology—A new discipline. *Mycologist*, 6, 64–65.

Cheung, P. C. K. (1996). The hypocholesterolemic effect of two edible mushrooms: *Auricularia auricula* (tree-ear) and *Tremella fuciformis* (white jelly-leaf) in hypercholesterolemic rats. *Nutrition Research*, *16*(10), 1721–1725.

Choi, Y., Lee, S. M., Chun, J., Lee, H. B., and Lee, J. (2006). Influence of heat treatment on the antioxidant activities and polyphenolic compounds of shiitake (*Lentinus edodes*) mushroom. *Food Chemistry*, *99*(2), 381–387.

Devece, C., Rodríguez-López, J. N., Fenoll, L. G., Tudela, J., Catalá, J. M., de los Reyes, E., and García-Cánovas, F. (1999). Enzyme inactivation analysis for industrial blanching applications: Comparison of microwave, conventional, and combination heat treatments on mushroom polyphenoloxidase activity. *Journal of Agricultural and Food Chemistry*, *47*(11), 4506–4511.

Editorial. (1997). The magic of mushrooms. *Nature*, *388*, 340.

Funebo, T., and Ohlsson, T. (1998). Microwave-assisted air dehydration of apple and mushroom. *Journal of Food Engineering*, *38*(3), 353–367.

García-Pascual, P., Sanjuán, N., Bon, J., Carreres, J. E., and Mulet, A. (2005). Rehydration process of *Boletus edulis* mushroom: Characteristics and modelling. *Journal of the Science of Food and Agriculture*, *85*(8), 1397–1404.

García-Segovia, P., Andrés-Bello, A., and Martínez-Monzó, J. (2011). Rehydration of air-dried shiitake mushroom (*Lentinus edodes*) caps: Comparison of conventional and vacuum water immersion processes. *LWT-Food Science and Technology, 44*(2), 480–488.

Giri, S. K., and Prasad, S. (2007). Drying kinetics and rehydration characteristics of microwave-vacuum and convective hot-air dried mushrooms. *Journal of Food Engineering*, *78*(2), 512–521.

Giri, S. K., and Prasad, S. (2009). Quality and moisture sorption characteristics of microwave-vacuum, air and freeze-dried button mushroom (*Agaricus bisporus*). *Journal of Food Processing and Preservation*, *33*, 237–251.

Gothandapani, L., Parvathi, K., and John Kennedy, Z. (1997). Evaluation of different methods of drying on the quality of oyster mushroom (*Pleurotus sp.*). *Drying Technology*, *15*(6–8), 1995–2004.

Guo, X.-H., Xia, C.-Y., Tan, Y.-R., Chen, L., and Ming, J. (2014). Mathematical modeling and effect of various hot-air drying on mushroom (*Lentinus edodes*). *Journal of Integrative Agriculture*, *13*(1), 207–216.

Gurtas Seyhan, F., and Evranuz, Ö. (2000). Low temperature mushroom (*A. bisporus*) drying with desiccant dehumidifiers. *Drying Technology*, *18*(1–2), 433–445.

Hernando, I., Sanjuán, N., Pérez-Munuera, I., and Mulet, A. (2008). Rehydration of freeze-dried and convective dried *Boletus edulis* mushrooms: Effect on some quality parameters. *Journal of Food Science*, *73*(8), E356–E362.

Jaworska, G., and Bernas, E. (2010). Effects of pre-treatment, freezing and frozen storage on the texture of Boletus edulis (Bull: Fr.) mushrooms. *International Journal of Refrigeration*, *33*(4), 877–885.

Jeong, S. C., Jeong, Y. T., Yang, B. K., Islam, R., Koyyalamudi, S. R., Pang, G., Cho, K. Y., and Song, C. H. (2010). White button mushroom (*Agaricus bisporus*) lowers blood glucose and cholesterol levels in diabetic and hypercholesterolemic rats. *Nutrition Research*, *30*(1), 49–56.

Jin, X., van der Sman, R. G. M., van Straten, G., Boom, R. M., and van Boxtel, A. J. B. (2014). Energy efficient drying strategies to retain nutritional components in broccoli (*Brassica oleracea var. italica*). *Journal of Food Engineering*, *123*, 172–178.

Kantrong, H., Tansakul, A., and Mittal, G. (2014). Drying characteristics and quality of shiitake mushroom undergoing microwave-vacuum drying and microwave-vacuum combined with infrared drying. *Journal of Food Science and Technology*, *51*(12), 3594–3608.

Kemp, I. C. (2005). Reducing dryer energy use by process integration and pinch analysis. *Drying Technology*, *23*(9–11), 2089–2104.

Kiranoudis, C. T., and Markatos, N. C. (2000). Pareto design of conveyor-belt dryers. *Journal of Food Engineering*, *46*(3), 145–155.

Kotwaliwale, N., Bakane, P., and Verma, A. (2007). Changes in textural and optical properties of oyster mushroom during hot air drying. *Journal of Food Engineering*, *78*(4), 1207–1211.

Krokida, M. K., Kiranoudis, C. T., Maroulis, Z. B., and Marinos-Kouris, D. (2000). Effect of pretreatment on color of dehydrated products. *Drying Technology*, *18*(6), 1239–1250.

Kumar, A., Singh, M., and Singh, G. (2013). Effect of different pretreatments on the quality of mushrooms during solar drying. *Journal of Food Science and Technology*, *50*(1), 165–170.

Lewicki, P. P. (2006). Design of hot air drying for better foods. *Trends in Food Science and Technology*, *17*(4), 153–163.

Loch-Bonazzi, C., Wolf, E., and Gilbert, H. (1992). Quality of dehydrated cultivated mushrooms (*Agaricus bisporus*): A comparison between different drying and freeze-drying processes. *LWT-Food Science and Technology*, *25*, 334–339.

Lombraña, J. I., Rodríguez, R., and Ruiz, U. (2010). Microwave-drying of sliced mushroom. Analysis of temperature control and pressure. *Innovative Food Science and Emerging Technologies*, *11*(4), 652–660.

Madamba, P. S. (1997). Optimization of the drying process: An application to the drying of garlic. *Drying Technology*, *15*(1), 117–136.

Martínez-Soto, G., Ocanña-Camacho, R., and Paredes-López, O. (2001). Effect of pretreatment and drying on the quality of oyster mushrooms (*Pleurotus ostreatus*). *Drying Technology*, *19*(3–4), 661–672.

Mattila, P., Könkö, K., Eurola, M., Pihlava, J.-M., Astola, J., Vahteristo, L., Hietaniemi, V., Kumpulainen, J., Valtonen, M., and Piironen, V. (2001). Contents of vitamins, mineral elements, and some phenolic compounds in cultivated mushrooms. *Journal of Agricultural and Food Chemistry*, *49*(5), 2343–2348.

Minato, K., Mizuno, M., Terai, H., and Tsuchida, H. (1999). Autolysis of lentinan, an antitumor polysaccharide, during storage of *Lentinus edodes*, shiitake mushroom. *Journal of Agricultural and Food Chemistry*, *47*(4), 1530–1532.

Mishkin, M., Saguy, I., and Karel, M. (1984). A dynamic test for kinetic models of chemical changes during processing: Ascorbic acid degradation in dehydration of potatoes. *Journal of Food Science*, *49*(5), 1267–1270.

Mizuno, T. (1995). Shiitake, *Lentinus edodes*: Functional properties for medicinal and food purposes. *Food Reviews International*, *11*(1), 109–128.

Mujumdar, A. S. (1997). Drying Fundamentals. In C. G. J. Baker (Ed.), *Industrial Drying of Foods*, (pp. 7–30). London: Blackie Academic and Professional.

Mullin, J. (1995). Microwave processing. In G. W. Gould (Ed.), *New Methods of Food Preservation*, (pp. 112–134): Springer US.

Nagaya, K., Li, Y., Jin, Z., Fukumuro, M., Ando, Y., and Akaishi, A. (2006). Low-temperature desiccant-based food drying system with airflow and temperature control. *Journal of Food Engineering*, *75*(1), 71–77.

Nijhuis, H. H., Torringa, H. M., Muresan, S., Yuksel, D., Leguijt, C., and Kloek, W. (1998). Approaches to improving the quality of dried fruit and vegetables. *Trends in Food Science and Technology*, *9*(1), 13–20.

Obodai, M., Cleland-Okine, J., and Vowotor, K. A. (2003). Comparative study on the growth and yield of *Pleurotus ostreatus* mushroom on different lignocellulosic by-products. *Journal of Industrial Microbiology and Biotechnology*, *30*(3), 146–149.

Pal, U. S., and Chakraverty, A. (1997). Thin layer convection-drying of mushrooms. *Energy Conversion and Management*, *38*(2), 107–113.

Pappas, C., Tsami, E., and Marinos-Kouris, D. (1999). The effect of process conditions on the drying kinetics and dehydration characteristics of some mw-vacuum dehydrated fruits. *Drying Technology*, *17*(1–2), 158–174.

Prajapati, V., Nema, P., and Rathore, S. (2011). Effect of pretreatment and drying methods on quality of value-added dried aonla (*Emblica officinalis* Gaertn) shreds. *Journal of Food Science and Technology*, *48*(1), 45–52.

Qi, L.-L., Zhang, M., Mujumdar, A. S., Meng, X.-Y., and Chen, H.-Z. (2014). Comparison of drying characteristics and quality of shiitake mushrooms (*Lentinus edodes*) using different drying methods. *Drying Technology*, *32*(15), 1751–1761.

Rajarathnam, S., Bano, Z., and Steinkraus, K. H. (1989). Pleurotus mushrooms. Part III. Biotransformations of natural lignocellulosic wastes: Commercial applications and implications. *Critical Reviews in Food Science and Nutrition*, *28*(1), 31–113.

Rhim, J.-W., and Lee, J. (2011). Drying kinetics of whole and sliced shiitake mushrooms (*Lentinus edodes*). *Food Science and Biotechnology*, *20*(2), 419–427.

Severini, C., Baiano, A., De Pilli, T., Carbone, B. F., and Derossi, A. (2005). Combined treatments of blanching and dehydration: Study on potato cubes. *Journal of Food Engineering*, *68*(3), 289–296.

Smith, J., Rowan, N., and Sullivan, R. (2002). Medicinal mushrooms: A rapidly developing area of biotechnology for cancer therapy and other bioactivities. *Biotechnology Letters*, *24*(22), 1839–1845.

Srivastava, B., Singh, K. P., and Zimik, W. (2009). Effects of blanching methods on drying kinetics of oyster mushroom. *International Journal of Food Engineering*, *5*, 1–13.

Strumillo, C., Jones, P. L., and Zylla, R. (1995). Energy aspects in drying. *Handbook of Industrial Drying*, *2*, 1241–1275.

Tadayyon, A., Hill, G. A., Mike Ingledew, W., and Sokhansanj, S. (1997). Contact-sorption drying of *Penicillium bilaii* in a fluidized bed dryer. *Journal of Chemical Technology and Biotechnology*, *68*(3), 277–282.

Tian, Y., Zhao, Y., Huang, J., Zeng, H., and Zheng, B. (2016). Effects of different drying methods on the product quality and volatile compounds of whole shiitake mushrooms. *Food Chemistry*, *197, Part A*, 714–722.

Torringa, E., Esveld, E., Scheewe, I., van den Berg, R., and Bartels, P. (2001). Osmotic dehydration as a pre-treatment before combined microwave-hot-air drying of mushrooms. *Journal of Food Engineering*, *49*(2–3), 185–191.

Tunde-Akintunde, T. Y. (2010). Effect of pretreatment on drying time and quality of chilli pepper. *Journal of Food Processing and Preservation*, *34*(4), 595–608.

Turło, J., Gutkowska, B., Herold, F., Krzyczkowski, W., Błażewicz, A., and Kocjan, R. (2008). Optimizing vitamin B12 biosynthesis by mycelial cultures of *Lentinula edodes* (Berk.) Pegl. *Enzyme and Microbial Technology*, *43*(4–5), 369–374.

Tutova, É. G., and Fel'dman, R. I. (1976). Influence of contact mass exchange on the process of dehydration in a vacuum. *Journal of Engineering Physics*, *30*(6), 702–705.

Vadivambal, R., and Jayas, D. S. (2007). Changes in quality of microwave-treated agricultural products—A review. *Biosystems Engineering*, *98*(1), 1–16.

Vega-Mercado, H., Marcela Góngora-Nieto, M., and Barbosa-Cánovas, G. V. (2001). Advances in dehydration of foods. *Journal of Food Engineering*, *49*(4), 271–289.

Walde, S. G., Velu, V., Jyothirmayi, T., and Math, R. G. (2006). Effects of pretreatments and drying methods on dehydration of mushroom. *Journal of Food Engineering*, *74*(1), 108–115.

Wang, H., Zhang, M., and Mujumdar, A. S. (2014). Comparison of three new drying methods for drying characteristics and quality of shiitake mushroom (*Lentinus edodes*). *Drying Technology*, *32*(15), 1791–1802.

Wang, H.-C., Zhang, M., and Adhikari, B. (2015). Drying of shiitake mushroom by combining freeze-drying and mid-infrared radiation. *Food and Bioproducts Processing*, *94*, 507–517.

Yu, M., Xu, X., Qing, Y., Luo, X., Yang, Z., and Zheng, L. (2009). Isolation of an anti-tumor polysaccharide from *Auricularia polytricha* (Jew's ear) and its effects on macrophage activation. *European Food Research and Technology*, *228*(3), 477–485.

Zhang, N., Chen, H., Zhang, Y., Ma, L., and Xu, X. (2013). Comparative studies on chemical parameters and antioxidant properties of stipes and caps of shiitake mushroom as affected by different drying methods. *Journal of the Science of Food and Agriculture*, *93*(12), 3107–3113.

Zhang, R., Huang, C., Zheng, S., Zhang, J., Ng, T., Jiang, R., Zuo, X., and Wang, H. (2007). Strain-typing of *Lentinula edodes* in China with inter simple sequence repeat markers. *Applied Microbiology and Biotechnology*, *74*(1), 140–145.

Zhang, Y., Li, S., Wang, X., Zhang, L., and Cheung, P. C. K. (2011). Advances in lentinan: Isolation, structure, chain conformation and bioactivities. *Food Hydrocolloids*, 25 (2), 196–206.

Zhang, Z., Song, H., Peng, Z., Luo, Q., Ming, J., and Zhao, G. (2012). Characterization of stipe and cap powders of mushroom (*Lentinus edodes*) prepared by different grinding methods. *Journal of Food Engineering*, *109*(3), 406–413.

Zhang, M., Tang, J., Mujumdar, A. S., and Wang, S. (2006). Trends in microwave-related drying of fruits and vegetables. *Trends in Food Science and Technology*, *17*(10), 524–534.

Zhang, M., and Xu, Y. Y. (2003). Research developments of combination drying technology for fruits and vegetables at home and abroad. *Journal of Wuxi University of Light Industry*, *22*(6), 103–106.

Section III

Changes in Properties during Vegetable Drying

12 Pigments and Nutrients during Vegetable Drying Processes, Dried Products Storage, and Their Associated Color Changes

Barbara Sturm and Oliver Hensel
University of Kassel

CONTENTS

12.1 INTRODUCTION

Drying increases the shelf life of vegetables significantly. However, drying also reduces the content of valuable components (e.g., vitamin C, pigments, and polyphenols) substantially. Simultaneously, the content of, mostly unwanted, reaction products further increases degradation within the product, and thus reduces the nutritional value and the achievable shelf life.

A thorough understanding of the changes of chemical components within the vegetable throughout the drying process and storage is necessary for the development of processing setups and strategies that target a reduction of detrimental changes within the product.

Commonly used technologies in industrial drying are convection, conduction, and, for valuable products, freeze-drying (FD) (Gehrmann et al., 2009). Sun, solar, and shade drying also play an important role, particularly in developing countries. Different types of processing setups and conditions have different effects on the resulting quality of foodstuff. In conventional drying applications, process settings, technologies, and control systems are generally based on experimentally found parameters from decades ago. This has far-reaching implications on the product quality and in energy efficiency extension (Mujumdar, 2007).

12.2 QUALITY OF DRIED VEGETABLES

There are hardly any products that do not change their characteristics during drying. Shape, size, and arrangement of macroscopic and microscopic elements are subject to change. Initially, the distribution of components within the raw material is balanced. During drying, subsequent removal of water takes place, and this balance is lost through the movement of components and their concentration. In combination with high temperatures, this leads to a number of chemical and physical changes in the product, which are detrimental for its quality, e.g., loss of valuable components, cell rupture leading to degradation of pigments and other components, as well as reduced rehydration capability (Timoumi et al., 2007; Santos and Silva, 2008; Miranda et al., 2009). The speed of these changes is increased by the concentration of soluble components, in addition to the influx of oxygen into the product.

For the development of drying strategies that allow high retention of necessary components, it is imperative to understand the mechanisms and processes that lead to product damage. Further, the knowledge of critical temperatures for different reactions is necessary to ensure minimization of losses. In addition, in the case of certain substances such as some enzymes, a destruction of these components might be desirable to maintain the product quality.

12.2.1 Quality Aspects, Perception, and Their Relation to Drying Processing

The quality of a dried vegetable can roughly be classified into microbial, chemical, physical, and nutritional values. The most important quality criteria of dried foodstuff are color, general appearance, shape, taste, microbial load, retention of valuable compounds, density, consistency, rehydration behavior, water activity, absence of pests and other contaminants, and absence of unwanted odors (Ratti, 2009).

Schuchmann and Schuchmann (2005) summarized the goals of traditional food dehydration as prevention of decay, retention of valuable components, and reduction of weight. Shewfelt (1999) noted that there is a big difference between the perception of producers and consumers when it comes to product quality, despite the fact that the main goal of processing agricultural products is customer satisfaction. Producers relate product quality to relatively easily measurable characteristics such as sugar content, firmness, vitamin C content, and color. In contrast, the consumers' understanding of quality is much more difficult to quantify. They tend to relate color and other optical characteristics such as shape with taste, hygiene, shelf life, and the nutritional value of the product (Ragaert et al., 2003; Pedreschi et al., 2006; Chen, 2008). Therefore, for production, it is desirable to preserve the optical appearance of a product as well as possible. Increasingly, however, consumers demand that products also meet other criteria related to the nutritional value. This in turn means that the processes need to be understood better by the producers, and novel measurement and control systems need to be developed and applied that take the dynamic changes within the product into account.

The quality of dried vegetables is influenced by a multitude of factors, which can lead to contradictory demands in the drying process. For the best possible retention of the cell structure, and therefore its reconstitution characteristics, low drying temperatures should be chosen (Vega-Gálvez et al., 2009). For the best retention of aromas, however, Thijssen (1979) suggested to dry at very high temperatures at least during the first phase of the process. Similarly, sometimes mutually exclusive conditions can be formulated for other components. For the suppression of Maillard reaction, for example, the temperature at the end of the drying process needs to be set very low (Miranda et al., 2009), whereas Kröll and Kast called for high temperatures at the end of drying for an increased retention of vitamin C. Thus, for the gentle drying of sensitive products and a high retention of valuable characteristics, the knowledge of the mechanisms involved in degradation is necessary.

Besides the main components in a dried product, there are a multitude of trace components present, such as minerals, vitamins, and fatty acids (Cheung and Mehta, 2015), which can differ greatly in concentration. These substances can significantly influence the course of quality-determining reactions. A quantitative prediction of possible conversion reactions is therefore difficult. In addition, spatial water content and temperature distributions within the product can vary greatly and, as a result, directly affect quality changes within the product. During subsequent storage, further degradation occurs, which greatly depends on storage conditions (temperature, light, presence of oxygen, etc.) and the packaging materials.

12.2.2 Mechanisms of Quality Changes in Relation to Water Activity

Foodstuff are made up of a multitude of components, which as a whole represent the nutritional value of a product. Despite their variety, chemically they can be grouped into a limited number of components or categories. The component that is represented at the highest level in fresh produce is water, and it varies highly depending on the products (Kröll and Kast, 1989). The presence of water in foodstuff is a key factor for product quality and consumability. In dried vegetables, the water content determines the degree of most of the degradation reactions and, therefore, the maximum achievable shelf life.

Although the generally accepted assumption was that a low water content is desirable, more recent investigations have shown that the optimum level of water activity (a_W) is different for every product and lies within a clearly defined area (Schuchmann and Schuchmann, 2005). Accordingly, a water activity level that is too high (>0.6) may increase the risks of microbial spoilage of the product, whereas overdrying leads to a reduction of product quality through lipid oxidation, enzymatic and nonenzymatic browning, and further loss of valuable contents such as etheric oils (Figure 12.1). Enzymatic spoilage is particularly high at high water activities. However, as opposed to microbial spoilage, it is not completely suppressed at low and very low a_W levels. The nonenzymatic browning has its highest reaction rate at medium a_W levels. At very high and very low water activity levels, it is almost negligible (Figure 12.1).

The optimum water content for stability of a product is where the sum of all degradation reactions is at its minimum. This water content often coincides with the so-called BET point (Brunauer–Emmett–Teller-theory) of the sorption isotherm, where the monolayer water binding state is reached. Thus, knowledge of the water vapor sorption behavior of a product allows for the prediction and evaluation of stability problems during drying, packaging, and storage.

In vegetable drying, important quality changes occur in the cell walls and the cytoplasm. Therefore, it is imperative to reduce cell rupture as far as possible (Lewicki and Pawlack, 2003). The loss of valuable components in a product is a function of temperature, moisture content, and duration of the process and presence of catalysts such as enzymes and trace metals (Rovedo and Viollaz, 1998).

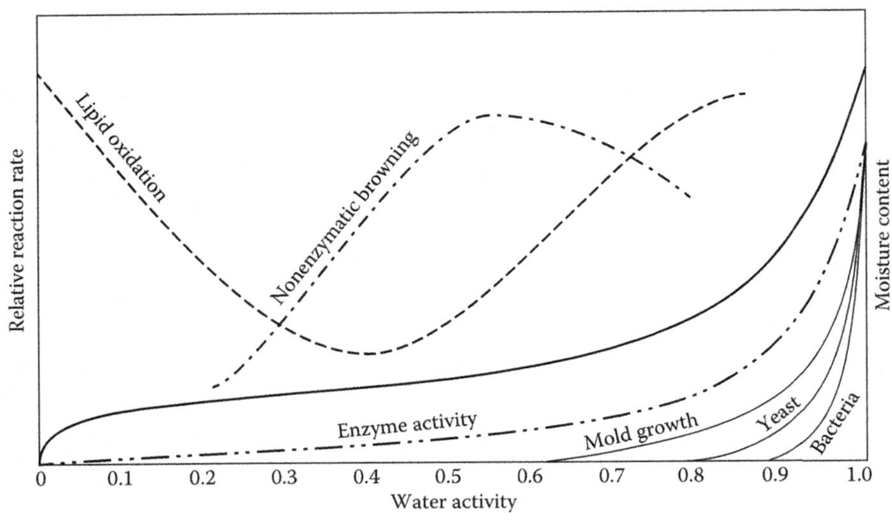

FIGURE 12.1 Water activity and stability diagram of foods.

Degradation of foodstuff, apart from microbiological degradation, is primarily caused by the following three types of reactions: (1) hydrolytic changes, (2) oxidative changes, and (3) enzymatic and nonenzymatic browning reactions. The most prominent sensorially tangible effects of chemical changes of a product during drying are color changes, loss of flavors and/or generation of off-flavors, and changes in the product's consistency (Tello-Ireland et al., 2011; Oliviero et al., 2013; Argyropoulos and Müller, 2014).

12.3 PIGMENTS AND NUTRIENTS IN VEGETABLES

Pigment and nutrient retention are important parameters to determine the quality of a dried foodstuff. As described earlier, a multitude of factors such as type of produce, type of processing, process settings, etc., influence the product quality. Numerous investigations have been undertaken to understand degradation pathways and the impact of the processing on different quality aspects of vegetables, of which Oliveira et al. (2016) give a comprehensive overview.

12.3.1 PIGMENTS

Pigments determine the color of vegetables and are very sensitive to the environmental conditions they are exposed to, particularly pH and temperature during drying and consequent storage (Marty-Audouin et al., 1992). Table 12.1 summarizes some of the most important pigments in vegetables and their color changes as well as acid/base conditions and heating (Bennion, 1980).

TABLE 12.1
Stability of Pigments against Physical and Chemical Conditions

Pigment	Natural Color	Acid	Base	Sustained Heat Treatment
Chlorophylls	Green	Olive green	Intensive green	Dull olive green, gray-brown, olive-yellow
Carotenoids	Yellow, red, orange, pink	Less-intensive color	Hardly any effect	Less-intensive color
Betalains	Purple-red, sometimes yellow-orange	Hardly any effect	Hardly any effect	Pale when pigments bleed out of tissue
Anthocyanins	Red, purple, blue	Red	Purple or blue	
Anthoxantins	White	White	Yellow	Dark, when temperatures are too high

Source: Adapted from Bennion, M., *The Science of Food.* John Wiley & Sons, New York, 1980. With permission.

12.3.1.1 Chlorophyll

Chlorophylls (Chl) exist in two forms in green plants: Chlorophyll a (Chl a), a blue green pigment and Chlorophyll b, (Chl b) a yellow green pigment, which are the most common plant pigments and are lipid soluble. Naturally, they occur in chloroplasts together with carotenoids and are both bound to a lipoprotein carrier (Eisenbrand and Schreier, 1995). The average ratio between Chl a and Chl b is 3:1 but can vary depending on the growing conditions of the product, with produce exposed to the sun having a higher ratio and produce grown in the shade, a lower ratio. Chl is highly susceptible to degradation during processing and storage, and degrades into gray-brown, dull olive green, or olive-yellow reaction products. Chl is sensitive to high temperature, acidic and alkaline surroundings, time, enzymes, oxygen, and light (Heaton and Marangoni, 1996). Degradation is initiated through tissue rupture, which leads to chemical and enzymatic changes, and eventually results in products of Chl catabolism (Heaton and Marangoni, 1996). Two main pathways for Chl degradation have been identified as follows: (1) acidically catalyzed degradation to pheophytins (pheophorbide-phytylesters), which are Mg-free derivatives of Chl; the pheophytins further react to pheophorbides through segregation of the phytol group; in the final step, the pheophorbides react to prophyridin carbon acid and aetioporphyrin; (2) oxidation-catalyzed lipoxygenase through which the Ch bound at C-10 is oxidized and which leads to subsequent bleaching (López-Ayerra et al., 1998). Chl degradation in dehydrated foods is directly linked to a_W. At high a_W, water is available for chemical reactions, and at very low a_W the mechanisms are linked to free radicals or lipid oxidation (Lajollo and Lanfer Marquez, 1982; López-Ayerra et al., 1998).

12.3.1.2 Carotenoids

Approximately 600 carotenoids are known to be found in foodstuff. Out of these, 50 show vitamin A activities. Carotenoids are lipid-soluble orange and yellow pigments and can be divided into carotenes (hydrocarbons) and xanthophylls (derivatives containing oxygen). Carotenoids containing hydroxyl groups are often bound to long-chain fatty acids. They are sensitive to light and oxygen. The degradation of carotenoids is accelerated through radical intermediates in foodstuff, such as unsaturated fatty acids, which are produced as a result of autoxidation or lipoxygenases-catalyzed co-oxidation (Eisenbrand and Schreier, 1995). Carotenoids isomerize from 5,6-exposides to 5,8-exposides, which are less color-intensive.

The most commonly found carotenes are α- and β-carotene (the major pigments in carrots) and lycopene (in tomatoes). Carotenes are able to quench singlet oxygen, which is the reason for their antioxidant characteristics. Among the carotenes, β-carotene is the most important precursor of vitamin A, and in green plants it occurs together with chlorophyll.

The beneficial characteristics of β-carotene include: color, antioxidant capacity, pro-vitamin A, and other aspects related to nutrient value. Thermo-oxidation leads to formation of colorless, lower-molecular-weight reaction products, leading to color loss and off-flavors (e.g., β-ionone and β-damascenone). Several studies have investigated the effects of different drying methods on the retention of β-carotene (King et al., 2001; Kidmose et al., 2007; Lefsrud et al., 2008).

12.3.1.3 Anthocyanins, Anthoxanthines, and Betalains

Anthocyanins are blue, purple, and red glycosides of anthocyanidin, which are water-soluble vacuole pigments. Their appearance and stability depends strongly on pH, temperature, light, oxygen, metal ions, and intramolecular and intermolecular association with other compounds (Andersen and Jurdheim, 2008). Anthocyanins are odorless, nearly flavorless, and act as antioxidants. The most frequently occurring anthocyanins in nature are cyanidin, delphinidin, malvidin, pelargonidin, peonidin, and petunidin, which occur in vegetables such as purple corn, red cabbage, and sweet potatoes. Anthocyanin degradation has a significant impact on color and nutritional properties of foodstuff.

Anthoxanthins are flavonoid pigments and are water-soluble. In acidic environments, they tend to be whiter, and in alkaline medium, yellow. They occur naturally in vegetables such as cauliflower. Similar to anthocyanins, anthoxanthins are very susceptible to color changes in the presence of minerals and metal ions, and they darken significantly when heated (Bennion, 1980).

Betalains are red-violet (betacyanins) and yellow-orange (betaxanthins), indole-derived pigments. The most important betacyanin is betanin, which is the main colorant in beetroots. Betalains are sensitive to heat treatment, which causes leaching. Anthocyanins and betalains have not been found to coexist in nature (Stafford, 1994).

12.3.1.4 Pigment Retention during Drying

A great number of publications have investigated the dependency of chlorophyll retention on the drying technology and the process settings. King et al. (2001) showed that the Chl retention of FD spinach samples was 12.4% higher than that of controlled low-temperature vacuum dehydration (CLTVD)-dried samples in case of Chl a and 4.3% for Chl b. Lefsrud et al. (2008) found that the Chl a and Chl b retention in both kale and spinach decreased significantly with increasing temperature (–25°C to 75°C). They further showed that these changes are not linear but there is an increase in losses for both components in defined temperature spans (25°C–50°C for Chl a and 50°C–75°C for Chl b).

Negi and Roy (2000) investigated the impact of drying methods (sun, solar, shade, cabinet) (65°C) and low temperature (30°C) on the retention of Chl acid in savoy beet, amaranth, and fenugreek. After drying, the Chl retention was 47%–60% in savoy beet, 35%–51% in amaranth, and 40%–50% in fenugreek, respectively. In all three products, a significant difference between different drying methods was noted. Although cabinet drying was the best method for amaranth, it was the worst for savoy beet in terms of retention, whereas for fenugreek, it resulted in an average retention. Conversely, low-temperature drying performed well in both savoy beet and fenugreek, but poorly in amaranth. These results indicate that the optimum choice of drying technology and process settings significantly depend on the products in question.

Kowalski et al. (2013) investigated the influence of cycling heating and cooling on the β-carotene retention of carrots. Regular and irregular changes in drying conditions led to retention of β-carotene of 60%–92%. Because the reaction products are less colorful than carotene, an increase in retention can be directly related to the decreased color changes. Lefsrud et al. (2008) found that drying temperature had a significant impact on retention of β-carotene and lutein in kale and spinach, showing that losses of β-carotene at the highest applied temperature (75°C) were 75% and 83% higher for kale and 78% and 75% higher in spinach in comparison to FD at −25°C. However, the highest change in retention occurred between 25°C and 50°C process temperature, showing that the critical degradation temperatures lie in this area.

A reduction of approximately 21% in all the analyzed trans-β-carotenes was reported by Kidmose et al. (2007), when orange flesh sweet potato (OFSP) chips were dried by shade drying to a final moisture content of 6%. Milling the chips into flour further reduced the content of pro-vitamin A. In another study, the reduction percentages in all trans-β-carotene content for forced air oven drying (at 57°C for 10 h), solar drying (SD), and open air sun drying (to MC <10%) were 12%, 9%, and 16%, respectively (Bengtsson et al., 2008). It is evident that the percentage of pro-vitamin A compounds was rather low in open air/natural sun drying processes, which could be explained by direct sunlight exposure and the extended duration of the drying process. Percentage losses of total carotenoids and trans-β-carotene for hot air, sun, and solar drying ranged from 13% to 33% and 16% to 34%, respectively (Bechoff et al., 2009). Bechoff et al. (2009) concluded that there was a significant difference in retention of pro-vitamin A between sun drying and SD. However, in their next study, Bechoff et al. (2010) mentioned that low-cost direct sun drying can be as effective and efficient as SD as far as pro-vitamin A retention is concerned. The mean losses in two OFSP varieties from Uganda were as low as 7.3%–10.7%.

Leong and Oey (2012) investigated the impact of FD on the retention of selected carotenoids (α-carotene, β-carotene, lycopene, lutein) in carrots and peppers. In carrots, they found losses of carotenoids was between 10.5% and 12.1%, while the losses of carotenoids in peppers varied between 5.7% and 14.3%.

Negi and Roy (2000) found a retention level for carotene between 28% and 40% for savoy beet, with low temperature leading to the highest retention and shade drying to the lowest retention. For amaranth, the differences were much higher at 18%–56%, with cabinet drying being the best and sun drying being the worst options. Retention in fenugreek was highest at 41%–85%. In fenugreek, low-temperature drying performed best, whereas sun drying performed worst. In line with the findings for Chl a and Chl b, these results show clearly β-carotene retention and thus, the preferred drying method, depend strongly on the product processed.

Vega-Gálvez et al. (2009) found that pigment discoloration increased with increasing temperatures in the drying of sweet pepper, and the discoloration of carotenoids could be linked to enzymatic and nonenzymatic browning.

In the drying of purple flesh sweet potatoes, levels of 71%, 75%, and 54% retention of anthocyanins were found for freeze-, vacuum-, and spouted bed-assisted microwave drying. The authors argued that the low retention for spouted bed drying was caused by the increased access of oxygen into the product (Liu et al., 2012).

Blanching has reportedly led to higher carotenoid retention of 37%–85% (Negi and Roy, 2000) in drying. However, it needs to be noted that in this study, the initial β-carotene content before drying was lower in blanched samples than in their nonblanched counterparts, and thus the actual concentration in the blanched samples after drying was lower than that in the fresh samples.

12.3.2 Phenolic Compounds

Polyphenol is an umbrella term for chemical compounds containing more than 2 phenol or phenol-ether groups, which belong to different groups. Some polyphenols (e.g., flavonols and anthocyanins) are red or yellow in the monomer state, but others are colorless (e.g., hydroxyl cinnamonacid, and catechines). Polyphenols are important food components because of their antioxidant activities. They act as reducing agents, hydrogen donors, singlet oxygen quenchers, and metal chelators (Deepa et al., 2007). However, all polyphenols are substrates of phenol-oxidases and can be converted into brown compounds when oxygen is present. These conversion reactions lead to changes in the color and taste of a product.

The amount of phenolic compounds in a drying product is temperature-dependent, and the amount of phenolic components decreases with increasing drying temperature. However, Vega-Gálvez et al. (2009) observed an increase in phenolic contents at very high temperatures and conclude that this might be due to the increased availability of precursors of phenolic components caused by nonenzymatic interconversions between phenolic components (Vega-Gálvez et al., 2009).

Mrcic et al. (2006) investigated the impact of process conditions of convection drying (air temperature and velocity) on kaempferol and total free phenolic compounds. At high temperatures and air velocities, an increase in phenolic contents was observed, which is attributed to the increased release of compounds from the matrix and more accessible to extraction.

12.3.3 Vitamins

Vitamins are among the heat-sensitive components in foodstuff. Their degradation during heat treatment is a complex process, which depends on further factors such as presence of oxygen, light, and water solubility (Awuah et al., 2007). Furthermore, the destruction of vitamins is influenced by pH and can be catalyzed by chemical components present in metals, other vitamins, and enzymes (Lewis and Heppell, 2000). The fat-soluble vitamin A (only when oxygen is present), D, E, and β-carotene, as well as the water-soluble vitamins C, B1 (thiamin), and B2 (riboflavin) in sour surroundings, are considered to be particularly heat-sensitive (Ryley and Kajada, 1994).

Vitamin C retention is commonly used as a basis for the estimation of overall nutritional quality of food products, particularly vegetables (Goula and Adamopoulos, 2006). Vitamin C loss in vegetables is the highest during heat treatment (Fennema, 1985). However, water solubility plays an important role in this context. For determination of vitamin C loss during food processing, the content of ascorbic acid is usually used (Rovedo and Viollaz, 1998; Timoumi et al., 2007; Santos and Silva, 2008; Miranda et al., 2009). Shortening of the drying process is shown to have a positive influence on the reduction of vitamin C losses (Goula and Adamopoulos, 2006). This allows for the conclusion that time of exposure to increased temperatures is as important as the actual temperature. It should to be noted that polyphenol oxidase (PPO) activity is inhibited when vitamin C is present. With an increasing degree of vitamin C degradation, therefore, PPO activity increases until its own natural deactivation temperature is reached. The mechanism behind degradation of vitamin C during drying may be associated with oxidation of ascorbic acid to dehydroascorbic acid, which is followed by further hydrolysis and oxidation (Santos and Silva, 2008).

Vitamin C degradation is significant in sweet potato, in which it was reduced by 33%–39% during cabinet drying at a temperature of 50°C and 70°C for 6h (Woolfe, 1992). Degradation of vitamin C in various fruits and vegetables during drying was discussed by Santos and Silva (2008). Compared with sun and SD, it was mentioned that shade drying might be a better way to get a final product rich in vitamin C.

In a study, ascorbic acid retention levels were found to be 2%–7% for savoy beet, 2%–8% for amaranth, and 13%–50% for fenugreek (Negi and Roy, 2000). Among the investigated drying techniques, low-temperature drying (30°C) performed best for all three products. Only for fenugreek, cabinet (47%) and SD (44%) resulted in similarly high retention rates.

Vitamin C in red peppers was shown to be highly susceptible to degradation in hot air drying and was strongly influenced by temperature. While drying at 50°C led to a vitamin C retention of 16.6%, at 90°C the vitamin C was almost completely destroyed (1.8%) (Vega-Gálvez et al., 2009).

The aforementioned results demonstrate that retention of vitamin C strongly depends on the product being considered and the technology used.

Vitamin C loss during drying can be reduced through the application of pretreatments such as blanching and additives (SO_2, and $CaCl_2$). Korus (2011) found a significant influence of drying method [FD and air drying (AD)] and pretreatment by blanching on vitamin C retention in kale. Nonblanched samples lost 25% (FD) and 31% (AD) of vitamin C, while blanched samples lost 16% (FD) and 23% (AD), respectively. However, the blanched samples had a lower vitamin C content prior to drying than the samples that were not pretreated. Thus, despite the lower losses during drying, the overall vitamin C contents in pretreated samples were 5% (AD) and 10% (FD) lower after drying than in the untreated samples. In carrots, the thermal pretreatment through blanching led to 80% loss in vitamin C (Gamboa-Santos et al., 2013).

12.3.4 Lipids

Lipid is an umbrella term for structurally different substances with common characteristics such as nonsolubility in water and includes actual fats and fat-like substances (Eisenbrand and Schreier, 1995). Lipids can be split into two general groups, (1) simple lipids such as alcohols (e.g., cholesterol), hydrocarbons (e.g., carotenoids), carbon acids (fatty acids), ether, esters (e.g., mono-, di-, triacyl glycerine) and (2) complex lipids.

Lipoperoxidation describes the complex reaction in which acyl-lipids that contain unsaturated fatty acids (oil, linol, and linolenic acid) are degraded into a multitude of volatile and nonvolatile products through oxidation. Lipolysis is the enzymatic hydrolysis of triacyl glycerines catalyzed by lipases, producing fatty acids. During vegetable processing (cutting, slicing, chopping, etc.), e.g., in potatoes, membrane lipids are degraded to unsaturated fatty acids (linol-, linolenic acid). Through further enzymatic (lipoxygenase, hydroperoxidlyase) oxidative degradation, odorous and flavoring substances are created (Eisenbrand and Schreier, 1995).

Lipid oxidation is responsible for the development of off-flavors, rancidity, and loss of fat-soluble vitamins and pigments in dehydrated foods. Moisture content plays a crucial role in this context. At high a_W, lipids can undergo enzymatic hydrolysis (e.g., soapy taste), and at very low a_W (<0.2), auto-oxidation of unsaturated fatty acids (e.g., rancidity) occurs, causing off-flavors (Perera, 2005).

Lipid oxidation can be decreased by the reduction of oxygen content and use of antioxidants, synergistic agents (e.g., citric and phosphoric acids), and chelating agents (e.g., EDTA, citric, and malic acid) (Barbosa-Cánovas and Vega-Mercado, 1996).

Muncia et al. (1992) found high levels of polyunsaturated fatty acids in spinach. King et al. (2001) discovered that levels of lipid peroxide in dried spinach (FD and CLTVD) increased with storage time and porosity. They concluded that a higher level of peroxide indicated increased peroxidation of fatty acids and, therefore, an increase of free radicals, which in turn contributed to Chl degradation.

12.3.5 Proteins

Proteins are macromolecules and are amino acid polymers. A general distinction can be made based on the different structural levels of proteins: primary (amino acid sequence of polypeptide chain), secondary (folding of polypeptide chain: α-helix, β-sheet structure etc.), tertiary (folding of the entire polypeptide chain), and quaternary (aggregation of several polypeptide chains into functional protein chains) structures.

Proteins are highly susceptible to denaturation through reversible or irreversible change of the ternary structure without destruction of the covalent bindings (except sulfide bonds). Denaturation is commonly caused by heat, change in pH levels, salt, increase of phase boundaries, etc. (Eisenbrand and Schreier, 1995) and is often followed by the release of amino acids from the protein, which can then further react with other chemical compounds via the Maillard reaction (Di Scala et al., 2011).

Most information regarding damage of proteins during drying given in the literature is based on quantification of lysin, which is often used in nutritional physiology as an indicator of protein damage.

Microwave–vacuum-dried carrots showed a significant loss of proteins (Chaughule and Thorat, 2011). The authors attributed this to the deactivation of heat-sensitive proteins such as peroxidases and polyphenol oxidases during both blanching and drying. Minimum protein retention was found to be 49%–53%, which the authors attributed to high heat generation through microwave heating and low sample thickness. When AD (50°C, 60°C, 70°C, and 80°C) okra retained 82%–89%, 82%–93%, 82%–84%, and 75%–79% protein, respectively (Pendre et al., 2012). Air and sun-dried tete, soko, and igbagba were investigated for their crude protein content. In air drying, 97%, 89%–96%, and 93%–99% were retained in tete, soko, and igbagba, respectively. Although retention in sun-dried samples was higher, these had not been blanched as opposed to the air-dried ones. Therefore, higher retention could be attributed to the pretreatment rather than drying (Onemani and Badifu, 1987).

In conclusion, AD seems to be gentler in regards to protein retention than microwave drying. However, inactivation of certain proteins might be desired i.e., PPO inactivation for the prevention of enzymatic browning.

12.3.6 Sugars

Sugars are organic compounds that fall either into the group of polyhydroxy aldehydes (aldoses) or polyhydroxy ketone (ketoses). They occur naturally as monosaccharides, disaccharides (e.g., sucrose), trisaccharides, and higher oligosaccharides. Sugars play a major role in the taste of produce and are highly susceptible to heat damage.

In vegetables, glucose and sucrose are the most commonly found sugars. When these sugars are heated, they undergo caramelization or Maillard reactions (see the following paragraphs for nonenzymatic browning).

Microwave–vacuum-dried carrots were found to have a glucose retention rate of 98%–99% (Chaughule and Thorat, 2011). The authors attributed this high retention to absence of oxygen, low drying temperature, and sugar concentration. Fante and Norena (2015) investigated the impact of AD (50°C, 60°C, and 70°C) and FD on inulin (a polysaccharide belonging to the group of fructans), glucose, and fructose for nonblanched and blanched garlic. Blanched samples before drying had 7% inulin, 23% glucose, and 19% fructose less than the nonblanched samples. This is because of the leaching of solids in steam blanching by evaporation and condensation of water. Both FD and AD decreased the inulin content while glucose and fructose contents increased. This could be due to hydrolysis of the inulin to reducing sugars. The FD samples showed significantly higher inulin contents than the air-dried ones. In air drying, the inulin content was reduced with increasing temperature. Glucose concentration increased with increasing temperature for both blanched and nonblanched samples, whereas fructose did not significantly increase in nonblanched samples but significantly increased in blanched samples.

12.4 ENZYMATIC AND NONENZYMATIC DEGRADATION MECHANISMS

Enzymatic and nonenzymatic browning are two major degradation mechanisms affecting vegetable product color. It is difficult to distinguish between the two unless the enzymes responsible for browning are inactivated. However, some nonenzymatic color changes can be the result of browning of intermediates, which are the products of enzyme-mediated oxidation that took place before the enzymes are inactivated (Wedzicha et al., 1984).

12.4.1 Enzymatic Browning

Enzymatic browning in plant food products is the enzymatically catalyzed degradation of phenolic components (particularly polyphenols) into polymeric products (phlobaphenes), which are brown in color. The oxidative browning is initiated by the collapse of intercellular membranes of the plant tissue. In the presence of oxygen,

PPOs degrade flavonoids such as catechin, hydrocinnamic acid, and pro-anthocyanins into o-chinons, which are consequently condensed to phlobaphenes (Eisenbrand and Schreier, 1995). Hydroxylation is a relatively slow process leading to colorless products, while oxidation processes are relatively quick and the produced chinons are colored. Further reaction of the chinons lead to the accumulation of melanin (through nonenzymatic polymerization), the brown or black pigments. These reactions often also result in negative changes of flavor and the denaturation of proteins (Eisenbrand and Schreier, 1995).

The most important factors influencing enzymatic browning are PPO and phenolic component concentrations, pH, temperature, and oxygen availability. PPOs are relatively heat-labile, and their heat resistance depends on the species and cultivars (Severini et al., 2003). Temperatures of higher than 50°C combined with the duration of the drying process reduce PPO activity. Above 80°C, PPOs are completely destroyed through denaturation.

PPOs can be deactivated and therefore enzymatic browning suppressed using several strategies such as heat inactivation, removal of substrates (O_2, phenols), reduction of pH, and addition of antioxidants (Severini et al., 2003). Weemaes et al. (1998) hypothesized that thermal treatment (blanching) is the most effective method for controlling enzymatic browning.

The main advantage of blanching in comparison to the use of ascorbic acid and citric acid or treatment with sulfur is that these compounds stay unaltered in terms of the introduction of other chemical components. Further, oxygen-free processing, e.g., vacuum drying, can be applied to suppress enzymatic browning. Another alternative is the thermal inactivation of the enzymes involved in browning. Several studies have shown that sensitive agricultural products may be exposed to high temperatures at the beginning of the drying process without being thermally damaged in terms of increased losses in valuable components (Chua et al., 2000; Sturm et al., 2012), while color can be better maintained, which might be due to the deactivation of enzymes on the surface, which leads to a reduction of enzymatic browning (Sturm et al., 2012).

12.4.2 Nonenzymatic Browning

Nonenzymatic browning is an umbrella term for many different reactions, among others chemical oxidation of phenols, Maillard reaction, caramelization, and Maderization (Manzocco et al., 2000). Nonenzymatic browning is intensified by heat.

In caramelization, anomeric shifts, ring size alteration, breakage of glycoside bonds, and formation of new bonds occur in sugars. Double bonds in the sugar rings produced intermediates to unsaturated rings, which absorb light. Other factors that can influence the browning reaction are the pH and the presence of metals, oxygen, phosphates, and SO_2 (Barbosa-Cánovas and Vega-Mercado, 1996).

At the beginning of a Maillard reaction, a concentration of soluble carbohydrates, proteins, and protein components (amino acids) leads to reactions between reducing sugar (particularly aldoses) and amino acids. Moderate heat treatment also can initiate Maillard reaction during the drying process. In this process, proteins are damaged and amino acids such as lysine, L-arginine, and L-histidine are lost. In particular, the loss of lysine is crucial due to its importance for the nutrition value of the product (Fennema, 1985).

The amadori-compounds (1-*N*-amino-acid-1-desoxtetoses) created from aldoses are sensorially not detectable (colorless and tasteless) primary reaction products of the Maillard reaction. These can further degrade and result in the browning of products during drying as well as storage (Kröll and Kast, 1989; Awuah et al., 2007; Baltes, 2007).

In summary, it can be proposed that nonenzymatic browning has a great influence on the resulting product quality in terms of color changes and changes in nutrition value and flavor. Therefore, process control development needs to target a reduction of nonenzymatic browning. This in turn will lead to an increase of the nutritional value, as well as the sensory properties of a product.

Nonenzymatic browning is significantly increased with an increase in drying temperature due to a number of degradation reactions in pigments and nutrients as described earlier. While drying of red peppers at 50°C led to a total color difference ΔE of 5.8 ± 2.9, drying at 90°C resulted in a difference of 7.8 ± 2.2 (Vega-Gálvez et al., 2009). This result was predominantly attributed to nonenzymatic browning.

12.5 CHANGES IN PIGMENTS AND NUTRIENTS DURING STORAGE

Some quality deterioration reactions such as browning and oxidation that occurred during drying can continue during storage, although at a lower speed. Thus, the drying process and storage conditions of dried vegetables need to be considered together. One prominent example is the reaction products of the Maillard reaction. To increase the shelf life of the product, it is important to dry the vegetables to a water content at which the reaction reaches a minimum (see also nonenzymatic browning and a_W).

Pigments, lipids, and vitamins (vitamin C) are highly vulnerable toward oxidation. Therefore, it is important to package dried products in airtight packaging. In some cases, e.g., products containing highly unsaturated fatty acids or etheric oils, it is optimal to package these in a nitrogen atmosphere to prevent oxidation. Many chemical components (e.g., lipids, vitamins, and amino acids) are additionally sensitive toward light and undergo photooxidation when exposed to light. Thus, products need to be packaged in nontranslucent packaging and/or stored in the dark.

Bechoff et al. (2010) showed that the losses of carotenoids during the drying of orange flesh sweet potatoes was not significant but were extremely high during storage of the products. The losses in the stored dried chips after 4 months were more than 70%, irrespective of sunlight exposure or the use of opaque or clear packaging material. Storage losses are also significant because dried products packed in polyethylene bags for 120 days at 24°C lost 50% of vitamin C (Woolfe, 1992).

King et al. (2001) investigated the impact of pretreatment (blanching, and nonblanching), drying method (FD, CLTVD), and storage conditions (–5°C, –20°C, –30°C, and –45°C) on the retention of Chl a and b. Results showed that for both drying strategies, Chl degradation was reduced in nonblanched samples. Further, the Chl degradation rates were lower in the CLTVD samples than in the FD samples, and Chl retention was the best at the highest freeze storage temperature (–5°C). The authors concluded that the higher porosity in FD samples increased the accessibility of O_2 into the product, which could explain the heightened degradation for FD samples. In addition, they hypothesized that blanching might have led to a loss of antioxidative characteristics of the product. They further showed that the levels of lipid peroxidases increased with both porosity of the material and duration of storage. Dehydrated green vegetables are susceptible to photo oxidation, which increases the singlet oxygen level and promotes peroxidation of fatty acids. The peroxidation of fatty acids leads to the production of free radicals, which then are active in the degradation of chlorophyll.

Nonblanched (nb) and blanched (b) FD and AD kale samples were stored for 1 year in refrigerated and ambient conditions. Vitamin C losses were higher in non-blanched samples at 22%–37%, while blanched samples had lost 9%–24%. Vitamin C levels were 15% higher in FD samples than in AD samples. Retention was improved by cold storage by 15% (b) and 22% (nb) for AD, and 12% (b) and 18% (nb) (Korus, 2011). Numerous authors have stated that an increase in storage temperature leads to a decrease in vitamin C retention (Negi and Roy, 2001; Uddin et al., 2002).

A study on storage conditions (ambient: 15°C–36.5°C, 45%–85% RH; cold: 7.5°C–8.5°C, 70%–75% RH) and packaging (single and double layered HDPE) after hot AD and SD of savoy beet and amaranth showed that β-carotene retention for both cases was significantly higher at lower temperatures, and in case of cold storage, retention was higher in double layer HDPE packaging. A continuous decline of ascorbic acid was observed at all storage temperatures. However, no significant relevance of storage temperatures and packaging was observed. Chlorophyll retention did not show significant differences for the different conditions, with the exception of SD savoy beet stored at ambient temperature where a drastic loss of chlorophyll was observed. On average, the hot air-dried samples had higher retention for all chemical components before storage than the solar dried ones, which is because of the shorter drying times and more constant temperatures applied (Negi and Roy, 2001).

Ishiguro et al. (2010) investigated the dependency of polysaccharide hydrolysis on storage conditions (0°C–20°C) for burdock roots. They found an increase in reducing sugar concentration and a decrease of inulin, which could be related to the residual activity of inulinase in the product.

Lipid oxidation during storage can be reduced through the use of antioxidants (e.g., polyphenolic compounds), vacuum, or inert gas packaging; in the former cases, alternative oxidation paths are provided, and in the latter, oxygen is completely removed.

12.6 MECHANISMS OF COLOR AND SPECTRAL CHANGES DURING DRYING AND STORAGE AND THEIR APPLICATIONS IN QUALITY ASSESSMENT

Color is an important characteristic of dried vegetable products (Kays, 1991; Chen, 2008). This is due to the correlation between color, taste, and aroma (Morris et al., 1953). During the drying of foodstuff, a multitude of chemical and biochemical changes can be observed as was described by Crapiste (2000). Color is often chosen as an indirect measurement for other quality criteria, such as pigments, aroma, etc., whose correlations are principally known and where measurement of color is the easiest and the quickest way of determination (Pathare et al., 2013). Fernandez et al. (2011) summarized the characteristics that change the color of a product during drying as follows: degradation of pigments (particularly chlorophyll and carotenoids), browning reactions (enzymatic and duration, variety, pretreatment, and concentration of compounds), and contamination with heavy metals. However, the color change is not directly related to the actual moisture content of the drying product (Mujumdar, 2000; Maskan, 2006).

In plant tissues, high rates of browning reactions occur during drying and storage (Krokida et al. 1998). Nonenzymatic browning and enzymatic phenol-oxidation are the two main categories of these reactions (Manzocco et al., 2000).

A multitude of newer publications has investigated the impact of drying conditions on product quality in terms of color (Vega-Gálvez et al., 2009; Sagar and Kumar, 2010; Guiné and Barroca, 2012). Maskan (2006) gave an overview on research in the area of color change during thermal treatment of foodstuff. He concluded that the quality of the final product depends significantly on the process chosen, but every technology and strategy leads to significant changes in color. However, the changes in color do not automatically indicate which reactions are taking place in the product. It is vital to correlate chemical changes to visual signals. Thus, in recent years, several new approaches were added to classical color determination using colorimeters. Nicolaï et al. (2014) gave an overview of the current most frequently used noninvasive methods for the investigation of inner and outer quality criteria of vegetables. Among them, computer-aided vision (CAV), thermography (TI), laser backscattering (LB), and hyperspectral imaging (HSI) have shown a great potential for the analysis of chemical and physical characteristics, color, defects, porosity, water content, content of soluble components, texture, firmness, shape, and shrinkage. Romano et al. (2011), Udomkun et al. (2014), and Quing et al. (2008) detected color, soluble components, and other attributes via LB. Liu et al. (2014) used HSI for the noninvasive simultaneous spatial and spectral detection of process and product characteristics. Huang et al. (2014) gave a comprehensive overview of work currently conducted in this field, which shows the wealth of recent work conducted. The changes that a product undergoes are detected noninvasively, and hence, the results are more accurate concerning the process, because the samples do not have to be removed from the system at any time.

12.7 CONCLUDING REMARKS

Vegetable products are highly susceptible to changes in pigments, nutritional value, and visual appearance during drying. Process temperatures and technologies applied have a major impact on these changes. In addition, the products themselves are significantly different in their appearance (size, shape, structure, etc.) and chemical composition. Thus, they behave extremely differently during processing and necessitate individual evaluation in terms of optimum technological setups and process settings during drying.

From the studies presented, it can be concluded that (1) nutrient destruction during drying is temperature-dependent; (2) there is a significant difference within and between nutrient classes regarding their sensitivity toward thermal destruction; (3) for each component, the critical temperatures/temperature ranges differ; and (4) the selection of optimum processing methods is product-dependent.

Many of the reaction pathways are not yet fully identified, and/or there is a significant lack of appropriate technological solutions and techniques for their noninvasive real-time determination during the drying process. Consequently, the dynamic changes in the composition of the product cannot be determined continuously and, thus, the dynamics of changes the products undergo throughout the process are not known. This information, however, is vital for the development of tailored control strategies and the selection of appropriate technologies.

Therefore, for the development of advanced drying processes regarding optimum retention of chemical components, the reaction pathways need to be understood, and technologies and techniques need to be developed, which can detect the dynamic changes within the product during the drying process. The availability of advanced visual systems in combination with chemometrics and powerful computer solutions allows for the targeted development of improved drying processes in terms of product quality and resource efficiency through model-based dynamic control systems.

As many degradation reactions in dried vegetable products are a consequence of biochemical changes during the drying, quality-optimized drying may also result in extended shelf life.

REFERENCES

Andersen, O. M., Jurdheim, M. (2008). Anthocyanins: Food applications. In *5th Pigments in Food Congress for Quality and Health*, University of Helsinki, Helsinki, Finland.

Argyropoulos, D., Müller, J. (2014). Changes of essential oil content and composition during convective drying of lemon balm (*Melissa officinalis* L.). *Industrial Crops and Products*, 52, 118–124.

Awuah, G. B., Ramaswamy, H. S., Economides, A. (2007). Thermal processing and quality: Principles and overview. *Chemical Engineering and Processing*, 46 (6), 584–602.

Baltes, W. (2007). *Lebensmittelchemie*. 6. Auflage, Springer, Berlin and Heidelberg.

Barbosa-Cánovas, G. V., Vega-Mercado, H. (1996). *Dehydration of Foods*. Chapman & Hall, New York.

Bechoff, A., Dhuique-Mayer, C., Dornier, M., Tomlins, K. I., Boulanger, R., Dufour, D., Westby, A. (2010). Relationship between the kinetics of β-carotene degradation and formation of norisoprenoids in the storage of dried sweet potato chips. *Food Chemistry*, 121 (2), 348–357.

Bechoff, A., Dufour, D., Dhuique-Mayer, C., Marouzé, C., Reynes, M., Westby, A. (2009). Effect of hot air, solar and sun drying treatments on pro-vitamin A retention in orange-fleshed sweet potato. *Journal of Food Engineering*, 92, 164–171.

Bengtsson, A., Namutebi, A., Larsson Alminger, M., Svanberg, U. (2008). Effects of various traditional processing methods on the all-trans-β-carotene content of orange-fleshed sweet potato. *Journal of Food Composition and Analysis*, 21, 134–143.

Bennion, M. (1980). *The Science of Food*. John Wiley & Sons, New York.

Chaughule, V. A., Thorat, B. N. (2011). Microwave vacuum drying of shredded carrots and its nutritional evaluation. *International Journal of Food Engineering*, 7 (4), 1556–3758. doi:10.2202/1556-3758.2235.

Chen, X. D. (2008). Food drying fundamentals. In Chen, X. D., Mujumdar, A. S. (Eds) *Drying Technologies in Food Processing.* Blackwell Publishing, Oxford, 1–54.

Cheung, P. C. K., Mehta, B. M. (2015). *Handbook of Food Chemistry*, Springer, Berlin and Heidelberg.

Chua, K. J., Mujumdar, A. S., Chou, S. K., Ho, J. C., Hawlader, M. N. (2000). Convective drying of banana, guave and potato pieces: Effect of cyclic variations of air temperature on convective food drying. *Drying Technology*, 18 (4–5), 907–936.

Crapiste, G. H. (2000). Simulation of drying rates and quality changes during the dehydration of foodstuffs. In Lozano, J. E., Anon, C., Parada-Arias, E., Barbosa-Cánovas, G. V. (Eds) *Trends in Food Engineering.* Technomic Publishing Co. Inc., Lancaster, PA.

Deepa, N., Kaura, C., George, B., Songh, B., Kapoor, H. (2007). Antioxidant constituents in some sweet pepper (*Capsicum annuum* L.) genotypes during maturity. *LWT-Food Science and Technology*, 40 (1), 121–129.

Di Scala, K., Vega-Gálvez, A., Uribe, E., Oyandale, R., Miranda, M., Vergara, J., Quispe, I., Lemus-Mondaca, R. (2011). Changes of quality characteristics of pepino fruit (*Solanum muricatum Ait*) during convective drying. *International Journal of Food Science and Technology*, 45 (9), 746–753.

Eisenbrand, G., Schreier, P. (1995). *Römpp Lexikon Lebensmittelchemie.* Georg ThiemeVerlag, Stuttgart, Germany.

Fante, L., Norena, C. P. Z. (2015). Quality of hot air dried and freeze-dried of garlic (*Allium sativum* L.). *Journal of Food Science and Technology*, 52 (1), 211–220.

Fennema, O. R. (Ed.). (1985). *Food Chemistry.* 2. Auflage, Marcel Dekker, New York.

Fernandez, F. A. N., Rodrigues, S., Law, C. L., Mujumdar, A. S. (2011). Drying of exotic tropical fruits: A comprehensive review. *Food and Bioprocess Technology*, 4, 163–185.

Gamboa-Santos, J., Soria, A. C., Perez-Mateos, M., Atanasio Carrasco, J., Montilla, A., Villamiel, M. (2013). Vitamin C content and sensorial properties of dehydrated carrots blanched conventionally or by ultrasound. *Food Chemistry*, 136, 782–788.

Gehrmann, D., Esper, G. J., Schuchmann, H. (2009) *Trockungstechnik in der Lebensmittelindustrie.* Behr's Verlag, Hamburg, Germany.

Goula, A.M., Adamopoulos, K.G. (2006). Retention of ascorbic acid during drying of tomato halves and tomato pulp. *Drying Technology*, 24 (1), 57–64.

Guiné, R. P. F., Barroca, M. J. (2012). Effect of drying treatment on texture and color of vegetables (pumpkin and green pepper). *Food and Bioproducts Processing*, 90 (1), 58–63.

Heaton, J. W., Marangoni, A. G. (1996). Chlorophyll degrardation in processed foods and senescent plant tissue. *Trends in Food Science and Technology*, 7 (1), 8–15.

Huang, H, Liu, L., Ngadi, M. O. (2014). Recent developments in hyperspectral imaging for assessment of food quality and safety. *Sensors*, 14, 7248–7276.

Ishiguro, Y., Onodera, S., Benkeblia, N., Shiomi, N. (2010). Variation of total FOS, total IOS, inulin and their related metabolizing enzymes in burdock roots (*Arctium lappa* L.) stored under different temperatures. *Postharvest Biology and Technology*, 56, 232–238.

Kays, S. J. (1991). *Postharvest Physiology of Perishable Plant Products.* Van Nostrand Reinhold, New York.

Kidmose, U., Christensen, L. P., Agili, S. M., Thilsted, S. H. (2007). Effect of home preparation practices on the content of provitamin A carotenoids in coloured sweet potato varieties (*Ipomoea batatas* Lam.) from Kenya. *Innovative Food Science and Emerging Technologies*, 8, 399–406.

King, V. A. E., Liu, C. F., Liu, Y. J. (2001). Chlorophyll stability in spinach dehydrated by freeze-drying and controlled low-temperature vacuum dehydration. *Food Research International*, 34, 167–175.

Korus, A. (2011). Effect of preliminary processing, method of drying and storage temperature on the level of antioxidants in kale (*Brassica oleracea* L. var. *acephala*) leaves. *LWT-Food Science and Technology*, 44, 1711–1716.

Kowalski, S., Szadzińska, J., Łechatańska, J. (2013) Non-stationary drying of carrot: Effect on product quality. *Journal of Food Engineering*, 118 (4), 393–399.

Krokida, M. K., Tsami, E., Maroulis, Z. B. (1998). Kinetics on color changes during drying of some fruits and vegetables. *Drying Technology*, 16 (3–5), 667–685.

Kröll, K., Kast, W. (1989). *Trocknen und Trockner in der Produktion*, Bd. 3. Springer Verlag, Berlin and Heidelberg.

Lajollo, F. M., Lanfer Marquez, U. M. (1982). Chlorophyll degradation in spinach system at low and intermediate water activities. *Journal of Food Science*, 47 (6), 1995–1998.

Lefsrud, M., Kopsell, D., Sams, C., Wills, J., Both, A. J. (2008). Dry matter content and stability of carotenoids in kale and spinach during drying. *HortScience*, 43 (6), 1731–1736.

Leong, S. Y., Oey, I. (2012). Effects of processing on anthocyanins, carotenoids and vitamin C in summer fruits and vegetables. *Food Chemistry*, 133, 1577–1587.

Lewicki, P. P., Pawlak, G. (2003). Effect of drying on microstructure of plant tissue. *Drying Technology*, 21 (4), 657–683.

Lewis, M., Heppell, N. (2000). *Continuous Thermal Processing of Food*. Aspen Publications, Gaithersburg, MD.

Liu, D., Sun, D. W., Zeng, X. A. (2014). Recent advances in wavelength selection techniques for hyperspectral image processing in the food industry. *Food and Bioprocess Technology*, 7, 307–332.

Liu, P., Zhang, M., Mujumdar, A. S. (2012). Comparison of three microwave-assisted drying methods on the physiochemical, nutritional and sensory qualities of re-structured purple-fleshed sweet potato granules. *International Journal of Food Science and Technology*, 47 (1), 141–147.

Lóopez-Ayerra, B., Murcia, M. A., Garcia-Carmona, F. (1998). Lipid peroxidation and chlorophyll levels in spinach during refrigerated storage and after industrial processing. *Food Chemistry*, 61 (1–2), 113–118.

Manzocco, L., Calligaris, S., Mastrocola, D., Nicoli, M. C., Lerici, C. R. (2000). Review of non-enzymatic browning and antioxidant capacity in processed foods. *Trends in Food Science and Technology*, 11 (9–10), 340–346.

Marty-Audouin, C., Lebert, A., Rocha-Mier, T. (1992). Influence of drying on the color of plant products. In Mujumdar, A. S. (Ed.) *Drying of Solids*. Oxford and IBH Publishing Co., New Delhi, India, 326–346.

Maskan, M. (2006). Effect of thermal processing on tristimulus colour changes of fruit. *Steward Postharvest Reviews*, 2 (5), 1–8.

Miranda, M., Maureira, H., Rodríguez, K., Vega-Gálvez, A. (2009). Influence of temperature on the drying kinetics, physicochemical properties, and anti-oxidant capacity of *Aloe Vera* (*Aloe Barbadensis* Miller) gel. *Journal of Food Engineering*, 91 (2), 297–304.

Morris, N. J., Lohmann, I. W., O'Connor, R. T., Freeman, A. F. (1953). Peanut butter, 4. Determination of color of peanut butter by spectral reflectance method. *Food Technology*, 7, 393–396.

Mrcic, V., Cocci, E., Dalla Rosa, M., Sacchetti, G. (2006). Effect of drying conditions on bioactive compounds and antioxidant activity of broccoli (*Brassica oleracea* L.). *Journal of the Science of Food and Agriculture*, 86 (10), 1559–1566.

Mujumdar, A. S. (2000). *Drying Technology in Agriculture and Food Science*. Science Publishers, Inc., Enfield, NH.

Mujumdar, A. S. (Ed.) (2007). *Handbook of Industrial Drying*. CRC Press, Boca Raton and New York.

Muncia, M.A., Vera, A., Garcia-Carmona, F. (1992). Effect of processing methods on spinach: Proximate composition in fatty acid and soluble protein. *Journal of the Science of Food and Agriculture*, 59, 473–476.

Negi, P. S., Roy, S. K. (2000). Effect of blanching and drying methods on β-carotene, ascorbic acid and chlorophyll retention in leafy vegetables. *LWT-Food Science and Technology*, 33, 295–298.

Negi, P. S., Roy, S. K. (2001). Effect of drying conditions on quality of green leaves during long term storage. *Food Research International*, 34, 283–287.

Nicolaï, B. M., Defraeye, T., de Ketelaere, B., Herremans, E., Hertog, M. L. A. T. M., Saeys, W., Torricelli, A., Vandendriessche, T., Verboven, P. (2014). Nondestructive measurement of fruit and vegetable quality. *Annual Review of Food Science and Technology*, 5, 285–312.

Oliveira, S. M., Brandão, T. R. S., Silva, C. L. M. (2016). Influence of drying processes and pre-treatments on nutritional and bioactive characteristics of dried vegetables: A review. *Food Engineering Reviews*, 8(2), 134–163. doi:10.1007/s12393-015-9124-0.

Oliviero, T., Verkerk, R., Dekker, M. (2013). A research approach for quality based design of healthy foods: Dried broccoli as a case study. *Trends in Food Science and Technology*, 30 (2), 178–184.

Onemani, O., Badifu, G. I. O. (1987). Effect of blanching and drying methods on nutritional and sensory quality of leafy vegetables. *Plant Foods Human Nutrition*, 37 (4), 291–298.

Pathare, P. B., Opara, U. L., Al-Said, F. A. J. (2013). Colour measurement and analysis in fresh and processed foods: A review. *Food and Bioprocess Technology*, 6, 36–60.

Pedreschi, F., León, J., Domingo, M., Moyano, P. (2006). Development of a computer vision system to measure the color of potato chips. *Food Research International*, 39 (10), 1092–1098.

Pendre, N. K., Nema, P. K., Sharma, H. P., Rathore, S. S., Kushwah, S. S. (2012). Effect of drying temperature and slice size on quality of dried okra (*Abelmoschus esculantus* (L.) Moench). *Journal of Food Science and Technology*, 49 (3), 378–381.

Perera, C. O. (2005). Selected quality attributes of dried foods. *Drying Technology*, 23 (4), 717–730.

Quing, Z., Ji, B., Zude, M. (2008). Non-destructive analyses of apple parameters by means of laser-induced light backscattering. *Postharvest Biology and Technology*, 48 (2), 215–222.

Ragaert, P., Verbekeb, W., Devliegherea, F., Debevere, J. (2003). Consumer perception and choice of minimally processed vegetables and packaged fruits. *Food Quality and Preference*, 15, 259–270.

Ratti, C. 2009. *Advances in Food Dehydration*. CRC Press, Boca Raton, FL.

Romano, G., Nagle, M., Argyropoulos, D., Müller, J. (2011). Laser light backscattering to monitor moisture content, soluble solid content and hardness of apple tissue during drying. *Journal of Food Engineering*, 104 (4), 657–662.

Rovedo, C. O., Viollaz, P. E. (1998). Prediction of degrading reactions during drying of solid foodstuff. *Drying Technology*, 16 (3–5), 551–578.

Ryley, J., Kajada, P. (1994). Vitamins in thermal processing. *Food Chemistry*, 49 (2), 119–129.

Sagar, V. R., Kumar, S. (2010). Recent advances in drying and dehydration of fruits and vegetables: A review. *Journal of Food Science and Technology*, 47 (1), 15–26.

Santos, P. H. S., Silva, M. A. (2008). Retention of vitamin C in drying processes of fruits and vegetables: A review. *Drying Technology*, 26 (12), 1421–1437.

Schuchmann, H. P., Schuchmann, H. (2005). *Lebensmittelverfahrenstechnik: Rohstoffe, Prozesse, Produkte*. Wiley-VHC, Weinheim, Germany.

Severini, C., Baiano, A., De Pilli, T., Romaniello, R., Derossi, A. (2003). Prevention of enzymatic browning in sliced potatoes by blanching in boiling saline solutions. *LWT-Food Science and Technology*, 36, 657–665.

Shewfelt, R. L. (1999). What is quality? *Postharvest Biology and Technology*, 15 (3), 197–200.

Stafford, H. A. (1994). Anthocyanins and betalains: Evolution of the mutually exclusive pathways. *Plant Science*, 101 (2), 91–98.

Sturm, B., Hofacker, W., Hensel, O. (2012). Optimizing the drying parameters for hot air dried apples. *Drying Technology*, 30 (14), 1570–1582.

Tello-Ireland, C., Lemus-Mondaca, R., Vega-Gálvez, A., López, J., Di Scala, K. (2011). Influence of hot-air temperature on drying kinetics, functional properties, colour, phycobiliproteins, antioxidant capacity, texture and agar yield of alga *Gracilaria chilensis*. *LWT-Food Science and Technology*, 44 (19), 2112–2118.

Thijssen, H. A. C. (1979). Optimization of process conditions during drying with regards to quality factors. *Lebensmittel-Wissenschaft und Technologie*, 12, 308–317.

Timoumi, S., Mihoubi, D., Zagrouba, F. (2007). Shrinkage, vitamin C degradation and aroma losses during infra-red drying of apple slices. *LWT-Food Science and Technology*, 40 (9), 1648–1654.

Uddin, M. S., Hawlader, M. N. A., Luo, D., Mujumdar, A. S. (2002). Degradation of ascorbic acid in dried guava during storage. *Journal of Food Engineering*, 51, 21–26.

Udomkun, P., Nagle, M., Mahayothee, B., Müller, J. (2014). Laser-based imaging system for non-invasive monitoring of quality changes of papaya during drying. *Food Control*, 42, 225–233.

Vega-Gálvez, A., Di Scala, K., Rodríguez, K., Lemus-Mondaca, R., Miranda, M., López, J., Perez-Won, M. (2009). Effect of air-drying temperature on physico-chemical properties, antioxidant capacity, colour and total phenolic content of red pepper (*Capsicum annuum*, L. var. Hungarian). *Food Chemistry*, 117 (4), 647–653.

Wedzicha, B., Lamikanra, O., Herrera, J., Panahi, S. (1984). Recent developments in the understanding of the chemistry of sulfur (IV) oxospecies in dehydrated vegetables. *Food Chemistry*, 15 (2), 141–155.

Weemaes, C. A., Ludikhuyze, L. R., van der Broeck, I., Hendrickx, M. E., Tobback, P. P. (1998). Activity, electrophoretic characteristics and heat inactivation of polyphenoloxidases from apples, avocados, grapes, pears and plums. *LWT-Food Science and Technology*, 31 (1), 44–49.

Woolfe, J.A. (1992). *Sweet Potato: An Untapped Food Resource*. Cambridge University Press, Cambridge.

13 Instant Controlled Pressure Drop as a Process of Texturing/Sterilizing Vegetables, Improving upon Conventional Drying Methods

Karim Allaf
University of La Rochelle

Tamara Allaf
ABCAR-DIC Process

Sabah Mounir
Zagazig University and University of La Rochelle

Farid Zerrouq
Sidi Mohamed Ben Abdellah University

CONTENTS

13.1 INTRODUCTION

In a very relevant study, The Swedish Institute for Food and Biotechnology (SIK) suggests that about 1.3 billion tons per year of vegetables, or roughly one-third of food produced for human consumption, are lost or globally wasted (Jenny Gustavsson et al., 2011). This is mainly due to the lack of relevant preservation operations capable of taking into account the specificity of various products and their required quality. Performances of postharvesting and food processes are different, specific, and numerous, depending on the infrastructure of the production area. Thus, although simple, freezing is not easily accessible in the harvesting zones and painfully applicable in the area of crop production. It is also economically disadvantageous in terms of transport to consumption regions. It implies possibilities of ruptures of cold chain, which results in weak quality of products and real risks in terms of bacterial development. Moreover, thawing of frozen fresh vegetables results in a very weak texture quality mainly because of wall rupture of full-of-water cells, even using high velocity freezing. Partial dehydration of dehydrofreezing processes should reduce such risks.

On the contrary, conventional hot air drying remains the most effective preservation operation and the simplest drying process. Furthermore, it is the easiest procedure that can be applied by the farmers in the vicinity of the harvesting zone. Moreover, since internal temperature remains at a relatively low level during a long time of the hot air drying process, this greatly preserves the biochemical quality of products. However, two main weaknesses dramatically reduce both drying performance and final product quality. The first is the great amount of energy needed for evaporating water. The second concerns the shrinkage, which the product undergoes, making it a more compact structure, often coupled to case-hardening. Shrinkage can be understood as a significant structural deformation in shape and size of a sample occurring along with simultaneous heat and mass transfer (Mayor and Sereno, 2004). The product has a rubbery behavior because of water content and temperature being higher than the glass transition level, the polymer loses water and collapses under this action. This results in dramatic decreasing of both drying rate and final product quality. Moreover, many authors indicated that case-hardening, normally due to excessive acceleration of product surface dehydration, inevitably slows down the drying process (Orikasa et al., 2005, 2008).

Some technologies such as heat pump or solar energy or both can contribute to provide the advantage of a substantial reduction in the normally high energy cost, however, possibly contributing to a high presence of bacteria and spores, which would imply sanitary problems. Moreover, the possible preservation of insect larvae greatly involves shorter preservation and storage time.

Dried product quality in terms of flavor, color, texture, and nutrients greatly depends on drying methods and treatment conditions (Vadivambal and Jayas, 2007; Head et al., 2008). Normally, the higher the drying temperature and time, the lower the chemical, functional, and sensorial quality. Compared with conventional hot air drying, the lower drying temperature in a lower oxygen environment used in vacuum drying leads to better retention of color, texture, and bioactive compounds (Suvarnakuta et al., 2005; Lewicki, 2006; Wu et al., 2007; Alibas, 2009).

Whatever the drying process, shrinkage usually results in low organoleptic quality and nutrition value. Compact texture makes it much more difficult for numerous posttreatment processes such as grinding, etc. It also implies low functional quality such as rehydration, water holding capacity (WHC), etc.

Other drying methods (freeze-drying, superheated steam, microwaves, etc.) have aimed at improving the dried product quality. However, freeze-drying is still too expensive for food applications, and it is neither able to control the texture nor meet the needs of a perfect microbiological decontamination. The use of superheated steam usually allows great thermal degradation, while vacuum superheated steam results in weak reliable equipment. Since microwaves imply internal heating, the temperature of the main part of the product should be much higher than what conventional hot air drying applies, leading to too large biochemical and nutritional degradation.

The lack of quality mainly correlated to the thermal degradation of hot air-dried products often takes place during the last stage of the process when the texture becomes more compact, and cementation results in a lower diffusivity and weaker drying rate.

Thus, texturing operations have been proposed as a method to improve airflow drying operation at both dried product quality and operation performance. The "swell drying" method has been defined as a very effective way to couple hot air drying with texturing by instant controlled pressure drop (DIC: Détente Instantanée Contrôlée; French for instant controlled pressure drop). While puffing can effectively be adopted mainly for cereal-based materials, DIC can be used to treat a wide range of products, despite their possible thermal fragility, greatly increasing drying rate and improving the textural, functional, and healthy quality of the final dried product. Thus, the finished product has better nutrition value than conventionally dried products. Moreover, DIC allows better color preservation, often considerably enhancing flavonoid availability and antioxidant activity. In addition, through a Ultra-High Temperature UHT-type microbiological decontamination and a complete elimination of insects and larvae, DIC treatment ensures a significantly longer lifetime.

13.2 PHENOMENOLOGICAL ANALYSIS OF AIRFLOW DRYING

Although the airflow drying process was one of the first unit operations of various cultures of human civilization, it still generates great scientific curiosity. Indeed, drying plays a key role in numerous industries, with inevitably growing importance for a huge potential of applications. Many scientific and technological studies and macroeconomic analyses have indicated requirements to improve the quality of the final product and increase process performance. This highlights the importance of achieving the objective of relevant intensification of this operation. Thus, empirical and/or kinematic models are not adequate for such an objective. Only a real phenomenological modeling can define the different fundamental processes and designate which is the highest resistant (limiting) process.

Since drying is a dehydration operation allowing specific interaction between exchange surface and water vapor, water is removed from the product exclusively by evaporation. The drying rate is dependent on the difference of vapor pressure from the product surface to the outside airflow. The higher the temperature and water activity of the exchange surface, the higher the vapor value at this surface. Vapor transfer at the surface depends in the present case of airflow drying and on the main characteristics of air (velocity, temperature, and relative humidity). The higher the drying rate, the colder the internal matrix; the thermal degradation of the material is very low for a long period of time during the main part of airflow drying.

13.2.1 Mathematical Phenomenological Modeling

13.2.1.1 Heat and Mass Transfers as Airflow/Solid Surface Interaction

The most important aspect of drying technology is the phenomenological model (Kavak Akpinar et al., 2006). Knowledge of the drying kinetics of biological materials is essential to intensify, design, optimize, and control the drying process (Sacilik et al., 2006). The principle of modeling is based on defining the fundamental phenomena before translating them on a set of mathematical equations that can adequately characterize the system. In particular, the solution of these equations must allow not only prediction of the process parameters (Gunhan et al., 2005) but also determination of the highest resistance process. Many mathematical models were proposed to describe the drying evolution mainly depending on the product (Özdemir and Onur Devres, 1999). Although there have been many researches on the mathematical modeling of experimental studies of various vegetables and fruits—such as parsley leaves (Kavak Akpinar et al., 2006), organic tomato (Sacilik et al., 2006), bay leaves (Gunhan et al., 2005), golden apples (Menges and Ertekin, 2006), banana (Dandamrongrak et al., 2002), berberis (Aghbashlo et al., 2008), Asian white radish (Lee and Kim, 2009), pea (Pardeshi et al., 2009), grape (Çağlar et al., 2009), and tomato (Demir and Sacilik, 2010)—the kinematic and/or empirical drying models neither allow to determine the highest resistance process nor even define appropriate intensified solutions.

The presence of water at the exchange surface combined with the heat transferred at this surface allows the generation of vapor. Later, the vapor is transported toward the surrounding environment usually through mass convection depending on airflow velocity, temperature (T_{air}), and relative humidity (HR). Thus, drying operation requires two tightly coupled processes: transferring heat from airflow to the exchange surface and establishing a vapor pressure difference from the exchange surface toward the surrounding environment.

Since the heat is transferred to the surface by convection, convection heat transfer rate $\dot{Q}$ achieved at the surface A should be the function of the heat convection coefficient h and the difference of temperatures of airflow and product surface T_{air} and T_s, respectively:

$$\dot{Q} = hA(T_{\text{air}} - T_s). \tag{13.1}$$

Vapor transport should be carried out through a mass convection rate depending on the mass convection coefficient k, the vapor density ρ_v, and the difference of vapor pressures at the product surface p_{ws} and in the surrounding medium $p_{w,\text{air}}$, respectively. One can express the evolution rate of the water content dry basis of the product, versus the difference of vapor pressures as following (Nguyen et al., 2016):

$$\dot{W} = -k\rho_v \frac{A}{m_d} \frac{\left(p_{ws} - p_{w,\text{air}}\right)}{P_{\text{total}}}. \tag{13.2}$$

Since the superficial vapor pressure p_{ws} depends on both temperature and water activity at the surface, $\left[p_{ws} = p_{w,T_s} a_{ws}\right]$ and the vapor pressure of air $p_{w,\text{air}}$ is correlated to the air temperature and its relative humidity HR, $\left[p_{w,\text{air}} = p_{w,T_{\text{air}}} \text{HR}\right]$

$$\dot{W} = -k\rho_v \frac{A}{m_d} \frac{(p_{w,T_s} a_{ws} - p_{w,T_{\text{air}}} \text{HR})}{P_{\text{total}}}. \tag{13.3}$$

Both heat and mass convection rates strictly depend on airflow velocity through coefficients h (expressed in W/m^2/K) and k (expressed in m/s), temperature, and water activity at the surface, respectively.

$$\dot{Q} = \dot{W}\ m_d\ L. \tag{13.4}$$

Internal wet bulb temperature $T_{w,s}$ can be achieved with the heat–mass balance system (Equations 13.2 and 13.4). $T_{w,s}$ is given by

$$T_{w,s} = T_{\text{air}} - \frac{k\rho_v}{h} \frac{L(p_{w,T_s} a_{ws} - p_{w,\text{air}})}{P_{\text{total}}} \tag{13.5}$$

This value of wet temperature $T_{w,s}$ is so low that internal heat transfer during the main part of the drying operation is equally low. Once the evaporation is achieved inside the internal pores and the water activity is low, the material temperature becomes relatively high, close to air temperature.

Both surface water activity and exchange surface (A) normally decrease during the drying operation because of the solute concentration and the shrinkage phenomenon.

Even when the drying operation is controlled by the external process, the drying rate $\dot{W}$ generally decreases because of the decreasing values of both exchange surface A and water activity of the surface $a_{w,s}$.

The main part of water vapor usually occurs at the exchange surface, which has the highest product temperature. Thus, during the main part of the operation, since internal temperature is lower, the "internal" evaporation process in the pores (water) of porous product stays very weak. Only at the final stage of the operation would it become relatively important. Consequently, during the main part of the airflow drying, water transfer inside the volume of the product toward its exchange surface takes place in the liquid phase. Therefore, airflow supplies the heat required for liquid/vapor vaporization change at the exchange surface. At the final stage of drying, internal temperature increases as does within-pore evaporation.

Airflow drying unit operation generally has five specific processes (Figure 13.1) as follows:

1. Heat transfer convection from the external medium to the exchange surface; the main part of this heat amount is mainly used for evaporating liquid water at the exchange surface.
2. Heat transfer conduction within the body.
3. Liquid water transfer within the body from its core to its exchange surface.
4. Vapor transfer within the body, much more relatively important at the last stage of drying.
5. Vapor transport from the exchange surface toward the surrounding medium; it is normally achieved through mass convection.

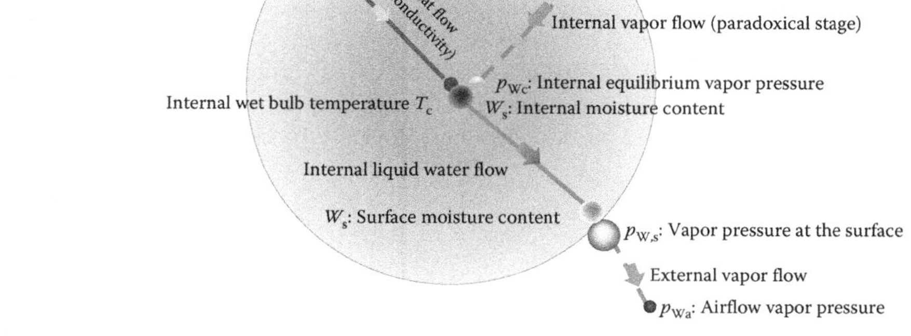

FIGURE 13.1 Schematic presentation of transfer phenomena of airflow drying (external heat transfer; heat for superficial evaporation; internal heat transfer; internal liquid water diffusion; external vapor transport; internal paradoxical vapor/heat transfer).

In terms of kinetics, these five processes get a "serial causality," implying the unit operation to follow the slowest process rate.

Other phenomena are normally indicated. They are as follows:

- The structural modification of the material (shrinkage phenomenon) depending on the drying conditions of temperature and water content and taking place during rubber-behavior state.
- The "biochemical" degradation issued from the temperature impacts (biochemical thermal degradation, oxidation, etc.).

13.2.2 Phenomenological Kinetic Model and Drying Steps

"Usually," hygroscopic porous media products undergo three drying phases as discussed in the following sections:

13.2.2.1 First Phase: Evaporation of Superficial Water

The first phase concerns the interaction between the surface of the product and the airflow, which usually has higher temperature and relative humidity HR < 1. This results in superficial transfer of vapor towards the surrounding medium, independently of any diffusion effect. Drying rate during this starting stage was reflected by Allaf and Allaf (2014), using the starting accessibility δW_s.

$$\delta W_s = -k\rho_v \frac{A}{m_d} \frac{\left(p_{w,T_s} a_{ws} - p_{w,air}\right)}{P_{total}} \delta t \tag{13.6}$$

The airflow characteristics (temperature, relative humidity, and velocity) as well as the value of exchange surface A take place in this stage. By increasing A through grinding or cut products and by getting higher value of water activity at the exchange surface, operators can greatly increase the value of starting accessibility.

13.2.2.2 Phase 2 of Superficial Evaporation Coupled to Liquid Water Diffusion within the Hygroscopic Porous Medium

The second phase starts once a gradient of water content is established in the product volume zone. Thus, liquid water transfer which take place includes various phenomena of capillarity, Fick's diffusion, etc. Then, drying process should be revealed by coupling the superficial vapor transport (mass convection in the surrounding medium) and internal liquid water diffusion by Fick's-type law:

$$\frac{\partial W_s}{\partial t} = -k\rho_v \frac{A}{m_d} \frac{\left(p_{w,T_s} a_{ws} - p_{w,air}\right)}{P_{total}}. \tag{13.7}$$

$$\frac{\rho_w}{\rho_m}\left(\overrightarrow{v_w} - \overrightarrow{v_m}\right) = -D_{eff}\vec{\nabla}\left(\frac{\rho_w}{\rho_m}\right). \tag{13.8}$$

With D_{eff} is the effective diffusivity.

This process exists in all the drying stages but cannot be reflected by the drying kinetics without getting the external superficial airflow interaction at a higher rate than "diffusion." Once airflow velocity is adequately increased, in addition to temperature and dryness, the drying process becomes limited by the diffusion process. Therefore, the internal water transfer resistance becomes high enough to be the limiting process of the entire drying operation. This is much more evident when one can assume that the thermal diffusivity is substantially greater than the diffusivity of liquid water within the material.

13.2.2.3 Phase 3 of Paradoxical Stage Coupling Internal Conduction Heat and Vapor Diffusion Transfers

By decreasing the amount of liquid water within the porous materials and by increasing the internal temperature much higher than its initial value of $T_{w,s}$, the vapor generated inside the internal pores becomes relatively more and more dominant and the internal mass transfer becomes a vapor transfer process in the porous material. This reveals the paradoxical stage of coupled temperature and vapor pressure gradients, both directed from the surface toward the core of material. Drying can continue and occur by involving a "progressive front" kinetic type. Thus, a considerable decrease in the drying rate occurs. However, the water content can keep decreasing until a balance limit value is achieved, which depends on the drying air conditions (air temperature [T_a] and relative humidity [HR]) and thermodynamic properties of the material.

Since vapor diffusion transfer within a porous medium has the gradient of vapor pressure as the driving force, similarly Fick's transfer law of diffusion can be used with $D_{v,eff}$ as the effective vapor diffusivity:

$$\frac{p_v/T}{\rho_m}\left(\vec{v_v} - \vec{v_m}\right) = -D_{v,eff}\vec{\nabla}\left(\frac{p_v/T}{\rho_m}\right). \tag{13.9}$$

Heat transfer within this volume implies both heating and evaporation of the water in the medium, more possibly within the pores. Conductivity to be considered should be dynamic effective conductivity λ_{eff}, whose value is usually much higher than the static conductivity value. Numerous authors assumed this high value of λ_{eff} to be due to various condensation/evaporating processes inside the pores. By assuming the main part of transferred heat is used for internal evaporating process, at constant temperature, heat flow can be defined by

$$-\lambda_{eff}\ \vec{\nabla}\cdot\left(\vec{\nabla}T\right) + \varepsilon_{abs} M_w L_v \frac{\partial p_v/RT}{\partial t} = 0. \tag{13.10}$$

Since, especially during this last phase of drying, we can often neglect the shrinkage impact and assume the density of dry material ρ_m to be constant and its proper velocity v_m to be null ($v_m = 0$), we can write this as

$$\frac{p_v}{T}\vec{v_v} = -D_{v,eff}\vec{\nabla}\left(\frac{p_v}{T}\right). \tag{13.11}$$

The two-equation system 13.10 and 13.11 revealed at one-dimension r gives

$$-\lambda_{\text{eff}} \frac{\partial^2 T}{\partial r^2} + \varepsilon_{\text{abs}} M_{\text{w}} L_{\text{v}} \frac{\partial p_{\text{v}}/RT}{\partial t} = 0. \tag{13.12}$$

$$\left(\frac{p_{\text{v}}}{T}\right)\overrightarrow{v_{\text{v}}} = -D_{\text{v,eff}} \frac{\partial (p_{\text{v}}/T)}{\partial r}\overrightarrow{e_r}. \tag{13.13}$$

The gradient of temperature within the porous medium is necessarily directed from the surface to the core of the sample. Thus, as (p_{v}/T) does not depend on the water concentration, but increases with the temperature, we can write

$$\overrightarrow{v_{\text{v}}} = -\frac{D_{\text{v,eff}}}{(p_{\text{v}}/T)} \frac{\partial (p_{\text{v}}/T)}{\partial r}\overrightarrow{e_r}. \tag{13.14}$$

Equation 13.14 leads to a movement of vapor directed from the surface toward the core, thus defining a paradoxical movement completely opposite to that required by the drying process. Only water activity should reduce the vapor pressure at the zone close to the external surface, and the whole kinetics of this phase should be an extremely slow "progressive front" kinetics.

13.3 HOW TO USE KINETIC ANALYSIS TO IDENTIFY EXTERNAL VAPOR CONVECTION COEFFICIENT *k* OR INTERNAL LIQUID WATER EFFECTIVE DIFFUSIVITY

In the case of many porous hygroscopic materials, such as minerals, food, or biological products, analysis of the drying kinetics first requires identifying the limiting process.

13.3.1 Conditions to Use Drying Kinetics for Identification of Effective Diffusivity

It is obvious that the kinetics of drying operation reveals the internal diffusion process only when the internal diffusion of liquid water within the porous material is the limiting process. However, this should be adopted only after proving that external mass transfers are not the limiting process. Two ways can be proposed to prove such an assumption:

- *Calculation way*: airflow velocity used in the operation is higher than a critical airflow velocity (CAV) depending on airflow temperature and humidity, size, and shape of the material:

$$\mathrm{CAV}=\left(\frac{-D_{\mathrm{eff}}P_{\mathrm{total}}\left[\dfrac{d\rho_{w}}{dr}\right]_{r=r_{\mathrm{surface}}}}{\rho_{v}\left(p_{v,T_s}a_{w,s}-p_{v,\mathrm{air}}\right)}\right)^{1.78}. \quad (13.15)$$

- *Experimental way*: various values of airflow velocity should be used. As long as the operation kinetics depend on the velocity, it means airflow velocity is lower than the CAV and the drying operation is translating the external airflow/surface interaction. Only when drying kinetics becomes independent of airflow velocity, this last converts in the range of higher than the CAV. Thus, the diffusion process can be observed through the drying kinetics.

13.3.2 Method to Determine the Effective Diffusivity from Drying Kinetics Data

13.3.2.1 Airflow Velocity Range

The validity of the condition of airflow velocity must be confirmed through its value (v > CAV) or impact (various values of airflow velocity do not imply modification in terms of drying kinetics). This should be done as the preliminary stage authorizing to analyze the diffusive model from experimental data of airflow drying. The strict verification of this condition of airflow velocity is the prerequisite to support not only the presence of the internal effective diffusion process of liquid water in the porous material, but also the possibility of the experimental data of drying curves of water content dry basis versus time to reveal this diffusion process. This is indispensable to consider the diffusion of liquid water within the porous material as the limiting process of drying.

13.3.2.2 Experimental Data Range

Once the limiting conditions (airflow velocity >> CAV) are satisfied, the main phase of drying is controlled by liquid water diffusion within the porous material coupled with superficial vapor generation and external transport of vapor toward the surrounding medium. Diffusion occurs by gradient of water content dry basis as the driving force, which is similar to Fick's first law, with the effective diffusivity as diffusion constant. Various solutions to Fick's second law can be obtained depending on the specific boundary conditions. Mathematical solutions are also available: series or error functions for less air drying time or trigonometric series or Bessel functions for higher air drying time. When it is possible to assume that shrinkage is also negligible and D_{eff} is constant, the solution proposed by Crank (1975) according to the geometry of the solid matrix is usually the most adopted one.

However, only the diffusion-part of the experimental data of water content dry basis versus time can be used. In this phase, internal vapor generation and transfer are negligible, and the diffusive transfer of liquid water within the porous material is preeminent. The values related to both start-up stage (neighboring time $t = 0$) and paradoxical stage (where vapor transfer within the porous material can be assumed to be preponderant) must be excluded from this liquid water diffusion study. Thus, the experimental data to be used in this liquid water diffusion model belong to the time range from $t = t_1$, distinct from the time $t = 0$ linked to the starting process and $t_{\text{paradoxical stage}}$.

Indeed, the theoretical value W_o that generally is obtained by extrapolation of the diffusional model toward $t = 0$ should be distinct from the initial real value of moisture content W_i. The difference between these values of W_i and W_o should reveal the amount of water easily removed from the surface, regardless of the internal liquid water diffusion process. The starting accessibility δW_s expressed as g H_2O/g dry basis db is given by

$$\delta W_s = W_i - W_o. \tag{13.16}$$

13.4 DRYING INTENSIFICATION: APPLICATION

13.4.1 Disadvantages of Airflow Drying

Airflow drying is always considered as the easiest and the most natural drying process. Since the internal level of temperature during the main part of the operation is much lower than the surface temperature, the biochemical quality of the product is preserved till the final stage of the operation. However, the weaknesses of this operation have been identified in various sectors which is discussed in the following sections.

13.4.1.1 Process Performance Drawbacks

- A very feeble internal diffusivity of moisture (liquid water) results in very low process kinetics. The drying rate is too weak, much more during the postshrinkage stage.
- The consumption of energy is very high because of the heat evaporation level.

13.4.1.2 Drawbacks of Final Product Quality Attributes

- Biochemical thermal degradation mainly during the final stage of drying.
- Poor functional behavior in terms of rehydration kinetics, levels, and WHC.
- Compactness and out-of-crispness texture.
- Possibility of microbiological contamination. A relevant decontamination method seems to be one of the most important requirements.

13.4.1.3 Drawbacks at Equipment Reliability

- Airflow production at high velocity and low relative humidity.
- Although very interesting in terms of heat and mass transfers, fluidized bed–type driers are too difficult for large scale equipment. Indeed, a 100-m^2 drier needs more than 15 m/s airflow velocity. Ventilator capacity should be higher than 5,400,000 m^3/h.
- Moreover, adequate distribution and interaction of the airflow at the whole product surface is always very difficult to be homogeneous and appropriately effective.

13.4.2 Intensification Ways

The intensification of the airflow drying process can be carried out by performing adequate modifications of the external medium (airflow temperature, velocity, and humidity), the type of drier (vibro-fluidized bed, etc.), or the product (shape, size, thickness, etc.).

13.4.2.1 First Intensification of External Transfers

Heat is normally transferred toward the matrix surface from the surrounding medium by convection, infrared (IR) radiation, and/or contact. In the opposite direction, vapor is removed from the exchange surface using airflow transport. Both flux of heat transfer and vapor transport should be perfectly correlated. The heat flow depends on the temperature and velocity of the surrounding medium (air). Equation 13.6 shows that to increase the flux of water vapor from the exchange surface toward the medium, it is necessary to:

- Increase the airflow velocity to obtain the highest possible value of vapor convection coefficient k.
- Increase the value of exchange surface A. This would be possible through grinding (to reduce granule size) or cutting, or any other possibility to obtain thinner sliced products. Another very relevant way to get both high airflow velocity and high exchange surface area is to use fluidized bed driers.
- Increase the values of both temperature and water activity of exchange surface in order to increase vapor pressure generated at the exchange surface. It is worth noting that the impact of high airflow temperature can be greatly reduced by establishing a high drying rate. This may greatly reduce material degradation.
- Reduce the relative humidity of airflow.

13.4.2.2 Second Intensification: Thermomechanical Texturing

The intensification of external heat and vapor transfer operations is relatively simple to perform and easy to control, unlike internal transfers that strongly depend on the structural characteristics of each material.

Since the first starting phase does not present any limitation, the higher the airflow velocity, the higher is the external heat and vapor transfers.

However, the second stage of the drying operation should combine external heat and vapor transfers with internal heat and liquid water diffusion within the porous material. This becomes the limiting process once the airflow velocity is higher than the CAV (Equation 13.15). Therefore, usually, the technological aptitudes of vegetable structure vis-à-vis drying operation are too weak, and drying often becomes limited by the internal transfer processes.

Moreover, the drying process is frequently associated with the shrinkage phenomenon; external heat transfer and vapor transport must inevitably decline with the exchange surface A_{eff}, and the technological aptitudes of material also become weaker. Only freeze-drying can remain unaffected by such shrinkage. It is well known that, generally, the thermal and liquid water diffusivities of many porous materials directly depend in an opposite direction on the degree of compactness. Higher porosity and more adequate pore distribution, etc. can become the best way to act against shrinkage impact. Numerous experimental data and various studies have often highlighted the importance of texturing as a relevant way to significantly improve mass diffusivity of porous materials.

The expansion of the porous structure improves water diffusivity regardless of the phase, liquid or vapor. Thus, the expansion by puffing, extrusion, DIC, etc. is a particularly suitable intensification at this stage of the drying operation.

"Swell drying" is a specific drying unit operation, which implies inserting a DIC expansion, often just after the first step of conventional hot air drying to overcome the compactness of the product via controlled expansion. The moisture content required for performing this expansion step should allow crossing the glass transition of the material just after the expanding process. This is strictly linked to the final equilibrium temperature and severely depends on the final total pressure. Only DIC permits to modify and control the level of this final pressure.

Once the expanded structure is performed, achieved, and preserved, the drying kinetics are amply intensified with a consequent reduction of the energy and a good improvement in the quality of the final product.

13.4.2.3 Third Intensification: Address the Paradoxical Phase Drying

During the final stage of drying, the residual water, whether bound or free, is purely evaporated in the volume inside the pores, and transferred as vapor within the porous matrix. These two transfers of heat (by conduction) and mass (vapor diffusion) are driven by gradients of temperature and vapor pressure, respectively. Both are directed from the surface towards the core, which results in too weak drying kinetics (paradoxical phase). Since porous material induces low conductivity, this paradoxical phenomenon is even more important in the case of DIC-treated polymers. To address this paradox and intensify the drying operation, three solutions can be adopted after the second intensification by DIC texturing has been conducted:

- Microwave heating
- Overheated steam drying
- Integral Starting Accessibility Drying (ISAD), often toward a vacuum.

These operations use Darcy's law partially or completely, with gradient of total pressure as a relevant way to solve this problem.

13.4.3 Texturing Operations

Since the natural structure of many plant-based materials implies too weak technology aptitudes regarding various processes of drying, extraction, chemical or enzymatic reactions, etc., texturing/restructuring processes are becoming increasingly important for vegetal features of various plants. Such texturing allows modifying the physical properties, leading to new functional behaviors, and usually is a very relevant way to intensify mass transfer phenomena.

Moreover, various types of drying operations except freeze-drying increasingly result in more compact solid matrix because of the shrinkage phenomenon. Indeed, the high moisture content of natural vegetal implies a glass transition temperature much lower than freezing level. During the main part of drying, fruits and vegetables, seaweeds and microalgae, etc. present rubberlike rheological behavior. Only after removing the main part of water content can these matrices cross the glass transition T_g border.

The more the compactness of the solid matrix, the weaker the drying rate and the higher the temperature within the solid matrix. Indeed, such a compact matrix issued from the shrinkage phenomenon reduces the diffusivity of water (liquid or vapor) while increasing thermal conductivity of the material. On the other hand, it is worth noting that decreasing drying rate should mean that water within the matrix is becoming more and more trapped without necessarily designating that drying is reaching the linked water stage.

Only a structural alteration (expansion) can then be adopted as an intensification means to improve the kinetics of drying during this stage of the operation.

13.4.3.1 Texturing Conditions

Texturing techniques are numerous and are based on various phenomena. Whatever the type (thermal, chemical, and microbiological, etc.) is adopted, wherever they are used, and whenever they take place, they normally include three stages: (1) generation of expanding element, usually high-pressure gas; (2) expansion of dough-similar material; and (3) preservation of the new structure. Thus, any texturing technique requires three coupled circumstances, during the expansion process:

1. A viscoelastic rheological behavior of rubbery materials during the expansion process. This is normally stretched because of adequate thermo-rheological chemical conditions (chemical composition, water/solvent content, and temperature).
2. A great amount of gas generated inside the matrix in short time. This great rate of gas assures the expansion of the material. Thus, a relevant mechanical impact allows creating a new and more porous structure. The gas phase can be engendered through thermal evolution (evaporation, autovaporization, etc.), thermochemical reactions, or microbiological/fermentation phenomena, etc. When the material has a specific appropriate viscoelastic rheological behavior adapted to the operation, porosity can be obtained thanks to the action of stresses generated by the pressurized gas.
3. Crossing the glass transition (T_g, W_g) border through decreasing temperature, water content, etc. to preserve the new matrix structure.

13.4.3.2 Main Types of Texturing

13.4.3.2.1 Thermal Texturing Techniques

The techniques of texturing are divided into chemical and thermal techniques (Linden, 1994). This chapter discusses only the thermal technique.

Making bread and some types of cakes can be cited as examples of thermal-texturing processes. These processes need the generation of a large amount of expanding gas such as CO_2 acting on dough material to assure an expansion process based on pore formation, which is followed by a dehydration stage till a vitreous stage is reached after crossing the glass transition border.

Frying should also be regarded as a thermal texturing operation. It implies high-speed generation of vapor using high-temperature vegetal oil as the heating source. More than the high temperature level of the surrounding medium, the specificity of such an operation is also an exceptionally high heat exchange convection coefficient, especially during the expanding phase.

13.4.3.2.2 Thermo-Mechanical Texturing Ways by Releasing toward the Atmospheric Pressure

Popping is a very particular thermo-mechanical texturing operation where the specific structure of the grain plays an exceptional role in high-pressure vapor growing within the grain owing to a high heating rate. Once this internal pressure reaches a value permitting to "explosion" the external membrane, the system undergoes an instant autovaporization, thus generating a great volume of vapor and letting the product temperature to considerably decrease till the final equilibrium level of the liquid water at atmospheric pressure (100°C). This temperature is sufficient to cross the glass transition border of the starch system.

A similar situation is seen with other starch-based polymers in the cases of puffing, extrusion cooking, steam explosion, etc. An initial high-temperature and high-pressure stage is achieved within an adequate vessel before mechanically expulsing the dough toward the external atmospheric pressure. The superheated liquid water in dough undergoes an immediate autovaporization process, promptly leading to a very fast, flash, or instantaneous expansion and cooling toward a level close to 100°C, which allows preserving the new porous expanded structure of this particular composition.

13.4.3.2.3 Texturing by Releasing toward a Vacuum: DIC

On the basis of fundamental approaches and theoretical studies concerning the expansion process (Allaf, 1988), DIC technology was defined, used, and optimized and was greatly developed at both laboratory and industrial scales. This operation has specific characteristics at the three levels of (1) the amount of expanding vapor produced by autovaporization, more increased due to the laws of the thermodynamics of instant transformations, (2) the large possible range of treatment temperature level to let the product get the most appropriate rheological characteristics, and (3) the perfectly controlled final temperature depending on the final pressure. This allows treating, expanding, and preserving the new structure of a very large domain of sensitive materials for many polymers used in various sectors of food, cosmetics, pharmaceuticals, etc. DIC technology is capable of coupling high processing performances and final product-controlled quality, while reducing energy cost.

13.4.4 Analysis of DIC

DIC is a high-temperature short-time (HTST) treatment that involves subjecting a wet product (usually neighboring water content of 0.3 g H_2O/g db) to a heat source (steam, microwave, hot gas, etc.). For food applications, treatment temperature (<180°C) is applied for a short time for a few seconds. The most important stage is the abrupt pressure drop toward a vacuum (absolute pressure from dozens of Pa to dozens of kPa). The instant pressure drop is performed in a few milliseconds (about 20 ms but always <200 ms).

13.4.4.1 Various Stages of DIC

The main stages of DIC are defined in Figure 13.2. It usually has two main stages:

1. Injection of high saturated steam pressure stage, which results in heating of the product within the processing vessel while the desired vacuum level is established in the large volume vacuum tank. Treatment time is very short (few seconds or dozens of seconds), normally defined to let the polymer get the best homogeneity at both temperature and water content. This first stage can be intensified by
   a. Establishing an initial step of preliminary vacuum before injection of steam to improve the effective exchange surface and increase the heating kinetics
   b. Gradually increasing the steam pressure within the processing vessel. This may increase the gradual/progressive diffusion of water issued from the steam condensed at the external exchange surface
   c. Using a rotating processing vessel. Such a dynamic system should allow more homogeneity and higher capacity, more especially in the case of dispersed materials, grains, powders, etc.
2. Instantaneously opening the great section instant opening valve between the treatment vessel and the vacuum tank. This very short time stage

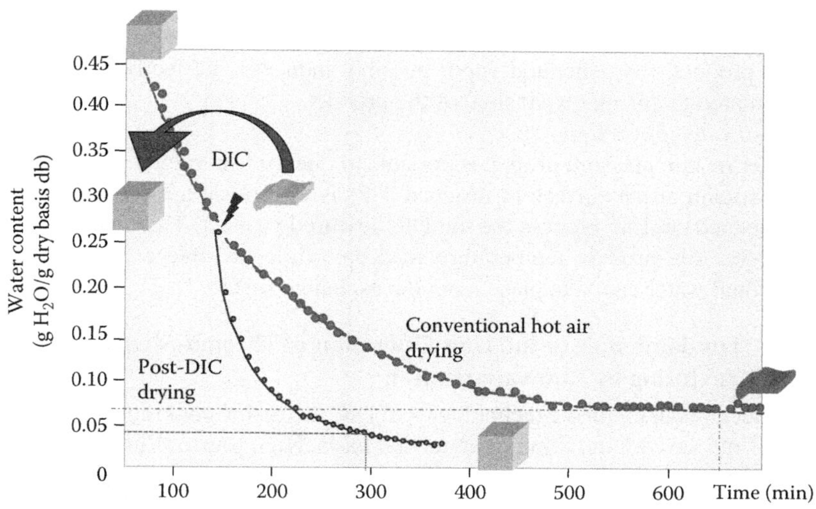

FIGURE 13.2 Intensification of airflow drying by inserting DIC texturing, thus defining the "swell drying" process.

(5–40 ms) ends by maintaining the vacuum for some seconds before releasing toward the atmospheric pressure. This stage may be intensified through

a. Direct expulsion of the product toward an external sterilized packaging.
b. Injection directly on the product surface of low-volume, high-velocity ambient airflow within the vacuum tank. This is to intensify the cooling of product.

13.4.4.2 Principal Technical Steps of Industrial DIC Treatment

1. *A first noninstantaneous vacuum*
   By using steam as heating and high pressuring tool, a first stage of establishing vacuum is performed after inserting the product within the treatment vessel. This is usually obtained through a direct connection with the vacuum pump.
2. *A gradual increase of steam pressure within the processing vessel*
   This is to perform gradual diffusions of the heat and of the water issued from the steam condensed at the outer exchange surface. Steam pressure should be performed in a few seconds and kept constant for a few seconds or tens of seconds to ensure a controlled product pressure and homogeneous temperature and moisture ($P < 2$ MPa, $T < 200°C$, $W < 1$ g H_2O/g dry basis).
3. *Pressure-drop into the vacuum*
   After adequate high temperature (usually up to 180°C), high pressure (typically up to 2 MPa), and short time (frequently between seconds and dozens of seconds), an abrupt decompression is achieved by instantly (<100 ms) connecting the treatment vessel to the vacuum high-volume tank (between 100 and 1000 Pa) using a large section instantaneously opening valve. This process, which can be achieved within a few thousandths or hundredths of a second, provides an instant autovaporization coupled to a similarly instantaneous cooling process. Depending on the thermo-rheological behavior of the product, the generated vapor possibly induces a perfectly controlled/regulated texturing (expansion) of the product.
4. *Final stage of vacuum*
   After instant pressure drop, it is possible to inject a low-volume, high-speed air stream at the surface of product. This is to provide a more intensified deepened cooling process the the DIC-textured product. This immediately triggers the product temperature to drop, while possible evaporation of residual water can take place from the exchange surface.

13.4.4.3 Fundamentals of the Unit Operation of Thermo-Mechanical Texturing by Autovaporization

Thermomechanical texturing technologies of DIC, extrusion cooking, puffing, steam explosion, etc. have all the same fundamental basis. High pressure/high temperature stage is systematically followed by a rapid release toward the atmospheric pressure for all these processes except DIC. This last process is characterized by its instant relaxation toward a vacuum. The final pressure of a few hundred kilopascal (usually up to 4 kPa) is the final stage of DIC. This is the most decisive aspect that highly discriminates DIC from the other thermo-mechanical texturing technologies such as puffing.

This difference is crucial. On one hand, it allows DIC to be performed at a relatively lower processing temperature. Thus, DIC can be used to treat and expand a large range of heat-sensitive products. On the other hand, final temperature just after decompression immediately reaches about 30°C instead of 100°C for the other thermo-mechanical texturing technologies. This results in the possibility of preserving the new structure of numerous polymers and expanding large range of starch-free materials whose glass transition occurs at temperature below 35°C, such as apple (Voegel-Turenne et al., 2001) and numerous other fruits and vegetables.

Since it is possible to regulate and adjust the numerous and various operating parameters, DIC has been used as well-controlled texturing process capable of improving performances of many unit operations such defining new processes of swell-drying, (Louka, 1996; Sahyoun, 1996; Juhel, 2000; Rakotozafy, 2001); parboiling cereals, thus defining a new postharvest treatment of rice (Habba, 1997); thermo-mechanical elimination of vegetative bacteria and spores, thus defining a new manner of microbiological decontamination (Debs-Louka, 2000); treatment of the archaeological wood (Sanya, 2000) vacuum-controlled spraying (VCS) process (Delgado-Rosas, 2002), specific processing of leguminous (Haddad, 2002), microwave-assisted DIC (Klima, 2006), extraction of essential oils (Kristiawan, 2006; Mellouk, 2007), three-stage spray drying capable of producing the new concept of expanded-granule powders (Mounir, 2007), extraction of vegetable oils (Nguyen van, 2010), extraction of antioxidants and active molecules (Ben Amor, 2008; Albitar et al., 2011; Berka-Zougali, 2011), thus defining the new operation of Tripolium, which is a three-pole intermittent solvent extraction of three-Pole intermittent solvent extraction (Allaf, 2014).

Since the specifically high pressure-drop rate is >0.5 MPa/s, toward a vacuum of absolute pressure of 5 kPa, DIC processing allows exceptionally respecting the laws of thermodynamics of instantaneity, when with the pressure decreasing, the temperature will be lower than balance level (short-time over-cooling process).

- *How to estimate the initial water content of raw material?*
  By using saturated steam for increasing product temperature, heating is performed by convection coupled with steam condensation (Figure 13.3). The water condensed at the exchange surface should allow an increase of water content dry basis by a global amount of ΔW_o, which is estimated to be closely linked to the difference in product temperature before A_o (initial temperature T_i) and after heating B_o (thermal treatment temperature T_t):

$$\Delta W_o = \frac{\left(c_{p,d} + W_i \ c_{p,H_2O}\right)}{L_v}\left(T_t - T_i\right). \tag{13.17}$$

By adiabatically dropping the pressure, the autovaporization engenders an amount of vapor (expanding gas), which should be proportional to the difference in product temperature before B_o (at thermal treatment temperature T_t) and after decompression C_o (T_d). It is worth noting that the final equilibrium

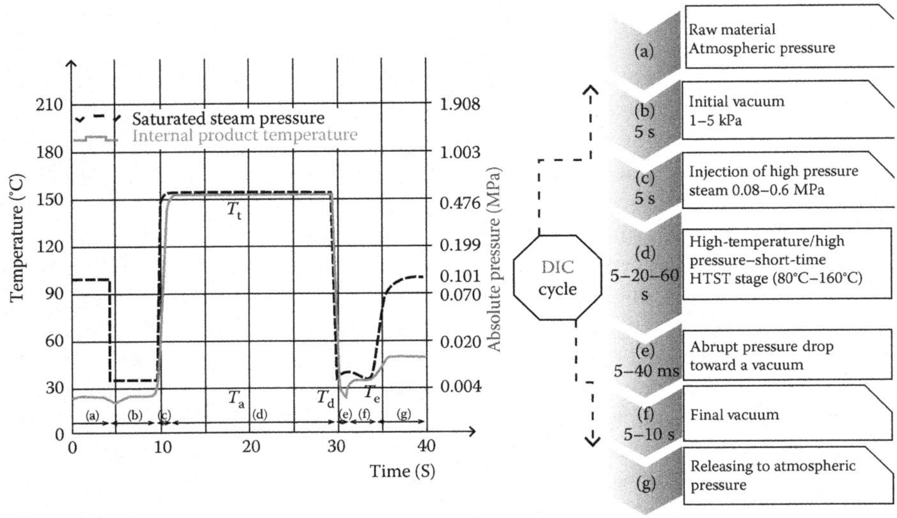

FIGURE 13.3 Evolution of the temperature and pressure during a DIC treatment: (a) atmospheric pressure; (b) initial vacuum; (c) saturated steam injection; (d) maintaining the saturated vapor pressure and temperature constant treatment; (e) steep relaxation to the empty; (f) maintaining the vacuum; and (g) returning to atmospheric pressure. HTST, high-temperature short-time.

temperature (T_e) at D_o is generally the boiling temperature of water at the final total pressure at vacuum stage (usually 5 kPa, approximately):

$$\Delta W_1 = \frac{\left(c_{p,d} + \left(W_i + \Delta W_o\right)\, c_{p,H_2O}\right)}{L_v}\left(T_d - T_t\right). \tag{13.18}$$

- *How to estimate the vacuum level of pressure-drop?*
  The thermal processing temperature T_t and the water content level (W_i + ΔW_o) are normally considered to allow the material to get visco-elastic behavior. The final pressure (vacuum level), and thus the thermal equilibrium temperature T_e, is typically considered to allow the material at a water content level of (W_i + ΔW_o + ΔW_1) to get vitreous behavior, which normally depends on glass transition level (generally determined from Gordon–Taylor relationship (Gordon and Taylor, 1952), which was formulated by Allaf (1982) as

$$T_{g(d,W)} = \frac{T_{g,d} + kWT_{g,w}}{1 + kW}, \tag{13.19}$$

where k is Gordon–Taylor correlation parameter, which depends on the material, W is the moisture content (dry basis), and T_g, $T_{g,d}$, and $T_{g,w}$ are the glass transition temperatures of the material concerned (at W water content dry basis), dry material, and pure water, respectively. By using the glass transition temperature $T_{g,w}$ of water as suggested by Orford et al. (1990) to be $T_{g,w} = -139°C$ and for a specific value $k = 1$, the schematic curve of the glass transition is given in Figures 13.4 and 13.5.

- *How to estimate the heating time?*
  As explained in Figure 13.3, the step of rapid or gradual high pressure-saturated steam injection results in a heating process mainly achieved by condensation. This possibly takes place after an initial vacuum step, which can increase the exchange surface between the steam and the material. Therefore, superficial temperature immediately reaches the steam temperature T_t and heat transfer takes place within the solid essentially by conduction:

$$\vec{\nabla} \cdot \vec{\varphi} + \rho c_p \frac{\partial T}{\partial t} = \vec{\nabla}\left(-\vec{\nabla}\lambda_{\text{eff}} T\right) + \rho c_p \frac{\partial T}{\partial t} = 0. \tag{13.20}$$

Physical characteristics (density ρ, specific heat c_p, effective conductivity λ_{eff}) of the matrix depend on both temperature and water content values. This heterogeneity can be simplified by using average values. (Equation 13.20) becomes

$$\frac{\partial T}{\partial t} = \langle a_{\text{eff}} \rangle \Delta T. \tag{13.21}$$

In the present case of wet porous material, the effective global conductivity $<\lambda_{\text{eff}}>$ value greatly exceeds its "static" value because of the presence

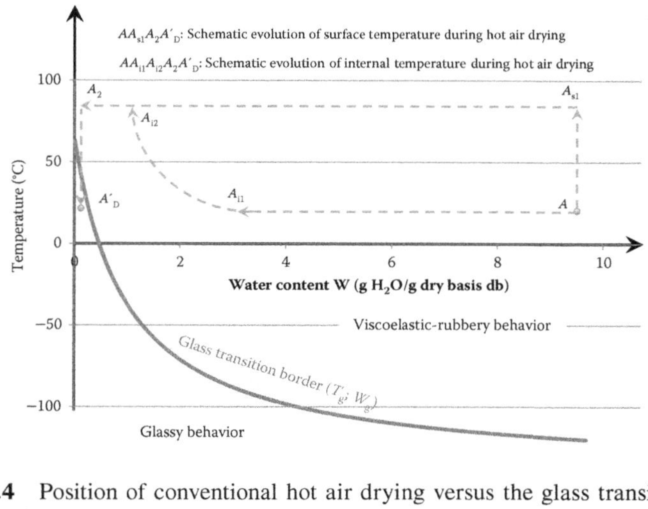

FIGURE 13.4 Position of conventional hot air drying versus the glass transition (T_g, W_g) border. Temperature of the surface exchange with airflow is represented by (A-A_{S1}-A_2- A'_D). Internal temperature (A-A_{i1}-A_{i2}-A_2-A'_D) keeps relatively low for a main part of drying process. Cooling process (A_2 to A'_D) often allows the product to pass through the glass transition boundary.

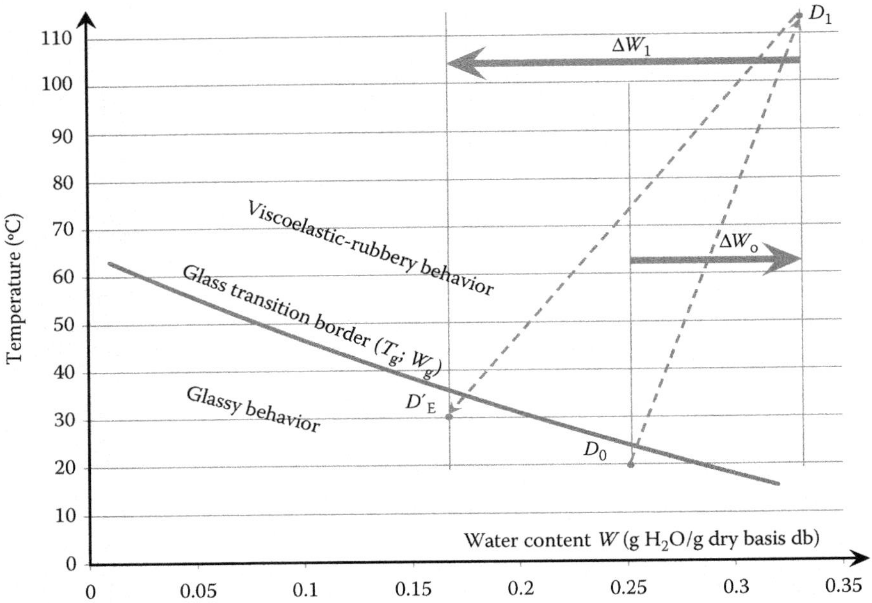

FIGURE 13.5 Evolution of DIC treatment versus the glass transition border (T_g, W_g): $D_0 \rightarrow D_1$ (crossing T_g border from glassy state toward viscoelastic-rubbery state, by increasing steam condensation, both temperature T and water content W); $D_1 \rightarrow D'_E$ (crossing T_g border from viscoelastic-rubbery state toward glassy state, by decreasing instant autovaporization, both temperature T and water content W). In the present case of DIC processing carried out using saturated steam, W changes from initial value at D_0, (W_o = 0.20–0.30 g H_2O/g db) to increase by condensation at D_1 ($\Delta W_o \approx$ 0.1 g H_2O/g db) and decrease by instant autovaporization ($\Delta W_1 \approx -0.2$ g H_2O/g db). (From Mounir, S., *Studies of New Manufacturing Process of Powders by Inserting the Instant Controlled Pressre Drop DIC within Spray-Drying, Swell-Drying and Controlled Vacuum Atomization of Dairy Products.* Université de La Rochelle, La Rochelle, France, 2007.)

of some local micro-phenomena of evaporation–condensation within the pores. Thus, temperature homogenization within the material must be conducted in a relatively rapid manner, while water condensed at the exchange area broadcasts within the material, according to Fick-similar diffusional law as developed by Allaf and Allaf (2014):

$$\frac{\rho_W}{\rho_d}\left(\overrightarrow{v_W} - \overrightarrow{v_d}\right) = -D_{\text{eff}}\vec{\nabla}\left(\frac{\rho_W}{\rho_d}\right). \tag{13.22}$$

At this stage, the diffusion phenomenon is conducted with negligible, and even in the absence of any, modification of the structure. It results in $\overrightarrow{v_d} = 0$ and ρ_d = constant. Equation 13.22 becomes

$$\rho_W \overrightarrow{v_W} = -D_{\text{eff}}\vec{\nabla}\rho_W. \tag{13.23}$$

Combined with the mass balance, assuming the main part of this operation is conducted after the homogenization of temperature, Equation 13.23 presents a similar form of the second Fick law:

$$\frac{\partial \rho_W}{\partial t} = \langle D_{\text{eff}} \rangle \nabla^2 \rho_W. \tag{13.24}$$

Since many studies and experiments have confirmed that the mass diffusivity D_{eff} of liquid water within a porous material is considerably lower than the effective thermal diffusivity α_{eff}, the whole thermal and water content homogenization of material subsequently to saturated steam condensation is well-ordered through liquid water transfer within the material. Consequently, the estimation of the heating processing time can be conducted to adequately satisfy (Equation 13.24) simultaneously at its shortest value so as to greatly reduce any possible thermal degradation.

- *How to perform a really instant pressure drop?*
  Popping of corn grains is a sudden pressure drop operation meaning that it occurs in an unexpected and uncontrolled manner. On the contrary, pressure drop of DIC treatment must take place and be performed only once both temperature and humidity become almost homogeneous in the material. Thanks to this specific aspect, and as the heating processing time of DIC is appropriately controlled, thermal degradation may often be much lower than other expansion processes and texturing autovaporization.

  To be instantaneous, pressure drop cannot be conducted using direct pumping, whatever the type, the rate, and the capacity of the vacuum pump. An intermediate, relatively high-volume (usually 70–100 times as a ratio) vacuum tank should be connected between the treatment vessel and the vacuum pump, using a large-section, perfectly controlled, abruptly opening valve. Moreover, since the conditions of the thermodynamics of instantaneity imply a highly organized one-dimension of initially fluctuated particle velocity, specific forms of vacuum tank would improve these specific conditions for a larger period of time.

BOX 13.1 FUNDAMENTAL: THERMODYNAMICS OF INSTANT TRANSFORMATIONS

INTRODUCTION

Allaf (2002) studied various instant processes. When the decompression of an ideal gas is performed through a quasi-static transformation (Figure 13.6: CT1: conventional thermodynamics), the random velocity of thermal fluctuation movement should uphold the same level whatever the spatial repartition. Temperature keeps the same level before and after such a decompression (Joule decompression of ideal gases).

(Continued)

BOX 13.1 (Continued) FUNDAMENTAL: THERMODYNAMICS OF INSTANT TRANSFORMATIONS

However, for an abrupt pressure drop, a sufficiently instantaneous, asymptotic transformation approach presented in Figure 13.6: CT3 should take place. Thus, instead of the well-defined temperature in the case CT1, which is

$$\frac{3}{2}RT = \frac{1}{2}M\sum_{i=1}^{3}\langle v_i^2\rangle, \qquad (13.25)$$

where v_i is the fluctuated velocity of the particles (molecules); during the very short time of instant decompression, the temperature should asymptotically reach a situation where the Velocity component along the decompression one-axis, will not be randomly distributed. Its value should not be implied in the determination of the temperature. Asymptotically, Figure 13.6: CT3 reveals

$$\frac{3}{2}RT = \frac{1}{2}M\sum_{i=1}^{2}\langle v_i^2\rangle. \qquad (13.26)$$

During this very short-time period, temperature should be

$$T_{\text{instantaneity}} = \frac{2}{3}T_{\text{initial}} = \frac{2}{3}T_{\text{final}}. \qquad (13.27)$$

Some specific experimental studies allowed determining this approach. For an ideal gas at initial temperature of 27°C = 300 K, during any decompression,

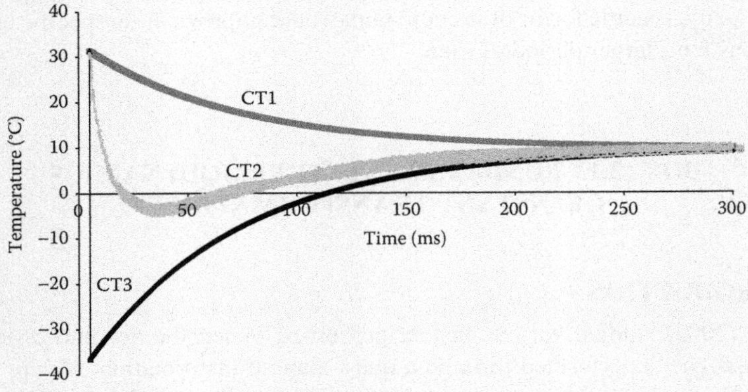

FIGURE 13.6 Impacts of instantaneity in ideal gas decompression: CT1 (quasi-static decompression); CT2 (rapid decompression); CT3 (Asymptotic evolution of instant pressure-drop).

(Continued)

BOX 13.1 (Continued) FUNDAMENTAL: THERMODYNAMICS OF INSTANT TRANSFORMATIONS

final temperature reaches the same value of 300 K. However, during the very short time of instant decompression, temperature should be: $200\,K < T < 300\,K$.

FUNDAMENTAL PRINCIPLES OF INSTANTANEOUS TRANSFORMATIONS

First Basis

Thermodynamics of quasi-static transformations systematically generate positive total entropy of the system and of the external medium. Moreover, in both instant and conventional phenomena, the rate of total entropy generation must be finite.

First Principle

Since the generation rate of the total entropy is a finite amount, total generated entropy of instantaneous transformation is always zero.

Main Implications

There are two types of thermodynamic system transformations: one-parameter processing (single transformation) and processing with several coupled parameters (complex transformation).

Single Transformation (One-Parameter Thermodynamic Process)

- Systematically, a single transformation is trying to reach the total thermodynamic equilibrium; the considered parameter must inevitably evolve toward finding the thermodynamic balance.
- A single transformation cannot be achieved instantaneously. It systematically has kinetics (evolution versus time). As examples, it is worth noting that this is the case of simple heat transfer, with temperature as the only thermodynamic parameter (Fourier laws, etc.) and the case of isothermal mass transfer (Fick laws, etc.).

Complex Transformation (Several Thermodynamic Parameters)

- A complex transformation can be performed instantaneously; thus, the total entropy is null.
- The evolution toward equilibrium of one of the parameters must inevitably generate an opposite evolution of another (other) parameter(s), which moves away far from the equilibrium state.
- Some cases were studied, such as
  - Instant pressure drop of gas or vapor
  - Rupture of an elastic material, which inexorably generates a sound wave
  - Specific chemical reactions

(*Continued*)

BOX 13.1 (Continued) FUNDAMENTAL: THERMODYNAMICS OF INSTANT TRANSFORMATIONS

Hierarchy of Thermodynamic Parameters

During a complex transformation (several thermodynamic parameters), a hierarchy is established between the thermodynamic parameters so as to reach the multi dimensional physical balance of the system. Example:

- Mechanical equilibrium is initially established
- Followed by the equilibrium of phase (vapor pressure)
- And finally the equilibrium of temperature

APPLICATIONS: INSTANT PRESSURE DROP

For an Ideal Gas (Joule Decompression)

The equilibrium state of instant pressure drop is the same as the quasi-static adiabatic transformation. However, the evolution of the operation implies a decrease of temperature before an ulterior evolution toward total equilibrium.

Progressive Autovaporization and Instant Autovaporization

- Since instant pressure drop operation can be "technically stopped" at an intermediate stage before reaching the same temperature of gradual decompression, instant autovaporization can consequently generate higher amount of vapor because of the greater difference between initial and final temperatures; thus, higher volume of expanding gas is then generated (Equation 13.18).
- During the expansion phenomenon caused by the stress of generated vapor, although the temperature should be much lower, the thermo-hydro-rheological behavior should follow the high temperature comportment.
- The preservation of the new structure takes place from glass transition when temperature is increasing from the lowest to equilibrium levels.

13.5 PRACTICAL SPECIFICITIES OF DIC

13.5.1 DIC Technology: An Intensification Basis of Innovative Industrial Unit Operations

DIC is an innovative high-performance manufacturing process focused on agro-industrial fields modifying and controlling plant structure/texture, implying an instant autovaporization, thus allowing highly relevant extraction (1-min/4-min essential oil extraction, high bioavailability of active molecules and ingredients), preserving nutritional and sensorial/taste quality, etc.

DIC technologies are very relevant:

1. As innovative technology in different industrial sectors, such as
   - Food industry
   - Nutraceutical industry
   - Cosmetic industry
   - Pharmaceutical industry
2. In large industrial processes of
   - Drying
   - Extraction
   - Decontamination
   - Natural antioxidant and antimicrobial activities
   - Energy generation
3. In different materials, such as
   - Fruits
   - Vegetables
   - Aromatic and medicinal herbs, flowers, and other parts of plants
   - Meat and sea-foods
   - Milk and dairy products (cheese, caseinates, etc.)
   - Seaweeds and microalgae

DIC technologies have specific impacts on industrial scales

- Low cost
- Low energy consumption
- Environment friendly
- High kinetics and great yield
- Easy scaling up for industrialization
- Perfectly adapted with food items having
  - High and controlled quality (needs for improving and controlling classic food product attributes in terms of nutritional content, safety, palatability, etc.)
  - High capability of long-time/easy storage at ambient temperature
  - Convenient to use and facility of transport

13.5.2 General Approaches of DIC Texturing Products

13.5.2.1 What Is DIC Technology?

DIC is a specific thermo-mechanical texturing operation, which aims at expanding all types of plant-based materials. Since it implies numerous sources of heating/pressuring systems (high-pressure hot steam; high-pressure neutral gas or air combined with microwaves; etc.), and perfectly regulated operating parameters of treatment pressure (usually ranging from 0.1 to 1 even tens of MPa) and temperature (frequently ranging between 80°C and 280°C), vacuum pressure level (commonly ranging from 0.1 to 10 kPa), decompression time (from a few milliseconds for instant pressure drop to several seconds for classical ways) implying perfectly synchronized pressure drop rate between 10 and few thousands of MPa/s; it results in expanding large types of products at perfectly defined ratios.

13.5.2.2 What Kind of Raw Materials Can Be Used in DIC Technology?

While popping by conventional or microwave heating only concerns corn grains, and puffing and cooking-extrusion only treat and expand mainly starch-based raw materials, DIC can texture and expand at adjusted levels all types of plant-based products (fruits, vegetables, tobacco, beans and seeds, coffee and cocoa beans, pellets, roots, leaves, woods, etc.) as well as mushroom, seaweeds and microalgae, meat, sea-foods, dairy products and cheese, etc. leading to starch-free snacking, textured and puffed materials, expanded-granule powders, etc.

13.5.2.3 Why Is DIC So Specific Compared with All Other Forms of Puffing, Extrusion, Popping, etc.?

At the three crucial levels of texturing parameters: (a) the amount of expanding gas, (b) thermo-rheological behavior, and (c) glass transition temperature, DIC is greatly specific because: (a) it produces a higher amount of expanding gas (vapor) without necessarily reaching too high temperature levels, (b) temperature range is very important, thus being able to perfectly respect both rheological characteristics during expansion phenomenon and thermochemical preservation of sensitive materials, (c) since DIC easily reaches suitably low final temperature with regard to glass transition temperature T_g of numerous polymers, it can preserve their new textured state. Conventional puffing, popping, extrusion, frying, etc. can principally be used for starch-based materials whose T_g is not so far from the equilibrium temperature 100°C of water at atmospheric pressure.

13.5.2.4 What Is the Difference between Flash-Expansion and DIC?

Treating fresh fruits such as grapes with vacuum is a process for improving wine quality, which increases the availability of grape polyphenols and aroma (Sebastian and Nadeau, 2002). A flash evaporator is used in the wine industry for its high cooling efficiency and its ability to improve the quality of wine. The flash-expansion is used following the harvest to immediately achieve a heating stage of 70°C–90°C coupled to a pressurization followed by a decompression achieved in a few seconds in a low-pressure tank, thus ensuring rapid cooling of both liquid and solid (skin, seeds, and pulp).

13.5.2.5 What Is the Difference between Vacuum Puffing and DIC?

Conventional vacuum puffing started in China by the end of 2006. It is made with the use of indirect external heating systems, which result in large heating time (2–10 min), against a few seconds to dozens of seconds for DIC. Consequently, vacuum puffing operation results in lower yield and capacity, with possibly higher degradation of quality. Moreover, DIC technology uses a direct hot/high pressure steam stream, with fully regulated operating parameters. This results in higher energy yield and fully controlled final product quality.

13.5.2.6 What Are the Properties of DIC Swell-Dried Final Product versus Conventional Air-Dried or Freeze-Dried Products?

Swell drying is coupling conventional or innovative airflow drying processes with DIC texturing. Airflow drying, as the easiest and the most natural drying process, is used when the internal temperature is much lower than the surface temperature. The biochemical composition quality of the product is preserved. Instead of a highly compact structure, optimized DIC treatment generally leads to porous product, regaining its original volume before shrinkage, to an expansion enlarging up to 33 times (in the case of cheese, etc.). The new, expanded product results in a very high internal diffusivity of moisture (liquid water), greatly improving the process performance in terms of kinetics and great decrease of the operation time. While the energy consumption is about 0.15 kWh/kg of dry matter for DIC treatment, the airflow drying of new restructured product implies about 0.12 kWh instead of 0.35 kWh/kg dry matter. Swell-dried product has a dry level that is able to reach up to 2%–3% instead of the usual 8% for un-restructured conventional hot air-dried fruits and vegetables. Grinding of these expanded dried products is much easier and the functional quality of expanded granule powders is much better. The final product has higher water absorption kinetics, with higher WHC. They are used as snacking, ready-to-use products, etc. The most important quality attributes are based on the microbiologic decontamination provoked by the thermal and mechanical impacts of DIC on the vegetative bacteria and spores (Allaf and Allaf, 2014). This allows DIC to be an exceptionally relevant process of decontamination of dried pieces and powders. It ensures microbiological safety, eliminates microflora growth, destroys insect larvae, and implies a much higher shelf-life of years instead of months for conventionally hot air-dried products (Figure 13.9).

13.5.2.7 What Is the Nutrition Value of DIC Textured Fruits, Vegetables, Seaweeds, Microalgae, etc.?

Swell-dried products are characterized by an exceptional preservation of fiber, minerals, and vitamins much higher than hot-dried products. It is worth noting that the porous structure results in much higher antioxidant availability, and great improvement of extraction process in terms of both yields and kinetics.

13.5.2.8 What Are Advantages of DIC Applications in Confectionery?

Swell-dried products can be used as baby foods, instant soups, ingredients of well-balanced bars, etc.

13.5.3 Additional Aspects

13.5.3.1 Between Puffing and Extrusion Cooking, and DIC

The main differences between puffing and extrusion-cooking processes, on one hand, and DIC treatment, on the other, may be understood through explanation of the hydro-thermo-mechanical expansion unit operations. In all these operations, expansion ratio closely depends on the amount, the effective volume, and the generation rate of vapor issued from decompression. It also depends on the thermo-rheological

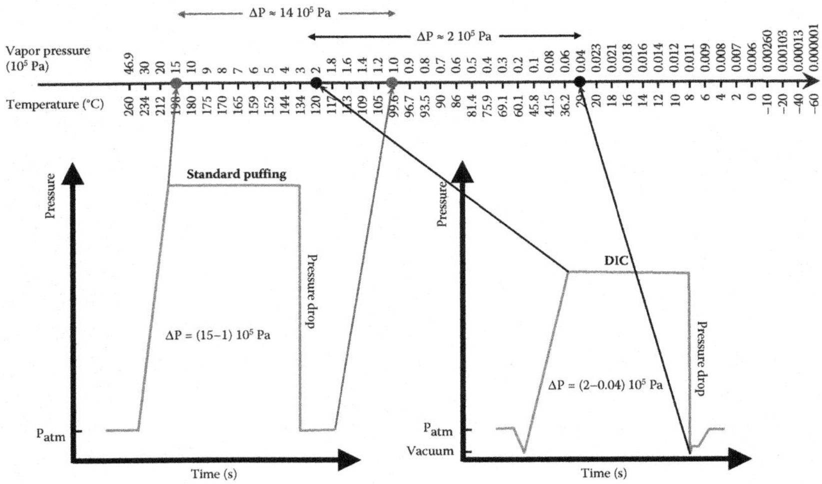

FIGURE 13.7 Evolution of the temperature and pressure during puffing and DIC. To get the same amount of vapor generated by autovaporization, DIC requires much lower steam pressure difference ΔP, much lower processing temperature inducing the possibility to texture sensitive materials. Final DIC temperature is so low that it results in crossing the glass transition, thus preserving the expansion of numerous starch-free compounds.

behavior of the considered materials (viscoelasticity) and the glass transition. They, respectively, imply the expansion and preservation of the new structure.

Therefore, Figure 13.7 shows that, compared to expansion operations using autovaporization toward atmospheric pressure, the treatment DIC is not a simple displacement of the level of depression, but an essential change making texturing autovaporization much more adapted to heat-sensitive and/or low-T_g materials. This last aspect (glass transition) is of particular importance. While only starch-based products are affected by puffing and cooking-extrusion, DIC concerns a particularly large number of products (fruits, vegetables, algae and microalgae, meat products, dairy products, cheese, etc.).

13.5.3.2 Drying by Successive Decompression Dehydration DDS

This other version of DIC treatment involves exploiting the autovaporization phenomenon as a step of gradually removing a certain amount of water, with the transfer of internal water and vapor. Similar to overheating, steam drying, or microwave drying, DDS uses Darcy's law whose driving force is the gradient of total pressure. Thus, a series of cycles implying gas (air, N_2, CO_2, etc.) pressurization, possibly at low temperature, followed each by an instant pressure drop toward a vacuum, ensures a very interesting intensification of the paradoxical stage (Al Haddad et al., 2008). DDS particularly suits heat-sensitive products such as fish (Haddad et al., 2001), baker's yeast (Rakotozafy, 2001), vaccines (Rakotozafy et al., 2000), etc.

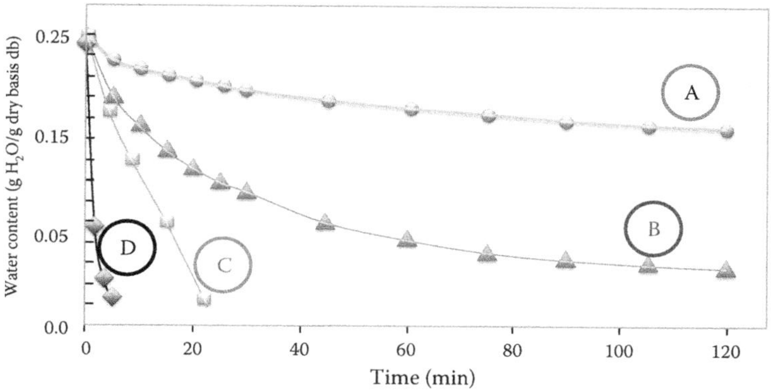

FIGURE 13.8 Comparative drying kinetics (A) by hot air, (B) by "swell drying" coupling the hot air drying with DIC expansion, (C) "swell drying" coupling DIC texturing with dehydration by successive decompressions DDS, and (D) "swell drying" with DIC coupled to a post-drying step by microwaves.

13.5.4 Comparison of the Performance of the Process and Quality of the Finished Product

Figure 13.8 reveals the great importance of expanded structure in reducing the drying time. It also highlights the specific impacts of DDS and microwave drying operations once they occur after DIC expansion process.

13.5.4.1 Decontamination

In terms of microbial inactivation, it is well recognized that dry heat is less effective than moist heat as proteins, which are an important component in maintaining cell viability are more stable in a low-moisture environment (Archer et al., 1998; Doyle and Mazzotta, 2000). Moreover, a few authors (Murphy et al., 2002; Chiewchan et al., 2007) found that cells attached to a tissue are more heat resistant than those unattached or dispersed throughout food or broth. Some previous studies illustrated that changes of the surface characteristics of food during drying led to better entrapment and hence protection of bacteria from direct heating at prolonged drying time (Chiewchan and Morakotjinda, 2009; Hawaree et al., 2009). All these mentioned factors significantly affect the heat resistance of bacteria during drying. However, Hawaree et al. (2009) proved that during the air drying of cabbage, surface characteristics such as water activity and shrinkage did not have a significant effect on the susceptibility of Salmonella attached on its surface.

From the point of maintaining quality attributes, using lower drying temperature to avoid chemical and physical deteriorations caused by heat may not always be effective in terms of microbial decontamination (Phungamngoen et al., 2011).

Mounir et al. in Allaf and Allaf (2014) studied thermal and mechanical impacts of DIC as a decontamination process. DIC as a high temperature–short-time saturated

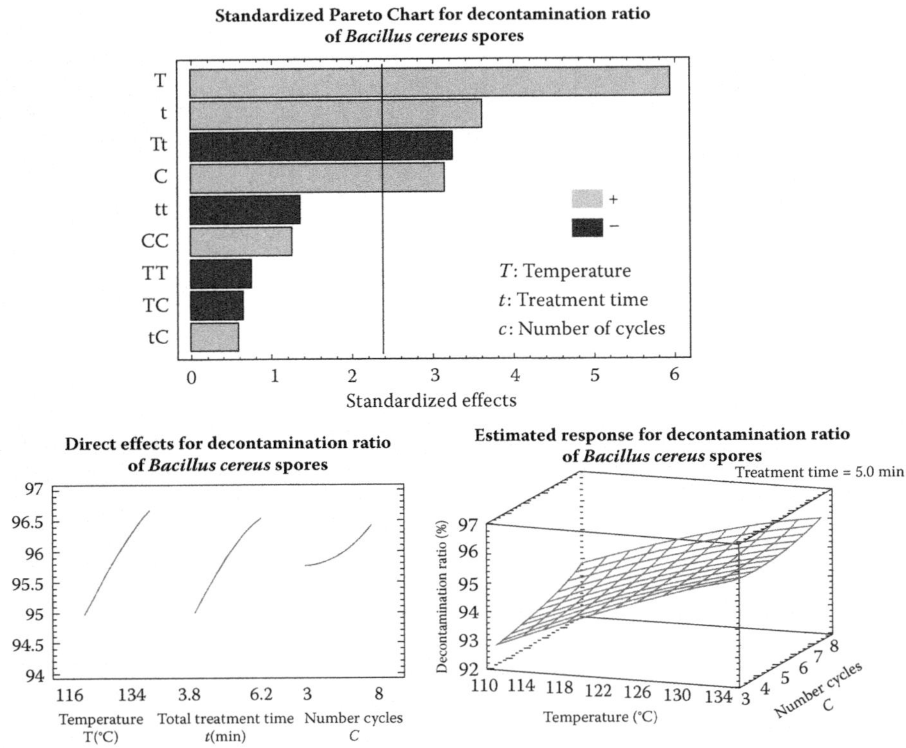

FIGURE 13.9 Pareto Charts for the treatment of dairy powders versus DIC parameters; impacts of temperature, total treatment time, and number of cycles on decontamination ratio. (From Mounir, S. et al., *Food Processing*, New York, Springer, 2014.)

steam pressure implies a very effective decontamination of both vegetative bacteria and spores. Indeed, DIC is similar to a Ultra-High Temperature (UHT) treatment, particularly specific to solids and powders. It allows rapidly heating the material while the instant cooling is made by dropping the pressure toward a vacuum (4–5 kPa; final temperature between 25°C and 35°C). A mechanical impact of the amount of generated vapor by autovaporization results in explosion of spores and specific importance of the number of cycles C (Figure 13.9).

13.5.4.2 Comparative Quality

The overall analysis of various materials issued from the different operations of hot air drying, microwave-drying, freeze-drying, and swell-drying coupling controlled hot air drying and DIC texturing allows us to draw a general trend in terms of the quality of the finished product and own performance operation. The following table (Table 13.1) provides an overall approach of these various operations; the precise comparison inevitably needs specific consideration of the material, processing conditions and requirements, and uses of the desired end product.

TABLE 13.1
Comparative Analysis of the Main Drying Processes to the Two Planes of the Process Performance and Overall Quality of the Finished Product to 4% Moisture

	Hot Air	Freeze-Drying	Microwave	Swell Drying: Airflow + DIC + (Airflow; DDS, or MW)
Shape and texture	Compact hard	Preserved shape Weak texture	Lightly compact Mid-hard	Controlled expansion (1–30 times); crispy.
Taste, flavor, color quality	1st stage: good Final: low quality	Low aroma preservation	Overheating risks	High preservation
Nutrient preservation	1st stage: good Final: degradation	Perfect preservation	Good/low preservation	Good/very good preservation
Hygienic content	Bad	Bad	Low	Perfect microbiology decontamination
Rehydration kinetics	Bad (dozens of min)	Instant (<1 min)	Few minutes	Very good (1–2 min)
WHC	Good	Very bad	Lightly good	Very good
Energy consumption (kWh/tonne db)	8,000	13,700	7,600	1st stage: 5000 DIC: 125 3rd stage: 1700

Source: Allaf, K., Vidal, P., Feasibility study of a new process of drying/swelling by instantaneous decompression towards vaccum. University of Technology of Compiègne, France, Chemical Engineering Department, 1989; Allaf, K. et al., Swell-drying: séchage et texturation par DIC des végétaux. *Techniques de l'Ingénieur* F3005 (Opérations unitaires du génie industriel alimentaire), 2012; Téllez-Pérez, C. et al., *Proc. Eng.*, 42, 978–1003, 2012; Alonzo-Macías, M., Comparative study of different drying processes of hot air drying, freeze-drying, and instant autovaporization: Application to strawberry. PhD, La Rochelle University, 2013.

13.5.5 Energy Consumption and Technical Specificity

In the case of saturated steam as a compressing/heating means, the energy required for heating and for compressing was calculated and experimentally determined at both laboratory and industrial scales:

- Heating of 1 metric-ton of 0.3 kg H_2O/kg dry basis material from ambient temperature (20°C) toward 0.5-MPa saturated steam temperature (150°C) requires, for 90% energy yield:

$$E_{\text{heating}} \approx 90 \text{ kWh/tonne dry basis.}$$

- In a 6-m³ treatment vessel, working at 0.5 MPa and $T = 150°\text{C}$, it will be an amount of compressing vapor:

$$M_v = 0.5 \times 10^6 \text{ Pa} \times 6 \frac{\text{m}^3}{8.314} \text{J/mol/K} \times (150 + 273)\text{K} \times 0.018 \text{ kg/mol} = 15.35 \text{ kg.}$$

- Energy for producing this amount of vapor with a yield of 80%: 15 kWh

$$E_E \approx 15 \text{ kWh/tonne dry basis}$$

- Other: +20%
- Total energy consumption for DIC treatment:

$$E_{\text{specific for DIC}} \approx 125 \text{ kWh/tonne dry basis.}$$

NOMENCLATURE

Symbol	Definition
T_a	Airflow temperature
T_s	Temperature at the product surface, or wet bulb temperature
$a_{w,s}$	Water activity at the product surface
$p_{w,T}$	Vapor pressure at the temperature T
$p_{w,a}$	Vapor pressure of airflow
$p_{w,s}$	Vapor pressure just adjacent to the surface of the product
W	Water content at time t in the material (g H_2O/g DM);
W_∞	Final water content at ($t \to \infty$) between the solid and the external medium (g H_2O/g DM);
W_o	Water content in the solid by extrapolating the diffusion (g H_2O/g DM)
D_{eff}	Effective diffusivity (m²/s)
dp	Characteristic length depending on the shape of the product: radius in the case of spherical or infinite cylinder or half the thickness in the case of a plate (m)
τ	Number of Fick = $D_{eff} \times t/\mathrm{d}p^2$
c_{p_s} and c_{p_w}	Specific heat (J/kg/K) of the dry material and liquid water, respectively
λ_{eff}	Overall effective thermal conductivity of the medium (W/m/K)
$\langle a_{eff} \rangle$	Average of thermal diffusivity (m²/s)
t	Time (s)

(Continued)

(*Continued*)

Symbol	Definition
M_w	Water molar mass (kg/mol)
T	Temperature (K)
ρ	Bulk density (kg/m^3)
L_v	Latent heat of evaporation (J/kg)
ε_{abs}	Absolute growth rate (%)
A_{eff}	The effective exchange area between the product to dry and the outside air, expressed in m^2
L	The latent heat of evaporation at the temperature T expressed in J/kg.

13.6 CONCLUSION

"Swell drying" involves hot air drying with the specific texturing operation of DIC. It effectively addresses the shrinkage phenomenon. This also enables porosity to increase up to 30 times, and thus, improves kinetics of the second drying phase. This is the second way of intensification, which results in higher performances of drying unit operation in terms of kinetics, energy consumption, and controlled operation parameters. Moreover, the quality attributes of the final product are much higher; indeed, the shorter the drying time, the higher the product quality. Vitamin content and availability of flavonoids and antioxidant activity of the finished product are higher than that of conventionally dried products. Moreover, the accessibility of active molecules of plant-based products are often considerably higher than the one of raw material itself. The microbiological decontamination thus generated, the controlled texture, the preservation of taste, and the good presence of aromatic content allow the "swellified" product quality to be well defined. A fundamental analysis of the kinetics of drying allows the intensification generated by DIC to be revealed through higher effective diffusivity D_{eff} and starting accessibility δW_s.

REFERENCES

Aghbashlo, M., M. H. Kianmehr, and H. Samimi-Akhijahani (2008). Influence of drying conditions on the effective moisture diffusivity, energy of activation and energy consumption during the thin-layer drying of berberis fruit (Berberidaceae). *Energy Conversion and Management* **49**(10): 2865–2871.

Albitar, N., S. Mounir, C. Besombes, and K. Allaf (2011). Improving the drying of onion using the instant controlled pressure drop technology. *Drying Technology* **29**(9): 993–1001.

Al Haddad, M., S. Mounir, V. Sobolik, and K. Allaf (2008). Fruits and vegetables drying combining hot air, DIC technology and microwaves. *International Journal of Food Engineering* **4**(6). doi:10.2202/1556-3758.1491.

Alibas, I. (2009). Microwave, vacuum, and air drying characteristics of collard leaves. *Drying Technology* **27**(11): 1266–1273.

Allaf, K. (1982). *Transfer Phenomena and Industrial Applications*. Beirut, Lebanon: Lebanese University.

Allaf, K. (1988). Approche à l'analyse fondamentale de l'expansion par alvéolation selon différents procédés (puffing, cuisson-extrusion…). Compiègne, France: Université de Technologie de Compiègne.

Allaf, K. (2002). Analysis of instantaneity in thermodynamic processes; Fundamental laws. In *1st Franco-Lebanese Symposium on Technologies and Studies on Process Engineering and Biochemistry*. Lebanese University, Beirut, Lebanon.

Allaf, T. (2014). Intermittent extraction process via solvent (DIC-TriPolium). A.-D. Process. France, WO 2015/162393 Al.

Allaf, T. and K. Allaf (2014). *Instant Controlled Pressure Drop (D.I.C.) in Food Processing*. New York: Springer.

Allaf, K., S. Mounir, and T. Allaf (2012). Swell-drying: séchage et texturation par DIC des végétaux. *Techniques de l'Ingénieur* F3005 (Opérations unitaires du génie industriel alimentaire).

Allaf, K. and P. Vidal (1989). Feasibility study of a new process of drying/swelling by instantaneous decompression towards vaccum. University of Technology of Compiègne, France, Chemical Engineering Department.

Alonzo-Macías, M. (2013). Comparative study of different drying processes of hot air drying, freeze-drying, and instant autovaporization: Application to strawberry. PhD, La Rochelle University.

Archer, J., Emma, T., Jervis, E. T., Bird, J., and Gaze, J. E. (1998). Heat Resistance of Salmonella weltevreden in Low-Moisture Environments. *Journal of Food Protection*, **61**(8), 969–973. doi: 10.4315/0362-028x-61.8.969.

Ben Amor, B. (2008). Maîtrise de l'aptiture technologique de la matière végétale dans les opérations d'extraction de principes actifs; texturation par Détente Instantanée Contrôlée DIC, Université de La Rochelle.

Berka-Zougali, B. (2011). Etudes physico-chimiques des plantes médicinales algériennes, Université des sciences et de la technologie Houari Boumediene (USTHB).

Çağlar, A., İ. T. Toğrul, and H. Toğrul (2009). Moisture and thermal diffusivity of seedless grape under infrared drying. *Food and Bioproducts Processing* **87**(4): 292–300.

Chiewchan, N., and Morakotjinda, P. (2009). Effects of Acetic Acid Pretreatment and Hot Air Drying on Resistance of Salmonella on Cabbage Slices. *Drying Technology*, **27**(9), 955–961. doi: 10.1080/07373930902901992.

Chiewchan, N., Pakdee, W., and Devahastin, S. (2007). Effect of water activity on thermal resistance of Salmonella krefeld in liquid medium and on rawhide surface. *International Journal of Food Microbiology*, **114**(1), 43–49. doi: http://dx.doi.org/10.1016/j.ijfoodmicro.2006.10.037.

Crank, J. (1975). *The Mathematics of Diffusion*. Oxford, UK: Oxford University Press.

Dandamrongrak, R., G. Young, and R. Mason (2002). Evaluation of various pre-treatments for the dehydration of banana and selection of suitable drying models. *Journal of Food Engineering* **55**(2): 139–146.

Debs-Louka, E. (2000). Destruction des microorganismes par voie thermo-mécanique contrôlée dans des produits solides en morceaux ou en poudre. Application aux épices et aromates, Université de La Rochelle.

Delgado-Rosas, M. (2002). Analyse et conception d'un réacteur à pulvérisation contrôlée sous vide (VCS), Université de La Rochelle.

Demir, K. and K. Sacilik (2010). Solar drying of Ayas tomato using a natural convection solar tunnel dryer. *Journal of Food, Agriculture and Environment* **8**(1): 7–12.

Doyle, M. E., and Mazzotta, A. S. (2000). Review of Studies on the Thermal Resistance of Salmonellae. *Journal of Food Protection*, **63**(6), 779–795. doi: 10.4315/0362-028x-63.6.779.

Gordon, M. and J. S. Taylor (1952). Ideal copolymers and the second-order transitions of synthetic rubbers. i. non-crystalline copolymers. *Journal of Applied Chemistry* **2**(9): 493–500.

Gunhan, T., V. Demir, E. Hancioglu, and A. Hepbasli (2005). Mathematical modelling of drying of bay leaves. *Energy Conversion and Management* **46**(11–12): 1667–1679.

Gustavsson, J., C. Cederberg, U. Sonesson, R. van Otterdijk, and A. Meybeck (2011). *Global Food Losses and Food Waste: Extent, Causes And Prevention*. Rome: FAO.

Habba, A. (1997). Nouveaux procédés de précuisson et d'étuvage du riz par détente instantanée contrôlée (DIC): Analyse des opérations, optimisation des procédés et approche à l'industrialisation, Université de Technologie de Compiègne.

Haddad, J. (2002). Impacts des procédés hydro-thermo-mécaniques dans la valorisation des graines oléoprotéagineuses, Université de La Rochelle.

Haddad, J., N. Louka, M. Gadouleau, F. Juhel, and K. Allaf (2001). Application du nouveau procédé de séchage/texturation par Détente Instantanée Contrôlée (DIC) aux poissons: impact sur les caractéristiques physicochimiques du produit fini. *Sciences des Aliments* **21**(5): 481–498.

Hawaree, N., Chiewchan, N., and Devahastin, S. (2009). Effects of Drying Temperature and Surface Characteristics of Vegetable on the Survival of Salmonella. *Journal of Food Science*, **74**(1), E16–E22. doi: 10.1111/j.1750-3841.2008.01010.x.

Head, D. S., Cenkowski, S., Holley, R., and Blank, G. (2008). Effects of superheated steam on Geobacillus stearothermophilus spore viability. *Journal of Applied Microbiology*, 104(4), 1213–1220. doi: 10.1111/j.1365-2672.2007.03647.x.

Juhel, F. (2000). Etude de l'application de deux nouvelles opérations de séchage/texturation du poisson; analyse comparative du procédé et du produit, Université de La Rochelle.

Kavak Akpinar, E., Y. Bicer, and F. Cetinkaya (2006). Modelling of thin layer drying of parsley leaves in a convective dryer and under open sun. *Journal of Food Engineering* **75**(3): 308–315.

Klima, L. (2006). Chauffage des matériaux par micro-ondes, Université de La Rochelle.

Kristiawan, M. (2006). Traitement thermo-mécanique des fleurs d'ylang-ylang indonésienne par détente instantanée contrôlée (DIC), Université de La Rochelle.

Lee, J. H. and H. J. Kim (2009). Vacuum drying kinetics of Asian white radish (*Raphanus sativus* L.) slices. *LWT-Food Science and Technology* **42**(1): 180–186.

Lewicki, P. P. (2006). Design of hot air drying for better foods. *Trends in Food Science and Technology* **17**(4): 153–163.

Linden, G. and D. Lorient (1994). *Biochimie Agro-Industrielle*. Masson: Paris.

Louka, N. (1996). Maîtrise de la qualité des produits agro-alimentaires séchés; modification texturale et réduction du coût énergétique par détente instantanée contrôlée (DIC) vers le vide. Conception et réalisation d'un nouveau procédé industriel, Université de Technologie de Compiègne.

Mayor, L. and A. M. Sereno (2004). Modelling shrinkage during convective drying of food materials: A review. *Journal of Food Engineering* **61**(3): 373–386.

Mellouk, H. (2007). Extraction des volatils à partir du bois par détente instantanée contrôlée, Université de La Rochelle.

Menges, H. O. and C. Ertekin (2006). Mathematical modeling of thin layer drying of golden apples. *Journal of Food Engineering* **77**(1): 119–125.

Mounir, S. (2007). *Studies of New Manufacturing Process of Powders by Inserting the Instant Controlled Pressre Drop DIC within Spray-Drying, Swell-Drying and Controlled Vacuum Atomization of Dairy Products*. La Rochelle, France: Université de La Rochelle.

Mounir, S., N. Albitar, and K. Allaf (2014). DIC decontamination of solid and powder foodstuffs. In: Allaf, T. and K. Allaf (eds) *Instant Controlled Pressure Drop (D.I.C.) in Food Processing*. New York: Springer, 83–94.

Murphy, R. Y., Duncan, L. K., Johnson, E. R., Davis, M. D., and Marcy, J. A. (2002). Thermal Inactivation of Salmonella Senftenberg and Listeria innocua in Beef/Turkey Blended Patties Cooked via Fryer and/or Air Convection Oven. *Journal of Food Science*, **67**(5), 1879–1885. doi: 10.1111/j.1365-2621.2002.tb08740.x.

Nguyen, T. H., J. L. Lanoisellé, T. Allaf, and K. Allaf (2016). Experimental and fundamental critical analysis of diffusion model of airflow drying. *Drying Technology* **34**(15): 1884–1899.

Nguyen van, C. (2010). Maîtrise de l'aptitude technologique des oléagineux par modification structurelle: applications aux opérations d'extraction et de transestérification in-situ, Université de La Rochelle.

Orford, P. D., R. Parker, and S. G. Ring (1990). Aspects of the glass transition behaviour of mixtures of carbohydrates of low molecular weight. *Carbohydrate Research* **196**: 11–18.

Orikasa, T., A. Tagawa, S. Soma, M. Iimoto, and Y. Ogawa (2005). Hot air drying characteristics of fruits and vegetables and surface hardening of samples during drying. *Journal of the Japanese Society of Agricultural Machinery* **67**(6): 62–70.

Orikasa, T., L. Wu, T. Shiina, and A. Tagawa (2008). Drying characteristics of kiwifruit during hot air drying. *Journal of Food Engineering* **85**(2): 303–308.

Özdemir, M. and Y. Onur Devres (1999). The thin layer drying characteristics of hazelnuts during roasting. *Journal of Food Engineering* **42**(4): 225–233.

Pardeshi, I. L., S. Arora, and P. A. Borker (2009). Thin-layer drying of green peas and selection of a suitable thin-layer drying model. *Drying Technology* **27**(2): 288–295.

Phungamngoen, C., Chiewchan, N., and Devahastin, S. (2011). Thermal resistance of Salmonella enterica serovar Anatum on cabbage surfaces during drying: Effects of drying methods and conditions. *International Journal of Food Microbiology*, **147**(2), 127–133. doi: http://dx.doi.org/10.1016/j.ijfoodmicro.2011.03.019.

Rakotozafy, H. (2001). Application du nouveau procédé de déshydratation par détentes successives (DDS), dans le séchage de produits biologiques à haute valeur ajoutée, Université de La Rochelle.

Rakotozafy, H., F. Vye, and K. Allaf (2000). Séchage de levure boulangère par un nouveau procédé : La déshydratation par détentes successives (DDS). Comparaison et couplage avec la lyophilisation, Montpellier.

Sacilik, K., R. Keskin, and A. K. Elicin (2006). Mathematical modelling of solar tunnel drying of thin layer organic tomato. *Journal of Food Engineering* **73**(3): 231–238.

Sahyoun, W. (1996). Maîtrise de l'aptitude de matériaux agro-alimentaires aux procédés de séchage. Etude de l'adéquation entre les états structuraux, biochimiques, physiques et comportementaux sur les processus de déshydratation = Mastering the aptitude of food products to drying processes. Study of the adequacy between structural, biochemical, physical, and behavioural states on the dehydratation process, Université de Technologie de Compiègne.

Sanya, E. (2000). Analyse du traitement hydro-amido-thermique de bois gorgé d'eau; application à la restauration-préservation du bois archéologique, Université de La Rochelle.

Sebastian, P. and J. P. Nadeau (2002). Experiments and modeling of falling jet flash evaporators for vintage treatment. *International Journal of Thermal Sciences* **41**(3): 269–280.

Suvarnakuta, P., S. Devahastin, and A. S. Mujumdar (2005). Drying kinetics and β-carotene degradation in carrot undergoing different drying processes. *Journal of Food Science* **70**(8): s520–s526.

Téllez-Pérez, C., M. M. Sabah, J. G. Montejano-Gaitán, V. Sobolik, C. A. Martínez, and K. Allaf (2012). Impact of instant controlled pressure drop treatment on dehydration and rehydration kinetics of green Moroccan pepper (*Capsicum annuum*). *Procedia Engineering* **42**: 978–1003.

Vadivambal, R., and Jayas, D. S. (2007). Changes in quality of microwave-treated agricultural products—a review. *Biosystems Engineering*, **98**(1), 1–16. doi: http://dx.doi.org/10.1016/j.biosystemseng.2007.06.006.

Voegel-Turenne, C., S. Benammar, and K. Allaf (2001). Evolution of the mechanical, textural and rheological characteristics of the Granny Smith apple during drying. *Sciences des Aliments* **21**(1): 27–34.

Wu, L., T. Orikasa, Y. Ogawa, and A. Tagawa (2007). Vacuum drying characteristics of eggplants. *Journal of Food Engineering* **83**(3): 422–429.

14 Aroma Aspects of Fresh and Dried Vegetables

Narendra Narain, Gomathi Rajkumar, Maria Terezinha Santos Leite Neta, Romy Gleyse Chagas Barros, Saravanan Shanmugam, and Marina Denadai
Federal University of Sergipe

Arun Sadashiv Mujumdar
McGill University

CONTENTS

14.1 INTRODUCTION

Vegetables are considered to be protective for human health due to the presence of some beneficial compounds that function against various oxidative-stress-related diseases. Over the past decades, consumers have been looking for more and more convenient food products, either partly or completely prepared food items. Owing to their high moisture content, vegetables are easily perishable. Moreover, there is a large volume of postharvest losses that occur due to deficient infrastructure facilities, rough handling, harsh climatic conditions, small scale of operation, and limited automation in the postharvest operations. In fresh vegetables, these losses may vary between 20% and 40% w/w, leading to an enormous economic waste all over the world (African Postharvest Losses Information System 2012; Dev and Raghavan 2012). To prevent huge quantitative and qualitative losses of vegetables, one important aspect is to improve the methods for preservation of vegetables that retain the fresh inherent organoleptic and nutritional qualities of these commodities (Gustavsson et al. 2011; Dev and Raghavan 2012).

The most important goal of vegetable processing is to transform perishable vegetables into stable products that can be stored for prolonged periods, whereby reducing losses, making them available in times of shortage and at places far away from the site of production. Drying is one of the most common methods that can be used to extend the shelf life while maintaining the desired characteristics of a food product. Modern food industries still follow various techniques to dehydrate and to preserve the food materials. The conventional hot air-drying (HAD) and freeze-drying (FD) techniques continue to play vital roles in the dehydration process (Montoya-Ballesteros et al. 2014; Sette et al. 2015; Song et al. 2015). The extent of the changes depends on the care taken in preparing the material before dehydration and on the dehydration process (Palzer et al. 2012). Major quality parameters associated with dried vegetable products are color, appearance, texture, shape, flavor, microbial load, retention of nutrients, bulk density, rehydration properties, water activity, and absence of pests, insects, other contaminants, preservatives, taints, and off-odors (Chiewchan et al. 2010; Ramírez et al. 2012; Chena and Opera 2013; Baker et al. 2005).

During dehydration, aroma and flavor play important roles in determining the quality of dried vegetable products. Aroma forms the most important criterion in elucidating the quality of vegetables. However, the retention of the naturally occurring volatile compounds in dehydrated vegetables has been a major challenge in food processing industries. Taking all these factors into consideration, this chapter will review initially on the mechanisms of the formation of volatile compounds in vegetables, effect of packaging and drying on the aroma of vegetables, modern techniques used in aroma analysis, and finally concentrate on the changes in aromatic profile of carrots subjected to most prominent dehydration methods such as the conventional air- and freeze-drying.

14.2 BIOGENERATION OF AROMA COMPOUNDS IN VEGETABLES

The major volatile compounds present in vegetables are formed by anabolic or catabolic pathways. These mechanisms include fatty acid derivatives, terpenes, glucosinolates, amino acids, and phenolics. A general vision of the currently known or hypothesized biochemistry of most important aroma compounds in vegetables are described below.

14.2.1 Compounds Formed by Degradation of Fatty Acids

Fatty acids play an important role in the aroma of all foods. They are catabolized through two main oxidative pathways: β-oxidation and lipoxygenase (LOX) (Fitz et al. 1999). The metabolism of polyunsaturated fatty acids by LOX catalysis results in volatile products like aldehydes and alcohols, which are responsible for fresh and green sensorial notes (Ong et al. 2010). The pathway is shown in Figure 14.1.

The LOX mechanism starts with the liberation of free fatty acids by the action of lipases. The key reaction in this pathway is the hydroperoxidation of polyunsaturated fatty acids with a (*Z*,*Z*)-1,4-pentadiene moiety, such as linolenic acids that are typically oxidized into 9-, 10-, or 13- chains hydroperoxides, depending on the regioselectivity and stereoselectivity of the catalysis. The hydroperoxides can be further converted into a range of volatile products by the action of hydroperoxide lyases, to finally get metabolized by isomerases and dehydrogenases, thus producing esters and alcohols.

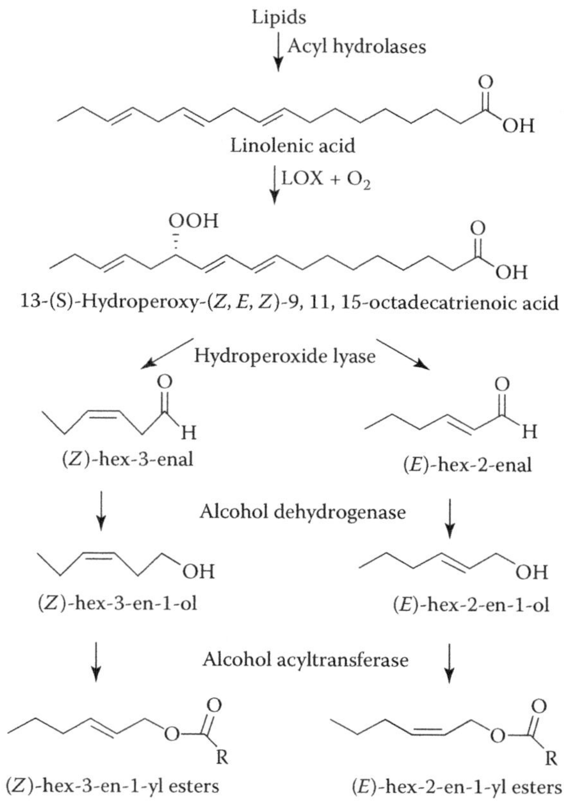

FIGURE 14.1 Scheme of degradation of linolenic acid via LOX pathway. (Adapted from Pérez, A. G. and Sanz, C., *Fruit and Vegetable Flavour*, Woodhead Publishing, Boca Raton, FL, 2008. With permission.)

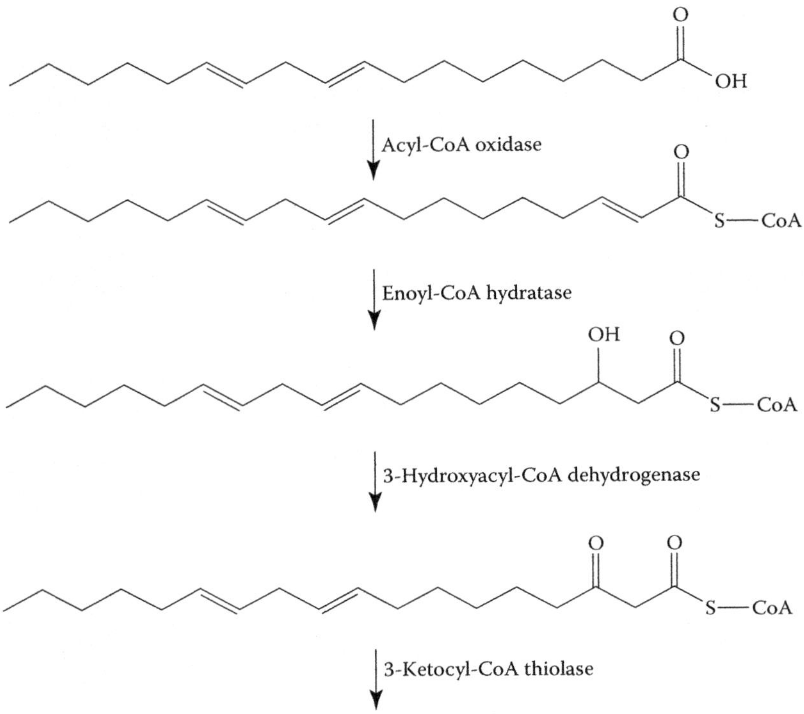

FIGURE 14.2 Cycle of β-oxidation of linolenic acid. (Adapted from Christensen, L. P. et al., *Flavours and Fragrances: Chemistry, Bioprocessing and Sustainability*, Springer, Berlin, Germany, 2007. With permission.)

The β-oxidation biogenesis requires the participation of acylcoenzimeA (acetyl-CoA) and the fatty acids metabolized are acyl-CoA derivatives. The loss of two carbons in every cycle of reaction is repeated several times until the complete breakdown of the compound. Figure 14.2 describes the main metabolism.

Other compounds derived from enzyme-catalyzed breakdown of unsaturated fatty acids may also be produced by autoxidation. This nonenzymatic oxidation of aliphatic chains yields a mixture of hydroperoxides that differ mainly in the geometrical isomerism of the double bonds, as well as in the position of the peroxide group. The aldehydes produced by this oxidation pathway may further lead to the formation of other volatile compounds (Christensen et al. 2007).

14.2.2 Amino Acid Metabolism

The metabolism of amino acids represents an important source of aliphatic, alcohols, carbonyls, acids, and esters. These volatile compounds are produced by three enzymatic activities: aminotransferase, decarboxylase, and alcohol dehydrogenase (Figure 14.3).

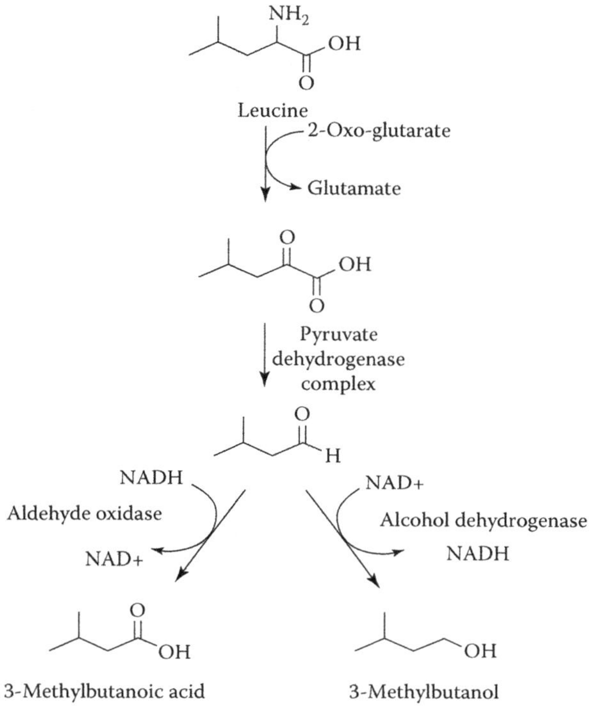

FIGURE 14.3 Scheme of formation of volatile compounds from amino acids. (Adapted from Fisher, C. and Scott, T. R., *Food Flavours: Biology and Chemistry*. The Royal Society of Chemistry, Cambridge, 1997. With permission.)

Enzymes remove both amine and carboxyl groups from the precursor amino acid, thus producing aldehydes. The aldehydes are oxidized or reduced to produce esters. The enzymatic activities can be affected by cultivar, maturity stage, and environmental conditions (Pérez and Sanz 2008) of the vegetables.

14.2.3 Glucosinolate Metabolism

Glucosinolates are thioglucosides that consist of a common basic skeleton containing a β-thioglucose group, a side chain, and a sulfonated oxime moiety. These are stable precursors to volatile compounds found in some vegetable species. When the plant tissue is damaged, glucosinolates are hydrolysed by the myrosinase enzyme, cleaving the glucose fragment to form unstable aglycones, which get rearranged into the respective isothiocyanates (Figure 14.4) (Fisher and Scott 1997; Christensen et al. 2007).

The products of this metabolism are glucose, sulfate, and aglycone intermediates, which are transformed into isothiocyanates, nitriles, glucose, and sulfate, depending on the structure of the glucosinolate and other conditions. When medium is acidic, production of hydroxynitrile derivatives are favored (Fisher and Scott 1997). Moreover, nitriles can be produced by the thermal decomposition of glucosinolates (Christensen et al. 2007).

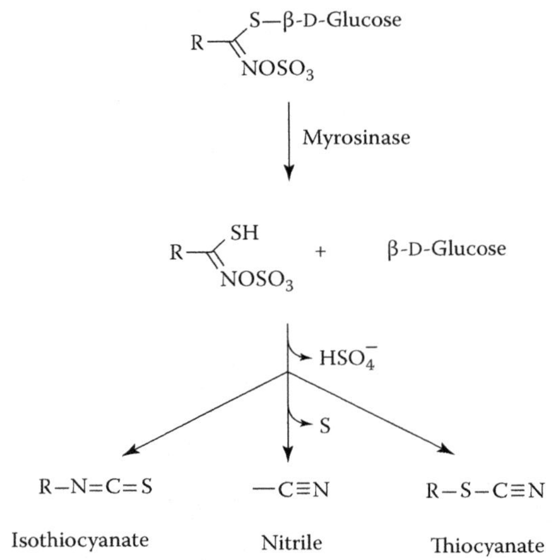

FIGURE 14.4 Scheme of glucosinolate hydrolysis. (Adapted from Fisher, C. and Scott, T. R., *Food Flavours: Biology and Chemistry*. The Royal Society of Chemistry, Cambridge, 1997. With permission.)

14.2.4 Biosynthesis of Terpenes

The terpenoids are metabolites formed from isoprene units and are widely distributed among vegetables. Monoterpenes and sesquiterpenes are important contributors to the flavor of vegetables. Both are present in intact plant tissues and are products of anabolic processes.

There are two routes of biosynthesis of terpenes in plastids: the mevalonate pathway, in cytosol, and the mevalonate-independent methylerythritol phosphate pathway (Mahmoud and Croteau 2002). Figure 14.5 describes both the mechanisms. The biosynthesis of monoterpenes starts with the production of terpenoid-building units (isopentenyl diphosphate and dimethylallyl diphosphate), followed by the condensation of these units by prenyltransferase, thus producing geranyl diphosphate. The next step is the conversion of geranyl diphosphate to the monoterpene parent skeleton. The transformation of the parent structure to various terpene derivatives ends the mechanism. Regular monoterpenes are derived exclusively from geranyl diphosphate, which is often cyclized to produce the backbone skeletons of the various monoterpene subfamilies (Mahmoud and Croteau 2002).

The volatile compounds of vegetables are usually labile and thus subject to rearrangements, cyclizations, oxidations, and degradations when submitted to processing procedures involving aeration and increasing temperature. Consequently, losses and modifications in the volatile profile are caused during industrial processing and storage, resulting in products that do not retain the original aroma of the fresh vegetables.

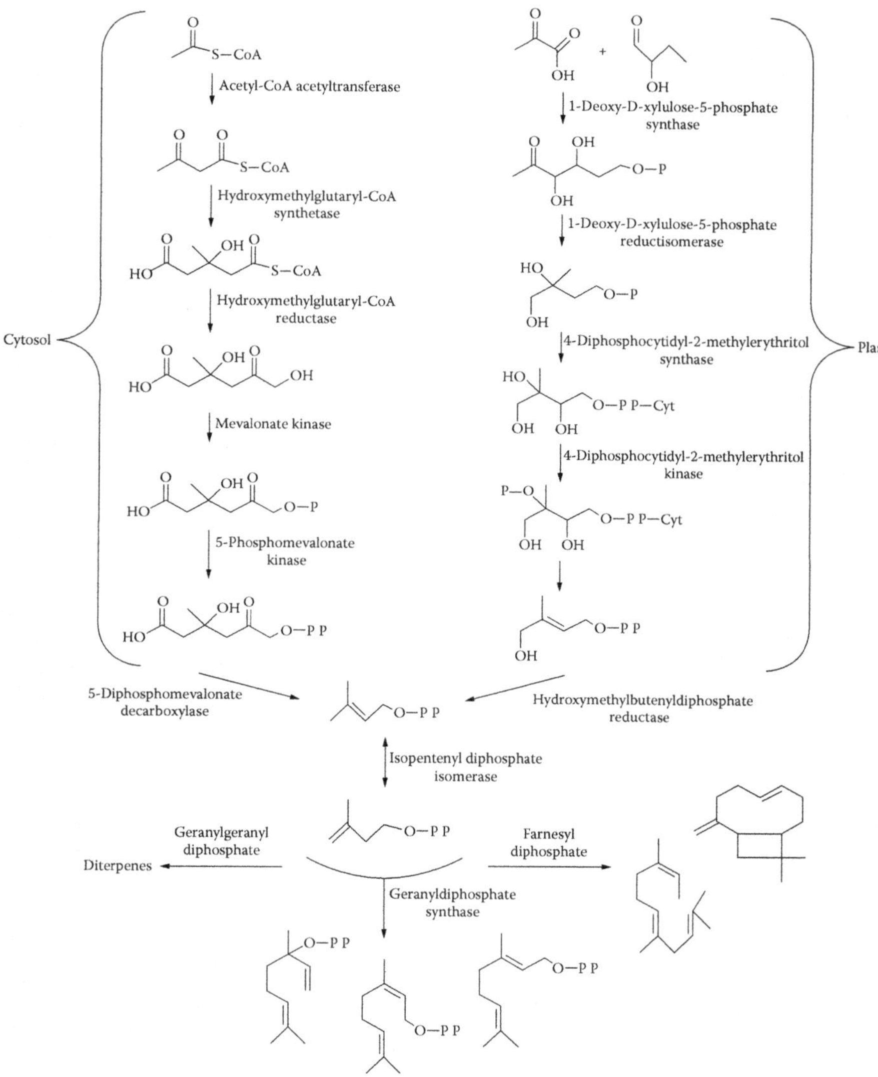

FIGURE 14.5 Scheme of biosynthesis of terpenoids. (Adapted from Pérez, A. G. and Sanz, C., *Fruit and Vegetable Flavour*, Woodhead Publishing, Boca Raton, FL, 2008. With permission.)

14.3 EFFECT OF PACKAGING ON AROMA COMPOUNDS IN VEGETABLES

Although there is extensive work on the evaluation of packaging material on aroma compounds in fruits, research work related to vegetables is quite limited. Effects of minimal processing and packaging on volatile compounds and other sensory aspects in carrots was studied by Edelenbos et al. (2010), and they reported that factors such as raw material, processing conditions, packaging material, and storage conditions affect the quality of carrots. Trimming and cutting into pieces and then storage for 2 days before processing resulted in a significant loss in monoterpenes and sesquiterpenes as compared to direct processing. The total terpene content decreased by 59%. Washing in water after cutting and then spin-drying further reduced the volatile content, especially the sesquiterpenes. Packaged carrots lost about 55% to 65% of the initial terpene content during the first 24 h of storage at 5°C, and the loss later increased to more than 88% on the 8th day in the carrots packaged in a needle-perforated polypropylene film in an atmosphere of more than 9% O_2 and less than 13% CO_2. However, when minimally processed carrots were packaged in a nonperforated film, the loss was about 70% of the initial terpene content and its atmospheric composition led to about 0% O_2 and up to 45% CO_2, which consequently results in poor sensory quality of carrots.

The effect of three modified atmosphere packaging (%O_2/%CO_2: 0/40; 2.5/40; and 2.5/60) on the quality preservation of several precooked vegetables such as cabbage, carrots, green beans, and bell peppers was studied by Barbosa et al. (2016). They evaluated the quality of the products for different storage periods up to 28 days. Only physicochemical parameters (pH, acidity, moisture and ash contents, antioxidant activity, color, and texture), microbial growth, organoleptic properties, and consumer acceptability were assessed. They observed only slight changes in physicochemical parameters and microbial growth, which was also confirmed by the trained panel that could not discriminate samples with different storage times. However, changes in volatile compounds were not determined in this study.

The volatile profile of two hybrids of "Radicchio di Chioggia," Corelli and Botticelli, stored in air or passive modified atmosphere during 12 days of cold storage, was determined by Cozzolino et al. (2016). The volatiles content of the samples varied depending only on the packaging conditions. Principal component analysis (PCA) showed that fresh product possessed a metabolic content similar to that of the modified atmosphere packaged samples after 5 and 8 days of storage. Specifically, 12 metabolites (3-nonanol, 4-nonanol, α-selinene, 2-ethylfuran, nonanal, octane, ethyl 3-methylpentanoate, tetradecane, ethyl tiglate, ethyl benzoate, 3-methylbutanal, and phenylethyl alcohol) describing the time evolution and explaining the effects of the different storage conditions were highlighted. Finally, a PCA analysis revealed that volatile organic compounds (VOC) profile significantly correlated with sensory attributes.

The influence of different preharvest and postharvest factors was investigated in fresh-cut iceberg lettuce packaged in low-O_2 modified atmospheres (Tudela et al. 2013). Fresh-cut iceberg lettuce developed undesirable off-odors under low O_2 and elevated CO_2 atmospheres. Higher CO_2 concentrations and higher accumulation of ethanol and acetaldehyde were detected in the headspace (HS) of modified atmosphere (MA)–packed lettuce from immature heads. Off-odor metabolites related with the LOX pathway such as hexanal, 1-hexanol, and cis-3-hexen-1-ol were generated. Volatile compounds such as cis-3-hexen-1-ol, elemene, ethyl acetate, and dimethyl sulfide increased their content more than 10 times compared with other volatiles.

14.4 EFFECT OF DRYING ON AROMA OF VEGETABLES

Several drying methods have been applied to vegetables, from the simplest ones such as solar and sun drying to the most expensive, applicable method of drying such as freeze, vacuum, osmotic, cabinet or tray, fluidized bed, spouted bed, Ohmic, microwave, and a combination thereof (George et al. 2004; Sagar and Suresh Kumar 2010). Figure 14.6 shows the various dehydration techniques employed with the most common processing parameters, their conditions, and quality control of dried products. To get dried products with high nutritional and sensorial attributes, nonconventional

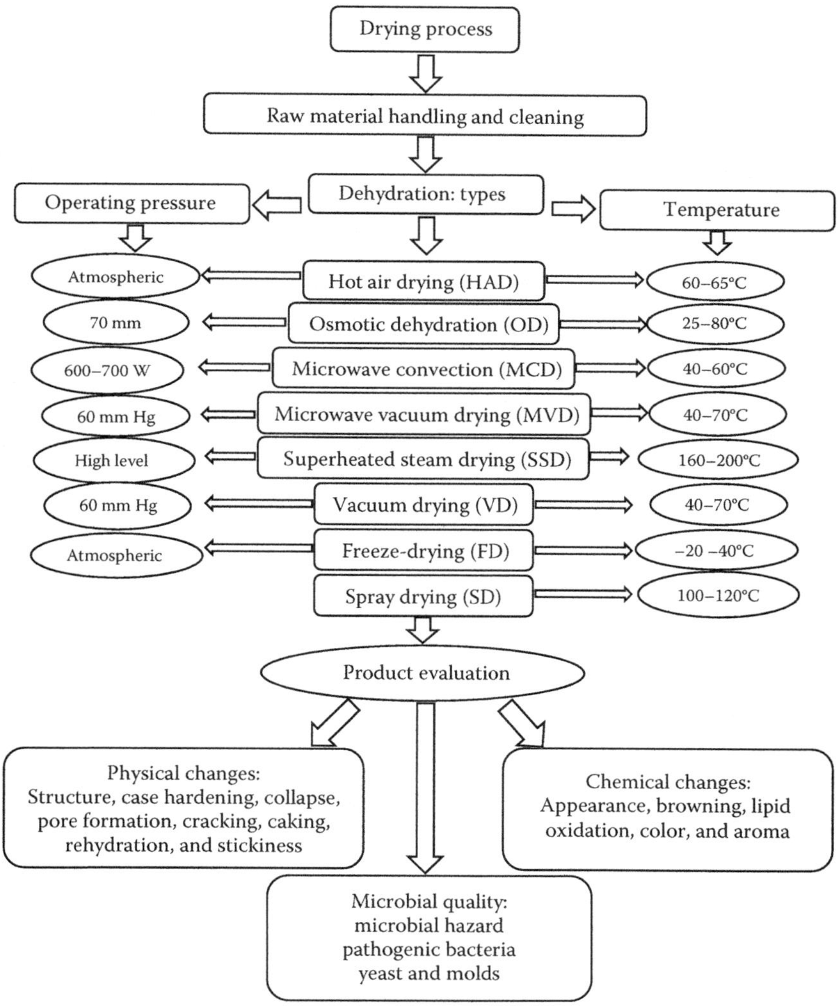

FIGURE 14.6 Various dehydration techniques demonstrating the most common processing parameters and product evaluation.

drying methods like modified atmosphere techniques have also been used (Santos and Silva 2008; Wanga et al. 2010; Verma and Vir Singh 2015). The drying process undoubtedly affects the quality of vegetables, and it is important that there be minimum adverse effects such as decrease in nutritional value, color, and flavor aspects.

Gaware et al. (2010) and Castoldi et al. (2015) studied drying of tomato using different methods of dehydration and rehydration kinetics. The results revealed that microwave vacuum drying showed the highest moisture diffusivity when compared to other drying techniques such as HAD and FD. Jorge et al. (2015) described the analytic hierarchy process used to identify the most suitable drying system for tomatoes among three available systems (FD, conventional drying oven, and bench-scale heated air flow), with the lowest possible cost and loss of quality of the dehydrated product.

Narain et al. (2010) monitored drying of tomato juice and dried products on their volatiles retention. Fresh tomato juice and its two blends—one with 5% maltodextrin and another with 5% tapioca flour were dried at 60°C in a forced-air circulation dryer. The volatiles retention was the highest in tomato powder prepared with 5% maltodextrin, which retained most of the sulfur compounds. To study the volatile profile changes during drying of tomato juice, certain volatile markers such as dimethyl sulfide, 2-ethyl furan, 1-hexanol, 5-hydroxymethyl furfural (5-HMF), α-terpineol, and acetaldehyde were selected. Dimethyl sulfide and 2-ethyl furan were selected for their sulfury note in tomato flavor; 1-hexanol for its fatty and fruity odor; 5-HMF for its formation during drying, basically due to Maillard reaction; α-terpineol for its delicate floral and sweet note; and acetaldehyde for its ethereal and nauseating odor.

Figure 14.7 shows the change in concentration (mg·Kg^{-1}) of some prominent volatile compounds such as geranyl butanoate, terpineol, ethanol, linalool, and acetaldehyde in tomato juice during drying. The major decrease was in ethanol, whose concentration gets reduced by about 1/3 in first hour of drying. The aroma impact of

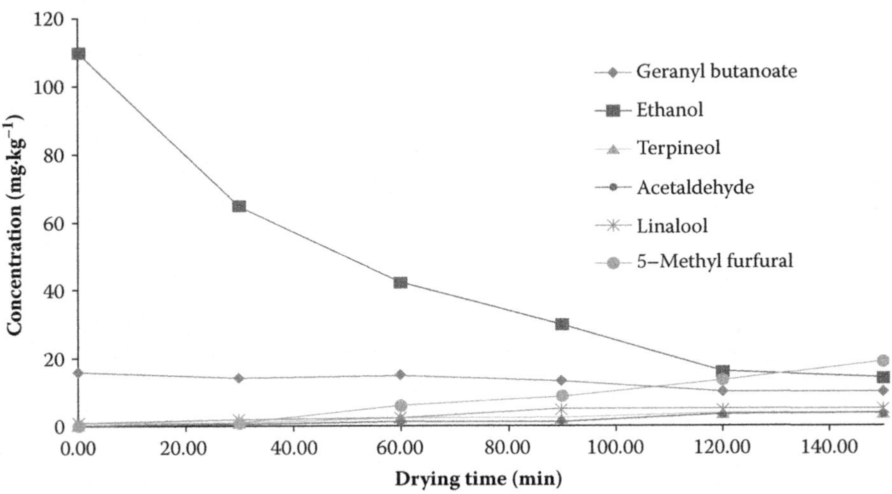

FIGURE 14.7 Concentration (mg/kg) of geranyl butanoate, terpineol, ethanol, linalool, and acetaldehyde in tomato juice during drying.

sulfur compounds in food products is very well-known. Generally, these compounds possess low odor thresholds; for example, dimethyl sulfide is reported to possess 0.3 ppb odor threshold in water (Buttery et al. 1990). There was an increase in concentrations of dimethyl sulfide, 2-ethyl furan, diethyl sulfide, and 2-acetyl furan in dried tomato powders, which could characterize for the sulfury note in dried products (Figure 14.2). Dimethyl sulfide and 3-methylthio propanal were reported to increase significantly in heated tomato products (Buttery 1996).

Linalool, 5-HMF, and α-terpineol concentrations also increased in the dried tomato powder. The concentrations of dimethyl sulfide and 2-ethyl furan increased with drying, and the dried tomato powder contained 11.8 and 7.6 mg/Kg of dimethyl sulfide and 2-ethyl furan, respectively (Figure 14.8). These compounds are known to possess strong sulfury aroma notes. The aroma attribute of dried tomato powder when rehydrated was found to be characteristic of tomato flavor, although it was quite mild.

Du et al. (2015) focused their study on identification of sulfur volatiles and GC-olfactometry aroma profiling in two fresh tomato cultivars ("Tasti-Lee" and "FL 47"). Dimethyl sulfide, dimethyl disulfide, dimethyl trisulfide 2-propylthiazole, and 2-s-butylthiazole were identified in fresh tomatoes. GC–olfactometry aroma profiling indicated that the most intense aroma category was earthy-musty, followed by fruity-floral, green-grassy, sweet-candy, and sweaty-stale sulfurous.

Montoya-Ballesteros et al. (2014) reported that among all conventional and unconventional drying techniques such as open-air sun drying, FD, microwave drying, and mechanical drying including ovens, tunnels, cabinets, spray dryers, heat pumps, and infrared drying chambers, FD and microwave drying resulted in reducing drying time while maintaining high concentration of bioactive compounds and good sensory attributes in chili peppers.

The quality parameters such as flavor or pungency, color, texture, and rehydration ratio of the garlic slices dried by the combination of MVD and air-drying

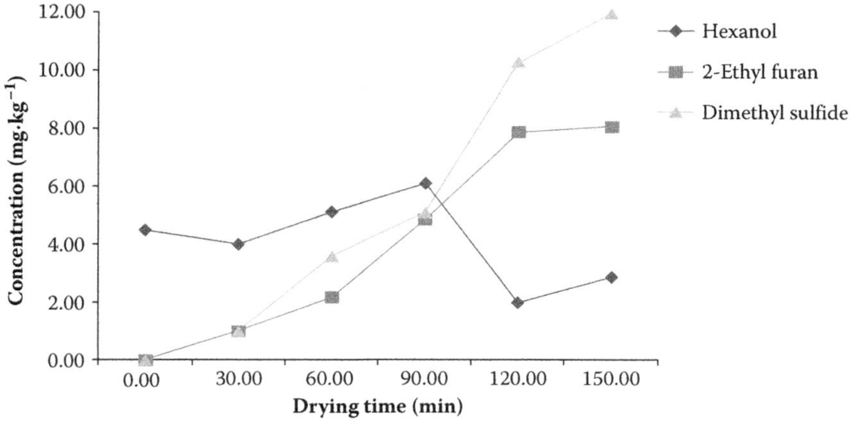

FIGURE 14.8 Concentration (mg/kg) of hexanol, 2-ethyl furan, and dimethyl sulfide in tomato juice during drying.

was close to that of freeze-dried products and much better than that dehydrated by conventional hot air-drying, indicating that the microwave–vacuum air-drying is a better way for drying garlic slices and other vegetables (Wei Cui et al. 2003). In the same way, Dev et al. (2011) studied that the effects of microwave-assisted HAD and conventional HAD on the drying kinetics, color, rehydration, and volatiles of drumstick (*Moringa oleifera*). They also suggested that the microwave-assisted HAD maintained the aroma and sensory qualities of the drumstick. Using microwave freeze dryer, some vegetable soups were successfully dried. Microwave power significantly influences the total drying time and sensory quality of final products. High microwave power resulted in shorter drying time but poorer product quality, whereas too low a microwave power leads to excessively long drying time (Wang et al. 2009).

During osmotic dehydration of potato cubes, the concentration of sucrose and salt mixtures, microwave power, and osmosis time were highly significant for water loss as well as solids gain. The combination of sucrose and salt solutions with pulsed microwave power and vacuum application during the first 4 min of osmotic dehydration of potato cubes reduced the osmosis time and resulted in greater mass transfer (Sutar et al. 2012). Rehydration is a complex process, and it indicates the physical and chemical changes induced by drying treatments in pumpkin slices, which also revealed that the rehydration ratio of pumpkin samples increases with the decrease in temperature and microwave power (Mayora et al. 2011; Seremet et al. 2016).

Drying methods and conditions have a profound effect on the quality and energy consumption of the dehydrated product. An et al. (2015) compared the various drying processes and reported that air-drying, FD, infrared drying, microwave drying, and intermittent microwave and convective drying had higher retention of chemical profiles, antioxidant activity, and cellular structures in Chinese ginger, which was attributed to their less intense heating. However, FD and infrared drying had relatively higher energy consumption and drying time, especially FD. Therefore, intermittent microwave and convective drying is a very promising technology for high sensitive products like fruits and vegetables due to its higher efficiency, good quality retention, and lower cost.

Mulet et al. (2000) reported that the drying process influences the shrinkage and other physical changes in vegetables like potato and cauliflower. Siucińska and Konopacka (2014) reviewed that ultrasound can enhance the dehydration process of vegetable tissue, both when applied during a conventional drying process and in preparation and/or enrichment prior to the drying. In continuation, the dehydration techniques have an influence on physicochemical, antioxidant, and microbial qualities of underutilized green leafy vegetable, *Ipomoea aquatica* (Shin et al. 2015). One of the solar dried methods also retained the physical quality and carotene content of green leafy and yellow succulent vegetables such as *Amaranthus gangeticus*, *Chenopodium album*, *Centella asiatica*, *Amaranthus tricolor*, and *Trigonella foenum graecum* (Mdziniso et al. 2006). These green leafy vegetables were steam blanched for 5 min followed by drying in an oven at 60°C for 10–12 h. The fresh and dried processed vegetables were analyzed for their nutrition properties, and the results suggested that the drying process helped in retention of the nutritional parameters like that present in fresh vegetables (Gupta et al. 2013).

Cabbage contains many flavor compounds, and isothiocyanates have been shown to be the major compounds that impart pungent flavor and sulfurous aroma to these vegetables (Banerjee et al. 2014). More than 16 compounds representing glucosinolates, sulfides, sugars, and some alcohols and aldehydes were reported to influence its flavor (Radovich et al. 2003; Radovich 2010). Sweet, hot, pungent, bitter, fruity, grassy/green, sulfurous, and earthy odors were reported to be the characteristic odors of cabbages (Radovich 2010; Lekcharoenkul et al. 2014). The important volatile contributors of these odors include 2-hexenal, 3-hexenol, 2,4-heptadienal, benzaldehyde, and phenylacetaldehyde (Valette et al. 2006). Lonchamp et al. (2009) reported the key volatile compounds of ready-to-use cabbages such as 1,4-dichlorobenzene, limonene, dimethyl sulfide, dimethyl disulfide, and allyl isothiocyanate during different storage periods. Hong and Kim (2013) analyzed hydrolysis products and other volatile constituents from Korean cabbages and its seeds, resulting in the identification of 16 and 12 volatile compounds, respectively. The primary volatile compound found in cabbage is ethyl linoleolate, while 4,5-epithiovaleronitrile is the primary volatile component in the seed. Dimethyl sulfide, dimethyl disulfide, dimethyl trisulfide, and methanethiol are off-odors in cabbage (Banerjee et al. 2014). Duan et al. (2007) reported that microwave freeze-drying of cabbage was better than FD for overall quality; however, they observed that aroma characteristics were lacking to make a definitive conclusion.

14.5 METHODS OF AROMA ANALYSIS

Aroma constitutes of a large number of volatile and semi-volatile organic compounds such as esters, alcohols, aldehydes, hydrocarbons, ketones, terpenes, sesquiterpenes, and sulfur compounds, these play an important role in flavor quality of vegetables. Aroma may be determined or evaluated with either instrumental or sensorial methods. Sensory evaluation has its own importance as this is directly related to the consumer acceptability of vegetables and vegetable products. The most common instrumental techniques used for identification and quantification of volatile compounds are provided by hyphenated instruments such as gas chromatograph coupled with various detectors, in particular, mass spectrometer and gas chromatograph attached to an olfactometer. Moreover, some other equipment such as electronic nose and electronic tongue have also been used lately but these still have some limitations.

14.5.1 Extraction of Volatile Compounds

The determination of aroma is often a very difficult task due to the complexity of the vegetable matrix and the presence of very low concentration of volatile compounds in comparison to the amount of other constituents such as lipids, proteins, and carbohydrates (Yang et al. 2013). The extraction techniques that allow the recovery of volatile compounds without the use of solvents are preferable to those that require solvents so as to not form unwanted compounds or artifacts, thereby retaining the authenticity of naturally occurring compounds.

In general, aroma chemical analysis involves four steps: (1) extraction or capture of volatile compounds, (2) concentration, (3) separation, and (4) identification and quantification. Direct extraction methods involve classical liquid–liquid extraction

(LLE), solid-phase extraction (SPE), and supercritical fluid extraction (SFE) procedures; distillation-based methods comprise mainly vapor distillation and simultaneous distillation–extraction (SDE) procedures, while the HS technique is based on static headspace (SHS) or dynamic headspace (DHS) capture of volatiles (Augusto et al. 2003). Thus, the chemical characterization of aroma ordinarily demands state-of-the-art techniques for sampling and sample preparation, which may require further concentration of volatile extracts before subjecting to separation, detection, and quantitative analysis of volatile compounds (Civille 1991; Sides et al. 2000). However, in this chapter, some of the prominent and currently used sample preparation techniques such as Purge and Trap, solid-phase microextraction (SPME), and stir bar sorptive extraction (SBSE) are described.

14.5.2 Purge and Trap Techniques

In DHS or Purge and Trap techniques, a flow of inert gas is bubbled into the sample and the volatile analytes are transferred to a trap containing an adsorbent polymer. Desorption of trapped analytes for subsequent analysis is performed either with small volume of adequate solvents (Rankin and Bodyfelt 1995) or by using an online automated thermal desorption unit (Valero et al. 1997), the latter being suitable for large number of samples analysis. Its main advantage is the high detectability of volatile compounds since the thermodynamic equilibrium between the sample, the analyte, and the HS is not necessarily required and the capture of volatiles in this method is increased by enrichment of the trap with the analytes of interest (Freire et al. 2008).

14.5.3 Solid-Phase Microextraction

Solid-phase microextraction (SPME) is an SPE sampling technique that involves the use of a fiber coated with an extracting phase, that can be a liquid (polymer) or a solid (sorbent), which extracts different kinds of analytes (volatile and nonvolatile) from different kinds of matrices. In recent years, this microextraction has attracted significant attention as this technique has high sensitivity, is simple to use, has short sample pretreatment time and a high enrichment factor, is solvent-free, and amenable to automation. Although it is an equilibrium technique, SPME was rapidly accepted as a simple, miniaturized, and green technique, which combines sampling, capture, concentration, cleanup, and sample introduction in a single step. In addition, this technique can be applied to gas, liquid, and solid sample matrices. Thus, this technique has become an attractive tool in the determination of volatile and semi-volatile compounds from plant matrices (Yang et al. 2013).

The SPME extraction has been widely used to determine volatile and semi-volatile compounds (volatile organic compounds including esters, alcohols, aldehydes, hydrocarbons, ketones, terpenes, sesquiterpenes, phenols, acids, etc.) and pesticides, including nicotinoids, carbamates, and fungicides (Melo et al. 2012) from various samples including plant matrices (fruits, vegetables, and medicinal plants) in the last 5 years. However, there is no standard protocol that can be adopted for different types of samples, thus leading to study of the best conditions for SPME by different researchers in different food matrices (Xu et al. 2016).

14.5.4 Stir Bar Sorptive Extraction

The extraction using stir bar was developed in 1999 by Baltussen et al. (1999). Like SPME, Stir Bar Sorptive Extraction (SBSE) also does not require any use of solvents. Moreover, SBSE is considered to be a technique with greater extraction efficiency when compared with SPME since this has a larger contact surface than the SPME fiber, which is an exceptional advantage (Lancas et al. 2009). However, the main disadvantage of the SBSE technique is that it requires large time for extraction and desorption (60 min for each step) when compared to SPME.

The stir bars are made of two different types of films, one with 24 μL of PDMS (known as Twister) and another 32 μL of PEG Silicon (known as GE Silicon Twister). The extraction using stir bars can be performed by placing a small magnetic bar directly in the solution to be analyzed, although NaCl may or may not be added. The desorption of volatile compounds is done by putting the stir bars in an appropriate thermal desorption unit that is installed directly on the injector of the gas chromatograph wherein the compounds are transferred onto the capillary column. The recovery range of analytes in SBSE could vary from SBSE 70% to 130% (Yang et al. 2013).

Due to the ease of handling and potential use of SBSE technique, it has become quite common and applicable to various types of food matrices. For vegetables, the SBSE technique has been used for the determination of volatile compounds in tomatoes (Pardo-García et al. 2013), organochlorine pesticides in lettuces (Barriada-Pereira et al. 2010), and of volatile organo seleno compounds (dimethylselenide and dimethyldiselenide) in onions (Duan et al. 2009).

14.5.5 Separation, Identification, and Quantification of Volatile Compounds

The volatile fraction of a vegetable, for whatever be the isolation method of obtaining it, consists of a complex mixture of volatile compounds, which requires a very powerful separation method. High-resolution gas chromatography meets these requirements as this technique is very selective, sensitive, and efficient (Franco and Janzantti 2004; Grob and Kaiser 2004). Of late, more stable columns with greater separation resolution efficiencies are commercially available. For analysis of aroma compounds, gas chromatographs are often coupled to detectors, such as thermal conductivity, flame ionization, electron capture, flame photometric, and mass spectrometric detectors. The detector senses the presence of the individual components as they leave the column, and their output, after suitable amplification, is acquired and processed by a computerized data system, resulting in a chromatogram containing peaks of the compounds.

The GC–MS is the combination of two powerful analytical tools wherein the GC separates efficiently the components of a complex mixture and the mass spectrometer provides qualitative information of the compounds such as mass spectra to identify them (Rood 1995; Van Ruth and O'Connor 2001).

Due to the complexity of aroma profiles, rapid analysis using a detector with fast mass spectral acquisition such as time-of-flight mass spectrometry (TOFMS) is recommended. Song et al. (1998) analyzed flavor volatiles in tomato and strawberry fruits and reported that an overlapping eluting compound extracted using SPME coupled to

TABLE 14.1
Principal Volatile Compounds of Some Vegetables

Vegetables	Volatile Compounds	Extraction Technique	Reference
Potato	6-Methyl-3,5-heptadien-2-one,decanal, (*E*)-2-hexenal	SPME	Kebede et al. (2014)
Tomato	1-Butanol, 2-pentanone, hexanal	Likens Nickerson Distillation	Narain et al. (2010)
Chilli pepper	Hexanal, 5-hexenol, β-caryophyllene	SPME	Bogusz Junior et al. (2012)
Cabbage	Ethyl linoleolate, benzenepropanenitrile, 2-phenylethyl isothiocyanate	Solvent extraction	Hong and Kim (2013)
Broccoli	Dimethyl pentasulfide, methyl (methylthio) methyl disulfide, 5-(methylthio)-pentanenitrile	SPME	Kebede et al. (2015)
Carrot	Ethanol, *trans*-caryophyllene, 3-hydroxy-2-butanone	SPME	Soria et al. (2008)
Red beet	2-Methylpropanal, 3-methylbutanal, 2-methylbutanal,	SPME	Kebede et al. (2014)
Onion	2-Mercapto-3,4-dimethyl-2,3-dihydrothiophene, allyl disulfide 2,4-dimethyl-thiophene,	SPME	Kebede et al. (2014)
Pumpkin	2,2,6-Trimethylcyclohexanone, 2,4-dimethyl-1-heptene, 1,2,5,5-tetramethyl-1,3-cyclopentadiene	SPME	Kebede et al. (2014)
Lettuce	2,3-Methylbutanal, 1-penten-3-ol, 2,3-butanedione.	Dynamic HS	Deza-Durand and Petersen (2014)
Spinach	Palmitic acid, ergost-5-en-3-ol, stigmasterol	Ultrasonic extraction	Shim and Baek (2012)

GC-TOFMS was still unresolved. Furthermore, flavor compounds present at extremely low concentrations (ppb, ppt) can still play a significant role in the key aroma profile. The comprehensive two-dimensional gas chromatography (GC–GC) is a powerful technique and is better suited for the analysis of complex mixtures. Using two orthogonal approaches on different combinations of apolar and polar columns, GC–GC enhances the separation efficiency as compared to the conventional one-dimensional GC. When coupled to a fast acquisition detector such as TOFMS or GC–GC-TOFMS, it has proven to be an established technique, providing better separation capabilities and a full mass spectrum with high-resolution two-dimensional contour plots (Dallüge et al. 2003).

Some of the main volatile compounds found in vegetables using different types of extraction procedures are presented in Table 14.1.

14.6 CARROT AROMA

The flavor of carrots is dominated by the presence of terpenes, sesquiterpenes, a few alcohols, styrene, and alkanes amounting to more than 90 volatile compounds

(Kjeldsen et al. 2003). Among its volatile profiles, monoterpenes and sesquiterpenes are of particular interest as their composition determines the sensory quality of the products. Alasalvar et al. (1999) observed that the storage temperature influences the content of monoterpenes and sesquiterpenes in raw, stored, and cooked carrots. An increase in the content of terpenoids during refrigerated storage and no change during frozen storage have been documented (Kebede et al. 2014). Kreutzmann et al. (2008) reported the sensory quality variations in raw carrots regarding bitterness, green flavor, and terpene flavor, which are characterized by volatiles such as terpinolene, β-pinene, sabinene, γ-terpinene, α-pinene, β-bisabolene, caryophyllene, and cuparene. Buttery and Takeoka (2013) identified linden ether as an important volatile in cooked carrots. An HS sorptive extraction technique employed by Fukuda et al. (2013) reported the role of terpenoids in influencing pleasant flavors such as fruity, fresh, and sweet notes to determine preferred carrot varieties. Duan and Barringer (2011) observed that the content of furan decreased during drying of carrot slices. Further, the impact of thermal and high pressure–high temperature processing had different effects on the volatile profile of blanched orange and yellow carrots (Duan and Barringer 2011; Fukuda et al. 2013). Li et al. (2010) developed a microwave drying system with temperature and power control and detected volatile signals employing α-pinene, β-pinene, α-terpinolene, and caryophyllene, with a fuzzy logic control system for drying carrots. Over the last two decades, although there have been many studies on the volatile profiles of carrots (Alasalvar et al. 1999; Kreutzmann et al. 2008; Li et al. 2010; Duan and Barringer 2011; Buttery and Takeoka 2013; Fukuda et al. 2013; Kebede et al. 2014), there is still lack of information on comparison of their aroma profiles under different drying conditions so as to enable a clear understanding of aroma stability and changes during dehydration.

Recently, a detailed study on physical properties and aroma profile of carrots as influenced by hot air and FD was performed by Rajkumar et al. (2016). The carrots were trimmed, scraped, washed, and cut into slices of 4.5 cm (length) × 1.5 cm (width) × 1.5 cm (thickness), and these were subjected to drying at 45°C by employing a conventional HAD, while FD was performed at −21°C at an absolute pressure of 85–90 Pa until constant moisture content was achieved. The final moisture content of the HAD and FD carrots was about 3.89% and 3.60%, respectively. HS volatile compounds were collected using an SPME fiber coated with 50/30 μm DVB/CAR/PDMS (Divinylbenzene/Carboxen/Polydimethylsiloxane). Separation, identification, and quantification of volatile compounds were achieved in a system of gas chromatography coupled with mass spectrometry (Rajkumar et al. 2016). The composition of the volatile compounds present in fresh, HAD, and FD carrots is presented in Table 14.2, and it presents several compounds including aldehydes, alcohols, terpenes, ketones, esters, acetates, and furans, identified in fresh and dehydrated carrots. It could be seen that most of the identified volatiles have been reported previously as constituents of carrots (Tatemoto and Michikoshi 2014; Tatemoto et al. 2014). Taking all the volatiles of carrot samples into consideration, which includes fresh, HAD, and FD samples, they were mainly comprised of terpenes (40) followed by alcohol (16), aldehydes (13), ketone (6), acetate (4), ester (2), ether (1), and furan (1) (Table 14.2).

TABLE 14.2
Principal Volatile Compounds of Fresh and Dehydrated (Hot Air- and Freeze-Dried) Carrots

No	Volatile Compounds	Concentration (μg/g)			Aroma Odor Description
		Fresh	Hot Air-Dried	Freeze Dried	
			Aldehydes		
1	Hexanal	8.467	16.471	1.872	Grass
2	Heptanal	4.655	2.024	2.534	Citrus
3	Octanal	13.349	2.698	—	Lemon, green
4	Benzene acetaldehyde	—	1.559	1.968	Floral, rose
5	(*E*)-2-nonenal	4.666	1.767	—	Cucumber, green
6	*n*-Decanal	—	—	1.286	Orange peel, tallow
7	β-Cyclocitral	4.291	2.274	2.881	Mint
8	Undecanal	1.727	—	—	Pungent, sweet
			Alcohols		
9	Ethanol	—	8.084	—	Sweet
10	2-Methyl-(S)-1-butanol	—	2.086	—	Roasted
11	n-Hexanol	—	1.622	—	Flower, green
12	1-Octen-3-ol	—	2.571	2.776	Earthy
13	(*Z*)-2-Octen-1-ol	—	0.390	—	Green
14	(*E*)-2-Nonenol	—	—	1.525	Fatty, melon
15	Borneol	—	—	4.376	Camphor
16	n-Nonanol	2.351	0.860	—	Fat, green
17	1-Octanol	5.979	—	—	Moss, nutty
18	*p*-Mentha-1,8-dien-6-ol	—	0.666	—	Caraway
19	*p*-Cymen-8-ol	—	—	2.451	Sweet, fruity
			Terpenes		
20	α-Thujene	11.714	2.323	3.851	Green, herb
21	α-Pinene	88.711	14.426	36.763	Pine, turpentine
22	Camphene	15.473	1.765	5.352	Camphor
23	β-Pinene	75.685	10.904	24.219	Pine, resin, turpentine
24	β-Myrcene	22.677	3.526	15.452	Balsamic, must, spice
25	α-Phellandrene	13.349	—	9.711	Turpentine, mint, spice
26	α-Terpinene	—	—	7.847	Lemon
27	*o*-Cymene	19.903	3.493	19.312	Citrus
28	Limonene	43.074	3.156	1.968	Lemon, orange
29	β-trans-Ocimene	—	—	1.389	Sweet herbal
30	(*E*)-β-Ocimene	5.154	—	—	Sweet, herbal
31	γ-Terpinene	21.018	6.273	43.302	Turpentine
32	Terpinolene	172.15	2.922	442.646	Sweet, fresh, piney citrus
33	α-Copaene	2.034	—	1.207	Woody, spicy

(*Continued*)

TABLE 14.2 (*Continued*)
Principal Volatile Compounds of Fresh and Dehydrated (Hot Air- and Freeze-Dried) Carrots

No	Volatile Compounds	Concentration (μg/g)			Aroma Odor Description
		Fresh	Hot Air-Dried	Freeze Dried	
34	α-Ylangene	10.790	—	—	Woody
35	α-Cedrene	8.315	—	—	Woody, cedar, sweet, fresh
36	(*E*)-Caryophyllene	163.410	46.003	1.845	Spicy, woody
37	α-Bergamotene	21.415	2.572	2.932	Woody, warm
38	α-Humulene	15.245	4.333	—	Woody
39	(*Z*)-β-Farnesene	28.440	3.775	2.986	Citrus, green
40	(*E*)-β-Farnesene	10.242	—	0.625	Woody, citrus, sweet
41	α-Acoradiene	20.893	—	—	Balsamic
42	α-Gurjunene	11.350	—	—	Woody, balsamic
43	γ-Himachalene	—	2.192	1.995	
44	α-Curcumene	37.802	5.344	4.076	Herb
45	(*Z,E*)-α-Farnesene	—	—	2.236	Woody, sweet
46	α-Zingiberene	11.804	2.012	—	Spice, fresh, sharp
47	β-Himachalene	10.941	—	—	
48	α-Bisabolene	12.517	1.557	1.669	Balsamic
49	β-Bisabolene	26.007	3.654	3.421	Balsamic
50	Patchoulene	1.654	—	—	Spicy
51	α-Dehydro-*ar*-himachalene	22.596	2.189	—	
52	β-Sesquiphellandrene	—	0.717	2.142	Woody
53	Bicyclogermacrene	73.988	—	—	Green, woody
			Ketones		
54	4-Methyl-5-hepten-2-one	2.740	2.506	2.864	Fruity, apple, musty
55	α-Ionone	—	2.635	2.312	Woody, violet
56	Geranyl acetone	—	—	4.457	Green
57	(*E*)-β-Ionone	—	3.803	3.452	Violet, flower
			Esters		
58	Ethyl ester octanoic acid	8.900	—	—	
59	Ethyl ester benzeneacetic acid	—	0.394	—	
			Acetates		
60	Ethyl acetate	—	2.160	—	Pineapple-like
61	*cis*-Chrysanthenyl acetate	—	—	1.785	
62	Bornyl acetate	229.931	22.754	66.396	Balsamic
			Furan		
63	*cis*-Linalool oxide	—	—	2.700	Flower, woody

Source: Rajkumar, G. et al., *Drying Technol.*, 2016. With permission.

14.6.1 Terpenes

An earlier report (Kreutzmann et al. 2008) suggests that the monoterpenes and sesquiterpenes accounted for almost 98% of the total volatiles of carrots, and these results agreed well with our findings. Compounds such as α-thujene, α-pinene, camphene, β-pinene, limonene, γ-terpinene, terpinolene, (*E*)-caryophyllene, α-bergamotene, and α-curcumene were the major carrot odor-related volatiles identified in all the samples including fresh carrots and in the HAD and FD ones. However, their occurrences varied between the samples (Table 14.2). The abovementioned compounds were observed to significantly contribute toward carrot aroma and flavor, and these compounds have been shown to correlate well with consumer acceptance of the vegetable (Buttery and Takeoka 2013).

Terpinolene and caryophyllene, responsible for "sweet" and "spicy" odor, respectively, were found to be the most abundant volatiles in fresh carrots, being 172.15 and 163.41 µg/g, respectively (Table 14.2). Terpinolene has been associated with cooked carrot aroma in processed products. Terpinolene, being a monoterpene and of light mass ($C_{10}H_{16}$), decreased by about 98% in HAD cabbage, while its concentration increased by 157% in FD cabbage (Table 14.2). Terpinolene is almost insoluble in water (9.5 mg/L at 25°C) and it is less dense than water. Thus, it is possible that this compound evaporates easily in HAD since the carrot structure and dimensions permit its slow evaporation even at 45°C, while in FD it is conditioned at a relatively lower temperature (–21°C), which inhibits its evaporation at these conditions. The increase in terpinolene concentration may also be due to the degradation of other sesquiterpenic compounds, leading to the formation of monoterpenes in the carrot volatile matrix (Figure 14.7). However, the caryophyllene concentration decreased in both dehydration processes. When compared to the fresh carrots, it decreased more (98.9%) in FD carrots and less (71.84%) in HAD carrots (Table 14.2). Caryophyllene is a higher mass compound ($C_{15}H_{24}$; 204.36 g·mol^{-1}) with a boiling point of 254°C–257°C at 760 mm Hg, and hence it gets reduced less in HAD; however, in FD, due to vacuum involvement, the compound gets removed during the dehydration process. Generally, terpinolene and caryophyllene are considered to be the predominant volatiles in carrots (Hagvall et al. 2008; Kebede et al. 2014). Reyes et al. (2010) reported that peroxidation products resulting from the oxidation of monoterpene terpinolene and autoxidation of caryophyllene degrade rapidly even at room temperature and/or rather severe conditions, leading to the formation of low molecular weight compounds.

Besides terpinolene and caryophyllene, several other key volatiles also participate in imparting characteristic aroma to carrots, and these undergo several changes during dehydration (Table 14.2). γ-Terpinene imparting turpentine odor to carrots increased considerably by about 106% in FD, while it decreased by 70% in HAD; the justification for this variation is similar to the compound terpinolene that also had the same behavior of an increase in FD and decrease in HAD carrots. (*Z*)-β-Farnesene, which possesses a fruity/green odor, decreased in both the samples (89.5% and 86.7% in FD and HAD, respectively). The "spicy" note of β-myrcene and (*E*)-caryophyllene showed a marked decrease of 84.4% and 71.8% in HAD, while in FD carrots, the decrease was 31.8% and 98.8%, respectively. The "balsamic" odor of β-bisabolene decreased almost the same extent (about 86%) in both processes of drying. In a similar way, other terpenes such as camphene (camphor odor) and limonene experienced decrease in their volatile contents after HAD and FD processes (Table 14.2). Further, it could be observed from the table that terpenes such as *o*-cymene and γ-himachalene were found only upon subjecting carrots to HAD and FD, while these were totally absent in the fresh ones. On the contrary, bicyclogermacrene (74 μg/g), α-acoradiene (21 μg/g), *o*-cymene (20 μg/g), α-gurjunene (11 μg/g), α-ylangene (11 μg/g), patchoulene (1.6 μg/g), α-cedrene (8 μg/g), (*E*)-β-ocimene (5 μg/g), and α-copaene (2 μg/g) existed only in fresh carrots, whereas, α-terpinene (8 μg/g), (*Z,E*)-α-farnesene (2.2 μg/g), and β-*trans*-ocimene (1.4 μg/g) were present in the FD carrots (Table 14.2). There is very little information on biosynthesis of terpene compounds in carrots, and there is no information available on volatile constituents in dehydrated products. Thus, the work reported by Rajkumar et al. (2016) revealed the main changes in aroma constituents when carrot slices were dried in HAD and FD processes, although the details of the mechanism of their formation in these processes are not yet elucidated.

Besides the major differences in compounds in the carrot samples, all possessed a characteristic odor such as green wood, citrus green, terpene-like, green, and fruity, which indicates that they exhibit the typical carrot odor contributing to consumer's preference. The presence of patchoulene exhibiting spicy odor in fresh carrots (1.65 μg/g) was reported for the first time by Rajkumar et al. (2016). It was absent in HAD and FD samples, which may indicate that the compound remains unstable during the dehydration process. It has been reported that odor sensations of carrots are divided into three distinct notes, namely, "carrot top," "fruity," and "spicy woody." α-Pinene, β-pinene, sabinene, α-phellandrene, β-myrcene, and *p*-cymene correspond to characteristic "carrot top" odor, whereas limonene, γ-terpinene, and terpinolene correspond to "fruity" notes, and β-caryophyllene, α-humulene, β-bisabolene, and (*E*)- and (*Z*)-γ-bisabolene were found to have "spicy" and "woody" notes (Alasalvar et al. 1999; Kjeldsen et al. 2003; Duan and Barringer 2011). Taking all compounds into consideration, the terpenoids of FD carrots retained several of the abovementioned characteristic aromas of fresh carrots when compared to HAD ones. Furthermore, the presence of some new terpenes such as β-*trans*-ocimene, neo-allo-ocimene, α-ylangene, α-cedrene, α-bisabolene, patchoulene, β-sesquiphellandrene, and bicyclogermacrene were also reported by Rajkumar et al. (2016).

14.6.2 Alcohols

The presence of alcohols in fresh, HAD, and FD carrots is unique and contribute partly to the aroma of the carrot. As can be seen from Table 14.2, 1-octanol and *n*-nonanol, which characterize aromatic and green odor, respectively, were detected exclusively in fresh carrots. On the other hand, ethanol, 2-methyl-1-butanol, *n*-hexanol, 1-octen-3-ol, (*Z*)-2-octen-1-ol, *n*-nonanol, and *p*-mentha-1,8-dien-6-ol were observed in HAD carrots as well. Even though the HAD sample showed some production of ethanol (8 μg/g), there was no formation of acetic acid, which indicates absence of off-flavors (Perera 2005). Further, the cumulative effects of all the other alcohols in the HAD sample could contribute to aromatic, grassy, earthy, and fruity odors, all depicting the characteristic aroma of fresh carrots. In the case of FD samples, 1-octen-3-ol, 2-(*E*)-nonenol, borneol, and *p*-cymen-8-ol were detected, and these characterized the presence of cucumber, green, camphor, and celery notes, offering an overall characteristic of fresh carrots.

Carrots contain 0.4%–0.8% of essential oil, which contains linoleic acid, and this on LOX oxidation leads to the formation of aldehydes, which consequently on the action of alcohol oxidoreductase form alcohols such as hexanol, *cis*-3-hexenol, *trans*-2-hexenol, etc. Alcohols present in fresh carrots such as 1-octanol, nonanol, and menthol at lower concentrations tend to evaporate on dehydration or form some small and low-molecular-weight alcohols. Thus, there is a wide variation in the concentration of alcohols found in fresh and dehydrated carrots. Moreover, the characteristic form of dehydration affects differently in alcohol concentration as in HAD dried carrots it was 8.084 μg/g while in FD carrots it was totally absent (Table 14.2).

14.6.3 Aldehydes and Ketones

Among the aldehydes formed, hexanal and 2-(*E*)-nonenal were reported to be the predominant components for the aroma of carrots (Vervoort et al. 2013). In general, the concentration of most of the aldehydes decreased the dehydration of carrots. However, the content of hexanal present in the fresh carrots (8.5 μg/g) was found to increased by about 94% after subjecting the carrots to HAD, while it was decreased by 77% in the FD process (Table 14.2). The explanation for this change is interpreted to the degradation of carotenes present in carrot since the contents of heptanal decreased by 56% in HAD and by 45% in FD samples. (*E*)-2-nonenal decreased by 62.13% in HAD, while it was completely absent in FD samples (Table 14.2). Soria et al. (2008) observed an increase in percentage composition of hexanal in dehydrated carrots; however, the reason for the increase in its concentration is not yet known. The contents of 2-nonenal in carrots were significantly ($p < .05$) more after processing in high pressure and high temperature when compared to their thermal equivalent by Grauwet et al. (2015). Furthermore, from the data presented in Table 14.2, β-cyclocitral was present in all the carrot samples, and its content decreased to a higher extent (47%) in HAD than in FD (33%) when compared to the fresh ones (4.3 μg/g).

Among the ketones present in fresh carrot was 4-methyl-5-hepten-2-one, which is considered to be a prominent aroma impact compound characterizing to fruity and musty odor in carrot; there was very little change in its concentration among fresh and dehydrated products. α-Ionone and (*E*)-β-ionone, which are formed from carotenoids, were present only in HAD and FD samples and were not seen in fresh samples. This reveals the degradation of various carotenoid compounds such as β-carotene, α-carotene, and lutein on dehydration of carrots, and these results are in agreement with the earlier findings of Vervoort et al. (2013) who reported that β-cyclocitral and 4-methyl-5-hepten-2-one are some of the carotenoid degradation products that are increasingly formed during thermal processing. All these above-mentioned compounds have been reported to be major contributors of off-odors in carrot products (Alasalvar et al. 1999). However, it is to be noted that most of these compounds were detected in low concentrations. Therefore, even though HAD and FD carrot samples showed traces of off-flavors, the overall effect of all other compounds contributes to retain the characteristic aroma of carrot. With regard to acetate, bornyl acetate loss in FD carrot samples (71%) is lower when compared to HAD carrot samples (91%). Thus, it was observed that the effect of volatiles after HAD and FD experienced various changes. Certain characteristic volatiles increased considerably; however, at the same time, certain volatiles such as limonene, terpinolene, (*E*)-β-farnesene, α-copaene, and several others were lost during HAD (Table 14.2).

Even though the content of certain key volatiles after FD was found to be slightly lower than HAD, FD retained most of the characteristic carrot aroma compounds. This could be attributed to an early report of Chiewchan et al. (2015), which stated that not only the total amounts of volatiles synthesized should be considered, sometimes presence of even small amounts of specific volatiles contribute greatly to the characteristic aroma of products. However, certain volatiles such as estragole, propanol, undecane, valencene, santalene, terpinyl acetate, (+)-cuparene, eugenol, and elemicin reported in carrots were not detected earlier by Rajkumar et al. (2016). This could be due to sample variations and differences in the method of volatile extractions.

Fukuda et al. (2013) reported the volatile profile of 12 carrot varieties, including 7 Kuroda (fresh) and 5 Flakee (processing). β-myrcene, terpinolene, sabinene, and 1,3,8-*p*-menthatriene were detected as odor-active volatiles by GC-olfactometry. Thymol methyl ether and caryophyllenes, especially caryophyllene oxide, were representative volatiles in Kuroda.

14.7 CONCLUSIONS

This chapter reviews the recent scientific information available on the aroma aspects of fresh and processed vegetables. There is very less work published on characteristic aroma impact compounds in vegetables and, in general, the data are quite limited. The various analytical techniques used for the isolation and identification of volatile compounds in vegetables are discussed in this chapter. The principal pathways for generating volatile organic compounds in vegetables and dried products are discussed in this chapter. Moreover, it updates the current knowledge on aroma of various vegetables, and finally it presents the changes in aromatic profile of carrots subjected to the most prominent dehydration methods such as the conventional air- and freeze-drying. Some prominent compounds contributing to the characteristic aroma of carrots are monitored and their effect on dried products are evaluated.

REFERENCES

African Postharvest Losses Information System. 2012. Estimated post harvest losses (%) 2003–2011. www.aphlis.net/index.php? form=losses_estimates (accessed February 10, 2012).

Alasalvar, C., Grigor, J. M., and Quantick, P. C. 1999. Method for the static headspace analysis of carrot volatiles. *Food Chemistry* 65: 391–397.

An, K., Zhao, D., Wang, Z., Wu, J., Xu, Y., and Xiao, G. 2015. Comparison of different drying methods on Chinese ginger (*Zingiber officinale* Roscoe): Changes in volatiles, chemical profile, antioxidant properties, and microstructure. *Food Chemistry* 197: 1292–1300.

Augusto, F., Lopes, A. L., and Zini, C. A. 2003. Sampling and sample preparation for analysis of aroma and fragrances. *Trends in Analytical Chemistry* 22: 160–169.

Baker, C. G. J. 2005. Energy efficiency in drying. *Stewart Postharvest Review* 1(4): 8–12.

Baltussen, E., Sandra, P., David, F., and Cramers, C. 1999. Stir bar sorptive extraction (SBSE), a novel extraction technique for aqueous samples: Theory and principles. *Journal of Microcolumn Separations* 11(10): 737–747.

Banerjee, A., Variyar, P. S., Chatterjee, S., and Sharma, A. 2014. Effect of post-harvest radiation processing and storage on the volatile oil composition and glucosinolate profile of cabbage. *Food Chemistry* 151: 22–30.

Barbosa, C., Alves, M. R., Rocha, S., and Oliveira, M. B. P. P. 2016. Modified atmosphere packaging of precooked vegetables: Effect on physicochemical properties and sensory quality. *Food Chemistry* 194: 391–398.

Barriada-Pereira, M., Serôdio, P., González-Castro, M. J., and Nogueira, J. M. F. 2010. Determination of organochlorine pesticides in vegetable matrices by stir bar sorptive extraction with liquid desorption and large volume injection-gas chromatography–mass spectrometry towards compliance with European Union directives. *Journal of Chromatography A* 1217(1): 119–126.

Bogusz Junior, S., Tavares, A. M., Filho, J. T., Zini, C. A., and Godoy, H. T. 2012. Analysis of the volatile compounds of Brazilian chilli peppers (*Capsicum* spp.) at two stages of maturity by solid phase micro-extraction and gas chromatography-mass spectrometry. *Food Research International* 48(1): 98–107.

Buttery, R. G. 1996. Quantitative and sensory aspects of flavor of tomato and other vegetable and fruits. In Teranishi, R., Flath, R. A., and Sugisawa, H (eds). *Flavor Science*. New York: Marcel Dekker. pp. 259–286.

Buttery, R. G. and Takeoka, G. R. 2013. Cooked carrot volatiles. AEDA and odor activity comparisons. Identification of linden ether as an important aroma component. *Journal of Agricultural and Food Chemistry* 38: 9063–9066.

Buttery, R. G., Teranishi, R., Flath, R. A., and Ling, L. C. 1990. Identification of additional tomato paste volatiles. *Journal of Agricultural and Food Chemistry* 38: 792–795.

Castoldi, M., Zotarelli, M. F., Durigon, A., Carciofi, B. A. M., and Laurindo, J. B. 2015. Production of tomato powder by refractance window drying. *Drying Technology* 33: 1463–1473.

Chena, L. and Opera, L. U. 2013. Approaches to analysis and modeling texture in fresh and processed foods—A review. *Journal of Food Engineering* 119: 497–507.

Chiewchan, N., Mujumdar, A. S., and Devahastin, S. 2015. Application of drying technology to control aflatoxins in foods and feeds: A review. *Drying Technology* 33: 1700–1707.

Chiewchan, N., Praphraiphetch, C., and Devahastin, S. 2010. Effect of pretreatment on surface topographical features of vegetables during drying. *Journal of Food Engineering* 101: 41–48.

Christensen, L. P., Edelenbos, M., and Kreutzmann, S. 2007. Fruits and vegetables of moderate climate. In Berger, R. G. (ed.). *Flavours and Fragrances: Chemistry, Bioprocessing and Sustainability*. Berlin, Germany: Springer. pp. 135–187.

Civille, G. V. 1991. Food quality: Consumer acceptance and sensory attributes. *Journal of Food Quality* 14: 1–8.

Cozzolino, R., Martignetti, A., Pellicano, M. P., Stocchero, M., Cefola, M., Pace, B., and De Giulio, B. 2016. Characterisation of volatile profile and sensory analysis of fresh-cut "Radicchio di Chioggia" stored in air or modified atmosphere. *Food Chemistry* 192: 603–611.

Dallüge, J., Beens, J., and Brinkman, U. A. T. 2003. Comprehensive two-dimensional gas chromatography: A powerful and versatile analytical tool. *Journal of Chromatography A* 1000: 69–108.

Dev, S. R. S., Geetha, P., Orsat, V., Gariépy, Y., and Raghavan, G. S. V. 2011. Effects of microwave-assisted hot air drying and conventional hot air drying on the drying kinetics, color, rehydration, and volatiles of *Moringa oleifera*. *Drying Technology* 29(12): 1452–1458.

Dev, S. R. S. and Raghavan, V. G. S. 2012. Advancements in drying techniques for food, fiber, and fuel. *Drying Technology* 30: 1147–1159.

Deza-Durand, K. M. and Petersen, M. A. 2014. Volatile compounds of modified atmosphere packaged cut iceberg lettuce: Effect of extremely low O_2, season, cultivar and storage time. *Food Research International* 62: 254–261.

Du, X., Song, M., Baldwin, E., and Rouseff, R. 2015. Identification of sulphur volatiles and GC-olfactometry aroma profiling in two fresh tomato cultivars. *Food Chemistry* 171: 306–314.

Duan, H. and Barringer, S. A. 2011. Changes in furan and other volatile compounds in sliced carrot during air-drying. *Journal of Food Processing and Preservation* 36: 46–54.

Duan, J., Li, X., Yu, C., and Hu, B. 2009. Headspace stir bar sorptive extraction combined with GC-ICP-MS for the speciation of dimethylselenide and dimethyldiselenide in biological samples. *Journal of Analytical Atomic Spectrometry* 24(3): 297–303.

Duan, X., Zhang, M., and Mujumdar, A. S. 2007. Studies on the microwave freeze drying technique and sterilization characteristics of cabbage. *Drying Technology* 25: 1725–1731.

Edelenbos, M., Balasubramaniam, M., and Pedersen, H. T. 2010. Effects of minimal processing and packaging on volatile compounds and other sensory aspects in carrots. *Acta Horticulturae* 876: 269–278.

Fisher, C. and Scott, T. R. 1997. Flavour compounds. In Fisher, C. and Scott, T. R (eds). *Food Flavours: Biology and Chemistry*. Cambridge: The Royal Society of Chemistry. pp. 15–55.

Fitz, W., Kerler, J., and Weenen, H. 1999. Lipid derived flavours. In Swift, K. A. D. (ed.). *Current Topics in Flavours and Fragrances*. Dordrecht, the Netherlands: Kluwer Academic Publishers. pp. 171–214.

Franco, M. R. B. and Janzantti, N. S. 2004. Avanços na metodologia instrumental da pesquisa do sabor. In Franco, M. R. B. (ed.). *Aroma e Sabor de Alimentos: Temas Atuais*. São Paulo, Brazil: Varela. pp. 17–27.

Freire, M. T. A., Bottoli, C. B. G., Fabris, S., and Reyes, F. G. R. 2008. Contaminantesvoláteisprovenientes de embalagensplásticas: Desenvolvimento e validação de métodosanalíticos. *Química Nova* 31(6): 1522–1532.

Fukuda, T., Okazaki, K., and Shinano, T. 2013. Aroma characteristic and volatile profiling of carrot varieties and quantitative role of terpenoid compounds for carrot sensory attributes. *Journal of Food Science* 78: S1800–S1806.

Fukuda, T., Tanaka, H., Ihori, H., Okazaki, K., Shinano, T., and Fukumori, T. 2013. Identification of important volatiles in fresh and processing carrot varieties: Using Kuroda and Flakee types. *Food Science Technology Research* 19(3): 497–504.

Gaware, T. J., Sutar, N., and Thorat, B. N. 2010. Drying of tomato using different methods: Comparison of dehydration and rehydration kinetics. *Drying Technology* 28(5): 651–658.

George, S. D., Cenkowski, S., and Muir, W. E. 2004. A review of drying technologies for the preservation of nutritional compounds in waxy skinned fruit. *North Central ASAE/CSAE Conference*, Winnipeg, Manitoba, Canada, September, 24–25, MB 04–104.

Grauwet, T., Kebede, B. T., Delgado, R. M., Lemmens, L., Manzoni, F., Vervoort, L., Hendrickx, M., Elmore, J. S., and Loey, A. V. 2015. Evaluating the potential of high pressure high temperature and thermal processing on volatile compounds, nutritional and structural properties of orange and yellow carrots. *European Journal of Food Research and Technology* 240: 183–198.

Grob, R. L. and Kaiser, M. A. 2004. Qualitative and quantitative analysis by gas chromatography. In Grob, R. L. and Berry, E. F. (eds). *Modern Practice of Gas Chromatography*. New York: John Wiley & Sons. pp. 403–460.

Gupta, S., Gowri, B. S., Jyothi Lakshmi, A., and Prakash, J. 2013. Retention of nutrients in green leafy vegetables on dehydration. *Journal of Food Science and Technology* 50: 918–925.

Gustavsson, J., Cederberg, C., Sonesson, U., van Otterdijk, R., and Meybeck, A. 2011. *Global Food Losses and Food Waste: Extent, Causes and Prevention*. Rome: Food and Agriculture Organization. www.fao.org/fileadmin/user_upload/ags/publications/GFL_web.pdf (accessed February 10, 2012).

Hagvall, L., Sköld, M., Bråred-Christensson, J., Börje, A., and Karlberg, A. T. 2008. Lavender oil lacks natural protection against autoxidation, forming strong contact allergens on air exposure. *Contact Dermatitis* 59: 143–150.

Hong, E. and Kim, G. 2013. GC-MS analysis of the extracts from Korean cabbage (*Brassica campestris* L. ssp. *Pekinensis*) and its seeds. *Preventive Nutrition and Food Science* 18: 218–221.

Jorge, A., Kubaski, E. T., Colmenero, J. C., Sequinel, T., Sequinel, R., and Tebcherani, S. M. 2015. Analytic hierarchy process applied to the choice of a long-life tomato (*Lycopersiconesculentum* Mill.) drying system. *Drying Technology* 33(10): 1180–1187.

Kebede, B. T., Grauwet, T., Magpusao, J., Palmers, S., Michiels, C., Hendrickx, M., and Van Loey, A. 2015. Chemical changes of thermally sterilized broccoli puree during shelf-life: Investigation of the volatile fraction by fingerprinting-kinetics. *Food Research International* 67: 264–271.

Kebede, B. T., Grauwet, T., Mutsokoti, L., Palmers, S., Vervoort, L., Hendrickx, M., and Van Loey, A. 2014. Comparing the impact of high pressure high temperature and thermal sterilization on the volatile fingerprint of onion, potato, pumpkin and red beet. *Food Research International* 56: 218–225.

Kjeldsen, F., Christensen, L. P., and Edelenbos, M. 2003. Changes in volatile compounds of carrots (*Daucuscarota* L.) during refrigerated and frozen storage. *Journal of Agricultural and Food Chemistry* 51: 5400–5407.

Kreutzmann, S., Thybo, A. K., Edelenbos, M., and Christensen, L. P. 2008. The role of volatile compounds on aroma and flavor perception in colored raw carrot genotypes. *International Journal of Food Science and Technology* 43: 1619–1627.

Lancas, F. M., Queiroz, M. E. C., Grossi, P., and Olivares. I. R. B. 2009. Recent developments and applications of stir bar sorptive extraction. *Journal of Separation Science* 32: 813–824.

Lekcharoenkul, P., Tanongkankit, Y., Chiewchan, N., and Devahastin, S. 2014. Enhancement of sulforaphane content in cabbage outer leaves using hybrid drying technique and stepwise change of drying temperature. *Journal of Food Engineering* 122: 56–61.

Li, Z., Raghavan, G. S. V., and Wang, N. 2010. Carrot volatiles monitoring and control in microwave drying. *LWT-Food Science and Technology* 43: 291–297.

Lonchamp, J., Barry-Ryan, C., and Devereux, M. 2009. Identification of volatile quality markers of ready-to-use lettuce and cabbages. *Food Research International* 42: 1077–1086.

Mahmoud, S. S. and Croteau, R. B. 2002. Strategies for transgenic manipulation of monoterpene biosynthesis in plants. *Trends in Plant Science* 7(8): 366–373.

Mayora, L., Moreirab, R., and Sereno, A. M. 2011. Shrinkage, density, porosity and shape changes during dehydration of pumpkin (*Cucurbitapepo* L.) fruits. *Journal of Food Engineering* 103: 29–37.

Mdziniso, P., Hinds, M. J., Bellmer, D. D., Brown, B., and Payton, M. E. 2006. Physical quality and carotene content of solar-dried green leafy and yellow succulent vegetables. *Plant Foods for Human Nutrition* 61: 13–21.

Melo, A., Aguiar, A., Mansilha, C., and Pinho, O. F. 2012. Monitoring pesticide residues in greenhouse tomato by combining acetonitrile-based extraction with dispersive liquid–liquid microextraction followed by gas-chromatography–mass spectrometry. *Food Chemistry* 130: 1090–1097.

Montoya-Ballesteros, L. C., González-León, A., García-Alvarado, M. A., and Rodríguez-Jimenes, G. C. 2014. Bioactive compounds during drying of chili peppers. *Drying Technology* 32: 1486–1499.

Mulet, A., García-Reverter, J., Bon, J., and Berna, A. 2000. Effect of shape on potato and cauliflower shrinkage during drying. *Drying Technology* 18(6): 1201–1219.

Narain, N., Galvão, M. S., Santana, K. L., and Moreira, J. J. S. 2010. Volatile compounds in tomato-based dried products. *Drying Technology* 28: 232–239.

Ong, P. K. C. and Liu, S. Q. 2010. Flavor and sensory characteristics of vegetables. In Sinha, N. K. (ed.). *Handbook of Vegetables and Vegetable Processing*. Oxford: Wiley-Blackwell. pp. 59–82.

Palzer, S., Dubois, C., and Gianfrancesco, A. 2012. Generation of product structures during drying of food products. *Drying Technology* 30: 97–105.

Pardo-García, A. I., Martínez-Gil, A. M., López-Córcoles, H., Zalacain, A., and Salinas, R. 2013. Effect of eugenol and guaiacol application on tomato aroma composition determined by headspace stir bar sorptive extraction. *Journal of Science Food and Agriculture* 93(5): 1147–1155.

Perera, C. 2005. Selected quality attributes of dried foods. *Drying Technology* 23: 717–730.

Pérez, A. G. and Sanz, C. 2008. Formation of fruit flavour. In Bruckner, B. and Wyllie, S. G. (ed.). *Fruit and Vegetable Flavour*. Boca Raton, FL: Woodhead Publishing. pp. 41–70.

Radovich, T. J. K. 2010. Cabbage flavor. In Hui, Y. H. (ed.). *Handbook of Fruit and Vegetable Flavors*. Hoboken, NJ: John Wiley & Sons. pp. 741–749.

Radovich, T. J. K., Kleinhenz, M. D., Sanchez-Vela, A., Scheerens, J. C., and Schult, B. 2003. Fresh cabbage sensory quality: Components and the impact of production factors. *Acta Horticulturae* 628: 787–795.

Rajkumar, G., Shanmugam, S., Galvâo, M. S., LeiteNeta, M. T. S., Sandes, R. D. D., Mujumdar, A. S., and Narain, N. 2016. Comparative evaluation of physical properties and aroma profile of carrot slices subjected to hot air and freeze drying. *Drying Technology*. doi.org/10.1080/07373937.2016.1206925.

Ramírez, J. R., Lagunas, L. M., Ortiz, A. L., and Torres, S. S. 2012. True density and apparent density during the drying process for vegetables and fruits: A review. *Journal of Food Science* 77: 145–154.

Rankin, S. A. and Bodyfelt, F. W. 1995. Solvent desorption dynamic headspace method for diacetyl and acetoin in buttermilk. *Journal of Food Science* 60(3): 1205–1207.

Reyes, A., Veja, R. V., and Bruna, R. D. 2010. Effect of operating conditions in atmospheric freeze drying of carrot particles in a pulsed fluidized bed. *Drying Technology* 28: 1185–1192.

Rood, D. 1995. *A Practical Guide to the Care, Maintenance and Troubleshooting of Capillary Gas Chromatographic Systems*. 2nd edn. Heidelberg, Germany: HuthigVerlag. pp. 323–325.

Sagar, V. R. and Suresh Kumar, P. 2010. Recent advances in drying and dehydration of fruits and vegetables: A review. *Journal of Food Science and Technology* 47(1): 15–26.

Santos, P. H. S. and Silva, M. A. 2008. Retention of vitamin C in drying processes of fruits and vegetables—A review. *Drying Technology* 26(12): 1421–1437.

Seremet (Ceclu), L., Botez, E., Nistor, O. V., Andronoiu, D. G., and Danut Mocanu, G. 2016. Effect of different drying methods on moisture ratio and rehydration of pumpkin slices. *Food Chemistry* 195: 104–109.

Sette, P. A., Franceschinis, L. E., Schebor, C., and Salvatori, D. 2015. Osmotic dehydrated raspberries: Changes in physical and bioactive compounds. *Drying Technology* 33: 659–670.

Shim, J. E. and Baek, H. H. 2012. Determination of trimethylamine in spinach, cabbage, and lettuce at alkaline pH by headspace solid-phase microextraction. *Journal of Food Science* 77: C1071–C1076.

Shin, L. E. R., Zzaman, W., Kuang, Y. T., and Bhat, R. 2015. Influence of dehydration techniques on physicochemical, antioxidant and microbial qualities of *Ipomoea aquatica* forsk. An underutilized green leafy vegetable. *Journal of Food Processing and Preservation* 39: 1118–1124.

Sides, A., Robards, K., and Helliwell, S. 2000. Developments in extraction techniques and their application to analysis of volatiles in foods. *Trends Analytical Chemistry* 19: 322–329.

Siucińska, K. and Konopacka, D. 2014. Application of ultrasound to modify and improve dried fruit and vegetable tissue: A review. *Drying Technology* 32(11): 1360–1368.

Song, C. F., Cui, Z. W., Jin, G. Y., Mujumdar, A. S., and Yu, J. F. 2015. Effects of four different drying methods on the quality-characteristics of peeled litchis (*Litchi chinensis* Sonn.). *Drying Technology* 33: 583–590.

Song, J., Fan, L., and Beaudry, R. M. 1998. Application of solid phase microextraction, and gas chromatography/time-of-flight mass spectrometry for rapid analysis of flavor volatiles in tomato and strawberry fruits. *Journal of Agricultural and Food Chemistry* 46: 3721–3726.

Soria, A. C., Sanz, J., and Villamiel, M. 2008. Analysis of volatiles in dehydrated carrot samples by solid-phase microextraction followed by GC-MS. *Journal of Separation Science* 31(20): 3548–3555.

Sutar, P. P., Raghavan, G. V. S., Gariépy, Y., Prasad, S., and Trivedi, A. 2012. Optimization of osmotic dehydration of potato cubes under pulsed microwave vacuum environment in ternary solution. *Drying Technology* 30: 1449–1456.

Tatemoto, Y., Haruna, Y., Harazaki, K., and Miyazawa, K. 2014. Evaporation of distilled water in a fluidized bed of various types of fluidizing particles under reduced pressure. *Drying Technology* 32(4): 476–483.

Tatemoto, Y. and Michikoshi, T. 2014. Drying characteristics of carrots immersed in afluidized bed of fluidizing particles under reduced pressure. *Drying Technology* 32(9): 1082–1090.

Tudela, J. A., Marín, A., Martínez-Sánchez, A., Luna, M. C., and Gil, M. I. 2013. Preharvest and postharvest factors related to off-odours of fresh-cut iceberg lettuce. *Postharvest Biology and Technology* 86: 463–471.

Valero, E., Miranda, E., Sanz, J., and Martínez-Castro, I. 1997. Automatic thermal desorption in the GC analysis of dairy product volatiles. *Chromatographia* 44: 59–64.

Valette, L., Fernandez, X., Poulain, S., Cuvelier, L. L., and Loiseau, A. M. 2006. Chemical composition of the volatile extracts from *Brassica oleracea* L. var. botrytis 'Romanesco' cauliflower seeds. *Flavour and Fragrance Journal* 21: 107–110.

Van Ruth, S. M. and O'Connor, C. H. 2001. Evaluation of three gas chromatography olfactometry methods: Comparison of odour intensity-concentration relationships of eight volatile compounds with sensory headspace data. *Food Chemistry* 74: 341–334.

Verma, A. and Vir Singh, S. 2015. Spray drying of fruit and vegetable juices—A review. *Critical Reviews in Food Science and Nutrition* 55(5): 701–719.

Vervoort, L., Grauwet, T., Njoroge, D. M., Van der Plancken, I., Matser, A., Hendrickx, M., and Loey, A. V. 2013. Comparing thermal and high pressure processing of carrots at different processing intensities by headspace fingerprinting. *Innovative Food Science and Emerging Technologies* 18: 31–42.

Wang, R., Zhang, M., and Mujumdar, A. S. 2010. Effect of food ingredient on microwave freeze drying of instant vegetable soup. *LWT-Food Science and Technology* 43: 1144–1150.

Wang, R., Zhang, M., Mujumdar, A. S., and Cai Sun, J. 2009. Microwave freeze–drying characteristics and sensory quality of instant vegetable soup. *Drying Technology* 27: 962–968.

Wei Cui, Z., Shi-Ying, X., and Da-Wen, S. 2003. Dehydration of garlic slices by combined microwave-vacuum and air drying. *Drying Technology* 21(7): 1173–1184.

Xu, C. H., Chen, G. S., Xiong, Z. H., Fan, Y. X., Wang, X. C., and Liu, Y. 2016. Applications of solid-phase micro extraction in food analysis. *TrAC Trends in Analytical Chemistry* 80: 12–29.

Yang, C., Wang, J., and Li, D. 2013. Micro extraction techniques for the determination of volatile and semi volatile organic compounds from plants: A review. *Analytica Chimica Acta* 799: 8–22.

Section IV

Others (Modeling, Measurements, Packaging, and Safety of Dried Vegetables and Vegetable Products)

15 Vegetable Dryer Modeling

Robert H. Driscoll and George Srzednicki
University of New South Wales

CONTENTS

15.1 INTRODUCTION

15.1.1 VEGETABLES

The daily availability of safe, fresh vegetables is essential to meet people's nutritional needs for healthy living. The annual production of vegetables worldwide exceeded 10^9 tonnes in 2010 (FAO, 2012). People who live on land where they can grow their own produce can be self-sufficient, and if their land area or farm is large enough, they may sell the produce to local people or to inhabitants of towns and cities. When produce is grown, harvested, and consumed on the day of harvesting, there is no need to apply any postharvest technologies. As farms are now located hundreds of kilometres away from the cities, it may take hours or days to transport them to the markets. As most vegetables are highly perishable, once they have been harvested, their nutritional quality and sensory characteristics may rapidly deteriorate. As opposed to grain crops, with very few exceptions, vegetables are harvested prior to the onset of senescence and dormancy, which makes them even more susceptible to deterioration.

Vegetables are mostly composed of water (generally over 80%), with carbohydrates (starch and sugars) being the second most abundant constituents, followed by dietary fibers including celluloses, hemicelluloses, pectin, and lignin. The latter group of compounds is essential for humans as it plays a significant role in the movement of food in the guts. Vegetables are also a major source of vitamins such as vitamin C and provitamin A (β-carotene) and also bioactive compounds, such as polyphenols (e.g., anthocyanins), which may act as antioxidants and thus protect the consumer from cancers or cardiovascular diseases. Last but not least, vegetables are an important source of minerals such as iron, potassium, and calcium.

Because of their importance in human nutrition, the production of vegetables on a commercial scale is constantly increasing, and thus a larger amount is to be harvested, handled, transported, and also preserved from deterioration. Because of their high moisture content after harvest, most vegetables are highly perishable. Although a large proportion of vegetables is consumed fresh, the seasonal nature of the production and the fact that large amounts are harvested at the same time result in the need to extend their shelf-life by using various methods of preservation and converting them into a more stable form. There are various ways of preserving them such as freezing, pickling, canning, or drying.

Drying belongs to the oldest ways of preserving food developed by human civilization. Its purpose is to reduce the moisture content to levels so as to prevent the dried product from deterioration due to microbial and biochemical activity. The safe levels of moisture are defined by the water activity of the dried product (a_w), which should be below 0.6 for storage of most of the dried products. The dried vegetables can be stored at ambient temperature. Other advantages of drying are the reduction in volume and weight of the dried products. This leads to the reduction in packaging, transportation, and storage cost.

Various drying techniques are used including use of high or low temperature, osmotic dehydration, low pressure, ultrasound, use of oxygen-free atmosphere, and many others. The minimization of the energy consumption in drying leads to increased use of renewable energy as a source of heat for dryers operating above ambient temperatures. The selection of the most suitable drying method depends on the type of the biological material, its price, the amount of material to be dried, and also on the capital and operating cost of the drying equipment.

Various parts of plants are consumed as vegetables.

Vegetables do not represent any particular botanical grouping; however, they can be grouped into three main categories: seeds and pods; bulb, roots, and tubers; flowers, buds, stems, and leaves. As the morphology and chemical composition of the above mentioned groups varies, the drying process will also be affected by the category to which a given vegetable belongs.

Plant parts are still alive after harvest and maintain respiration by using stored reserves of energy until, eventually, they senesce or ripen. They can be divided into vegetative and reproductive organs. The derivation of the various vegetables consumed by people is indicated in Figure 15.1.

A. Flower bud. Example: artichoke
B. Stems. Fleshy stems are important vegetables. Examples: bamboo shoots and asparagus.
C. Seeds and seed pods.
    - The seeds are small embryonic plants, usually with some stored food.
    - Formation of seed starts with pollination of the flower. The embryo is developed from the zygote and the seed coat from the integuments of the ovule.
    - After fertilization, the endosperm becomes the food.
    - Plant seeds used for food include rice, legume seeds (peanuts, peas, and beans), corn, and coconut.
    - Some seeds are consumed immature such as sweet corn and peas. Beans and peas (snow peas) are usually eaten when seed pods are fleshy.

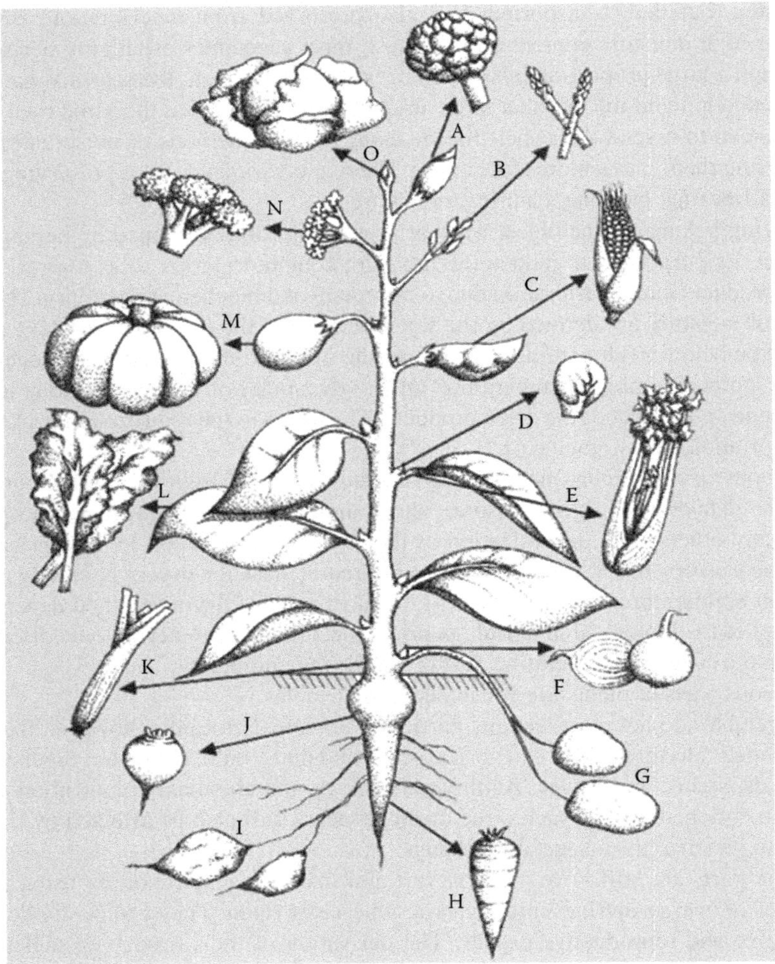

FIGURE 15.1 Plant parts consumed as vegetables (APEC, 2008).

D. Axillary buds. Example: brussels sprout.
E. Petioles. Example: celery.
F. Bulbs. Examples: onion and garlic.
G. Stem tubers and rhizome. Potato is a well-known example of a stem tuber, and ginger is an example of rhizome that is derived from swollen stem tissue.

H, I. Roots.

- The major function of roots is to anchor or support the aerial parts of the plant and to absorb water and inorganic nutrients from the soil.
- In many plant species the tap root functions as a vegetative reproductive organ that can maintain the species through winter or dry conditions until favorable growing conditions returns. These roots store large amounts of carbohydrates and usually have low rates of respiration.

Typical examples are carrots and radishes (H). Sweet potatoes are swollen and are lateral roots (I).

J. Swollen hypocotyls. In some species, swollen hypocotyls function as a storage organ. Examples are beetroot, kohlrabi, and turnip.

K. Swollen leaf bases. Example: leek.

L, O. Leaves and main buds.

- Leaves containing chlorophyll are the main sites of photosynthesis. Leaves are prominent in the human diet as leafy vegetables.
- Plants using leaves for food include leafy herbs, leafy vegetables including cabbages and lettuces.

M. Fruit. Many fruits are consumed as vegetables. These include cucurbits that are harvested immature as well as mature pumpkins and squashes.

N. Swollen inflorescence. Examples: broccoli, cauliflower, and broccolini.

O. Terminal buds. Example: cabbage.

15.1.2 Vegetable Preservation

Drying is a major method of preserving foods (Chen and Mujumdar, 2008). Since harvest tends to occur at particular times of the year for each product, preservation reduces product deterioration, which allows us to consume the product over longer periods. Drying is effective because the reduction in product moisture causes a reduction in the availability of water within the product, slowing down the transport of chemical reactants, reactions that require water and microbiological activity.

There are many other techniques for preservation, but drying has particular advantages:

1. Nothing is added to the product (as happens in brining or pickling).
2. Nothing is extracted from the product, except water. By reducing product water activity, the shelf-life can be extended considerably.
3. The product bulk is reduced, thus reducing transport costs and maximizing storage space.
4. Flavors and aromas tend to get concentrated, so dried foods generally have a more intense taste and smell. Colors may also be intensified.
5. Nutritional content is better preserved, for example, vitamins, fibers, antioxidants, and proteins. Some vitamins such as vitamin C may be damaged by heating during drying, and therefore low-temperature drying techniques are available to reduce susceptible vitamin loss.
6. Drying can be done at very low cost (for example, sun drying, heat pump, and in-store drying) and is a highly developed technology that is well understood.
7. Dried foods can generally be reconstituted by adding back the water removed to a composition similar to the original product. Dried vegetables used in soups may not need reconstituting. Drying vegetables to powders allows combination of the product with other powders, broadening the range of products available.

Many fruits and vegetables are commonly dried. Apart from preservation by reduction of water activity, drying intensifies taste and flavor and generates a range of new states such as sheets, purees, sauces, powders, flakes, pastes, and pulps (Akanbi et al., 2006).

Drying in the sun is possible, but many vegetables deteriorate significantly in the time when they are sun-dried, since they lack the natural preservative effects of acidity or sweetness possessed by foods such as fruits. For any substantial increase in shelf-life, vegetables should be dried to below the glass transition temperature (Ross et al., 1996; Matveev et al., 2000; Moraga et al., 2006), so that they become brittle. The amount of water to cause deterioration will normally be a large percentage of the initial product weight, for example, about 90% for onions. Typically, fresh vegetables are over 90% water and are dried to under 10%.

Fruits and vegetables may need pretreatment before drying, particularly to suppress enzymes that may be activated by exposure to oxygen and may affect product quality. Section 18.5 describes in detail the Various pretreatments, especially blanching, used in the industry. This process may add water to the product, yet conversely reduce drying time by opening up the internal cell structure so that water can escape more easily.

Examples of vegetables that are commonly dried:

1. Carbohydrate-rich foods (often staple foods) such as potatoes, turnips, pumpkin, beans, carrots, and beets
2. Fiber-rich foods such as cauliflower, broccoli, spinach, cabbage, asparagus, rhubarb, and celery
3. Others, such as tomatoes, herbs, onions, and mushrooms

15.1.3 Properties of Air

Since air drying is the most common method of drying vegetables commercially, the properties of air are important for studying drying. Vapor that evaporates from the surface does not dissolve in the air, but forms a chemical mixture where the vapor component and the dry gases components can be considered separately. The term *vapor* is used to describe a gas that is normally a liquid at air temperature. Water as gas in the air is called *water vapor*, but the remaining components present in air (nitrogen, oxygen, etc.) are simply called *gases*.

The study of a mixture of gases and vapor is called *psychrometry*, useful for understanding how air carries water away from the product. The following psychrometric properties are important for drying using heated air.

15.1.3.1 Absolute Humidity

Absolute humidity H is a measure of the air moisture content:

$$H = \frac{m_{\mathrm{V}}}{m_{\mathrm{G}}}. \tag{15.1}$$

This is the mass ratio of water vapor (m_{v}) to mass of dry gases (m_{G}) and is called a dry basis measurement because the denominator refers to the dry gases only, not the total mass of air. Why dry basis is used? The dry basis definition of Equation 15.1 is preferred for drying, because the denominator does not change during drying. The same will be taken into consideration for product moisture also.

Typically, air contains <1% water, so H is of the order 0.01. For this reason, charts of psychrometric properties often express H in terms of g/kg, so that 1% becomes 10 g/kg dry air.

Air properties are conveniently plotted as dry bulb temperature T against absolute humidity H. The resulting chart is called a psychrometric chart.

15.1.3.2 Relative Humidity

As the amount of water in the air increases, the saturation of the air also increases, until at 100% saturation the air cannot hold any more moisture in a stable state. Relative humidity r is defined as the ratio of the air absolute humidity to the air saturation humidity H_S:

$$r = \frac{H}{H_S}. \tag{15.2}$$

Completely dry air has a relative humidity of zero. As relative humidity increases, the capacity of the air to dry decreases.

15.1.3.3 Density

The pressure of each component adds up to the total pressure, typically 1 atmosphere (1 bar or 101325 Pa). By the law of partial pressures,

$$P_T = P_G + P_V, \tag{15.3}$$

where P indicates pressure, and the subscripts indicate total (T), gas (G), and vapor (V) components. Similarly, the number of moles of gas n can be written as

$$n_T = n_G + n_V. \tag{15.4}$$

The ideal gas equation states that

$$n_T = \frac{P_T V}{R T_K}, \tag{15.5}$$

where R is the gas constant, V is the gas volume, and subscript K indicates absolute temperature (degrees Kelvin).

Density ρ is defined as total mass per unit volume. Moles and mass are related by the molecular weight MW, and so we can calculate the density of the air as

$$\rho = \frac{m_T}{V} = (n_G + n_V)\frac{P_T}{R T_K}. \tag{15.6}$$

The molecular weight of water is about 18 kg/kmol and of air is about 28 kg/kmol. We can rewrite this equation in terms of the absolute humidity H. For 1 kg of dry air with H kg of water vapor:

$$\rho = \left(\frac{1}{\mathrm{MW_G}} + \frac{H}{\mathrm{MW_V}} \right) \frac{P_T}{RT_K}. \tag{15.7}$$

15.1.3.4 Enthalpy

Both the dry air and the water vapor contained in the air possess heat, and the total heat content is called as enthalpy. Since enthalpy is heat energy, we can measure enthalpy relative to any standard reference condition that we like. For drying, we choose the following reference conditions:

Dry air	Gas state at 0°C
Water vapor	Liquid state at 0°C

There is a reason for choosing these two reference conditions. We expect the air to remain as gas during drying, but we expect the water to change from liquid to vapor.

Calculating the mixture enthalpy h using the reference conditions given above:

$$h = m_a[c_a T + H(c_v T + \lambda_o)], \tag{15.8}$$

where the three terms come from the following sources:

Dry air	Sensible heat term	$m_a c_a T$
Water vapor	Sensible heat term	$m_a H c_v T$
Latent heat term	$m_a H \lambda_o$	

In Equation 15.8, m_a is the total mass of air, c_a is the specific heat of dry air (about 1.0 kJ/kg·K), c_v is the specific heat of water vapor (about 1.87 kJ/kg·K), and λ is the latent heat of evaporation of water. The latent heat λ_o is evaluated at 0°C and is equal to 2501 kJ/kg. On a psychrometric chart of temperature versus absolute humidity, Equation 15.8 is close to a straight line, with a negative slope.

During drying, the air exchanges heat with the product in such a way that the convection heat coming into the product exactly supplies the heat needed for evaporation. As a result, the enthalpy of the air and the product stays constant during drying. This fact is useful for analyzing hot air dryers.

15.2 THIN-LAYER DRYING CURVES

The main goal of dryer modeling is to determine the drying time of a product. This is done by determining the drying rate as a function of time and integrating over the required change in moisture.

A thin layer placed on a tray in a dryer will dry in a specific way, depending on many factors such as slice thickness and air temperature. To define a thin layer precisely is difficult. If the layer of product is too thick, then not all of the product will dry at the same rate. Some will be obscured from the drying air by other product, and the air will be increasingly humidified as it moves through the product. As a practical definition, a thin layer should be defined as sufficiently thin that the exit air is not significantly different in quality (temperature and humidity) than the inlet air to the layer. "Thin layer" does not mean a single layer of product. An example of a thin layer drying curve is shown in Figure 15.2.

For the following discussion, heated air dryers are specifically considered. These dryers work by heating air, which in turn transfers heat to the product allowing evaporation. Other forms of dryers will be discussed in later sections.

The product moisture content W (wet basis) is defined as

$$W = \frac{\text{Mass of water}}{\text{Total mass of product}}. \tag{15.9}$$

This is the definition most commonly used for commercial measures of moisture. For scientific work, an alternative definition M (dry basis) is used for greater precision:

$$M = \frac{\text{Mass of water}}{\text{Dry solids content}}. \tag{15.10}$$

The symbols W and M will be used to distinguish between the two definitions. For a practical manual on industrial drying, W is more useful. However, for a discussion on drying models, the second definition is preferred. One reason for this choice can be seen by differentiating the above two equations with respect to time. Let m_w be the mass of water and m_s the mass of nonwater components in a product. Then

$$\frac{dW}{dt} = \frac{d}{dt}\left(\frac{m_w}{m_s + m_w}\right) = \frac{dm_w}{dt}\left(1 - \frac{m_w}{(m_s + m_w)^2}\right), \tag{15.11}$$

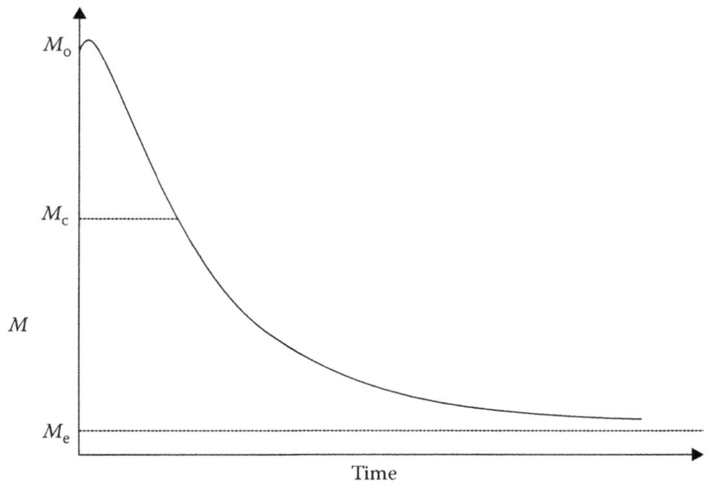

FIGURE 15.2 Thin-layer drying curve.

where t is time. However, when we differentiate the dry basis definition, we obtain a far simpler equation:

$$\frac{dM}{dt} = \frac{1}{m_s}\frac{dm_w}{dt}. \qquad (15.12)$$

The differential dM/dt is the rate of drying. Equation 15.12 (abovementioned) more clearly relates the rate of change of moisture content with the rate of evaporation of water from the product. The drying rate will typically be the highest at the start of drying, dropping to zero as the product approaches the equilibrium moisture content.

Both wet and dry basis definitions are important. To change one form to the other, the following equations may be used:

$$M = \frac{W}{1-W}. \qquad (15.13)$$

$$W = \frac{M}{1+M}. \qquad (15.14)$$

For example, wet basis moisture content (W) of 20% is equivalent to a dry basis moisture content (M) of 25%. This can be written as

$$20\%\ \text{wb} \equiv 25\%\ \text{d}b$$

where "wb" indicates wet basis, "db" indicates dry basis.

Vegetables are typically dried in 3–8 h. The rate at which a product dries is an important characteristic of each product and depends on many factors (such as thickness, cell structure, surface waxiness). Initially water may be close to the product surface and thus easily removable, so the drying rate is highest with fresh product. However, as surface moisture is depleted, the rate of diffusion of moisture through

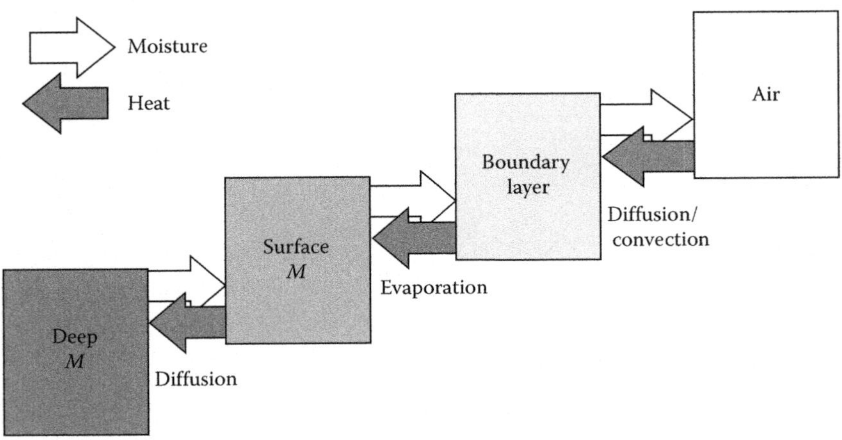

FIGURE 15.3 Moisture and heat transfer during drying.

the product becomes the dominant effect, and the drying rate reduces. As a result, the drying rate varies with time.

If a product was left in the dryer, it would continue to dry until the product moisture level comes to equilibrium with the adjacent air. This point is approached asymptotically, so theoretically is never reached—perfect equilibrium cannot be attained. However, this theoretical equilibrium state is a useful reference condition for studying drying. We define a quantity M_e, called the equilibrium moisture content of the product, as the lowest moisture content that a product can reach for given drying conditions. For example, in humid conditions, the equilibrium moisture content may be very high, and the product will not dry sufficiently to be safe for storage, just as on damp misty days, clothes on a line may not dry sufficiently. In hot dry conditions, drying is far easier.

From this discussion, it is apparent that there is a relationship between the concentration of moisture in a product and the air conditions (see Figure 15.3).

Two parameters are important for defining the air conditions, the air temperature (T_a) and the relative humidity (RH). We can formalize this by the following equation:

$$M_e = f(T_a, \text{RH}). \tag{15.15}$$

This equation is dependent on the product, especially its composition and structure. For historic reasons, Equation 15.15 is called the *isotherm* equation. Each product will have its own unique isotherm equation, relating to its moisture at equilibrium with the storage air conditions. The form of the equation (its shape) will vary, and many different forms are used to describe isotherms for different products.

Of the two parameters on the right hand side of the isotherm equation, relative humidity is far more important than dependence on temperature. So, to achieve a specific final product moisture, we need to be aware of the air conditions, especially the air relative humidity. A mechanical dryer is a means of controlling the air humidity so that adequate drying can be achieved (see Figure 15.4).

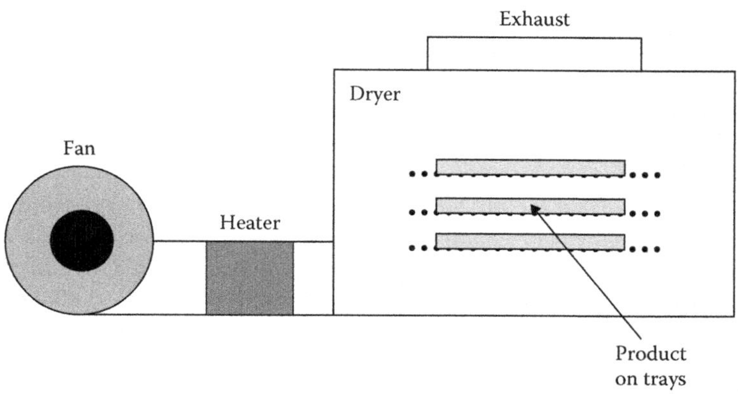

FIGURE 15.4 Concept of hot air dryer.

Note that air humidity primarily affects the final product state, but does not strongly affect the rate of drying. Here, temperature (or more correctly air enthalpy) is more important. Increasing the air temperature affects the drying rate by increasing the internal rate of diffusion, the rate of transfer of heat into the product, the capacity of the air to hold moisture, and reducing the viscosity of the air in the boundary layer around the product.

As a result, many factors affect the drying rate. We try to summarize these factors into a model of the rate of drying, called the *thin layer drying curve*. This is a model of the drying process, and, just like the isotherm equation, is a unique property of each product. In the present section, we are going to study this property.

15.2.1 Important Regions in the Drying Curve

A product drying curve may be plotted by recording (continuously) the mass of the product against time, and then plotting moisture content versus time. To convert the masses to moisture contents, a careful estimate of the final product moisture content is required. This can be done by taking a sample of the product, weighing it, and reweighing after drying completely in an oven (assumed to be at zero relative humidity). A few replicate measurements should be taken. This allows calculation of the product dry mass m_s:

$$m_s = m_f(1 - W_f), \tag{15.16}$$

where m_f is the final product mass, W_f is the final product moisture content, and subscript "f" indicates the final state. Now we can calculate the product moisture from the definition in Equation 15.10:

$$M = \frac{(m - m_s)}{m_s}. \tag{15.17}$$

The resulting plot of moisture versus time should look something like the drying curve in Figure 15.2. A thin-layer drying curve is not a straight line, but approximates an exponential decay curve. Four specific regions can be identified:

1. An initial region, where the product is not in thermal equilibrium with the drying air. In this region, heat transfer between the product and the air has not reached steady state. Depending on the difference in enthalpy between the drying air and the product, there may be an increase or a decrease in surface moisture. For example, if very cold tomatoes are put into warm humid air, moisture may deposit as droplets on the surface. Thermal equilibration is fast compared with moisture equilibration, so normally this region appears as a small anomaly in the shape of the drying curve at time zero, but within a few minutes is no longer apparent. However, when fitting drying models to data, the first 5 min of data should never be included in the curve fitting exercise, since the initial transient is not a true drying effect.

2. A constant rate period, where surface moisture is available and the product dries like a free water surface. The rate of drying will be approximately constant on a plot of dry basis moisture versus time, but will not be constant on a wet basis plot. The effect of product shrinkage may cause this region to deviate from a straight line, since the surface area available for drying may be changing. However, this effect is small, because the removal of free moisture has little structural effect, and so a straight-line approximation is normally used to describe this region.
3. A transitional region, where both surface evaporation and internal diffusion are important. This region is the most complex to model, but is rarely observable as a distinct region, and so is rarely (never?) modeled.
4. A falling rate region, where surface moisture is depleted and internal moisture diffusion controls the rate of drying. This region will be discussed in more detail in the following section.

These regions are not of equal importance. The most important drying region for food products is generally the falling rate period (FRP), although some fruits and vegetables (especially purees) may have significant constant rate periods.

15.2.2 Falling Rate Period

Drying occurs when sufficient heat energy is available to provide the latent heat of evaporation. Heat is transferred from the hot air to the product surface by convection, increasing the energy of surface water molecules which can then escape, taking heat with them as latent heat as they mix with the drying air. For moisture to reach the surface, it must first diffuse through the outer layers until it reaches the surface.

We can think of this as two barriers or resistances in series. The surface acts as a barrier, where the supply of heat, availability of air, and thickness of the boundary layer, all act to restrict the rate of drying. The product interior acts as a second barrier, where the rate of molecular diffusion through the product limits the rate at which water reaches the product surface. The total resistance to drying is the sum of these two resistances.

At the start of drying, the internal resistance due to diffusivity is not important, as moisture is available near the surface. The rate of drying will stay roughly constant. This is the constant rate period. However, after the surface moisture has been partially depleted, the internal diffusion resistance increasingly dominates, and the rate of drying starts to fall. This is the FRP. The rate of internal moisture diffusion depends on the characteristics of the product (especially internal structure and composition) and the local moisture gradient. In addition, as internal moisture becomes depleted, the internal moisture gradient also decreases and so the drying rate decreases. This is summarized by Fick's first law of diffusion:

$$\frac{1}{\rho}\frac{\mathrm{d}m}{\mathrm{d}t} = -DA\frac{\mathrm{d}M}{\mathrm{d}x}, \tag{15.18}$$

where ρ is the product density, D is the product moisture diffusivity, A is area, and x is distance. For convenience, Fick's equation has been stated in terms of mass flows and mass concentrations. Solutions to this equation provide a basis for many drying models. For many products, the approximation can be made that the internal part of the product is the only region of importance, and so drying models often concentrate on this region exclusively.

15.2.3 Summary of Drying Rates

Drying of vegetables can be approximated by

1. A falling rate model only, if the product has low surface water availability (most cases).
2. A constant rate period followed by an FRP (high-moisture products with free water available at the surface initially).
3. More complex models that specifically model both heat and mass transfer and allow inclusion of the initial transient region, constant rate region, and falling rate region in modeling the thin-layer drying rate.

Due to its simplicity, the first option is by far the most common choice for commercial drying calculations.

15.2.4 Pressure Drop

Choosing an appropriate size of fan and burner is important for dryer design. The two sizes are interrelated; to heat double the airflow to the same temperature requires double the heater size. Choosing the fan size is based on generating the correct volume flow of air, so that moisture evaporated from the product can be carried away:

$$\frac{dm_w}{dt} = \dot{m}_a(H_E - H_I), \tag{15.19}$$

where $\dot{m}_a$ is the air flow rate (dot indicates the time differential d/dt), H_E is the exit air absolute humidity, and H_I is the inlet air humidity. Doubling the air flow requires eight times the fan motor size.

To generate sufficient air flow, a fan converts electrical energy into mechanical energy of movement and potential energy (pressure). From Bernoulli's equation:

$$E = \dot{m}_a\left(\frac{1}{2}v^2 + \frac{P}{\rho}\right), \tag{15.20}$$

where E is the fluid mechanical energy, v is the air speed, g is gravitational acceleration, h is height, and P is the air pressure at a point. As the air moves through the dryer, energy is used up by friction with the walls of the dryer, corners, and moving around the product. For constant flow, the pressure energy is gradually depleted. As a result, the fan is designed to provide sufficient pressure energy to allow the dryer to operate at the correct flow rate. Since power increases as the cube of the required air speed, a small error in calculating the required air flow can be expensive, and so correct use of Equation 15.20 is important.

For a deep bed of material, the pressure drop through the bed can be a major component of energy loss. Energy use is dependent on the tortuosity of the path followed by the air as it moves between particles, and for deep-bed drying especially, the pressure generated at different speeds will be an important design factor. At low flow rates, a linear relationship between pressure drop and speed is observed, but at high airflows, the relationship becomes quadratic. Overall, the best description is given by Ergun's equation:

$$\frac{\Delta P}{L} = 150 \frac{(1-\varepsilon)^2}{\varepsilon^3} \frac{\mu}{D_P^2} v_s + 1.75 \frac{(1-\varepsilon)}{\varepsilon^3} \frac{\rho}{D_P} v_s^2, \tag{15.21}$$

where $\Delta P/L$ is the pressure drop within distance through the product bed, ε is the porosity of the bed (a measure of the available space between the particles), μ is the air viscosity, D_p is the particle average diameter, and v_s is the superficial or face speed of the air through the product. Since Ergun's equation was developed for spherical particles and ideal geometries, the constants 150 and 1.75 are usually replaced by values found empirically.

The actual airflow is affected by a large number of factors that may change during drying. For drying of product in-store, the following factors should be considered:

1. The degree of compaction. A product dropped from a low height results in a loose fill with a high porosity. However, over time, the particulates may move, reducing the interstitial volume and thus reducing the airflow through the bulk product.
2. Fine material may block the pore space between particles, reducing airflow. This may especially happen during loading, where fine dust may concentrate near the loading chutes.
3. High-moisture products may be rheologically plastic and deform under high loads, blocking the air path.
4. Insects or microbiological activity may generate detritus, which blocks the airflow.
5. Air introduced to a product store through ducts may not be uniformly distributed, so it may not dry all of the product.

If there are variations in porosity within a bulk store, regions of low porosity will receive very little air, as the pressure increases as the cube of porosity (Equation 15.21). Air will tend to take the path of least resistance and bypass these regions, compromising the whole vegetable store by creating a zone of increased biological activity, sometimes called a "hotspot."

15.2.5 Residence Time

Since a wet product loses moisture faster than a dry product, drying is inherently stable with respect to residence time, and a certain variation in initial moisture content between products is acceptable. This is not always the case, and the exceptions listed below are given as examples:

1. In bulk product stores, dry products must never be loaded on top of wet products. For example, during harvest, loads of products may enter a store having varying levels of moisture, and due to commercial pressures, the dryer operator might be tempted to load the products into an in-store dryer as it arrives. This can lead to a situation where the dried product receives high moisture air from the wet layer, and so is rewetted. The product will then be at microbiological risk, energy use will be poor, and the wetted product may expand and damage the dryer.
2. Many dryers use conveyor belts to carry products from one end of the equipment to its exit. The residence time in this case will be constant, and the entire product will receive the same drying effect. However, other dryers may allow product mixing; for example, mixing rolls between belts in a belt dryer, or the tumbling action of a fluidized bed dryer. These dryers will have a greater variation in residence time, and so some product may be over- or underdried.

The result of variation in residence time is to have products at different moisture contents being stored in the same container. This is not ideal if moisture can migrate from wet to the surface of the dried products, causing regions of high water activity, making the product susceptible to deterioration; for example, mold.

15.2.6 Effects of Main Parameters on Drying Kinetics

On the basis of our discussion so far, it is now possible to consider the main factors that affect the drying rates. For hot air dryers, air properties are important, so are considered first.

15.2.6.1 Air Temperature

The inlet air temperature of a dryer has a significant effect on the drying rate. Temperature affects the drying capacity of the air to hold moisture, and so the moisture from the dryer is removed by thermal advection of the moving fluid (air). The air temperature affects the product temperature, thus affecting the rate of diffusion of moisture through the product.

15.2.6.2 Air Relative Humidity

Relative humidity directly affects the final state of the product leaving the dryer, because a product cannot be dried to a lower moisture content than that corresponding to equilibrium with the air moisture content. Relative humidity also has a weak effect on drying rates, by increasing the driving force difference due to moisture concentration.

15.2.6.3 Air Speed

Most people expect that doubling the airflow will double the drying rate. However, for thin-layer dryers, air speed has less effect than expected. Evaporation can only be fast if the rate of heat supply is fast:

$$\dot{Q} = h_c A(T_s - T_a) = \dot{m}_w \lambda, \tag{15.22}$$

where h_c is the heat transfer coefficient, A is the surface area available for heat transfer, T_s is the surface temperature, T_a is the air temperature, and $\dot{m}_w$ is the rate of evaporation of water (mass per unit time). In effect, convection of heat from the air to the product is the bottleneck restricting the rate of evaporation.

At first glance, Equation 15.22 does not appear to depend on air speed at all. However, increasing air speed affects the boundary layer around a product, making it thinner, and so increasing the heat transfer coefficient h_c. From empirical measurement, the heat transfer coefficient increases as $v_s^{0.6}$, so that to double the heat transfer rate, we would need to triple the air speed, which may be expensive to achieve! Despite this, many dryer operators choose to use very high air speeds to achieve a more uniform final product, sacrificing the cost of the power for improved product uniformity.

Product properties will also be significant. Some key factors affecting drying kinetics are discussed further.

15.2.6.4 Product Composition

Internal structure and composition have strong effects on the rate of drying. These can be summarized in one product property, the moisture diffusivity D. For example, an open porous structure allows a faster rate of diffusion, and high oil content products such as oilseeds tend to have lower rates of diffusion. Diffusivity may vary within the product; for example, the oil-rich bran outer layer of rice acts like a dam wall holding back the internal moisture and preventing it from diffusing easily to the surface. Many products have waxy outer layers that reduce moisture transfer rates.

15.2.6.5 Product Surface

Few products have the geometrically ideal shapes often portrayed in drying texts. Many products have complex convoluted surfaces. Increase in surface area generally has a positive effect on drying rates, by allowing a greater surface for receiving heat from the air, although highly convoluted surfaces tend to have larger air boundary layers trapped at the product surface.

15.2.6.6 Product Thickness

The effect of product thickness has a nonlinear effect on drying rate. For example, for a cylindrical slab, where the thickness d is much less than the radius R, the time required for drying will vary between d and d^2. This effect is due to the distance that must be travelled by water molecules to reach the product surface, and the effect of the boundary layer around the product. For this reason, a thinner product will dry faster.

By the same mechanism, allowing the product to dry from both sides also reduces the drying time, since the effective thickness is halved. This can be done, for example, by using mesh belts for conveying product through the dryer.

15.3 THEORETICAL PREDICTIONS OF DRYING BEHAVIOR OF VEGETABLES

In this section, we will develop the main models used for representing vegetable drying. We should straightaway distinguish between two categories of models: models of the product and models of the dryer. Due to the complexity of the drying process, it is rare to have one model do both in any detail (Driscoll, 2004).

We will start by developing models of the product, by considering how the basic physical processes involved in drying can conveniently be applied to thin layer drying. The physical process can be separated into two main components, drying at the surface and diffusion of moisture within the product.

15.3.1 Empirical Models of Product Drying

Apart from the theoretical and semitheoretical models described before, several empirical models exist. Since the theory of drying is complex, a given model may not include all of the physical principles of importance. For example, a model based on diffusion principles only cannot describe the initial transient region well. So for a particular product, an empirical model may represent the shape of the drying curve more precisely than a theoretical model. In addition, the model might be constructed in a way that allows easy calculation of drying time.

15.3.2 Modeling Drying at Surface

To evaporate moisture from the product surface, heat must first be transferred from the drying air into the product. Heat travels by exchange of molecular vibration from the free air particles through an air boundary layer and then into the product particles. This process is well described by the convection equation based on Newton's law of cooling (see Equation 15.22).

The transferred heat supplies the energy required for evaporation. The rate at which mass leaves the surface can be modeled by an equation exactly analogous to the heat transfer equation:

$$\dot{m}_{\mathrm{w}} = \frac{\mathrm{d}m_{\mathrm{w}}}{\mathrm{d}t} = k_{\mathrm{y}} A \left(H_{\mathrm{s}} - H_{\mathrm{a}} \right), \qquad (15.23)$$

where H_s is the air absolute humidity in a region close to the product surface, H_a is the air absolute humidity, and k_y is called the mass transfer coefficient. Most of the heat transferred goes directly into evaporating moisture, although some may be used to change the sensible heat of the product and to bring the evaporated moisture into thermal equilibrium with the drying air. Assuming steady state (product temperature stays constant):

$$\dot{Q}_{\mathrm{w}} = \dot{m}_{\mathrm{w}}[\lambda + c_{\mathrm{v}}(T_{\mathrm{a}} - T_{\mathrm{s}})], \tag{15.24}$$

where λ is the latent heat of evaporation at the product temperature T and c_v is the specific heat of water vapor. The second term represents the sensible heat gained by the evaporated moisture, and is normally small compared with the latent heat of evaporation.

If we do not assume steady state, then Equation 15.24 should be modified to include sensible heat changes in the product. An overall heat balance for a thin layer of product then gives the rate of product heat gain as

$$\dot{Q} = m_{\mathrm{p}}c_{\mathrm{p}}\frac{\mathrm{d}T_{\mathrm{p}}}{\mathrm{d}t} = -\dot{m}_{\mathrm{a}}(h_{\mathrm{E}} - h_{\mathrm{I}}) = hA(T_{\mathrm{a}} - T_{\mathrm{p}}) - \dot{m}_{\mathrm{w}}[\lambda + c_{\mathrm{v}}(T_{\mathrm{a}} - T_{\mathrm{s}})], \tag{15.25}$$

where subscript p refers to the product. This form of equation allows us to model not just the two main drying periods but the initial transient as well, so is preferred for dryer simulation purposes. The mass transfer coefficient k_y can be conveniently estimated for air/water mixtures by means of the Lewis number, the ratio of thermal to mass diffusivity.

For the constant rate period and for practical purposes, we can write

$$\frac{\mathrm{d}M}{\mathrm{d}t} = -k_{\mathrm{o}}, \tag{15.26}$$

where k_o is the drying rate of a high moisture product. This constant can be determined experimentally from the slope of moisture content versus time in the constant rate region. The negative sign is chosen to reflect our interest in drying of the product. Note that Equation 15.26 must be written in terms of dry basis moisture content, and is not a valid equation if M is replaced by W (wet basis).

Many vegetables shrink substantially during drying, so the surface area of the product will decrease. Since heat transfer and surface evaporation are proportional to product surface area, the drying rate will decrease due to shrinkage. Shrinkage is caused by internal moisture movement by diffusion, and so the product should not change shape until surface water is depleted. For this reason, shrinkage should not occur in the constant rate period, and so will not invalidate Equation 15.26.

15.3.3 Modeling Drying within Product

Once there is no excess free water at the surface, drying becomes controlled by internal moisture movement. Equation 15.18 is the governing equation for this region. By looking at a control volume within the product and equating the mass fluxes in and out with the change in mass within the control volume, we obtain Fick's second law of diffusion:

$$\frac{\mathrm{d}M}{\mathrm{d}t} = -D\frac{\mathrm{d}^2M}{\mathrm{d}x^2}. \tag{15.27}$$

For constant air conditions, uniform initial state and constant product properties, solutions to this equation for ideal shapes (thin slab, infinite cylinder, and sphere) were obtained by Crank (1975), and have the form

$$M(x,t) = a_0 + \sum_{n=1}^{\infty} a_n \cdot f(b_n x) \cdot \exp c_n\, t, \tag{15.28}$$

where a_n, b_n, and c_n are constants, and f is a function such as sine. For example, for a spherical particle under constant drying conditions, and with no shrinkage, the mathematical solution to Fick's second law (Equation 15.27) is

$$\frac{M(x,t) - M_e}{M_o - M_e} = 1 + \frac{2R}{\pi r} \sum_{n=1}^{\infty} \frac{(-1)^n}{n} \sin\left(\frac{n\pi r}{R}\right) \exp\left(\frac{-Dn^2\pi^2 t}{R^2}\right), \tag{15.29}$$

where r is the radius at a point, R is outer radius of sphere, M_e is equilibrium moisture content, M_o is initial moisture content. Integrating over the interior volume for constant drying air conditions, this can be written as

$$\frac{M(t) - M_e}{M_o - M_e} = a_1 \exp(-k_1 t) + a_2 \exp(-k_2 t) + \cdots, \tag{15.30}$$

where a_i and k_i are constants.

Successive terms in Equation 15.30 will have decreasing importance. The first term will be the most important, the second term about half as important (due to the factor of $1/n$ in Equation 15.29), and so on. So, a truncated form of Equation 15.30 should give realistic approximations to drying, within the limits of the abovementioned assumptions.

Consider the case where only the first term is kept. Defining MR as the moisture ratio, as given by the left-hand side of Equation 15.30, then

$$\mathrm{MR} = \frac{M - M_e}{M_o - M_e} = a_1 \exp(-k_1 t) \tag{15.31}$$

is a first term approximation to the solution to Fickian diffusion. Surprisingly, despite all of the assumptions, this turns out to be a reasonable approximation to a thin layer drying curve under constant drying conditions.

We can make a further assumption by ignoring the initial transient effect and considering only internal diffusion. In this case, we would like the average moisture content M at time zero to be exactly the initial moisture content, and with this boundary condition, we obtain $a_1 = 0$. Then

$$\mathrm{MR} = \frac{M - M_e}{M_o - M_e} = \exp(-k_1 t). \tag{15.32}$$

Differentiating this equation with respect to time

$$\frac{dM}{dt} = -k_1(M - M_e), \tag{15.33}$$

which shows that the drying rate is proportional to the driving force, defined as the difference between the current product average moisture content and the equilibrium (final) moisture content. Equations 15.32 and 15.33 are the most commonly used equations for representing thin-layer drying of a product. Equation 15.31 (with the extra constant a_1) is better in most cases, as it compensates for the initial transient when the product initially comes in contact with the drying air. Typically, Equation 15.32 explains about 85% of the variation in moisture content with time, which is adequate for most commercial drying applications. This solution is independent of product shape.

Returning to Equation 15.30, we next consider a two-term truncation, giving

$$\text{MR} = \frac{M(t) - M_e}{M_o - M_e} = a_1\exp(-k_1 t) + a_2\exp(-k_2 t). \tag{15.34}$$

This is commonly called the two-compartment model, since it has been derived previously by considering a product made up of several layers or compartments, each with a separate diffusivity. However, the derivation shown is for a homogeneous material. This model is better at predicting the shape of the thin-layer drying curve, typically accounting for about 95% of the variation in moisture during drying.

For the two-compartment model, it is not advisable to set $a_1 + a_2 = 1$ (which would force the initial moisture content to agree with experimental values), because of the effect of the initial transient. No pure diffusion model can properly include this effect, since it neglects surface resistance.

Equation 15.34, although more precise, is more difficult to use in practice and cannot be explicitly solved for time. Despite this, the two-compartment model of thin-layer drying is very commonly used in the literature, because of its superior shape agreement to the drying curve and consequently better predictive capacity for drying times.

15.3.4 Constant vs. Changing Conditions

It is useful at this point to summarize the models for thin-layer drying so far. We have ignored the initial transient region, since it requires more complex modeling (see Equation 15.25). If the product exhibits a constant rate period, we can model this using Equation 15.26, being careful to discard the first few minutes of data that will be affected by the initial transient. Once the surface has dried, a single or two-term model (Equations 15.32 or 15.34) can be fit to the remaining data. Many products do not exhibit a constant rate period, in which case we can apply the single or two-term models directly to the thin-layer drying data, making curve fitting much easier.

To obtain the thin-layer equations given above, many assumptions were required:

- The initial transient region can be ignored.
- The drying conditions stay constant, particularly the air speed, temperature, and relative humidity.
- The product is homogeneous.
- The initial moisture distribution is uniform.
- Product shape does not change during drying.
- The product's thermophysical properties do not change during drying.
- Diffusion is the only significant mechanism for the FRP.
- A truncated series solution is a good shape approximation to the drying curve.

Whenever we use the models described above, we should be aware of these assumptions and not push the models beyond what they were designed to do. More complex methods of modeling drying exist, which allow some or all of these assumptions to be discarded, and these will be considered in Section 15.3.5.

In this section, we will consider the assumption of constant aeration conditions. This assumption is fine for the laboratory, but is rarely observed in practice. Can the thin-layer drying Equations 15.32 and 15.34 still be used?

For clarity, let us consider the single-term model of Equation 15.32. This was presented in two different forms, in integrated form (Equation 15.32) and in differential form (Equation 15.33). They are the same model, since Equation 15.32 can be obtained from 15.33 by integration over time. To integrate, we require that the rate constant k_1 and the equilibrium moisture content M_e are constants during drying. If we have varying drying conditions, this is not true.

Our conclusion is that for varying conditions, the differential form is valid, but integral forms of thin-layer drying equation are invalid and should not be used.

Equation 15.34 can also be differentiated, giving

$$\frac{dM}{dt} = -(M_o - M_e)[k_1 a_1 \exp(-k_1 t) + k_2 a_2 \exp(-k_2 t)]. \quad (15.35)$$

This form is valid for varying drying conditions, provided that we keep updating the variables as air condition changes.

15.3.4.1 Two-Layer Model

For varying conditions, other forms of thin-layer drying model exist. One form that models the initial transient, surface evaporation, and internal diffusion will be described. The model is relatively new.

The model assumes a product can be represented as an internal layer of mass m_1 and a surface layer of mass m_2. The internal layer does not come directly into contact with the drying air—it is totally buried within the product. The second layer interacts with the air and the "buried" layer. Using a simple linear approximation to Fick's first law,

$$\frac{dM_1}{dt} = -k_1 \frac{m_{s2}}{m_{s1}}(M_1 - M_2), \tag{15.36}$$

$$\frac{dM_2}{dt} = -k_1(M_2 - M_1) - k_e(M_2 - M_e), \tag{15.37}$$

where the average moisture content is

$$M = \frac{m_{s1}M_1 + m_{s2}M_2}{m_{s1} + m_{s2}}. \tag{15.38}$$

In Equations 15.36 through 15.38, k_1 and k_e are rate constants, assumed to exhibit an Arrhenius temperature dependence, m_s indicates dry mass (which is the product mass not including water), and subscripts 1 and 2 refer to the two layers.

Since the model is written in differential form, predictions of drying rate change with changing air conditions. Some initial experiments are required to determine the equation constants. The model can then be used in simulations of drying using changing air conditions such as real weather data.

Integrating Equations 15.36 and 15.37 for constant air conditions, successively eliminating M_1 and M_2, leads to an equation for the overall drying rate. Although the integrated form is not of practical usefulness, it is of interest to compare it with the one- and two-term models described before. We find that the final equation is similar to the two-term model of Equation 15.34, but has one less independent constant.

15.3.5 Finite Element/Finite Difference Models

The thin-layer drying models that we have developed so far have been based on many simplifying assumptions. The approaches described in this section do not need as many assumptions, but are therefore more complex. The two methods described here are called finite difference (FDM) and finite element models (FEM). Both methods are basically approaches for solving the fundamental heat and mass transfer differential equations that describe drying.

The heat and mass transfer equations are also called transport equations. The heat equation is derived by applying Fourier's equation for heat flow to a control volume. The derivation is as follows:

$$\text{Fourier's equation: } \dot{Q} = -kA\frac{dT}{dx}, \tag{15.39}$$

where $\dot{Q}$ is the rate of heat flow. For a one-dimensional cell, heat accumulation in the cell is the difference between the heat entering and leaving:

$$\rho cV\frac{dT}{dt} = \left(kA\frac{dT}{dx}\right)_{\text{left}} - \left(kA\frac{dT}{dx}\right)_{\text{right}}. \tag{15.40}$$

Taking the limit as the element volume drops to zero and extending to three dimensions gives

$$\nabla .\,(k\nabla T) = \rho c\dot{T}. \tag{15.41}$$

The equation for mass transfer can be developed in a similar way from Fick's first law or by analogy to the above heat transfer equation:

$$\nabla .\,(D\nabla M) = \dot{M}. \tag{15.42}$$

Comparing the two equations, $k/\rho c$ defines thermal diffusivity, which is analogous to the mass diffusivity D. The two equations are solved simultaneously across the interior of the product. The surface of the product is the physical and mathematical boundary. As air conditions change, the boundary conditions can also be modified. The boundary equations are provided by Equations 15.22 and 15.23.

Up to this point, the two methods are the same. However, finite difference and finite element methods differ in the way they approach solving the two transport equations.

15.3.5.1 Finite Difference Method

To solve the equations in the interior of the product, the product is approximated by a series of cells. Each cell must have a regular shape. For a long, thin rod, we could break the rod into a series of cells of length Δx and cross-sectional area A, resulting in rectilinear cells. At the center of each cell is positioned a node, and the main state variables (moisture M and temperature T) are specified at these nodes. The method is easiest to use if the product shape has a simple geometry. The cells should have uniform dimensions, except on the boundaries.

The boundary and initial conditions do not need to be uniform. Once the geometry has been specified, the governing equations are reformulated as difference equations. These equations are then programmed into a computer. The computer solves the equations by working from known conditions (such as the initial state of the product and the boundaries of the product) and integrating forward in time.

At the beginning of each time step, the thermophysical properties inside each cell are recalculated. For example, the product-specific heat changes with moisture content, so as moisture diffuses from a cell, the specific heat of the product in that cell must change. An added complexity is that heat and mass must be conserved, so the equations used to recalculate properties must satisfy the conservation requirements.

The heart of the FDM method is the approximation:

$$\frac{\mathrm{d}Y}{\mathrm{d}x} \approx \frac{Y_2 - Y_1}{x_2 - x_1}, \tag{15.43}$$

where Y is a dependent variable and x an independent variable. When applied to two consecutive nodes, this provides a method for the computer to calculate differentials of functions. Second-order derivatives are calculated by applying this method twice. So, for example, for a one-dimensional rod, the heat transport equation can be written as

$$-k\frac{T_{i+1}^{p}-2T_{i}^{p}+T_{i-1}^{p}}{\Delta x^{2}}=\rho c\frac{T_{i}^{p+1}-T_{i}^{p}}{\Delta t}, \qquad (15.44)$$

where Δx is the distance between nodes, Δt is the time step, superscripts represent indexed times, and subscripts represent indexed positions. This equation is a representation of the heat transport equation in a form that a computer can use, with the advantage that a computer can apply this calculation over all cells and at all times, integrating forward to obtain a solution to the drying problem. The final solution will present the moisture and temperature at points within the product at different times.

Boundary cells are handled differently. The equation for boundary nodes is worked out by replacing one of the heat transfer terms in Equation 15.40 with the convection equation, then dividing both sides by the boundary cell volume to obtain an equation similar to Equation 15.44 but involving a convection term.

This method is more powerful than the semiempirical equations described earlier, but still has limitations, especially that the exact product shape cannot be easily represented, the solution may not converge to the correct answer, and numerical inaccuracies may accumulate over each time step, leading the solution away from the correct answer. However, with care, clever time step algorithms and for simple drying situations, this method can give reasonable answers.

15.3.5.2 Finite Element Methods

FEMs use similar methods like FDM, except that the state variables are linearly interpolated over the interior of each cell. This allows both heat and mass balances to be exactly satisfied for each time step, whereas the FDM method does not guarantee exact mass and energy conservation.

Furthermore, the regular boundary shapes are no longer required, and the cells can take arbitrary shapes that very closely match to the true product boundary. Finally, the integrated error over a cell can be determined, giving solutions with greater precision.

The method is more difficult to set up initially, requiring detailed integrations of the interpolated equations for heat and mass transfer. Once the integrations have been done, however, the resulting equations are extremely simple linear equations, predicting new cell values in terms of the previous values. Figure 15.5 is an example of a finite element analysis of a peanut (done by the authors).

15.3.5.3 Summary of Modeling Methods

Models of drying rate allow predictions of drying time. How do we choose which model to apply? Since food products are highly variable, the single-compartment model of 15.31 is adequate for commercial work, where the rate constant k_1 is assumed to have an Arrhenius temperature dependence. Used in differential form, this equation can respond to changing aeration conditions (although in the authors' experience, not very well).

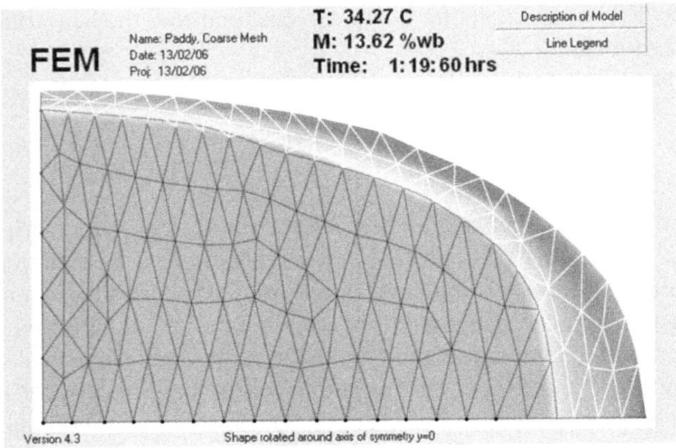

FIGURE 15.5 Finite element model of drying a rice grain under constant conditions.

Few vegetables are dried in deep beds, and so the more sophisticated models described in the text (especially 15.36 through 15.38) are not normally required. However, for theoretical studies, these models should be used. Here are some examples:

1. Two-layer model: useful for deep-bed drying, nonuniform drying, with a simple memory of the drying history of the product (how much it has been wetted or dried previously).
2. FEM: useful for studies of how moisture moves within a product; for example, the effect of oil-rich or low-moisture conduction zones within a product, so that better drying techniques can be developed.
3. FDM models: can be used for individual particles (and in a sense the two-layer model is a form of FDM model). FDM models are also very well suited to modeling complete dryers, even complex dryers with multiple stages and flow conditions (Bala, 1997).

15.4 MODELING OF SPECIFIC DRYERS

15.4.1 Modeling of Dryers

A generic hot air dryer has already been shown in Figure 15.4. Air is brought into the dryer using a fan, the air is heated, and then blown across the product before exiting. The exit air may be recirculated by mixing a portion (about 90%) with fresh inlet air.

15.4.1.1 The Main Equations

A product moves through the dryer at a mass flow of $\dot{m}_P$. The product is exposed to air flowing past the product at a mass flow of $\dot{m}_a$. The two streams interact with each other, and the resulting state of both streams can be predicted by means of four equations for continuous dryers (steady state assumed at all points):

1. Mass balance, specifically of the moisture content. The mass of water lost from the product is the mass of water gained by the air:

$$-\dot{m}_{\mathrm{p}}\frac{\mathrm{d}M}{\mathrm{d}x}=\dot{m}_{\mathrm{a}}\frac{\mathrm{d}H}{\mathrm{d}y}. \tag{15.45}$$

2. Heat balance. The heat gained by the product is the heat lost by the air:

$$\dot{m}_{\mathrm{p}}\left[c_{\mathrm{p}}+c_{\mathrm{w}}M\right]\frac{\mathrm{d}T_{\mathrm{p}}}{\mathrm{d}x}-\frac{\mathrm{d}m_{\mathrm{w}}}{\mathrm{d}t}\lambda_{T_{\mathrm{p}}}=-\dot{m}_{\mathrm{a}}\left[c_{\mathrm{a}}\frac{\mathrm{d}T_{\mathrm{a}}}{\mathrm{d}y}+c_{\mathrm{v}}\frac{\mathrm{d}HT_{\mathrm{a}}}{\mathrm{d}y}\right]. \tag{15.46}$$

3. Mass transfer, again specifically of moisture. This is the thin-layer drying equation described in the previous sections, and is specific to the product being dried:

$$\frac{\mathrm{d}M}{\mathrm{d}t}=f(T_{\mathrm{a}},H_{\mathrm{a}},M_{\mathrm{p}}). \tag{15.47}$$

4. Heat transfer:

$$\frac{\mathrm{d}Q}{\mathrm{d}t}=hA(T_{\mathrm{a}}-T_{\mathrm{s}}), \tag{15.48}$$

where x is the direction of product movement and y is the direction of air movement. These four equations are the fundamental equations for any air drying system. They will simplify a little if the following assumptions can be made:

1. The air and product are in thermal equilibrium (but not at the same temperature), in which

   case $\mathrm{d}Q/\mathrm{d}t=0$.

2. The product is stationary in the dryer.

The form of these equations, as noted before, suggest using FDM methods, so that a computer can do the repeated calculations required to solve the complete set of equations over the length of the dryer. There are other possible methods of solving the equation that are more sophisticated and precise, such as Runge–Kutta algorithms or forward–backward differences, but FDM method is the simplest and will give reasonable precision provided that the time interval for integration is small. Using this method, a differential such as $\mathrm{d}M/\mathrm{d}x$ can be written as

$$\frac{\mathrm{d}M}{\mathrm{d}x}\approx\frac{M_{\mathrm{new}}-M_{\mathrm{old}}}{\Delta t}. \tag{15.49}$$

There are many ways of classifying dryers, and the most important factors are considered in the following paragraphs.

15.4.1.2 Direction of Air Flow

If the dryer uses air to carry moisture from the product, then the four equations described in Section 15.4.1.1 apply, although there are some differences in how we use the equations for specific air dryers.

For air dryers, the direction of air flow is critical. The air flow may be concurrent (in the same direction as the product as in Figure 15.6), counter-current (in the opposite direction to the flow, as in Figure 15.7), crossflow (at right angles to the product flow direction, either across the product or through a supporting mesh), or a combination of all of these.

The counter current configuration is the most energy efficient, but tends to heat the product to higher air temperatures. The concurrent configuration is more gentle, with the product leaving at a lower temperature.

Examples of dryers that do not use air to heat the product include conduction dryers, freeze dryers, and flash (vacuum) dryers.

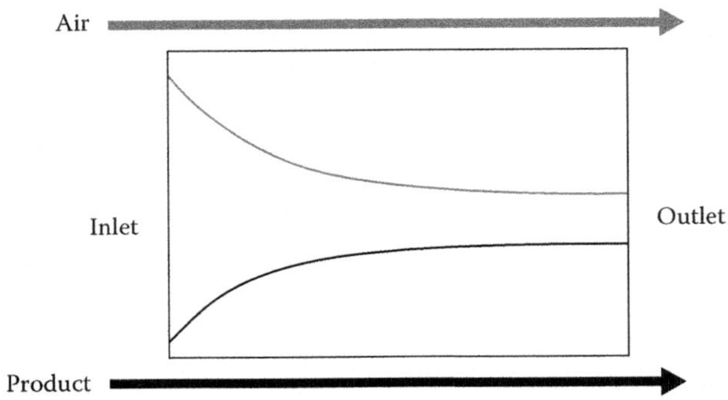

FIGURE 15.6 Temperature profile of product and air: concurrent flow.

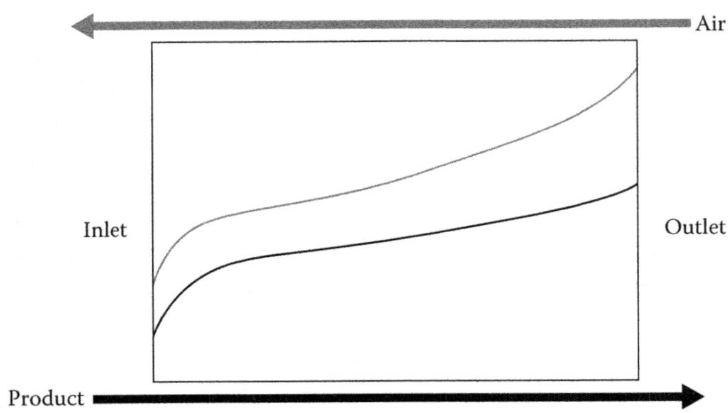

FIGURE 15.7 Temperature profile of product and air: counter-current flow.

15.4.1.3 Batch Dryer or Continuous Dryer?

Is the dryer operated in batch or continuous mode? A batch dryer requires loading, drying, then unloading, whereas a continuous dryer (e.g., a belt dryer) does all three operations simultaneously. There are advantages to both. Typically, a batch dryer is suited to small amounts of products, whereas a continuous dryer allows process control to obtain a more uniform product final moisture. A batch dryer normally has a smaller footprint (takes up less space), but a continuous dryer processes more quickly as unloading and loading times are not needed.

15.4.1.4 Type of Product Used?

The type of product is possibly the most important factor in choosing a suitable dryer. Typically, vegetables are large particulate solids, a fact that reduces the normal variation across food materials, and so makes this choice easier than for other food groups. For example, a tunnel dryer can dry a wide range of vegetables without needing to be modified.

Many other factors are important (direct heat versus indirect, recirculation), but the given four are the most critical for decision making of choice of dryer.

15.4.2 Hot Air Drying

Several forms of hot air dryers exist for vegetables. Generally, these dryers are versatile, suited to a large number of products.

15.4.2.1 Kiln Dryer

The simplest form is the kiln dryer, where heat produced by combustion is funneled into a drying chamber or through a drying bed containing the product. Provided all of the product receives similar air conditions (which can be difficult to achieve in practice), the four equations described earlier can be used once per time interval to develop an unsteady-state model of the kiln dryer. Typically, the integration time interval is chosen to be of the order of minutes.

To solve the differential equations, the air properties are required, and in fact a quick solution to the drying rate can be obtained using a psychrometric chart. Although basic, this solution method is beyond the scope of the current chapter (see, for example, Driscoll, 2004).

15.4.2.2 Tray Dryer

Slices of vegetables can be spread on trays, and the trays placed into a tray dryer (also called a cabinet dryer). The trays may be arranged onto trolleys or trucks, which can be quickly slid in or out of the dryer. The dryer consists of four main sections (see Figure 15.8):

1. Some air is exhausted from the dryer; the remainder enters the drying chamber.
2. After removing some moisture, the air exiting the drying chamber is mixed with fresh inlet air.

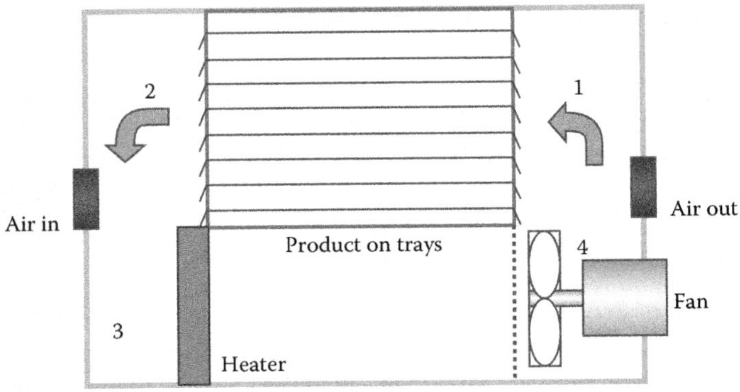

FIGURE 15.8 Tray dryer.

3. The mixed air is heated.
4. A fan provides propulsion energy.

A characteristic of the tray dryer is that they are normally set up to recirculate the air. As the air passes around the four sections, about 10% is exhausted through the outlets, and the remaining 90% is recirculated. This improves the air efficiency by retaining the air enthalpy (Heldman and Hartel, 1995).

To model recirculation, we can start with the kiln analysis but add a mixing stage, to represent 10% fresh air being mixed with the fan inlet air:

$$H_{\mathrm{M}} = R\dot{m}_{\mathrm{a}}H_{\mathrm{a}} + (1-R)\dot{m}_{\mathrm{a}}H_{\mathrm{A}}, \tag{15.50}$$

$$h_{\mathrm{M}} = R\dot{m}_{\mathrm{a}}h_{\mathrm{a}} + (1-R)\dot{m}_{\mathrm{a}}h_{\mathrm{A}}, \tag{15.51}$$

where R is the proportion by mass of air being recirculated, H is the air absolute humidity, h is the air enthalpy, subscripts are a for air and A for ambient conditions.

If the air passes over very long trays, it will tend to lose its drying effect as it cools and picks up moisture from the product, effectively becoming a deep bed. Products at the far end of the tray from the air entering the drying chamber will tend to dry more slowly. This effect can be reduced by increasing the air speed.

Typically, tray dryers are operated with 10 or 20 trays, at airspeeds of 10–30 m/min, and temperatures of 35°C–50°C.

15.4.2.3 Tunnel Dryers

One of the most common methods for drying vegetables is by tunnel drying. The product is loaded onto trays, which are then placed in small trolleys, and each trolley pulled by a linking chain to the next trolley. As each trolley enters the dryer it is exposed to hot air, which blows across the trays. Tunnel dryers are configured as concurrent or counter-current. The air is normally heated by a burner, with scope for supplementary heating of the air from solar collectors on the roof of the dryer.

Analysis of the tunnel dryer is similar to the tray dryer, except that the product is moving, although not all tunnel dryers use air recirculation.

15.4.2.4 Belt Dryers

Replacing the trays and trucks with a continuous belt gives us a belt dryer. The belt may be solid (for example, rubber) or mesh, which allows aeration from above and below. Vegetable belt dryers can be very long, and typically take about 3–4 h to dry product particles of around 7–12 mm thickness. Different stages of the belt dryer can operate at different temperatures and different flow configurations, mixing up cross-flow, concurrent flow, and counter current flow to achieve the best drying profile.

Simulation of belt dryers is useful for determining the best configuration for removing water without exceeding the product maximum allowable temperature. Since the aeration conditions are changing throughout the dryer, the two-layer model is the minimum standard required for modeling the thin-layer drying rate. This model has an inherent memory of moisture movement (the buried layer) allowing more precise prediction of the drying behaviour through successive sections.

Most vegetables can be dried on a belt dryer, either whole or sliced. Breaking rollers can be added between sections to separate components such as onion rings. To reduce the dryer length, belts may be arranged vertically or helically.

15.4.3 Solar Dryers

A solar dryer collects energy transmitted by the sun by radiation. The collection device might be a solar cell or a black surfaced collector. In turn, this energy is passed into air moving through the drying bed. The differences to an air dryer are that the heat source is the sun, the airflow is nonuniform (and close to zero at night), and that the product bed may pick up some direct solar energy.

The flow of air through the dryer may be generated by the temperature difference between the solar-heated air and ambient air, since hot air rises. Equating the energy input to the sensible heat change of the air,

$$\varepsilon \cdot \dot{Q}_{\mathrm{sol}} = \dot{m}_{\mathrm{a}} c_{\mathrm{a}} \Delta T, \tag{15.52}$$

where ε is the collector efficiency. For a given irradiation of the collector, Equation 15.52 shows an important relationship between the air speed v_{a} and the gain in temperature of the air ΔT,

$$v_{\mathrm{a}} \Delta T = \text{constant}.$$

We can design the solar dryer in such a way that we either favor air speed or air temperature rise. For example, a narrow air inlet will generate hotter, slower air flow, whereas a wider inlet will give comparably cooler, faster air. Increasing both is of course ideal, since higher temperatures will increase drying rates and air capacity, and air speed will allow more moisture to be carried away.

Generally with vegetables, oil contents are low, and so the maximum possible drying rate is desirable. This is not true for all food products. Initially, a faster air flow with small heating will give the best result, since the limiting factor for wet product is the air capacity. However, once the surface has dried, higher air temperatures are preferred to increase the rate of internal moisture diffusion and the air capacity. Trials may need to be done to find this optimum point. The prevailing wind direction may be a useful factor to consider. For example, coffee dryers in some countries consist simply of a parabolic arch supporting a plastic sheet spread over the product, allowing the wind open access to the drying bed. A fully enclosed system allows better control, but may not be as effective at using the wind's energy to supplement that from the sun.

Since from Bernoulli's equation (Equation 15.20) the airflow varies as the square root of the height difference between the inlet and the outlet, air speed can be controlled by changing this height. For an enclosed dryer, a draft tube or chimney can be added over the drying bed, and this will increase air speed. A T-junction at the top of the tube will allow the prevailing wind (again!) to induce an increased air flow by the Venturi effect.

Solar dryers use a low density energy source and require careful maintenance to protect collection surfaces, and as such do not have an advantage over spreading the product in the sun unless higher temperatures are required to adequately dry the product. However, they may provide other functions such as security, storage, and protection from contamination that improve their utility.

15.4.4 Fluidized Bed Dryers

Fluidized bed dryers use high air speeds through a granulated product to provide a buoyancy force. As air speeds increase, the drag force between the air and the product increases as the square of the speed, reducing the effective weight until the particles are able to move freely as a fluid (see Figure 15.9). Further increases in air speed will cause entrainment, where the particles are carried away from the drying bed by the air, which is not desirable. The fluidization region is perfect for

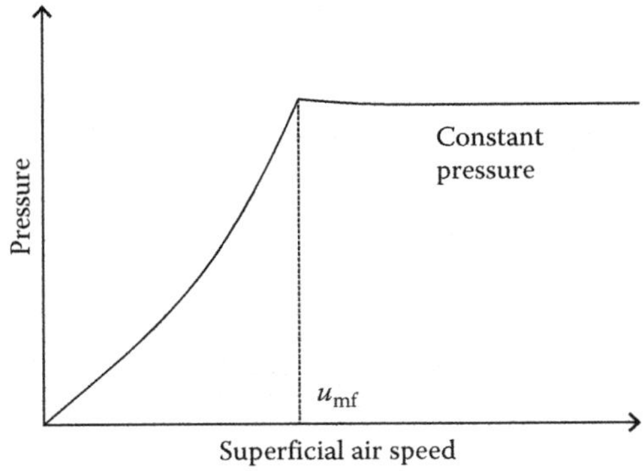

FIGURE 15.9 Pressure drop across fluidized bed.

tumbling the product, exposing all of its surfaces to the drying air and so increasing the drying rate.

Generally, vegetable products are not fluidized directly—they are too large to be fluidized easily, and anyway would be bruised by the tumbling action. However, small particles of vegetables up to the size of peas can be fluidized and this is useful for addition to mixes such as soups.

Since the air speed required for fluidization is relatively large (about 2 m/s), there will normally be excessive air, and the change in air humidity will be smaller than with a conventional hot air dryer, so there is less need to model the air properties. This simplifies the drying analysis to a heat and energy balance across the fluidized layer of the bed.

However, the internal structure of a fluidized bed of products is very distinctive, and the literature contains many attempts to analyse in detail what is happening within this region (Driscoll and Srzednicki, 2014). Under the best drying conditions, bubble structures are formed, with relatively dense particle concentrations between the bubbles. An analysis of the hydrodynamics of this flow requires computational fluid dynamics, which is beyond the scope of this chapter.

Particles suitable for fluidization are categorized according to the Geldart classification (Geldart, 1973):

1. Group A: (20–100 μm), low particle density (<1400 kg/m^3). Bed expands in size as it fluidizes, and then starts bubbling as described earlier.
2. Group B: (40–500 μm), medium particle densities between 1400 to 4000 kg/m^3. Bubbling starts as product fluidizes.
3. Group C: (20–30 μm), very fine and so cohesive particles. These are difficult to fluidize, and so may require vibratory fluidization. An example is drying of whole tea leaves as it is sometimes practiced in India.
4. Group D: (> 600 μm), high density particles, difficult to fluidize.

A fluidized bed can also provide an ideal medium for agglomeration of fine particles, by the addition of moisture by sprays. Fluidization occurs when the weight of the bed is sustained by the drag forces. This can be written in terms of the pressure drop across the bed:

$$\frac{\Delta P}{\Delta L} = (1-\varepsilon)(\rho_s - \rho_f)g, \tag{15.53}$$

where ΔP is pressure drop across the bed, ε is the fluidized bed porosity, ρ_s and ρ_f are the particle and fluid densities, respectively, and g is gravitational acceleration. This equation is useful for fan design for fluidized bed dryers.

15.4.5 Heat Pump Dryers

Drying is energy expensive, since the latent heat of water for evaporation is very high. In all hot air dryers, heat from combustion is added to the air, then transferred to the product, from where it evaporates the moisture into the air and is conveyed out to the environment. To achieve a high energy efficiency, we need to modify this process so that the latent heat of evaporation is not lost from the system. But for this to happen, we need to recondense the water in the air. A heat pump dryer achieves this by supercooling a small fraction of the air, so that condensation reduces the air moisture. This releases the latent heat of condensation, which is then returned to the air using a heat exchanger.

A refrigeration system is used to condense air moisture, then restore the heat. As hot moist air from the dryer passes over the refrigeration evaporator, the air is cooled, and the condensed water passes out of the dryer. The refrigerant is then compressed to a high temperature and pressure, and the heat transferred back to the air in the refrigerator condenser.

To model the heat pump dryer, a representation of refrigeration on a psychrometric chart is required. This can be done in several ways:

1. The air in the heat pump dryer is first split into the drying air and the bypass air. The proportion of drying air is equal to the recirculation rate R, and the bypass air has a proportion $1 - R$ compared with the total air through the fan.
2. A detailed model will start with the refrigerant properties, and assuming a saturation cycle, can then predict conditions in each stage of the refrigeration cycle given any two bits of information about the cycle. This level of modeling will allow predictions of the energy used by the cycle, but requires the refrigerant properties to be modeled.
3. The next step is to measure the evaporator and condenser temperatures, heat transfer coefficients, and areas for both the evaporator and the condenser.
4. Since the refrigerant properties are difficult to model precisely, a shortcut would be to measure conditions in a dryer, specifically the evaporator air exit temperature and the condenser air exit temperature. This still allows system modeling, although the power used by the refrigeration cycle may not be estimated theoretically, but would have to be measured.
5. Whichever method is used, the heat removed by the evaporator can be calculated allowing the temperature of the cooled bypass air to be calculated. Cooling the air below the dew point dehumidifies the air, so we can assume that the air leaving the evaporator is saturated, giving us sufficient information to calculate all of the air properties.
6. The air is then reheated via the refrigeration condenser, and the amount of heat being added calculated from the condenser properties. As a rough guide, the air should return to the same enthalpy as the bypass air entering the refrigeration system, and any reduction below this enthalpy is a loss in efficiency of the cycle. Since this is the addition of dry heat, the air moisture content (humidity) will not change, so a heat balance can be used to calculate the condenser exit air temperature.
7. The air which passes through the drying chamber must also be modeled, but this is an identical modeling exercise to the tray dryer and will not be discussed further here.

8. The bypass air is then mixed with the air exiting the drying chamber.
9. The water heat generated by the circulating fan is a significant contributor to the total enthalpy and must be included. In some cases the only heat input to a heat pump dryer is the waste heat from the fan.

Often a heat pump dryer is insulated, and the closed circuit operation (no fresh air enters the system) makes it ideal for working with a nitrogen atmosphere, helping to prevent product oxidation during drying but also reducing microbiological activity.

The main advantages of a heat pump dryer are

1. High thermal efficiency (typically close to 100% and theoretically in excess of 100% is possible), since the heat required for evaporation is recovered and reused many times (Perera and Rahman, 1997). This gain is offset to some extent by the use of higher cost electric power for the air conditioning system.
2. Gentle drying conditions, since the heat pump dryer can only operate at close to ambient conditions in order to adequately supercool the drying air.
3. Potential combination with other technologies such as microwave (Jia et al., 1993).

The main disadvantages are

1. Since the equipment requires a refrigeration system, and ideally is insulated, the equipment is more expensive than a conventional dryer.
2. The low drying air temperature causes long drying times, during which time the product is held in intermediate temperatures and moistures that may be ideal for microbial growth.

Typically, the conditions in a heat pump dryer (especially 40°C–45°C) are adequate to deactivate enzymes over the longer drying times required.

15.4.6 Spray Dryers

Spray dryers are used for liquids, slurries, and pulps such as vegetable pulps, pastes, juices, and purees. The liquid is pumped (positive displacement, to ensure uniform flow) from a supply tank into the top of the drying chamber. The product is then atomized into fine droplets using either a spinning disk or stationary nozzles. The droplets are allowed to fall through a heated air flow (Lee, 1983; Giovanelli et al., 1995), where the air temperature can be high (over 200°C) since the product is cooled by evaporative cooling to about 100°C. Each spherical droplet first forms a crust of dried material, and then as the internal temperature rises, vaporized water breaks through the crust to form the characteristic broken shell shapes. These shell fragments have a high area to volume ratio, so they rehydrate well.

Examples of vegetables currently spray dried are tomatoes, spinach, rhubarb, and peas. The high fiber content of vegetables tends to clog the spray dryer nozzles easily, so care must be taken to reduce the fiber content or use cleaning air bursts to keep the nozzle clear. Vacuum spray drying may be used for heat-sensitive materials.

The final powder must be free-flowing, and below the glass transition temperature, so additives (such as gums, starches, and maltodextrin) may be added to bind product water and so increase the glass transition moisture content. Otherwise the product will be too cohesive. Little research has been done on spray drying of vegetables (Verma and Singh, 2015), but there is a growing commercial trend to spray dry vegetables to brightly colored powders, increasing storage life and reducing transport costs for drinks when reconstituted.

15.4.7 Sublimation Drying

Sublimation is the process of direct conversion from solid to gas phase. The freeze dryer, also called a lyophilizer, uses sublimation instead of evaporation to remove water from the product (Di Matteo et al., 2003). Sublimation removes about 90% of the water (Liapis et al., 1996) exposed to the drying chamber vacuum, starting at the product surface and moving inward, so the product has a core of high moisture material and an outer layer that is almost dry. For this reason, freeze-dried products maintain good retention of shape, flavor, color, vitamins, and are easily rehydrated. The main components necessary for a freeze dryer are the vacuum chamber and pump, a vapor trap, and heaters for the product shelves (Karel, 1975). The product may first be frozen in an external freezer (e.g., individual quick freezing [IQF] or blast freezing), or simply placed on the drying shelves under vacuum so that sublimation cools the product to the required operating temperature.

The capital cost of a freeze dryer is normally higher than other types, and the operating costs may also be high (Chou and Chua, 2001; Mujumdar, 2001), and hence the freeze dryer is mainly used for high value foods. But combinations of vacuum drying with other technologies such as fluidization and microwaves offer promise.

15.5 CONCLUSIONS

This chapter has described the mechanism of drying of vegetables, ways of modeling the process, and the main dryers used in the industry. Models provide predictions of drying behaviour, allowing testing of various drying strategies and estimation of drying time. Various models are described and discussed with regard to the appropriateness of their use in relation to the type of biological product to be dried and the type of dryer.

Special consideration was given to simulation of changing drying conditions, allowing improved prediction under circumstances where the air conditions in the dryer vary during operation. This may be deliberate (as in belt, tunnel, or instore dryers) or simply due to normal ambient variation. The information provided in this chapter may be useful in design and development of dryers and improvement of processing techniques for vegetables.

REFERENCES

Akanbi, C. T., Adeyemi, R. S., and Ojo, A. 2006. Drying characteristics and sorption isotherm of tomato slices. *Journal of Food Engineering* 73:141–146.

APEC. 2008. *Postharvest: A technology for living produce.* In Kanlayanarat, S. (ed.) CD version. APEC Secretariat, Singapore.

Bala, B. K. 1997. *Drying and Storage of Cereal Grains.* New Delhi, India: Oxford and IBH Publishing Co. Pvt. Ltd. India.

Chen, X. D., and Mujumdar, A. S. 2008. Preface. In Chen, X. D., and Mujumdar, A. S. (eds) *Drying Technologies in Food Processing*, Chichester, UK: Blackwell Publishing Ltd.

Chou, S. K., and Chua, K. J. 2001. New hybrid drying technologies for heat sensitive foodstuffs. *Trends in Food Science and Technology* 12:359–369.

Crank, J. 1975. *The Mathematics of Diffusion*, 2nd edition. Oxford, UK: Oxford University Press.

Di Matteo, P., Donsì, G., and Ferrari, G. 2003. The role of heat and mass transfer phenomena in atmospheric freeze-drying of foods in a fluidized bed. *Journal of Food Engineering* 59:267–275.

Driscoll, R. H. 2004. Food dehydration. In Smith, J. S., and Hui, Y. H. (eds) *Food Processing: Principles and Applications*, 31–44. Oxford, UK: Blackwell Publishing Ltd.

Driscoll, R. H., and Srzednicki, G. 2014. Fluidized bed, spouted bed, and in-store drying of grain, Chapter 10. In Varzakas, T., and Tzia, C. (eds) *Food Engineering Handbook*, 438–487. London: CRC Press.

FAO. 2012. *FAO Statistical Yearbook 2012.* Rome: World Food and Agriculture, FAO.

Geldart, D. (1973). Types of gas fluidization. *Powder Technology* 7(5):285–292.

Giovanelli, G., Zanoni, B., Filkova, I., and Mujumdar, A. S. 1995. Industrial spray drying systems. In Mujumdar, A. S. (ed.) *Handbook of Industrial Drying*, 263–307. New York: Marcel Dekker.

Heldman, D. R., and Hartel, R. W. 1995. *Principles of Food Processing.* Gaithersburg: Aspen Publishers Inc., 196–197.

Jia, X., Clements, S., and Jolly, P. 1993. Study of heat pump assisted microwave drying. *Drying Technology* 11(7):1583–1616.

Karel, M. 1975. Heat and mass transfer in freeze-drying. In Goldblith, S. A., Rey, L., and Rothmayr, W. W. (eds) *Freeze Drying and Advanced Food Technology.* New York: Acadamic Press, 177–202.

Lee, D. A. 1983. Spray drying. In Bhatia, M. V. (ed.) *Transfer Operations in Process Industries*, 55. Lancaster, PA: Technomic.

Liapis, A. I., Pikal, M. J., and Bruttini, R. 1996. Research and development needs and opportunities in freeze-drying. *Drying Technology* 14(6):1265–1300.

Matveev, Y. I., Grinberg, V. Y., and Tolstoguzov, V. B. 2000. The plasticizing effect of water on proteins, polysaccharides and their mixtures. Glassy state of biopolymers, food and seeds. *Food Hydrocolloids* 14:425–437

Moraga, G., Martinez-Navarrete, N., and Chiralt, A. 2006. Water absorption isotherms and isosteric heats of sorption of potatoes. *Journal of Food Engineering* 35:267–280.

Mujumdar, A. S. 2001. Recent developments in the drying technologies for the production of particulate materials. In Levy, A., and Kalman, H. (eds) *Handbook of Conveying and Handling of Particulate Solids*, 533–545. Amsterdam, the Netherlands: Elsevier.

Perera, C. O., and Rahman, M. S. 1997. Heat pump dehumidifier drying of food. *Trends in Food Science and Technology* 8:75–79.

Ross, Y. H., Karel, M., and Kokinin, J. L. 1996. Glass transition in low moisture and frozen foods: Effects on shelf life and quality. *Food Technology* 50:95–108.

Verma, A., and Singh, S. V. 2015. Spray drying of fruit and vegetable juices: A review. *Critical Reviews Food Science and Nutrition* 55:701–719. doi:10.1080/10408398.2012.672939.

16 Numerical Modeling of Morphological Changes of Food Plant Materials during Drying

H. C. P. Karunasena
University of Ruhuna

Wijitha Senadeera
University of Southern Queensland

CONTENTS

16.1 INTRODUCTION

Food plant materials, particularly fruits and vegetables, when undergoing drying are subjected to higher levels of morphological changes, leading to alteration of various physical properties characterizing the dried food product. The main factors driving such morphological changes are the moisture content, drying temperature, atmospheric conditions, rate of moisture removal, and properties of the food plant variety. Prediction of such morphological changes is critical for improving the product quality and processing efficiency in food engineering. In that context, different modeling techniques are being researched, each having its own pros and cons depending on the fundamental nature of the technique and its level of advancement achieved, targeting a given application. Among these modeling techniques, numerical modeling has gained considerable attention since the recent past, and which holds true for the present too. In this background, this chapter initially presents an overview of the different modeling techniques used in the field, and then it specifically presents a novel numerical modeling technique available its key applications, limitations, and future prospects.

16.1.1 Characteristics of the Cellular Structure Existing in Food Plant Materials

Food plant materials are essentially made out of cells aggregated to form tissues. In those plant tissues, cells are mainly composed of two elements: cell fluid and cell wall. Cell fluid usually accounts for about 80%–90% of the whole cell's volume or mass. Cell fluid is mainly occupied by vacuole, which is basically a large storage of watery liquid constituents (Jangam 2011). The cell wall is a thin porous layer mainly composed of microfibers and has viscoelastic properties. The cell wall acts as a boundary to ensure the cell fluid is contained within the cell volume. When the food plant materials are in fresh conditions, cells tend to be in their fully turgid states, having higher turgor pressure values owing to the higher moisture content of the material. As a result, they become large in size and tend to be more firm.

In a generic tissue structure, cells are of nonuniform shapes and sizes. Particularly, there exist clearly identifiable regions or layers of different types of cells in a given food plant material such as cells in the skin, flesh, and the center core (Verboven et al. 2008). Additionally, not only cell shape and size, other cell properties such as stiffness, chemical composition, pigments and most of the other physical and chemical properties are also considerably different in such cell regions. Cells in the tissue are always bonded to the neighboring cells through intercellular bonds via a middle lamella, which is a layer mainly composed of pectin. Intercellular bonds are strong enough to keep cells aggregated together even when the tissue gets deformed under external forces or is subjected to shape alterations. Other than the cells, intercellular spaces are frequently observed in a plant tissue, which consist of heterogeneous mixtures of gaseous and liquid constituents (Konstankiewicz and Zdunek 2001). Furthermore, the cell walls, pectin layers, and the tissue as a whole are semipermeable structures, allowing the exchange of water and other chemical substances across cell layers from or to the outside environment (Taiz and Zeiger 2010). Accordingly, the cellular structure available in a given tissue is essentially in a discrete form because it is composed of cells and intercellular spaces, which behave as individual units. Further, those units are made out of liquid, solid, and gaseous substances in different proportions and therefore the tissue as a whole is essentially a multiphase system. Additionally, it is broadly a permeable structure, allowing diffusion of different constituents across the tissue.

16.1.2 What Happens to the Tissue Structures of Food Plant Materials during Drying

During drying, a fresh tissue structure is subjected to predetermined environmental conditions characterized mainly by the temperature, humidity, pressure, and the flow of the air used for drying in a given dryer. As described earlier, since the tissue being a moist permeable structure, water in the cell interior gradually diffuses from the central tissue regions to the exterior tissue regions and eventually into the tissue boundaries (Simal et al. 1994). The rate of such moisture transfer is usually high in the initial phase of drying because the tissues have higher moisture content at the beginning, assisting higher rate of diffusion. However, as the drying progresses and the moisture is gradually removed from the tissue, the rate of drying reduces (Martin et al. 2006). Furthermore, if the drying temperature is considerably high, a hard case is formed at the exposed outer boundary of the tissue, leading to a phenomenon called "case hardening", restricting the progress of drying caused by the outward moisture flow mainly driven by diffusion (Ratti 1994; Wang and Brennan 1995; Rahman et al. 2005).

Irrespective of the mode of moisture transfer, moisture removal is the key phenomenon leading to numerous benefits of dried food products such as reduced microbial reactions, extended shelf life of the product, and considerably reduced weight of the product. Accordingly, when moisture gets removed, a proportional volumetric contraction is experienced by the cells and the tissue, leading to extensive deformation or shrinkage phenomena corresponding to the volumetric contraction. As a norm, shrinkage causes the tissue structure to distort excessively and irregularly. Such cellular-level distortions of the food plant products during drying eventually result in bulk or macroscale material deformations. On the other hand, some of the bulk-level deformations or related physics can cause microscale deformations as well. Accordingly, there exist important multiscale relationships among these deformations. Further, compared to fresh food plant materials, dried forms of those materials have unique characteristics such as texture, mouth-feel, flavor, color, and nutritional content. As a result, in dried food plant products, there are wider opportunities available for altering those properties to match with different customer requirements than in fresh food plant products.

16.1.3 Importance of Predicting Physical and Morphological Changes of Food Plant Materials during Drying and Different Techniques Involved

In food engineering, it is critical to predict and control numerous physical changes of food plant materials. As done in other engineering fields, software-based simulations are ideal in this context, with the aid of an appropriate material model within a simulation software platform. Accordingly, one would be able to conveniently simulate the physical changes and quality variations of a given food variety during drying. Furthermore, it is even more beneficial to be able to simulate shrinkage as well. However, up to now, there is hardly any such effective method commercially available, which can be used for the purpose mentioned above. This is mainly due to the unavailability of effective modeling techniques capable of accounting for the critical changes of physical properties of plant tissues during drying. However, other than such software, there are two dominant types of models developed for practical use in food engineering in order to predict the behavior of food plant materials during drying which are empirical models and theoretical models.

Here, empirical models are the most frequently used type, which are based on experimentally observed physical parameters. However, these models are usually deficient in general applicability owing to constraints corresponding to the respective experimental processes involved in developing such models. Therefore, the relationships obtained are basically valid for the specific food plant variety under given drying conditions (Rahman 2003). The parameters or coefficients used in such models are generally weak in theoretic or realistic terms (Rahman 1994). Also, most of these experimental results are based on microscopy images, which are rather qualitative than quantitative (Hills and Remigereau 1997; Lewicki and Pawlak 2003; Rahman et al. 2005; Devahastin and Niamnuy 2010; Sansiribhan et al. 2010). For example, it is very challenging and almost impossible to measure forces, or stresses, in subcellular levels using experimental techniques, although such fine details are highly critical to explain the sophisticated morphological changes of the food structure during drying. Sometimes it is even very difficult to obtain a clear understanding of the structural properties owing to equipment limitations, particularly the limited imaging capabilities of microscopes (Bai et al. 2002). Therefore, although these empirical models are popular and have direct relationship with the experimental observations, their extensive use as a means of drying models for general applications are restricted due to highly controlled experimental conditions, limited availability of the technology, and high cost and time involved in sample preparation and experimentation.

On the other hand, the theoretical models are fully or partially based on fundamental heat and mass transfer concepts and are applicable for general plant materials under different processing conditions. However, owing to the use of sophisticated theories, which usually rely on oversimplified boundary conditions and approximated heat and mass transfer phenomena, these models have numerous practical limitations. For instance, many of such theoretical models are based on mass transfer fundamentals of Fick's laws, which rely on several unrealistic approximations such as homogeneous food materials, omission of heat transfer or shrinkage during drying, and oversimplified geometrical approximations with ideal boundary conditions (Wang and Brennan 1995). Furthermore, since nonlinear diffusion equations are frequently used in such theoretical drying models, computationally expensive numerical treatments are indispensable in solving them (Crapiste et al. 1988b; Yesilata and Aktacir 2009).

Tables 16.1 and 16.2 present a collection of such bulk-scale empirical and theoretical models available in the literature, along with the corresponding drying

TABLE 16.1
Empirical Bulk-Scale Models for Food Plant Material Drying

Food Product	Drying Technique	Model Parameter	Source
Apple	Convective air drying	Moisture content; porosity; shrinkage; bulk density; solid density	Lozano et al. (1980)
Apple	Convective air drying	Moisture content; bulk-scale geometrical parameters: area, diameter, perimeter, roundness, elongation	Mayor et al. (2005)
Apple; banana; carrot; garlic; potato	Convective air drying; osmotic dehydration	Moisture content; drying temperature; material density	Boukouvalas et al. (2006)
Apple; cabbage; carrot; potato	Forced convective air drying	Moisture content; porosity; bulk density; solid density; pore size; cumulative specific bulk surface area	Karathanos et al. (1996)
Bean; pea; potato	Heat pump–assisted fluidized bed drying	Moisture ratio; drying time; aspect ratio; drying temperature; effective diffusion coefficient	Senadeera et al. (2003)
Bean; pea; potato	Heat pump–assisted fluidized bed drying	Moisture ratio; shrinkage; aspect ratio	Senadeera et al. (2005)
Bean; pea; potato	Heat pump–assisted fluidized bed drying	Moisture content; drying rate; shrinkage rate	Senadeera (2008a)
Potato	Superheated steam drying; forced convective air drying	Moisture ratio; drying time; drying temperature	Leeratanarak et al. (2006)
Potato	Forced convective air drying	Moisture content; bulk density; shrinkage; porosity; air volume to total volume ratio; sample thickness, length, and width	Wang and Brennan (1995)
Spinach	Combined microwave and forced convective air drying	Moisture ratio; drying time	Karaaslan and Tunçer (2008)

TABLE 16.2
Theoretical Bulk-Scale Models for Food Plant Material Drying

Food Product	Drying Technique	Model Parameter	Source
Any cellular material in general	Any drying process (valid only until the cellular structure is intact)	Moisture content; shrinkage	Crapiste et al. (1988a)
Apple; corn; orange; pear; potato; tomato	Convective air drying (literature data)	Moisture content; thermal conductivity; drying temperature	Maroulis et al. (2002)
Apple; potato	Convective air drying (literature data)	Moisture content; porosity	Rahman (2003)
Apple; potato	One-dimensional convective air drying	Moisture content; shrinkage; mass flux through the cellular structure	Crapiste et al. (1988b)
Carrot	Fluidized bed drying	Moisture content; drying temperature	Jaros and Pabis (2006)
Carrot; celery	Fluidized bed drying	Moisture content; drying time	Pabis (2007)
Carrot; garden beet	Convective air drying	Moisture content; drying time; shrinkage; drying coefficient	Pabis and Jaros (2002)
Carrot; garlic; mushroom; onion; red beet	Convective air drying (literature data)	Moisture content; drying time	Stanisław (1999)
Chilli red pepper	Convective air drying	Moisture content; drying time; effective diffusion coefficient	Yesilata and Aktacir (2009)
Potato	Forced convective air drying	Moisture content; drying time; effective diffusion coefficient	Rosselló et al. (1992)
Potato	Convective air drying	Moisture content; drying time; effective diffusion coefficient; drying temperature; sample size	Simal et al. (1994)
Potato	Convective air drying	Moisture content; effective diffusion coefficient; drying temperature; shrinkage	Hassini et al. (2007)

techniques and physical parameters involved. Table 16.3 presents a collection of microscale drying experiments and related empirical relationships available in the literature on different food plant materials.

16.2 NUMERICAL MODELING OF MORPHOLOGICAL CHANGES OF FOOD PLANT MATERIALS DURING DRYING

In order to fill the gap of a widely applicable modeling approach, numerical modeling has emerged, which combines the capabilities of both empirical and theoretical modeling. Here, the physical characteristics of a given food material are simplified and mathematically represented by a system of equations, and drying

TABLE 16.3
Microscale Experimental Studies and Related Empirical Relationships Developed for Food Plant Material Drying

Food Product	Experimental Technique	Cellular Parameters Studied	Source
Apple	Forced convective air drying; puff drying; freeze-drying	Moisture content; cell area; shape factor; diameter; perimeter	Lewicki and Pawlak (2003)
Apple	Forced convective air drying	Moisture content; drying time; drying temperature; porosity	Bai et al. (2002)
Apple	Fluidized bed drying	Moisture content; drying time	Hills and Remigereau (1997)
Apple	Forced convective air drying	Moisture content; drying time; porosity; pore diameter; pore volume	Rahman et al. (2005)
Apple	Convective air drying	Moisture content; cell area; diameter; perimeter; roundness; elongation; compactness	Mayor et al. (2005) and Karunasena et al. (2014a)
Apple; apricot	Forced convective air drying	Moisture content; cell roundness; elongation	Bolin and Huxsoll (1987)
Grape	Forced convective air drying	Drying time; drying temperature; cell area; diameter; perimeter; elongation; compactness	Ramos et al. (2004) and Ramos (2010)
Carrot	Forced convective air drying	Moisture content; drying time; cell diameter; fractal dimension; material hardness	Sansiribhan et al. (2010)
Carrot	Forced convective air drying	Moisture content; cell diameter; drying time; drying temperature; fractal dimension; material hardness	Sansiribhan et al. (2012)
Potato	Forced convective air drying	Moisture content; drying time; cell area; fractal dimension	Campos-Mendiola et al. (2007)

mechanisms are also incorporated into the system, mathematically. The majority of parameters and coefficients used in the equations of a given numerical model are custom selections based on the food variety and drying conditions (not like theoretical models that oversimplify and eventually far deviate from the real physical nature of the food material). The numerical models are essentially simulated by solving the model equations in a computer program in an iterative manner, and the results are eventually visualized to obtain the model predictions for a given drying scenario. Not like in fundamental theoretical models that involve nonlinear equations or differential equations, in numerical models, the model equations are essentially converted into discretized set of equations and established in a computer code, facilitating simulations in digital computers. Further, in theoretical models or even in empirical models, the basic form is a relationship between two or several physical parameters (e.g., the relationship between the drying temperature and the moisture content). However, numerical models have a much broader perspective aiming to represent the actual physical form or the behavior of a given tissue during drying. For instance, a numerical model may potentially represent the real-time moisture transfer at different points within a given tissue while accounting for localized deformations. Further, such numerical models produce results enriched with much visual information rather than just a set of curves representing parameter relationships, as in the case of most of the empirical or theoretical models (Heredia et al. 1995; Hills and Remigereau 1997; Konstankiewicz and Zdunek 2001; Bruce 2003; Ramos et al. 2003; Loodts et al. 2006; Rahman 2008). In this background, numerical modeling has become more popular in recent times where more research efforts are reported on numerous numerical modeling approaches.

16.2.1 Conventional Grid-Based Numerical Methods and Their Limitations for Modelling Plant Materials during Drying

The most developed and conventionally used numerical modeling techniques are the grid-based ones such as the finite element method (FEM), finite difference method (FDM), or the finite volume method (FVM). The fundamental approach behind all these techniques is the use of a fixed grid or a set of nodes to estimate the properties of a given point of interest within the problem domain. These techniques have been widely researched and have become so popular that there are numerous commercial software platforms available today for different engineering and other scientific uses.

Among the numerous cellular level numerical models available in the literature, other than the very few models on microscale drying, the majority are dedicated to study the basic mechanical responses of plant cells and tissues to external interactions such as compression, tension, or impact (see Table 16.4 for a comprehensive collection of numerical models on plant microstructure). In Table 16.4, except for the most recent FEM-based tissue drying model (Fanta et al. 2014), almost all of the other numerical models involve oversimplified approximations of the cellular structure and do not sufficiently account for large deformations or considerable moisture content reduction during drying. Also, the models are mostly limited to basic force-deformation studies and represent only very limited morphological details. Therefore, in order to develop a comprehensive numerical model that can account for large deformations and moisture transfers of the heterogeneous cellular structure, the practical applicability of these models is very limited.

Even the most advanced and recently developed plant tissue drying model in the identified literature (Fanta et al. 2014; see Table 16.4) has considerable conceptual drawbacks that limit its applicability in modeling actual dried food plant materials. The model couples water transport phenomenon and deformations of a customized cellular structure generated using a vertex model that approximates cells as polygons of different shapes (Abera et al. 2013). This particular vertex model has a limited scope since it is dedicated to model plant tissues under slow dehydration conditions, which usually occur during storage, rather than actual convective drying processes involved in dryers or industrial drying setups. As a result, the model can only

TABLE 16.4
State-of-the-Art Numerical Models on Food Plant Material Microstructure

Key Features	Limitations	Source
FEM-based 2-D tissue drying model to study shrinkage during drying; accounted for cell wall permeability and basic mechanisms of intercellular spaces; tissue structure uses a separate cortex tissue geometric model	Critical dry states and cell wall wrinkling not accounted; complex time-consuming hybrid simulation approach involved	Fanta et al. (2014)
2-D tissue model for mass transfer during drying	A basic geometrical model; cellular deformations not accounted	Rotstein and Cornish (1978)
2-D tissue model for tensile responses of cells; cells approximated to fluid flied compartments with thin boundaries; to study basic mechanical responses of tissues	Only valid for fresh tissues; limited subcellular details incorporated	Pitt and Davis (1984)
2-D tissue model represented as a matrix of equilateral triangles; cells placed at triangle nodes; to study basic mechanical responses of tissues	Only valid for fresh tissues; limited sub-cellular details incorporated	Pathmanathan et al. (2009)
2-D tissue model with discretized cell elements; cell boundaries represented as a spring-mass network; to study basic mechanical responses of tissues	Only valid for fresh tissues; cell volume conserved	Loodts et al. (2006)
FEM-based 3-D single cell model; to study basic mechanical responses of cells	Only valid for fresh cells	Wu and Pitts (1999)
3-D tissue model; cells modeled as 3-D shapes with hexagonal sides; to study basic mechanical responses of tissues	Only valid for fresh tissues; cell wall permeability omitted	Zhu and Melrose (2003)
3-D single cell model; cells modeled as fluid-filled spheres with thin walls; to study basic mechanical responses of cells	Only valid for fresh cells	Wang et al. (2004)
3-D tissue model; cells modeled as tetrakaidecahedrons with hexagonal and square sides; turgor pressure and intercellular bonds accounted; to study basic mechanical responses of tissues	Only valid for fresh tissues; cell volume conserved	Gao and Pitt (1991)
FEM-based 3-D single cell model; cells modeled as fluid-filled spheres with thin walls; accounts for wall permeability and volume loss during compression	Only valid for fresh cells	Smith et al. (1998)
3-D tissue model; spherical cells used, cells bonded with elastic intercellular connections; to study basic mechanical responses of tissues	Only valid for fresh cells; cell volume conserved	Nilsson et al. (1958)

simulate very limited moisture content reduction (about 30%), which is not the case in actual drying processes where the overall moisture reduction can even be as high as 95%. Also, the model has another key limitation due to the use of polygon-shaped cells with straight cell wall segments. Since these polygon sides are maintained as straight lines during the full length of drying simulations, the deformed cell shapes

(a) (b) (c)

FIGURE 16.1 Scanning electron microscope (SEM) images of apple cells at different states of dryness: (a) = 1.0, (b) = 0.1, and (c) = 0.01.

do not replicate the highly wrinkled cell walls of actual dried tissues (Figure 16.1). Also, the model assumes that the cell wall mass is lumped only on polygon vertices of the cell, which is also not acceptable in reality, because cell wall mass is usually well distributed along the cell wall. To perform simulations, a separate MATLAB® code is used and the tissue geometry is numerically solved in each time step, which is then fed into a Comsol-based FEM model to compute water exchange parameters in each time step. This hybrid numerical simulation approach demands a considerable effort to implement, since it relies on the cosimulation of two independent codes in each time step. These limitations are broadly due to the fundamental limitation of the grid-based modeling techniques used in these models.

Further, other than these microscale numerical models, even the available grid-based macroscale numerical drying models face similar application limitations. For instance, a gel material model is reported in the literature for modeling wrinkled shapes of dehydrated plant leaves (Liu et al. 2010). The technique oversimplifies the plant cellular structure to a continuum material that is then approximated by a hypothetical gel material, which does not compare well with the actual discrete cellular structure of real plant materials. Furthermore, quantifying such unique gel material properties becomes problematic, particularly in the case of drying of realistic plant materials having different cellular and bulk-scale characteristics. Further, the relationship between material deformations and moisture content reduction is omitted, which mainly limits the applicability of this model in real drying simulations. Surface wrinkling of plant leaves under dehydration has also been recently modeled with FEM (Jeong et al. 2013). The work relates wrinkling of leaves with the moisture content and accounts for localized variations of the moisture content. However, due to the simplified two-layer thin structure specifically developed to suit the thin leaf geometry, its applicability for bulk-scale food plant materials is limited.

The general, the validity, accuracy, and computational performance of grid-based techniques have been adequately serving most of the problem domains involving continuum materials and their elastic or finite plastic deflations. However, in the case of plant tissue drying, which essentially involves large deformations, multiphase, or phase-change conditions of discrete materials, the use of conventional FEM, FDM, and FVM approaches are quite challenging due to the fundamental limitations of such mesh or grid-based techniques. For instance, in the case of large deformations or when a given material undergoes cracking (such as in the case of food materials during drying), the neighboring nodes of a given grid-based model can fundamentally be unavailable and the mesh or the grid can get distorted leading to instabilities of the simulations. To recover from such undesirable situations, mesh refining or remeshing methods have been proposed, which can extensively increase the overall computational cost as well as the effort of setting up the model. The technical challenges and the various limitations of these state-of-the-art plant material drying models, particularly the microscale models, highlight the limited applicability of the grid-based approaches. Therefore, in order to fill this research gap, grid-free or mesh-free methods have been proposed, which are dedicated to handling large deformations of multiphase noncontinuum systems, which can even undergo phase changes.

16.2.2 Potential of Mesh-Free Methods to Numerically Model the Dried Food Microstructure

As mentioned earlier, grid-based methods have limited applicability in modeling of dry cell and tissue because of their fundamental shortcomings in handling advanced physics such as multiphase materials with solid, liquid, and gas interactions (cell wall, fluid, and intercellular spaces), excessive boundary deformations (cell walls and tissue boundaries), discrete materials (aggregated cellular structure), and multiscale relationships (interrelated subcellular, cellular, and bulk-scale physics). To overcome such fundamental shortcomings, recently developed mesh-free methods are more applicable since they primarily do not use any interconnected grids as in the case of grid-based methods, which entirely rely on grids (Liu and Liu 2003). In mesh-free methods, a set of non-interconnected particles is used to discretize the problem domain. Therefore, compared to the grid-based techniques, mesh-free methods have a higher capability to model materials with large deformations. Further, discretization refinement can be easily achieved when compared to mesh refinement of grid-based methods, particularly for complex geometries (Liu and Liu 2003). In the literature, there are different mesh-free methods (Belytschko et al. 1996) such as the smoothed particle hydrodynamics (SPH) (Gingold and Monaghan 1977), the element-free Galerkin (EFG) method (Belytschko et al. 1994), the point interpolation method (PIM) (Liu and Gu 2001b, 2002), the mesh-less local Petrov–Galerkin (MLPG) method (Atluri and Zhu 1998), the local radial point interpolation method (Liu and Gu 2001a), and the boundary point interpolation method (Liu and Gu 2004). These methods have found many applications in engineering and science (Wu et al. 2005; Gu and Liu 2006; Gu et al. 2007).

Among these methods, SPH is a well-developed, particle-based mesh-free technique, which was initially developed for astrophysical applications (Gingold and Monaghan 1977; Lucy 1977). SPH defines a given problem domain as a set of non-interconnected particles that carry physical properties, which can evolve to represent new states of a problem domain in time and space. The Navier–Stokes equations are solved without a mesh, but with a localized weighting function, which is called the smoothing function or the kernel (Frank and Perré 2010; Perré 2011). SPH can handle large deformations easily and more efficiently than grid-based methods (e.g., free surface flows, deforming boundaries, and moving interfaces), because the stability and the performance of the method are not affected by the particle evolution (in grid-based techniques, as the material undergoes extreme deformations, the stability of the grid scheme is negatively affected). In this regard, modeling of explosions is one of the good examples of SPH capabilities to handle large deformations better than grid-based methods. Furthermore, problems related with a set of discrete particles rather than a continuum can be better solved with SPH (Liu and Liu 2003). Another distinguishing property of SPH is the capability to incorporate new physics and mechanisms into the basic formulation of SPH to cater for novel problem domains of interest (Morris et al. 1997; Liu and Liu 2003).

In this regard, there is much evidence of successful application of SPH for solving different fluid flow problems such as Poiseuille flow, shock tube problem, 2-D heat conduction, Couette flow, shear-driven cavity problem, free surface flows, gas expansion, 1-D detonation, 2-D gas explosion, and underwater explosion (Liu and Liu 2003). SPH solutions closely replicate the analytical and numerical solutions obtained from conventional modeling techniques such as FEM and FDM (Liu and Liu 2003). Some other specific SPH applications are multiphase flows in porous media (Li and Liu 2002; Liu and Liu 2003; Tartakovsky et al. 2007b, 2009; Frank and Perré 2010; Perré 2011), complex fluid flow in porous media (Morris et al. 1999; Zhu et al. 1999; Zhu and Fox 2001, 2002; Vakilha and Manzari 2008), wetting of different solid boundaries (Tartakovsky and Meakin 2005c; Tartakovsky and Meakin 2006; Wróblewski et al. 2008), conduction heat transfer in irregular geometries and boundaries (Cleary and Monaghan 1999; Jeong et al. 2003; Vishwakarma et al. 2011), microscopic liquid flows in porous media (Tartakovsky et al. 2009), surface tension and viscosity simulations (Tartakovsky and Meakin 2005c; Wróblewski et al. 2008), and fluid transportation in fractured porous media (Tartakovsky and Meakin 2005a,b; Tartakovsky et al. 2007a). SPH is also incorporated with commercial modeling and simulation software (e.g., LS-DYNA) (Vesenjak and Ren 2008).

Further, in the context of microscale material deformations, SPH is an ideal computational technique due to its proven performances in modeling: porous media and diffusion (Zhu and Fox 2001), non-Newtonian and viscoelastic fluids (Ellero and Tanner 2005), cellular details with large deformations and discontinuities (Van Liedekerke et al. 2010a) and subcellular details (Van Liedekerke et al. 2010b), multiscale problems with many small-scale elements (tissues consisting of many individual cells; Li and Liu 2004; Ghysels et al. 2009; Van Liedekerke et al. 2009, 2010a, 2011), and the problems with thermal fluctuations (Español and Revenga 2003; Vázquez-Quesada et al. 2009a,b). Particularly for plant cells and tissues, there are several models developed to study basic mechanical responses of a single cell and tissue subjected to external compression, tension, shear, and impact loads, by coupling SPH with a discrete element method (DEM) (Van Liedekerke et al. 2009, 2010a,b, 2011). DEM works well with cell wall models to represent elastic deformations accurately (Pathmanathan et al. 2009) and can sufficiently represent elastic and plastic deformations of intercellular interactions (Loodts et al. 2006).

16.2.3 Fundamentals of Mesh-Free Methods and Their Applications

Mesh-free methods can provide accurate and stable solutions for analytical problems defined through partial differential equations or integrals, even with complex boundary conditions. For this purpose, a set of arbitrarily distributed nodes are used that are non-interconnected, unlike in conventional grid-based techniques, which use connected grids. As a result, computational time and resources for the initial mesh generation can be saved in comparison with grid-based methods. Among the different mesh-free methods, the mesh-free particle methods are very popular, representing different states of the problem domain by using particles that can displace with the material deformation. These methods are applicable in any scale, ranging from atomistic scales to macroscopic scales and even to astrophysical scales (Liu and Liu 2003).

In mesh-free particle methods, any field variable of a particle at a given location is derived as a summed influence of the field variable values of neighboring particles, as shown in Equation 16.1. The influence is evaluated using a function called a smoothing function or kernel, which has non-zero values only for neighboring particles. Here, N represents the total number of particles in the domain and is the value of the field variable of particle.

$$u(x)=\sum_{i=1}^{N}\varphi_i(x)\,u_i. \tag{16.1}$$

SPH is frequently used for hydrodynamic problems, where the fluid is modeled with particles that do not originally have any interconnections. The particles can have fluid properties such as density, pressure, viscosity, energy, position, velocity, and acceleration, and these properties evolve with time during the time integration of governing equations.

To model any problem domain of interest with the mesh-free method, first, the fundamental physical laws of conservation need to be established: conservation of mass, momentum, and energy (Navier–Stokes equations). One way of representing the Navier–Stokes equations is with the Lagrangian method (based on the material description) (Liu and Liu 2003):

conservation of mass (continuity equation):

$$\frac{d\rho}{dt}=-\rho\frac{\partial \boldsymbol{v}^{\beta}}{\partial \boldsymbol{x}^{\beta}}, \tag{16.2}$$

conservation of momentum:

$$\frac{d\boldsymbol{v}^{\alpha}}{dt}=\frac{1}{\rho}\frac{\partial \sigma^{\alpha\beta}}{\partial \boldsymbol{x}^{\beta}}, \tag{16.3}$$

conservation of energy:

$$\frac{de}{dt}=\frac{\sigma^{\alpha\beta}}{\rho}\frac{\partial \boldsymbol{v}^{\alpha}}{\partial \boldsymbol{x}^{\beta}}, \tag{16.4}$$

where ρ, $\boldsymbol{v}$, e, σ, and t are the density, velocity vector, internal energy, total stress tensor, and time and position vector, respectively. α and β indicate coordinate directions.

SPH basically agrees with the Lagrangian concept since SPH particles carry material properties that can move with the material when it deforms. Accordingly, the most popular SPH versions of Equations 16.2 through 16.4 are (Liu and Liu 2003)

conservation of mass (continuity equation):

$$\frac{\mathrm{d}\rho_i}{\mathrm{d}t} = \sum_{j=1}^{N} m_j \boldsymbol{v}_{ij}^{\beta} \cdot \frac{\partial W_{ij}}{\partial \boldsymbol{x}_i^{\beta}}, \tag{16.5}$$

conservation of momentum:

$$\frac{\mathrm{d}\boldsymbol{v}_i^{\alpha}}{\mathrm{d}t} = \sum_{j=1}^{N} m_j \left(\frac{\sigma_i^{\alpha\beta}}{\rho_i^2} + \frac{\sigma_j^{\alpha\beta}}{\rho_j^2} \right) \frac{\partial W_{ij}}{\partial \boldsymbol{x}_i^{\beta}}, \tag{16.6}$$

conservation of energy:

$$\frac{\mathrm{d}e_i}{\mathrm{d}t} = \frac{1}{2} \sum_{j=1}^{N} m_j \left(\frac{P_i}{\rho_i^2} + \frac{P_j}{\rho_j^2} \right) \boldsymbol{v}_{ij}^{\beta} \frac{\partial W_{ij}}{\partial \boldsymbol{x}_i^{\beta}} + \frac{\mu_i}{2\rho_i} \varepsilon_i^{\alpha\beta} \varepsilon_i^{\alpha\beta}, \tag{16.7}$$

where P, μ, and m are pressure, viscosity, and mass, respectively. The i represents the particle on focus, j represents the surrounding particles, N is the total number of particles in the problem domain, ε is the shear strain rate, and W_{ij} is the smoothing function (kernel), which is used to estimate the properties of particle i by the use of the properties of the surrounding j particles.

For low Reynolds number flow simulations of incompressible fluids, SPH-based mass and momentum conservation equations are further refined as (Morris et al. 1997)

conservation of mass (continuity equation):

$$\frac{\mathrm{d}\rho_i}{\mathrm{d}t} = \sum_{j=1}^{N} m_j \boldsymbol{v}_{ij} \cdot \nabla_i W_{ij}, \tag{16.8}$$

conservation of momentum:

$$\frac{\mathrm{d}\boldsymbol{v}_i}{\mathrm{d}t} = -\sum_{j=1}^{N} m_j \left(\frac{P_i}{\rho_i^2} + \frac{P_j}{\rho_j^2} \right) \nabla_i W_{ij} + \sum_{j=1}^{N} \frac{m_j (\mu_i + \mu_j) \boldsymbol{v}_{ij}}{\rho_i \rho_j} \left(\frac{1}{r_{ij}} \frac{\partial W_{ij}}{\partial r_{ij}} \right) + \boldsymbol{F}_i, \tag{16.9}$$

where, ∇_i is the gradient with respect to the coordinates of particle i; r_{ij} is the position of particle i relative to particle j; $\boldsymbol{v}_{ij}$ is the velocity of particle i relative to particle

j; and F_i is the body force acting on particle i per unit mass. In model development, Equations 16.8 and 16.9 are mainly used, which are explained in the following sections. In addition to these computational techniques, sufficient understanding is essential on the basics of cellular mechanisms in order to develop numerical models for cellular drying, which is presented next.

16.2.4 Useful Insights Drawn from Fundamentals of Cellular Mechanisms

In order to ensure good intercellular contact, the majority of cells embedded in a natural plant tissue are polyhedrous in shape (Sterling 1963; Jarvis 1998). However, isolated cells can be approximated by spheres (Nilsson et al. 1958; Hills and Remigereau 1997; Jarvis 1998; Lewicki and Pawlak 2003), and the shape factor* can range from 0.6 to 1.2 (Bolin and Huxsoll 1987; Lewicki and Drzewucka 1998; Mayor et al. 2005). A cell is composed of a cell fluid volume surrounded by a flexible cell wall, and the cell fluid hydrodynamic pressure is counterbalanced by a stiff cell wall (Nilsson et al. 1958; Hettiaratchi and O'Callaghan 1978; Wu et al. 1985; Gao and Pitt 1991; Chaplain 1993; Jarvis 1998; Smith et al. 1998; Lewicki and Pawlak 2003; Wang et al. 2004; Taiz and Zeiger 2010; Karunasena et al. 2014f), which also has viscous characteristics (Lewicki and Pawlak 2003). Cellulose in the cell wall provides rigidity and strength (Lewicki and Pawlak 2003), while pectin provides plasticity characteristics (Heredia et al. 1995; Lewicki and Pawlak 2003). The cell fluid mainly exists as vacuole and cytoplasm, which contain water-based heterogeneous solutions (Lewicki and Pawlak 2003) and account for around 90% of the water content of the cell. The cell wall contains a comparatively less water (Hills and Remigereau 1997; Karunasena et al. 2014f). Therefore, for modeling purposes, the cell fluid is regarded as water having an augmented viscosity, and the cell wall is approximated by a solid boundary. Cells tend to maintain a spherical shape due to the turgor pressure (Jarvis 1998), which implies that the walls are under a pre-tension (Konstankiewicz and Zdunek 2001). Cell walls also can undergo plastic deformations due to irreversible movements of their cellulose fibers (Köhler and Spatz 2002; Keckes et al. 2003). The cell turgor pressure influences the stiffness of the tissue (Nilsson et al. 1958; Lin and Pitt 1986), as well as the pressure of the air–water mixture in intercellular spaces (Konstankiewicz and Zdunek 2001). Tissues are elastic structures (Mayor et al. 2005; Karunasena et al. 2014f) in which the cells are bonded by pectin substances of the middle lamella (Lewicki and Pawlak 2003). Cell separation is resisted by the middle lamella, which is free of liquid water (Verboven et al. 2008), and provides a continuum medium for moisture transfer from cell to cell during drying (Lewicki and Pawlak 2003). The intercellular addition is positively influenced by large cell diameter or higher turgor pressure (Jarvis 1998). Intercellular spaces are loosely arranged in the honeycomb-like cellular structure (Lewicki and Pawlak 2003), and create a network with incomplete connections (Verboven et al. 2008). These spaces are spherical in shape (Sterling 1963; Khan and Vincent 1990; Mayor et al. 2005), and the basic cellular structure

* $4\pi A/P^2$.

persists intact throughout the ripening life cycle of food plant materials (Verboven et al. 2008). These microscale changes in the cellular structure contribute strongly to the macroscale physical behaviors of the tissue (Nilsson et al. 1958; Murase et al. 1980; Lin and Pitt 1986).

16.2.5 Useful Insights Drawn from Fundamentals of Cellular Drying Mechanisms

In addition to the previously mentioned generic details on the cellular structure, cellular drying mechanisms and related insights are also very important in building a cell-level drying model. In that regard, a summary of the most related literature is presented from here on. The overall moisture transfer inside the tissue is mainly determined by the cell-to-cell moisture transfer via vacuole, cytoplasm, and cell wall. Cell-to-cell moisture transfer prevails initially in the outer cells and gradually propagates into the inner cell layers (Crapiste et al. 1988b). Water removal from the vacuole mainly contributes to the overall cellular moisture content reduction during drying (Hills and Remigereau 1997; Zogzas et al. 2002). The surface area of tissue reduces linearly with the moisture content reduction during drying (Mazza and Lemaguer 1980; Suarez and Viollaz 1991; Karathanos 1993; Mayor et al. 2005), and the macroscopic shrinkage is isotropic (Lee et al. 1967; Lozano et al. 1980; Lewicki and Drzewucka 1998; Moreira et al. 2000; Ramos et al. 2004; Mayor et al. 2005). Cell walls are sufficiently flexible for shrinkage-based deformations (Hills and Remigereau 1997), and the majority of cells remain intact during drying without considerable cell wall breakage (Bai et al. 2002). In such conditions, cells lose their original roundness and undergo excessive shrinkage (Bolin and Huxsoll 1987; Mayor et al. 2005; Rahman 2008). Until the material is extremely dried, the void network remains intact (Lozano et al. 1980). However, shrinkage is resisted up to a certain extent as a result of case-hardening (Mayor et al. 2005) and cell wall thinning (Weier and Stocking 1949).

The size of the intercellular spaces reduces linearly as moisture is removed from the structure (Mayor et al. 2005), while the temperature positively influences the formation of large intercellular spaces (Funebo et al. 2000). In the final stage of drying, shrinkage is resisted by the formation of air voids in the structure (Lee et al. 1967). As a result, the porosity increases (Lozano et al. 1980; Hills and Remigereau 1997; Bai et al. 2002; Askari et al. 2004; Rahman et al. 2005; Rahman 2008), eventually leading to intense crack formation in the cell structure (Lewicki and Pawlak 2003; Rahman 2008). The water loss–driven shrinkage is dominant compared to the expansion of air voids (Hills and Remigereau 1997) or pore formation (Bai et al. 2002). These pores are of two types: externally connected pores and locked-in pores (Lozano et al. 1980). Gases contained in these spaces can expand and escape due to heating (Weier and Stocking 1949). Pore formation is also affected by the pressure existing in the drying environment (Rahman 2008). The osmotic potential and cell wall permeability mainly influence moisture transfer in cells, which vary as the material undergoes drying (Bolin and Huxsoll 1987). The microscale and the macroscale changes of the food plant materials during drying are interrelated (Lozano et al. 1980; Bai et al. 2002; Mayor et al. 2005).

16.3 BASICS OF MESH-FREE–BASED NUMERICAL MODELS APPLICABLE FOR THE PREDICTION OF PLANT TISSUE MORPHOLOGICAL CHANGES DURING DRYING

16.3.1 Representation of Plant Cells in a Tissue

As shown in Figure 16.2a, a given plant tissue is simplified as an aggregate of uniform cylindrical cells and the top surface of each cell is taken as a 2-D representation of it. Here, the axial deformations along the Z axis is assumed to be uniform, and the XY plane stresses in the top and bottom surfaces are neglected so as any Z directional velocity components (Figure 16.2c and 16.2d). As shown in Figure 16.2b, cell is treated as composed of two main components: cell wall and cell fluid, where the cell wall behaves as a stretched boundary housing the pressurized cell fluid as observed in nature (Gao and Pitt 1991; Chaplain 1993; Jarvis 1998; Smith et al. 1998; Lewicki and Pawlak 2003; Wang et al. 2004; Taiz and Zeiger 2010).

16.3.2 Modeling the Cell Wall

As shown in Figure 16.2c and 16.2d, the cell wall is modeled using a DEM approximating it to a neo-Hookean elastic material model while accounting for viscous interactions (Van Liedekerke et al. 2010a), different drying-related mechanisms, and stability requirements (Karunasena et al. 2014b,c). Accordingly, as shown in Figure 16.3, the cell wall model has seven types of force interactions: cell wall stiff forces ($\boldsymbol{F}^e$),

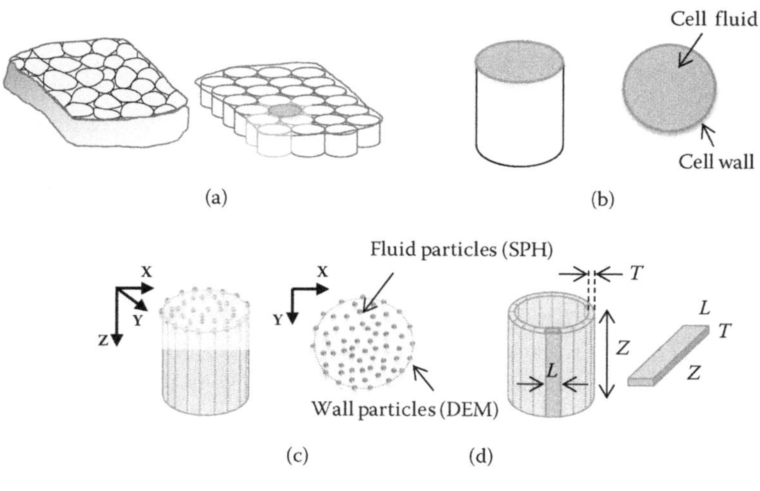

FIGURE 16.2 2-D representation of a plant tissue: (a) A plant tissue is approximated to an aggregate of individual cylindrical cells; (b) Top surface of each cylindrical cell is taken as a 2-D representation of the whole cell that consists of two basic components: cell fluid and cell wall; (c) The problem domain is discretized and represented as two sets of particles: SPH particles for the cell fluid and DEM particles for the cell wall; and (d) Individual elements of the discretized cell wall. (Reproduced from Karunasena, H.C.P. et al., *Soft Matter*, 10(29), 5249–5268, 2014e, with permission from the Royal Society of Chemistry.)

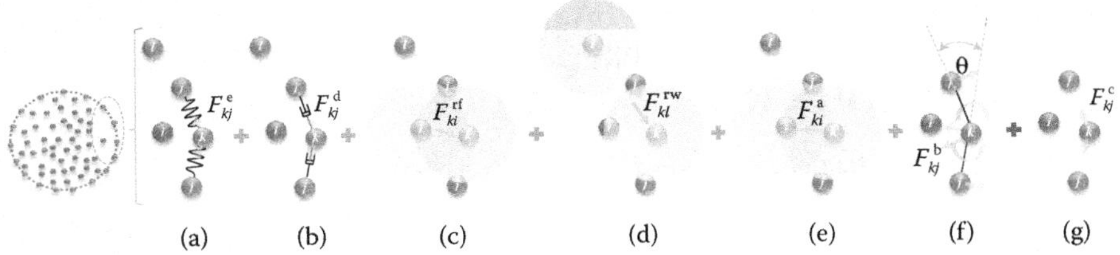

FIGURE 16.3 Force interactions used in a DEM-based cell wall model: (a) stiff forces, (b) damping forces, (c) wall–fluid repulsion forces, (d) non-bonded wall–wall repulsion forces, (e) wall–fluid attraction forces, (f) wall bending stiff forces, and (g) wall contraction forces. (*i*: fluid particles; *j*, *k*, *l*: wall particles). (Reproduced from Karunasena, H.C.P. et al., *Soft Matter*, 10(29), 5249–5268, 2014e. With permission from the Royal Society of Chemistry.)

wall damping forces ($\boldsymbol{F}^{d}$), wall–fluid repulsion forces ($\boldsymbol{F}^{rf}$), nonbonded wall–wall repulsion forces ($\boldsymbol{F}^{rw}$), wall–fluid attraction forces ($\boldsymbol{F}^{a}$), forces due to bending stiffness of the cell wall ($\boldsymbol{F}^{b}$), and forces to produce contractions of the cell wall during drying ($\boldsymbol{F}^{c}$). Therefore, the resultant force ($\boldsymbol{F}_k$) on any wall particle is defined as (Karunasena et al. 2014e)

$$\boldsymbol{F}_k = \boldsymbol{F}^{e}_{kj} + \boldsymbol{F}^{d}_{kj} + \boldsymbol{F}^{rf}_{ki} + \boldsymbol{F}^{rw}_{kl} + \boldsymbol{F}^{a}_{ki} + \boldsymbol{F}^{b}_{kj} + \boldsymbol{F}^{c}_{kj}, \tag{16.10}$$

where i is any neighboring fluid particle, j is any bonded wall particle, and l is any nonbonded k wall particle. Here, a nonlinear spring model is used to define the stiff force on any wall particle k due to any bonded wall particle j (Van Liedekerke et al. 2010a) as

$$\boldsymbol{F}^{e}_{kj} = GZ_0T_0\left(\lambda_\theta - \frac{1}{\alpha^2\lambda_\theta^5}\right), \tag{16.11}$$

where G is the shear modulus ($\approx E/3$), with E being the Young's modulus of the wall material, Z_0 is the height of the cell at the initial condition, T_0 is the initial thickness of the cell wall, $\lambda_\theta = L/L_0$ is the extension ratio computed individually for each of the cell wall elements at the current time step, L is the width of each of the cell wall elements at the current time step (distance between particle k and j), and L_0 is the corresponding initial width of the cell wall element of interest. Here, α is a dimensionless parameter calculated as follows for cylindrical cells using $\beta = 0.5$ (Van Liedekerke et al. 2010a):

$$\alpha = \sqrt{\frac{\beta + \sqrt{\beta^2 - 4(\beta - 1)/\lambda_\theta^6}}{2}}. \tag{16.12}$$

In Equation 16.10, a linear dashpot model is used to represent the viscous forces acting on any wall particle k due to the neighboring wall particles j (Van Liedekerke et al. 2010a):

$$\boldsymbol{F}^{d}_{kj} = -\gamma\,\boldsymbol{v}_{kj}, \tag{16.13}$$

where γ is the cell wall damping constant and $\boldsymbol{v}_{kj}$ is the velocity of particle k relative to particle j. In Equation 16.10, the $\boldsymbol{F}^{\mathrm{rf}}$ represents the repulsive forces required to prevent the possibility of fluid particle penetration through the cell wall, and $\boldsymbol{F}_{ki}^{\mathrm{rf}}$ on any wall particle from any other fluid particle is defined as (Karunasena et al. 2014e)

$$\boldsymbol{F}_{ki}^{\mathrm{rf}} = f_{ki}^{\mathrm{rf}} \boldsymbol{x}_{ki}, \tag{16.14}$$

where f_{ki}^{rf} is the magnitude of the repulsion force and $\boldsymbol{x}_{ki}$ is the position vector of particle k relative to particle i. This follows a Lennard-Jones (LJ) force type and is defined as (Karunasena et al. 2014e)

$$f_{ki}^{\mathrm{rf}} = \begin{cases} f_0^{\mathrm{rf}} \left[\left(\dfrac{r_0}{r_{ki}} \right)^8 - \left(\dfrac{r_0}{r_{ki}} \right)^4 \right] \left(\dfrac{1}{r_{ki}^2} \right) & \left(\dfrac{r_0}{r_{ki}} \right) \geq 1 \\ 0 & \left(\dfrac{r_0}{r_{ki}} \right) < 1 \end{cases}, \tag{16.15}$$

where r_0 is the initial gap between the two particles, r_{ki} is the current gap between the two particles, and f_0^{rf} is the strength of the LJ contact. Further, in Equation 16.10, a similar LJ repulsion force $\boldsymbol{F}_{kl}^{\mathrm{rw}}$ is used to prevent self-penetration of the nonbonded wall particles and the respective LJ contact strength is f_0^{rw}. Next, in Equation 16.10, an LJ-type wall–fluid attraction force $\boldsymbol{F}_{ki}^{\mathrm{a}}$ is used to limit any unphysical detachments between the cell wall and the cell fluid during drying, which is critical in simulating dried cells than fresh cells (Karunasena et al. 2014b).

In Equation 16.10, bending stiff forces $\boldsymbol{F}_{kj}^{\mathrm{b}}$ represent the cell wall bending stiffness, preventing sharp wrinkles forming on the cell wall during drying and are defined based on the cell wall curvature on any wall particle k within the k and j particle pair as (Hosseini and Feng 2009; Pan and Wang 2009; Shi et al. 2012; Karunasena et al. 2014b)

$$\boldsymbol{F}_{kj}^{\mathrm{b}} = \frac{k_{\mathrm{b}}}{L} \tan\left(\frac{\Delta\theta}{2} \right), \tag{16.16}$$

where k_{b} is the cell wall bending stiffness, L is the width of any given wall element at any given time step, θ is the external angle between the particular wall element and the adjacent wall element as shown in Figure 16.3, and $\Delta\theta$ is the change of the angle θ during time evolution.

Further, in Equation 16.10, cell wall contraction forces ($\boldsymbol{F}^{\mathrm{c}}$) are included in the wall model to incorporate experimentally observed cell perimeter reductions (Lewicki and Pawlak 2003; Mayor et al. 2005; Blum 2011; Karunasena et al. 2014a) and are defined as (Karunasena et al. 2014c,e)

$$\boldsymbol{F}_{kj}^{\mathrm{c}} = k_{\mathrm{wc}} \left[L - L_0' \left[1 - \frac{a}{b} \left(1 - \frac{X}{X_0} \right) \right] \right], \tag{16.17}$$

where k_{wc} is the force coefficient of cell wall contractions, L is the current width of any particular wall element (see Figure 16.2d), L_0' is the width of the wall element at fully turgid condition, a and b are empirical factors, and X/X_0 is the normalized moisture content of the cell to be simulated. The a and b factors are taken as 0.2 and 0.9 to align with experimental findings on cell perimeter (Karunasena et al. 2014a,e). Further, when simulating dry cells, the moisture content of the cell wall is assumed to be reducing proportionally with the overall cellular moisture content reduction (Karunasena et al. 2014e). Additional parameter values involved in modeling the cell wall are given in Table 16.5.

16.3.3 Modeling the Cell Fluid

Cell fluid is approximated to a high viscous incompressible Newtonian fluid with low Reynolds number flow characteristics and is modeled with SPH (Van Liedekerke et al. 2009, 2010a, 2011; Karunasena et al. 2014e). Further, the attraction forces between the cell fluid and the cell wall are used to prevent any detachments of the cell fluid from the cell wall during the simulation of cell drying (Karunasena et al. 2012c). Accordingly, the cell fluid model involves four types of force interactions as shown in Figure 16.4: fluid pressure forces ($\boldsymbol{F}^{\mathrm{p}}$), fluid viscous forces ($\boldsymbol{F}^{\mathrm{v}}$), wall–fluid repulsion forces ($\boldsymbol{F}^{\mathrm{rw}}$), and wall–fluid attraction forces ($\boldsymbol{F}^{\mathrm{a}}$). The resultant of these forces: $\boldsymbol{F}_i$ on any fluid particle i is defined as

$$\boldsymbol{F}_i = \boldsymbol{F}_{ii'}^{\mathrm{p}} + \boldsymbol{F}_{ii'}^{\mathrm{v}} + \boldsymbol{F}_{ik}^{\mathrm{rw}} + \boldsymbol{F}_{ik}^{\mathrm{a}}. \tag{16.18}$$

Here, following the standard Lagrangian-type SPH formulation for weakly compressible low Reynolds number fluid flows, the $\boldsymbol{F}_{ii'}^{\mathrm{p}}$ and $\boldsymbol{F}_{ii'}^{\mathrm{v}}$ for any given fluid particle i, are defined using the properties of the neighboring fluid particles i' as (Morris et al. 1997; Van Liedekerke et al. 2010a; Karunasena et al. 2014e)

$$\boldsymbol{F}_{ii'}^{\mathrm{p}} = -m_i \sum_{i'} m_{i'} \left(\frac{P_i}{\rho_i^2} + \frac{P_{i'}}{\rho_{i'}^2} \right) \left(\frac{1}{Z} \right) \nabla_i W_{ii'}, \tag{16.19}$$

$$\boldsymbol{F}_{ii'}^{\mathrm{v}} = m_i \sum_{i'} \frac{m_{i'} (\mu_i + \mu_{i'}) \boldsymbol{v}_{ii'}}{\rho_i \rho_{i'}} \left(\frac{1}{Z} \right) \left(\frac{1}{r_{ii'}} \frac{\partial W_{ii'}}{\partial r_{ii'}} \right), \tag{16.20}$$

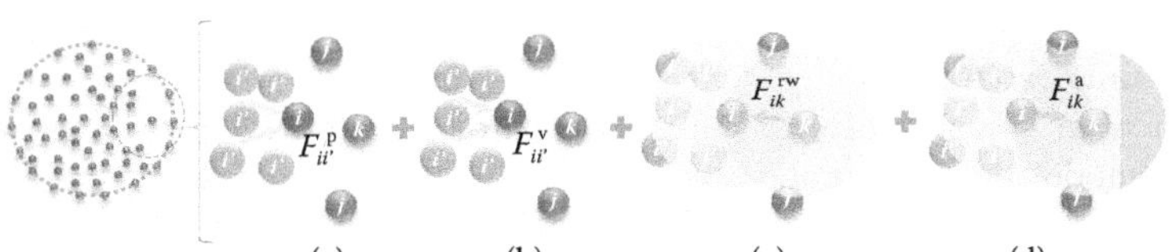

FIGURE 16.4 Force interactions used in the SPH-based cell fluid model: (a) pressure forces, (b) viscous forces, (c) wall–fluid repulsion forces, and (d) wall–fluid attraction forces. (i, i': fluid particles; j, k: wall particles). (Reproduced from Karunasena, H.C.P. et al., *Soft Matter*, 10(29), 5249–5268, 2014e, with permission from the Royal Society of Chemistry.)

where at any given time m, P, ρ, μ, Z, and W are the particle mass, pressure, density, dynamic viscosity, cell height, and the smoothing kernel, respectively. The smoothing kernel W is calculated for any given fluid particle i by using the quartic smoothing kernel as (Liu and Liu 2003; Karunasena et al. 2014b,e)

$$W_{ij} = \frac{15}{7\pi h^2}\begin{cases} \left(\frac{2}{3} - \frac{9}{8}S^2 + \frac{19}{24}S^3 - \frac{5}{32}S^4\right) & 0 \le S \le 2, \\ 0 & S > 2, \end{cases} \tag{16.21}$$

where h is the smoothing length at the current time step, S is the ratio of $r_{ii'}/h$, and $r_{ii'}$ is the gap between particle i and any surrounding fluid particle i' within the influence domain of the particle i $(0 \le S \le 2)$. To improve computational accuracy, the smoothing length h is dynamically adopted using a geometrical relationship as (Karunasena et al. 2014b)

$$h = \left(\frac{D}{D_0}\right)h_0, \tag{16.22}$$

where D is the average cell Feret diameter at the current time step, D_0 is the initial cell diameter, and h_0 is the initial smoothing length. During the time evolution of the model, standard SPH state equation is used to evolve the pressure of each of the fluid particles (Liu and Liu 2003; Van Liedekerke et al. 2010a)

$$P_i = P_T + K\left[\left(\frac{\rho_i}{\rho_0}\right)^7 - 1\right], \tag{16.23}$$

where P_T is the initial cell turgor pressure, K is the fluid compression modulus that ensures the fluid behaves in a incompressible manner (Van Liedekerke et al. 2010a), ρ_i is the current density of each fluid particle, and ρ_0 is its initial density assumed to be equal to the density of water. To obtain the current density of fluid particles, the density is evolved as (Van Liedekerke et al. 2010a)

$$\frac{d\rho_i}{dt} = \frac{1}{Z}\frac{d\rho_i^*}{dt} - \frac{\rho_i^*}{Z^2}\frac{dZ}{dt} + \frac{\rho_i}{m_i}\frac{dm_i}{dt}. \tag{16.24}$$

The first term in Equation 16.24 accounts for minor density fluctuations, which are defined using the standard SPH continuity equation as

$$\frac{d\rho_i^*}{dt} = m_i \sum_{i'} \mathbf{v}_{ii'} \cdot \nabla_i W_{ii'}, \tag{16.25}$$

where ρ_i^* is the 2-D density of fluid particle i defined as $\rho_i^* = Z\rho_i$. The second term in Equation 16.24 adjusts the density evolution by accounting for any cell volume changes due to cell height variations, and is defined as (Van Liedekerke et al. 2010a)

$$\frac{dZ}{dt} = \frac{Z_t - Z_{t-\Delta t}}{\Delta t}, \tag{16.26}$$

where at any given time, Z_t and $Z_{t-\Delta t}$ are the cell heights at the current and previous time steps, and Δt is the time step size. By considering the incompressibility of the cell wall material, the cell height is updated in each time step as (Van Liedekerke et al. 2010a)

$$Z = (\alpha\lambda_\theta)Z_0. \tag{16.27}$$

The third term in Equation 16.24 represents the minor variation of fluid density as influenced by the cell fluid exchange across the cell wall due to its permeability, which is defined as (Taiz and Zeiger 2010; Van Liedekerke et al. 2010a)

$$\frac{dm_i}{dt} = -\frac{A_c L_p \rho_i}{n_f}(P_i + \Pi), \tag{16.28}$$

where A_c, L_p, n_f, and Π represent the total surface area of the cylindrical cell at any given time, cell wall permeability, total number of fluid particles used, and the osmotic potential of the cell fluid at a given dry cell state, respectively. In Equation 16.18, the last two terms are the repulsion forces $\boldsymbol{F}_{ik}^{rw}$ and attraction forces $\boldsymbol{F}_{ik}^{a}$ acting on a given fluid particle i, which are defined using the LJ force type as

$$\boldsymbol{F}_{ik}^{rw} = \sum_k f_{ik}^{rw}\boldsymbol{x}_{ik}, \tag{16.29}$$

$$\boldsymbol{F}_{ik}^{a} = \sum_k f_{ik}^{a}\boldsymbol{x}_{ik}. \tag{16.30}$$

The numerical values used for different cell fluid properties are given in Table 16.5.

16.3.4 Modeling the Tissue

Modeling a tissue can be realized by bonding cells together. By having hexagonal cell shapes, the aggregation is much facilitated and it basically resembles the frequently observed honeycomb cell arrangement of tissues (Karunasena et al. 2014e). Further, to represent the middle lamella observed in actual plant tissues, a known gap is kept among cells, providing more realistic cellular arrangement compared to some of the initial mesh-free–based tissue models (Van Liedekerke et al. 2010a) or FEM-based tissue models (Fanta et al. 2014). Accordingly, the tissue is arranged as shown in Figure 16.5a and b, and as shown in Figure 16.5c and d, the cell–cell

TABLE 16.5
Key Physical Parameters Used for the Model

Parameter	Value	Source
Initial cell diameter (D_0)	150 μm	Hills and Remigereau (1997)
Initial cell height (Z_0)	100 μm	Van Liedekerke et al. (2010a)
Wall initial thickness (T_0)	6 μm	Wu and Pitts (1999)
Initial cell fluid mass	1.77×10^{-9} kg	Calculated Karunasena et al. (2012c)
Wall mass (10% of cell fluid mass)	1.77×10^{-10} kg	Calculated Van Liedekerke et al. (2010a)
Fluid viscosity (μ)	0.1 Pa s	Van Liedekerke et al. (2010a, 2011)
Initial fluid density (ρ_0)	1000 kgm^{-3}	Set
Fresh cell turgor pressure (P_T)	200 kPa	Wang et al. (2004) and Van Liedekerke et al. (2010a)
Fresh cell osmotic potential (Π)	−200 kPa	($\Pi = -P_T$) Van Liedekerke et al. (2010a, 2011)
Wall permeability (L_P)	2.5×10^{-6} m^{2}N^{-1}s	Set
Wall shear modulus (G)	18 MPa	Wu and Pitts (1999) and Van Liedekerke et al. (2010a)
Wall bending stiffness (k_b)	1×10^{-12} Nm rad^{-1}	Set
Wall damping ratio (γ)	5×10^{-6} Nm^{-1}s	Set Van Liedekerke et al. (2010a)
Fluid compression modulus (K)	20 MPa	Van Liedekerke et al. (2010a)
Wall contraction force coefficient (k_{wc})	4×10^4 Nm^{-1}	Set
LJ contact strength of wall–fluid repulsions $\left(f_0^{rf}\right)$	1×10^{-12} Nm^{-1}	Set
LJ contact strength of wall–wall repulsions $\left(f_0^{rw}\right)$	1×10^{-12} Nm^{-1}	Set
LJ contact strength of wall–fluid attractions $\left(f_0^{a}\right)$	2×10^{-12} Nm^{-1}	Set
LJ contact strength of cell–cell repulsions $\left(f_0^{rc}\right)$	1×10^{-10} Nm^{-1}	Set
Pectin layer stiffness (k_{pectin})	20 Nm^{-1}	Set
Pectin layer thickness (T_p)	8 μm	Set
Initial smoothing length (h_0)	1.2 × initial fluid grid spacing	Set
Time step (Δt)	2×10^{-9} s	Set

Source: Karunasena, H.C.P. et al., *Soft Matter,* 10(29), 5249–5268, 2014e.

contact is defined based on two intercellular forces: stiff forces representing stiffness of the intercellular pectin layer, and LJ-type repulsion forces to avoid interpenetrations of neighboring cells. A linear spring model is used to define the pectin layer stiff force on any wall particle due to its initial neighboring particle on the adjacent cell wall as (Van Liedekerke et al. 2010a; Karunasena et al. 2014d,e)

$$\boldsymbol{F}_{km}^{e\_pectin} = -k_{pectin}\Delta\boldsymbol{x}_{km}, \tag{16.31}$$

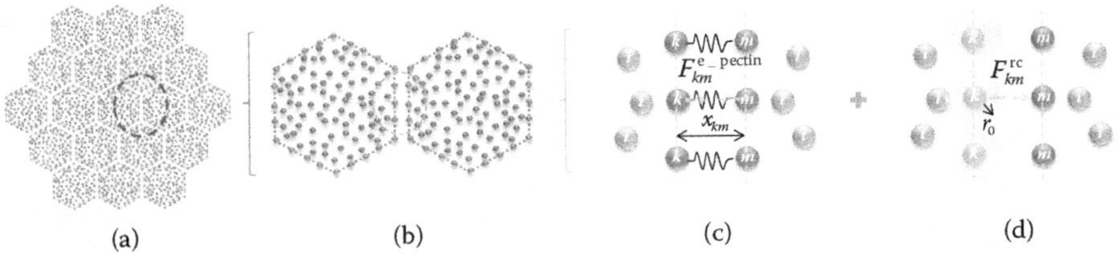

FIGURE 16.5 Tissue model and cell–cell force interactions: (a) hexagonal shaped cells are used for tissue initialization with a positive pectin layer gap, (b) interacting wall particle pairs of adjacent cells, (c) pectin layer stiff forces, and (d) cell–cell repulsion forces. (i: fluid particles; k, m: wall particles). (Reproduced from Karunasena, H.C.P. et al., *Soft Matter*, 10(29), 5249–5268, 2014e. With permission from the Royal Society of Chemistry.)

where k_{pectin} is the pectin layer stiffness and $\Delta \boldsymbol{x}_{km}$ is the difference of the gap between the two neighboring wall particles compared to their initial gap. In addition to this force, a secondary LJ-type force is incorporated to specifically ensure that the cells are not interpenetrated. This force acts on the cell wall particles as (Karunasena et al. 2014e)

$$\boldsymbol{F}_{km}^{\text{rc}} = \sum_{j} f_{km}^{\text{rc}} \boldsymbol{x}_{km}, \tag{16.32}$$

where f_{km}^{rc} is the strength of the LJ force field defined similar to that of the cell wall LJ force field, and $\boldsymbol{x}_{km}$ is the position vector of particle k relative to particle m. The numerical values of the intercellular force field parameters are given in Table 16.5.

Then the tissue model is implemented as a computer code and is time-evolved following a standard procedure to obtain different states of dryness defined by the normalized dry basis moisture content values X/X_0, where $X = m_{\text{water at any given condition}}/m_{\text{dry solid}}$ and $X_0 = m_{\text{water at fresh condition}}/m_{\text{dry solid}}$ (Karunasena et al. 2014e). The obtained qualitative results are visualized using the Open Visualization Tool (OVITO) (Stukowski 2010) and quantitative results are analyzed using a set of normalized cellular geometrical parameters (A/A_0, D/D_0, P/P_0, R/R_0, EL/EL_0 and C/C_0), where A (cell area), D (feret diameter*), P (perimeter), R (roundness†), EL (elongation‡), and (compactness§).

The simulation outcomes are initially compared with a number of experimental findings on apple tissue morphological changes during drying (Mayor et al. 2005; Karunasena et al. 2014a) and are then compared with the drying data on some other tissue varieties such as grape, potato, and carrot (Karunasena et al. 2015b). The key outcomes are briefly discussed in the next sections.

* $\sqrt{4A/\pi}$.

† $4\pi A/P^2$.

‡ $\sqrt{4A/\pi}$/(Major axis length).

§ Major axis length/Minor axis length.

16.4 KEY SIMULATION OUTCOMES FOR THE PREDICTION OF PLANT TISSUE MORPHOLOGICAL CHANGES DURING DRYING

16.4.1 Simulation of Morphological Changes of a Single Cell during Drying

Figure 16.6 shows the fundamental morphological changes of a single cell during drying, in the absence of any intercellular interactions. Figure 16.6a represents the initial hexagonal single cell, which gets inflated when simulated with fresh cell conditions as shown in Figure 16.6b. Next, Figure 16.6c through 16.6f present different dry cell conditions where, as the moisture content reduces, the cell simulations clearly indicate cell contractions. This phenomenon fundamentally agrees with realistic cellular shrinkage observed in plant cells during drying. Also, since the cell is not externally restricted by any other cells or intercellular influences, the circularity is uninterrupted and no cell wall wrinkling is observed, which is due to the positive turgor pressure and cell wall contraction forces discussed earlier (Karunasena et al. 2014e).

16.4.2 Simulation of Morphological Changes of a Food Plant Tissue during Drying

Simulation of a tissue composed of 37 cells is presented in Figure 16.7 and the corresponding enlarged views are presented in Figure 16.8. Here, compared to the circular

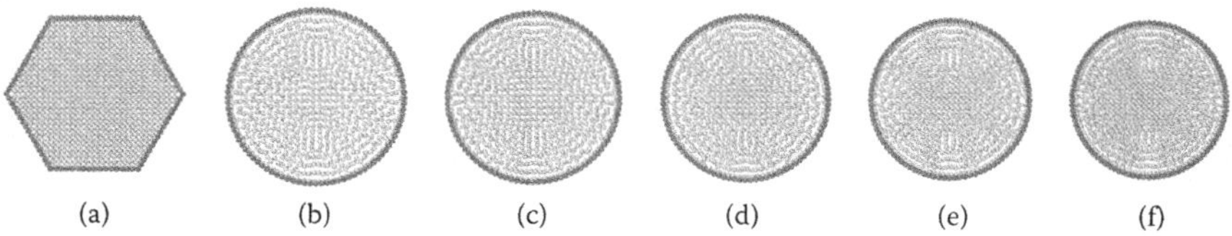

FIGURE 16.6 Single cell model undergoing drying: (a) initial condition before simulations; (b) turgid condition: $X/X_0 = 1.0$ and $P_T = 200\,kPa$; dried conditions: (c) $X/X_0 = 0.8$ and $P_T = 160\,kPa$, (d) $X/X_0 = 0.6$ and $P_T = 120\,kPa$, (e) $X/X_0 = 0.4$ and $P_T = 80\,kPa$, and (f) $X/X_0 = 0.3$ and $P_T = 60\,kPa$. (Reproduced from Karunasena, H.C.P. et al., *Soft Matter*, 10(29), 5249–5268, 2014e, with permission from the Royal Society of Chemistry.)

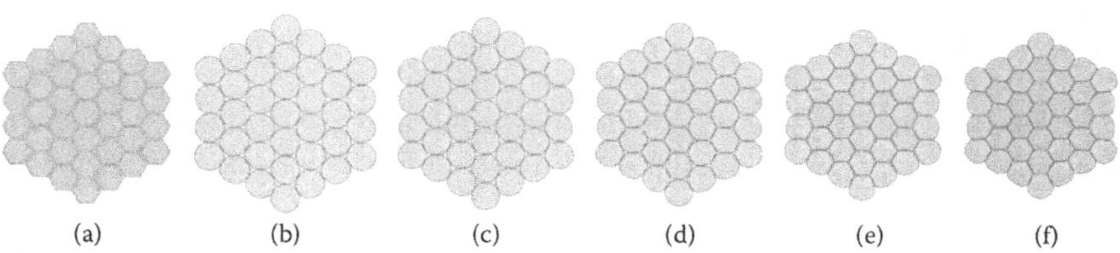

FIGURE 16.7 37-cell tissue model undergoing drying: (a) initial condition before simulations; (b) turgid condition: $X/X_0 = 1.0$ and $P_T = 200\,kPa$; dried conditions: (c) $X/X_0 = 0.8$ and $P_T = 160\,kPa$, (d) $X/X_0 = 0.6$ and $P_T = 120\,kPa$, (e) $X/X_0 = 0.4$ and $P_T = 80\,kPa$, and (f) $X/X_0 = 0.3$ and $P_T = 60\,kPa$. (Reproduced from Karunasena, H.C.P. et al., *Soft Matter*, 10(29), 5249–5268, 2014e, with permission from the Royal Society of Chemistry.)

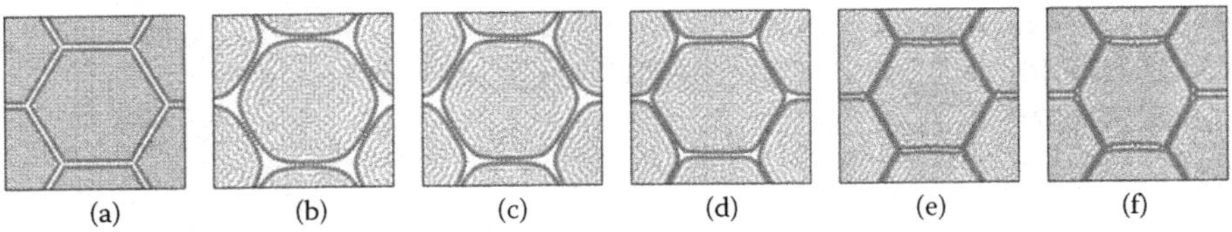

FIGURE 16.8 37-cell tissue model undergoing drying (enlarged view): (a) initial condition before simulations; (b) turgid condition: $X/X_0 = 1.0$ and $P_T = 200$ kPa; dried conditions: (c) $X/X_0 = 0.8$ and $P_T = 160$ kPa, (d) $X/X_0 = 0.6$ and $P_T = 120$ kPa, (e) $X/X_0 = 0.4$ and $P_T = 80$ kPa, and (f) $X/X_0 = 0.3$ and $P_T = 60$ kPa. (Reproduced from Karunasena, H.C.P. et al., *Soft Matter*, 10(29), 5249–5268, 2014e, with permission from the Royal Society of Chemistry.)

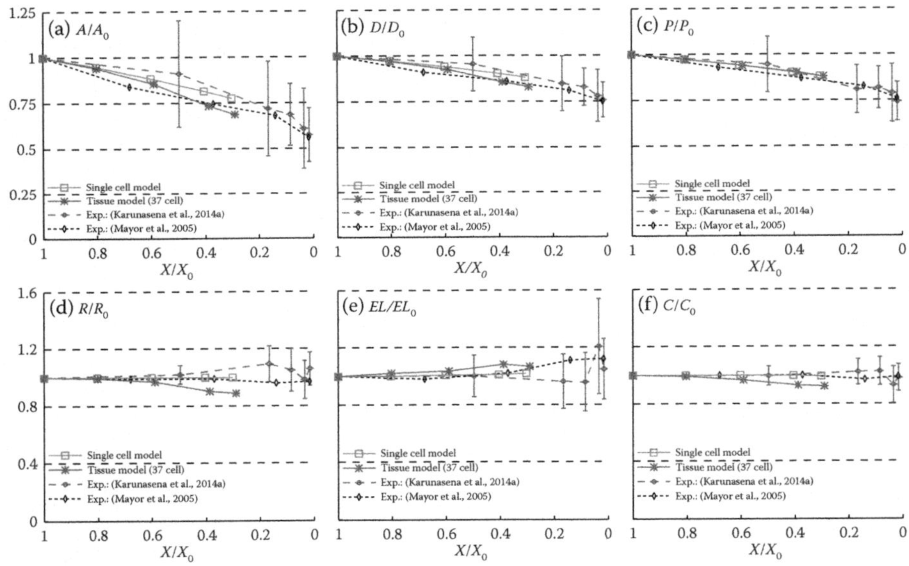

FIGURE 16.9 Influence of drying for cellular geometric property variations of dense tissue: (a) A/A_0, (b) D/D_0, (c) P/P_0, (d) R/R_0, (e) EL/EL_0, and (f) C/C_0. (Error bars indicate one standard deviation.)

shape observed in the single cell simulations mentioned earlier, the cells in the tissue are fairly hexagonal, as a result of the intercellular interactions. Further, as the tissue gets dried, the pectin layer tends to experience lower stretch conditions compared to the fresh samples. This indicates that the tissue is more prone to deformations mainly due to the lower moisture content and turgor pressure at dry conditions (Karunasena et al. 2014e). Also, from Figure 16.8e and f, it is clearly observed that the cell wall has experienced some local wrinkling or warping at extremely dried conditions. This clearly mimics the wrinkled cell walls observed in actual cells as given in Figure 16.1. When the simulation outcome is quantitatively compared with experimental findings as presented in Figure 16.9, a good agreement is observed (Karunasena et al. 2014e).

16.4.3 Simulation of Morphological Changes of Different Types of Food Plant Tissues during Drying

Other than apple cell and tissue shrinkage simulations, to demonstrate the general applicability of this mesh-free–based modeling technique, tissues of different food plant varieties such as potato, grape, and carrot have been studied using custom model parameter values (Karunasena et al. 2015b). Figure 16.10 shows a qualitative comparison between the simulation outcomes indicating clear differences of morphological changes among the tissue varieties. The difference is further evident when referring to Figures 16.11 through 16.14, where enlarged interior of each of the tissues are presented. The differences are mainly based on the tissue size, level of shrinkage, shapes of the cells at different cell layers, wrinkled patterns of the cell wall, and overall shape of the dried tissues. The main causes for such morphological differences are the cell size, cell wall stiffness, turgor pressure, pectin layer thickness and stiffness, and cell wall contraction parameters (Karunasena et al. 2015b).

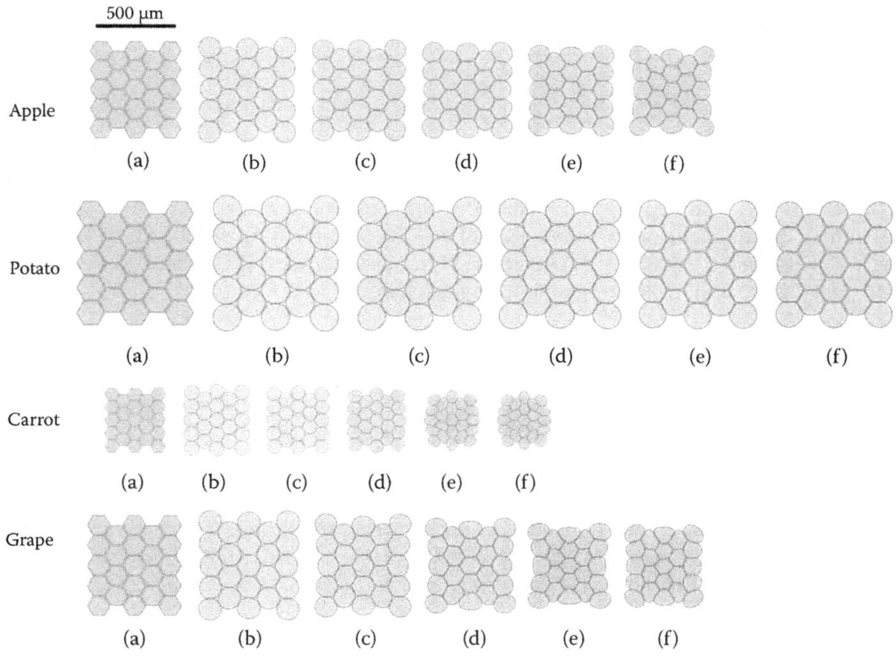

FIGURE 16.10 Tissue simulations at different states of dryness: (a) initial condition before simulations, (b) $X/X_0 = 1.0$, (c) $X/X_0 = 0.8$, (d) $X/X_0 = 0.6$, (e) $X/X_0 = 0.4$, and (f) $X/X_0 = 0.25$. (Reproduced from Karunasena, H.C.P. et al., *J. Food Eng.*, 146, 209–226, 2015b, with permission from Elsevier.)

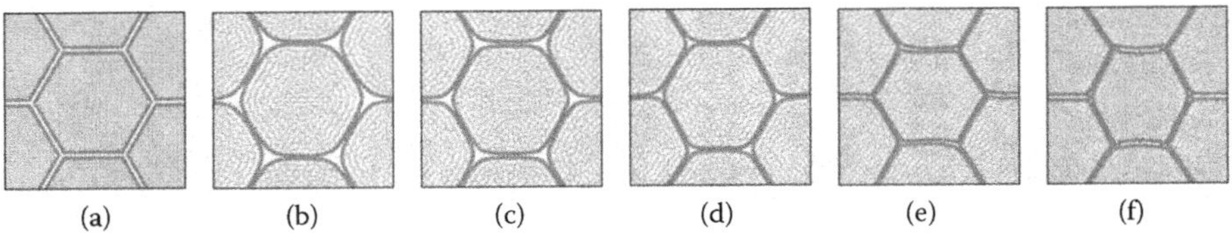

FIGURE 16.11 Apple tissue simulations at different states of dryness (enlarged view): (a) initial condition before simulations, (b) $X/X_0 = 1.0$, (c) $X/X_0 = 0.8$, (d) $X/X_0 = 0.6$, (e) $X/X_0 = 0.4$, and (f) $X/X_0 = 0.25$. (Reproduced from Karunasena, H.C.P. et al., *J. Food Eng.*, 146, 209–226, 2015b, with permission from Elsevier.)

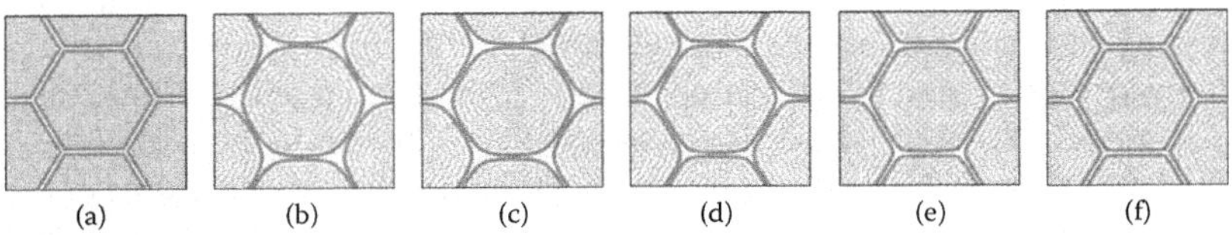

FIGURE 16.12 Potato tissue simulations at different states of dryness (enlarged view): (a) initial condition before simulations, (b) $X/X_0 = 1.0$, (c) $X/X_0 = 0.8$, (d) $X/X_0 = 0.6$, (e) $X/X_0 = 0.4$, and (f) $X/X_0 = 0.25$. (Reproduced from Karunasena, H.C.P. et al., *J. Food Eng.*, 146, 209–226, 2015b, with permission from Elsevier.)

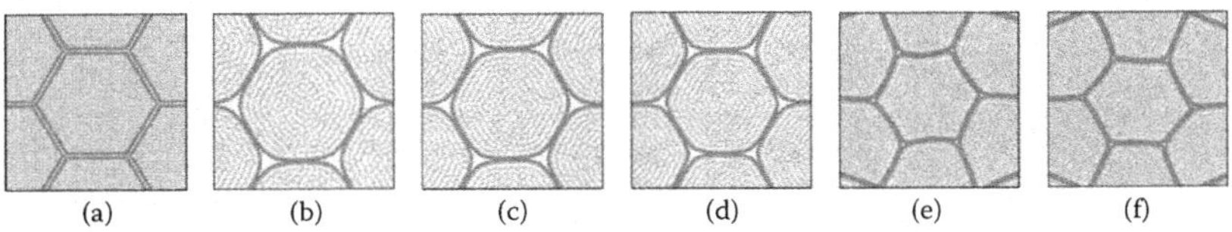

FIGURE 16.13 Carrot tissue simulations at different states of dryness (enlarged view): (a) initial condition before simulations, (b) $X/X_0 = 1.0$, (c) $X/X_0 = 0.8$, (d) $X/X_0 = 0.6$, (e) $X/X_0 = 0.4$, and (f) $X/X_0 = 0.25$. (Reproduced from Karunasena, H.C.P. et al., *J. Food Eng.*, 146, 209–226, 2015b, with permission from Elsevier.)

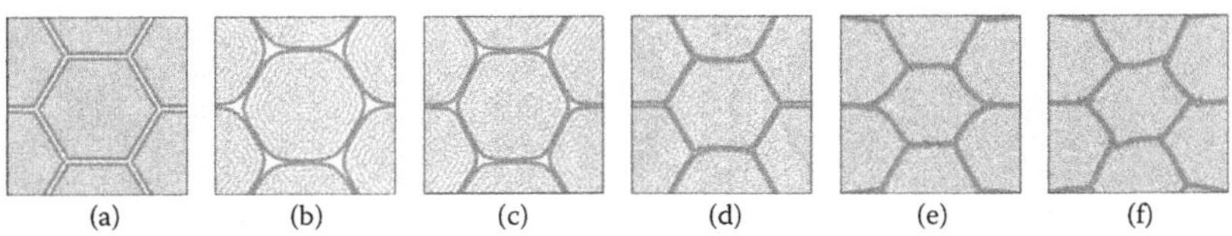

FIGURE 16.14 Grape tissue simulations at different states of dryness (enlarged view): (a) initial condition before simulations, (b) $X/X_0 = 1.0$, (c) $X/X_0 = 0.8$, (d) $X/X_0 = 0.6$, (e) $X/X_0 = 0.4$, and (f) $X/X_0 = 0.25$. (Reproduced from Karunasena, H.C.P. et al., *J. Food Eng.*, 146, 209–226, 2015b, with permission from Elsevier.)

16.4.4 Simulation of Porosity Development of Food Plant Tissues during Drying

Realistic tissues have cellular voids consisting of moist gaseous constituents. Using the same modeling approach discussed earlier, tissue porosity has been studied as it directly influences the deformation of a given tissue (Karunasena et al. 2015a). In Figure 16.15, both apple and grape tissues with 5%, 10%, 15%, and 20% porosity are comparatively presented. Here, the pores are assumed to be open pores connected with the atmosphere and simply modeled as cellular voids without

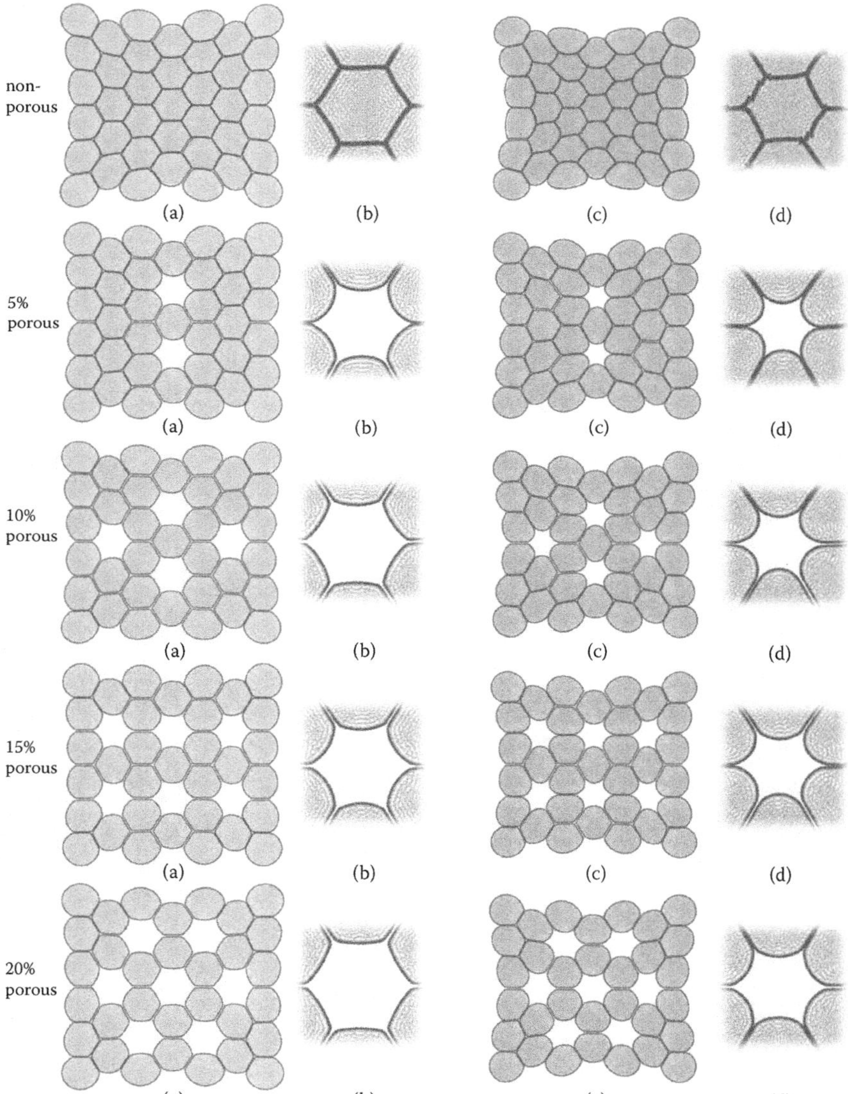

FIGURE 16.15 Comparison of shrinkage of dried porous tissues at $X/X_0 = 0.3$: (a) apple tissue, (b) a cell/ pore of apple tissue (enlarged), (c) grape tissue, and (d) a cell/pore of grape tissue (enlarged). (Reproduced from Karunasena, H.C.P. et al., *Biosyst. Eng.*, 132, 71–87, 2015a, with permission from Elsevier.)

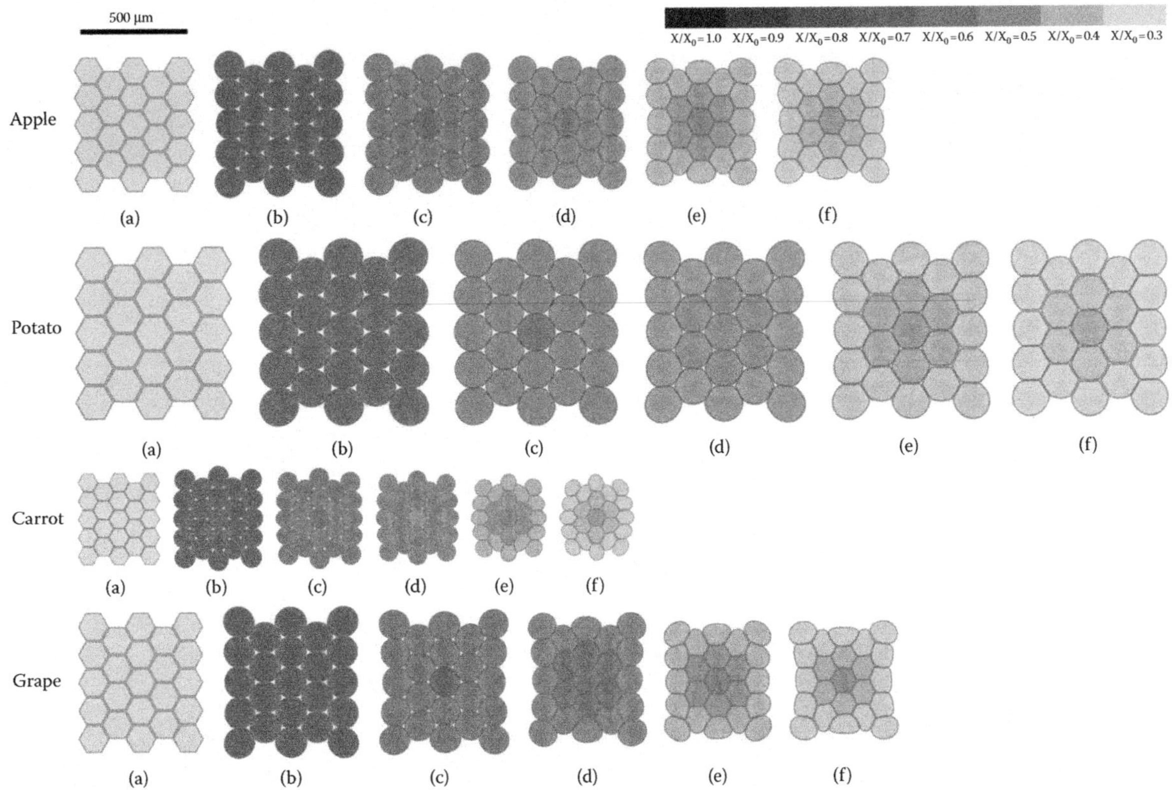

FIGURE 16.16 Tissue simulations at different states of dryness, incorporating case hardening: (a) initial condition before simulations, (b) $X/X_0 = 1.0$, (c) $X/X_0 = 0.8$, (d) $X/X_0 = 0.6$, (e) $X/X_0 = 0.4$, and (f) $X/X_0 = 0.3$ (These values correspond to any cell at the tissue boundary). (Reproduced from Karunasena, H.C.P. et al., *Drying Technol.*, 33 (6), 713–734, 2014g, with permission from Taylor & Francis Ltd.)

containing any constituents. In the case of tissues without pores, shrinkage of grape tissue is comparatively higher than the apple tissues, showing clearly identifiable cell wall wrinkling effects. The cell wall wrinkling is more intense toward the center of the tissue, mainly as a result of the larger number of neighboring cells undergoing contractions during drying. Also, when considering the porous tissues, it is observed that the pores of grape tissues are smaller in size when compared with the apple pores.

16.4.5 Simulation of Case Hardening of Food Plant Tissues during Drying

The same modeling technique has been used to study the case hardening of tissues during drying with the aid of tissues of different food plant materials as shown in Figure 16.16 (Karunasena et al. 2014g). Since case hardening resists moisture removal from the core of a given tissue, interior cell layers tend to have higher moisture contents and the level of shrinkage of those is limited compared to a tissue that is not subjected to case hardening during drying. Further, the simulation results clearly indicate that the tissues of different food varieties tend to undergo distinct deformations under case hardening.

16.5 FUTURE PROSPECTS OF NUMERICAL MODELING

For the purpose of predicting the morphological changes of food plant materials during drying, mainly grid-based and mesh-free-based approaches are available at present. Considering the current level of developments of both the approaches, there are numerous advantages as well as limitations. For future development purposes, knowing the limitations of the existing methods are important. In that regard, the key limitations of the existing grid-based numerical modeling techniques are difficulties in solving multiphase interactions, simulating extremely dried tissues (at present, simulations are possible only until about 30% moisture reduction), addressing the complex physics of intercellular spaces and middle lamella, solving discrete tissue systems, and large deformations with multiscale relationships such as cell wall wrinkling and related tissue deformations.

When it comes to mesh-free methods, the state-of-the-art plant tissue drying models of that kind are found to be superior compared to grid-based counterparts in treating large deformations, excessive moisture content reduction, representing the discrete tissue structures and accounting for subcellular details. Further, the mesh-free models have demonstrated favorable agreements with drying experiments when simulated for individual cell mechanisms so as tissue mechanisms such as different tissue types, porosity, shrinkage, and case hardening. Further, the mesh-free models are more flexible in incorporating new physics into the main formulae of the model. However, the current SPH–DEM–coupled mesh-free models for food plant tissue drying have a number of limitations, and therefore the below improvements are recommended as future work.

At present the model is basically 2-D, and therefore 3-D models need to be developed for studying real cellular deformations, which are 3-D by nature. The proposed SPH–DEM approach readily supports such modifications.

Other than simply treating the cell fluid as a Newtonian fluid, real cell interior constituents and their unique characteristics and physical properties have to be incorporated into the model.

Turgor pressure variation and cell wall drying are simply assumed to be directly proportional to the cell moisture content, which has to be experimentally verified, and the actual behavior needs to be incorporated into the model.

So as to minimize the complexity of the model and avoid excessive computational cost, the influence of temperature and time for cellular deformations are currently omitted (a moisture domain–based simulation approach has been used (Karunasena et al. 2014b) for all the simulation works), which have to be incorporated into the model.

Cell wall permeability, pectin layer properties, and other cell wall and cell fluid physical properties are assumed to be constant during drying. These properties need to be found experimentally for each case and those custom values need to be incorporated into modelling.

In the tissue model, intercellular water diffusion is not considered and all of the cells are assumed to undergo similar moisture removal during drying, which does not fully replicate actual moisture transfer phenomena of tissues undergoing drying. Therefore, diffusion and related physics would better be incorporated into modelling in future.

In tissue modeling, all the cells are assumed to be identical for the convenience of automatically generating the initial tissue geometry. However, the computer source code should be updated to accommodate realistic tissue structure to be modeled having cells of different sizes and shapes.

At present, the intercellular spaces in the tissue are simply represented as cellular voids with no material in it. Since moist gaseous mediums exist in real intercellular spaces, such material models have to be developed and incorporated to improve the model predictions.

The cell wall model can be improved by incorporating the plastic nature of cell wall deformations and having a multilayered cell wall with solid and fluid particles to replicate the physical thickness and moist nature of the cell wall.

Advanced wall–fluid interaction models may be used for improving the model performance. For instance, rather than using separate techniques to model the fluid and wall, a common technique could be used. SPH is recommended for this purpose, and such improvements would demand several alterations to the fundamental SPH formulation.

Since the model predictions highly depend on the turgor pressure hypothesis, single cell-based drying experiments may need to be conducted to obtain further details on actual turgor pressure variations of plant cells during drying.

Bigger tissue models could be developed by aggregating a large number of cells and even incorporating localized variations of cell moisture content and the turgor pressure. Such modifications would assist advanced bulk tissue morphological studies such as case hardening, bulk tissue wrinkling, porosity development, and density variation.

To further assist bulk material modeling, multiscale material models can be developed by incorporating these cellular level deformation characteristics. Such material models may be adapted to commercial software to suit a wider user community.

16.6 CLOSING REMARKS

Numerical modeling is a significant research area of interest in the field of food engineering. This chapter mainly focused on introducing different modeling techniques used for predicting morphological changes of food plant materials during drying, with more emphasis on most recent developments in the direction of microscale numerical models. For the purpose, different methods of numerical modeling are available, which can broadly be categorized into grid-based and mesh-free-based methods, and both have higher future expansion potentials. Among these two methods, the SPH–DEM coupled mesh-free approach elaborated in this chapter is comparatively better than the grid-based models in the current context. However, irrespective of the numerical modeling technique of interest, the current limitations existing in any of such techniques need to be identified to assist future developments. Such developments will essentially end up in comprehensive software tools readily available for food engineers so that physical properties and morphological changes can effectively be predicted to improve the product quality and process efficiency in the food industry.

REFERENCES

Abera, M. K., S. W. Fanta, P. Verboven, Q. T. Ho, J. Carmeliet, and B. M. Nicolaï. 2013. Virtual fruit tissue generation based on cell growth modelling. *Food and Bioprocess Technology* 6 (4):859–869. doi:10.1007/s11947-011-0775-4.

Askari, G. R., Z. Emam-Djomeh, and S. M. A. Mousavi. 2004. Effect of drying method on microstructural changes of apple slices. Paper Read at 14th International Drying Symposium (IDS 2004), São Paulo, Brazil.

Atluri, S. N. and T. Zhu. 1998. A new meshless local Petrov-Galerkin (MLPG) approach in computational mechanics. *Computational Mechanics* 22 (2):117–127.

Bai, Y., M. S. Rahman, C. O. Perera, B. Smith, and L. D. Melton. 2002. Structural changes in apple rings during convection air-drying with controlled temperature and humidity. *Journal of Agricultural and Food Chemistry* 50 (11):3179–3185. doi:10.1021/jf011354s.

Belytschko, T., Y. Krongauz, D. Organ, M. Fleming, and P. Krysl. 1996. Meshless methods: An overview and recent developments. *Computer Methods in Applied Mechanics and Engineering* 139 (1–4):3–47. doi:10.1016/s0045-7825(96)01078-x.

Belytschko, T., Y. Y. Lu, and L. Gu. 1994. Element-free Galerkin methods. *International Journal for Numerical Methods in Engineering* 37 (2):229–256. doi:10.1002/nme.1620370205.

Blum, A. 2011. Plant water relations, plant stress and plant production. In *Plant Breeding for Water-Limited Environments* 26. Springer, New York.

Bolin, H. R. and C. C. Huxsoll. 1987. Scanning electron microscope/image analyzer determination of dimensional postharvest changes in fruit cells. *Journal of Food Science* 52 (6):1649–1650. doi:10.1111/j.1365-2621.1987.tb05897.x.

Boukouvalas, C. J., M. K. Krokida, Z. B. Maroulis, and D. Marinos-Kouris. 2006. Density and porosity: Literature data compilation for foodstuffs. *International Journal of Food Properties* 9 (4):715–746. doi:10.1080/10942910600575690.

Bruce, D. M. 2003. Mathematical modelling of the cellular mechanics of plants. *Philosophical Transactions of the Royal Society of London. Series B: Biological Sciences* 358 (1437):1437–1444. doi:10.1098/rstb.2003.1337.

Campos-Mendiola, R., H. Hernández-Sánchez, J. J. Chanona-Pérez, L. Alamilla-Beltrán, A. Jiménez-Aparicio, P. Fito, and G. F. Gutiérrez-López. 2007. Non-isotropic shrinkage and interfaces during convective drying of potato slabs within the frame of the systematic approach to food engineering systems (SAFES) methodology. *Journal of Food Engineering* 83 (2):285–292. doi:10.1016/j.jfoodeng.2007.02.027.

Chaplain, M. A. J. 1993. The strain energy function of an ideal plant cell wall. *Journal of Theoretical Biology* 163 (1):77–97. doi:10.1006/jtbi.1993.1108.

Cleary, P. W. and J. J. Monaghan. 1999. Conduction modelling using smoothed particle hydrodynamics. *Journal of Computational Physics* 148 (1):227–264. doi:10.1006/jcph.1998.6118.

Crapiste, G. H., S. Whitaker, and E. Rotstein. 1988a. Drying of cellular material—I. A mass transfer theory. *Chemical Engineering Science* 43 (11):2919–2928. doi:10.1016/0009-2509(88)80045-9.

Crapiste, G. H., S. Whitaker, and E. Rotstein. 1988b. Drying of cellular material—II. Experimental and numerical results. *Chemical Engineering Science* 43 (11):2929–2936. doi:10.1016/0009-2509(88)80046-0.

Devahastin, S. and C. Niamnuy. 2010. Invited review: Modelling quality changes of fruits and vegetables during drying: A review. *International Journal of Food Science and Technology* 45 (9):1755–1767. doi:10.1111/j.1365-2621.2010.02352.x.

Ellero, M. and R. I. Tanner. 2005. SPH simulations of transient viscoelastic flows at low Reynolds number. *Journal of Non-Newtonian Fluid Mechanics* 132 (1–3):61–72. doi:10.1016/j.jnnfm.2005.08.012.

Español, P. and M. Revenga. 2003. Smoothed dissipative particle dynamics. *Physical Review E* 67 (2):026705.

Fanta, S. W., M. K. Abera, W. A. Aregawi, Q. T. Ho, P. Verboven, J. Carmeliet, and B. M. Nicolaï. 2014. Microscale modeling of coupled water transport and mechanical deformation of fruit tissue during dehydration. *Journal of Food Engineering* 124:86–96. doi:10.1016/j.jfoodeng.2013.10.007.

Frank, X. and P. Perré. 2010. The potential of meshless methods to address physical and mechanical phenomena involved during drying at the pore level. *Drying Technology* 28 (8):932–943. doi:10.1080/07373937.2010.497077.

Funebo, T., L. Ahrné, S. Kidman, M. Langton, and C. Skjöldebrand. 2000. Microwave heat treatment of apple before air dehydration—Effects on physical properties and microstructure. *Journal of Food Engineering* 46 (3):173–182. doi:10.1016/s0260-8774(00)00080-7.

Gao, Q. and R. E. Pitt. 1991. Mechanics of parenchyma tissue based on cell orientation and microstructure. *Transactions of ASAE* 34:232–238.

Ghysels, P., G. Samaey, B. Tijskens, P. Van Liedekerke, H. Ramon, and D. Roose. 2009. Multi-scale simulation of plant tissue deformation using a model for individual cell mechanics. *Physical Biology* 6 (1):016009.

Gingold, R. A. and J. J. Monaghan. 1977. Smoothed particle hydrodynamics—Theory and application to non-spherical stars. *Monthly Notices of the Royal Astronomical Society* 181:375–389.

Gu, Y. T. and G. R. Liu. 2006. Meshless techniques for convection dominated problems. *Computational Mechanics* 38 (2):171–182. doi:10.1007/s00466-005-0736-8.

Gu, Y. T., Q. X. Wang, and K. Y. Lam. 2007. A meshless local Kriging method for large deformation analyses. *Computer Methods in Applied Mechanics and Engineering* 196 (9–12):1673–1684. doi:10.1016/j.cma.2006.09.017.

Hassini, L., S. Azzouz, R. Peczalski, and A. Belghith. 2007. Estimation of potato moisture diffusivity from convective drying kinetics with correction for shrinkage. *Journal of Food Engineering* 79 (1):47–56. doi:10.1016/j.jfoodeng.2006.01.025.

Heredia, A., A. Jiménez, and R. Guillén. 1995. Composition of plant cell walls. *Zeitschrift für Lebensmittel-Untersuchung und -Forschung A* 200 (1):24–31. doi:10.1007/bf01192903.

Hettiaratchi, D. R. P. and J. R. O'Callaghan. 1978. Structural mechanics of plant cells. *Journal of Theoretical Biology* 74 (2):235–257. doi:10.1016/0022-5193(78)90074-7.

Hills, B. P. and B. Remigereau. 1997. NMR studies of changes in subcellular water compartmentation in parenchyma apple tissue during drying and freezing. *International Journal of Food Science and Technology* 32 (1):51–61. doi:10.1046/j.1365-2621.1997.00381.x.

Hosseini, S. M. and J. J. Feng. 2009. A particle-based model for the transport of erythrocytes in capillaries. *Chemical Engineering Science* 64 (22):4488–4497. doi:10.1016/j.ces.2008.11.028.

Jangam, S. V. 2011. An overview of recent developments and some R&D challenges related to drying of foods. *Drying Technology* 29 (12):1343–1357. doi:10.1080/07373937.2011.594378.

Jaros, M. and S. Pabis. 2006. Theoretical models for fluid bed drying of cut vegetables. *Biosystems Engineering* 93 (1):45–55. doi:10.1016/j.biosystemseng.2005.08.011.

Jarvis, M. C. 1998. Intercellular separation forces generated by intracellular pressure. *Plant, Cell and Environment* 21 (12):1307–1310. doi:10.1046/j.1365-3040.1998.00363.x.

Jeong, J. H., M. S. Jhon, J. S. Halow, and J. van Osdol. 2003. Smoothed particle hydrodynamics: Applications to heat conduction. *Computer Physics Communications* 153 (1):71–84. doi:10.1016/s0010-4655(03)00155-3.

Jeong, S. H., S. H. Park, and C. H. Kim. 2013. Simulation of morphology changes in drying leaves. *Computer Graphics Forum* 32 (1):204–215. doi:10.1111/cgf.12009.

Karaaslan, S. N. and I. K. Tunçer. 2008. Development of a drying model for combined microwave-fan-assisted convection drying of spinach. *Biosystems Engineering* 100 (1):44–52. doi:10.1016/j.biosystemseng.2007.12.012.

Karathanos, V. 1993. Collapse of structure during drying of celery. *Drying Technology* 11 (5):1005–1023. doi:10.1080/07373939308916880.

Karathanos, V. T., N. K. Kanellopoulos, and V. G. Belessiotis. 1996. Development of porous structure during air drying of agricultural plant products. *Journal of Food Engineering* 29 (2):167–183. doi:10.1016/0260–8774(95)00058-5.

Karunasena, H. C. P., Y. T. Gu, R. J. Brown, and W. Senadeera. 2014g. Numerical investigation of case hardening of plant tissue during drying and its influence on the cellular level shrinkage. *Drying Technology* 33 (6):713–734. doi:10.1080/07373937.2014.982759.

Karunasena, H. C. P., Y. T. Gu, R. J. Brown, and W. Senadeera. 2015a. Numerical investigation of plant tissue porosity and its influence on cellular level shrinkage during drying. *Biosystems Engineering* 132:71–87. doi:10.1016/j.biosystemseng.2015.02.002.

Karunasena, H. C. P., P. Hesami, W. Senadeera, Y. T. Gu, R. J. Brown, and A. Oloyede. 2014a. Scanning electron microscopic study of microstructure of gala apples during hot air drying. *Drying Technology* 32 (4):455–468. doi:10.1080/07373937.2013.837479.

Karunasena, H. C. P., W. Senadeera, R. J. Brown, and Y. T. Gu. 2014c. Simulation of plant cell shrinkage during drying—A SPH–DEM approach. *Engineering Analysis with Boundary Elements* 44:1–18. doi:10.1016/j.enganabound.2014.04.004.

Karunasena, H. C. P., W. Senadeera, R. J. Brown, and Y. T. Gu. 2014e. A particle based model to simulate microscale morphological changes of plant tissues during drying. *Soft Matter* 10 (29):5249–5268. doi:10.1039/C4SM00526K.

Karunasena, H. C. P., W. Senadeera, R. J. Brown, and Y. T. Gu. 2014f. A novel approach for numerical simulation of plant tissue shrinkage during drying. In *International Research Symposium on Postharvest Technology*, edited by Thilakarathne, B. M. K. S., H. M. A. P. Rathnayake, W. A. H. Champa, and W. M. C. B. Wasala. Research and Development Centre, Institute of Post Harvest Technology, Anuradhapura, Sri Lanka.

Karunasena, H. C. P., W. Senadeera, R. J. Brown, and Y. T. Gu. 2015b. Application of meshfree methods to numerically simulate microscale deformations of different plant food materials during drying. *Journal of Food Engineering* 146:209–226. doi:10.1016/j.jfoodeng.2014.09.011.

Karunasena, H. C. P., W. Senadeera, Y. T. Gu, and R. J. Brown. 2012c. A particle-based micromechanics approach to simulate structural changes of plant cells during drying. Paper Read at 4th International Conference on Computational Methods (ICCM 2012), November 25–28, 2012, Gold Coast, Australia.

Karunasena, H. C. P., W. Senadeera, Y. T. Gu, and R. J. Brown. 2014b. A coupled SPH-DEM model for micro-scale structural deformations of plant cells during drying. *Applied Mathematical Modelling* 38 (15–16):3781–3801. doi:10.1016/j.apm.2013.12.004.

Karunasena, H. C. P., W. Senadeera, Y. T. Gu, and R. J. Brown. 2014d. A meshfree model for plant tissue deformations during drying. *ANZIAM Journal* 55:C110–C137.

Keckes, J., I. Burgert, K. Fruhmann, M. Müller, K. Kölln, M. Hamilton, M. Burghammer, S. V. Roth, S. Stanzl-Tschegg, and P. Fratzl. 2003. Cell-wall recovery after irreversible deformation of wood. *Nature Materials* 2 (12):810–813. doi:10.1038/nmat1019.

Khan, A. A. and J. F. V. Vincent. 1990. Anisotropy of apple parenchyma. *Journal of the Science of Food and Agriculture* 52 (4):455–466. doi:10.1002/jsfa.2740520404.

Köhler, L. and H.-C. Spatz. 2002. Micromechanics of plant tissues beyond the linear-elastic range. *Planta* 215 (1):33–40. doi:10.1007/s00425-001-0718-9.

Konstankiewicz, K. and A. Zdunek. 2001. *Influence of Turgor and Cell Size on the Cracking of Potato Tissue*. Vol. 15. Polish Academy of Sciences, Institute of Agrophysics, Lublin, Poland.

Lee, C. Y., D. K. Salunkhe, and F. S. Nury. 1967. Some chemical and histological changes in dehydrated apple. *Journal of the Science of Food and Agriculture* 18 (3):89–93. doi:10.1002/jsfa.2740180302.

Leeratanarak, N., S. Devahastin, and N. Chiewchan. 2006. Drying kinetics and quality of potato chips undergoing different drying techniques. *Journal of Food Engineering* 77 (3):635–643. doi:10.1016/j.jfoodeng.2005.07.022.

Lewicki, P. P. and J. Drzewucka. 1998. Effect of drying on tissue structure of selected fruits and vegetables. Paper Read at Proceedings of the 11th International Drying Symposium 1998, Greece.

Lewicki, P. P. and G. Pawlak. 2003. Effect of drying on microstructure of plant tissue. *Drying Technology* 21 (4):657–683. doi:10.1081/drt-120019057.

Li, S. and W. K. Liu. 2002. Meshfree and particle methods and their applications. *Applied Mechanics Reviews* 55 (1):1–34.

Li, S. and W. K. Liu. 2004. *Meshfree Particle Methods*. Springer, Berlin, Germany.

Lin, T. T. and R. E. Pitt. 1986. Rheology of apple and potato tissue as affected by cell turgor pressure. *Journal of Texture Studies* 17 (3):291–313. doi:10.1111/j.1745-4603.1986.tb00554.x.

Liu, G. R. and Y. T. Gu. 2001a. A local radial point interpolation method (LRPIM) for free vibration analyses of 2-D solids. *Journal of Sound and Vibration* 246 (1):29–46. doi:10.1006/jsvi.2000.3626.

Liu, G. R. and Y. T. Gu. 2001b. A point interpolation method for two-dimensional solids. *International Journal for Numerical Methods in Engineering* 50 (4):937–951. doi:10.1002/1097-0207(20010210)50:4<937::aid-nme62>3.0.co;2-x.

Liu, G. R. and Y. T. Gu. 2002. Comparisons of two meshfree local point interpolation methods for structural analyses. *Computational Mechanics* 29 (2):107–121. doi:10.1007/s00466-002-0320-4.

Liu, G. R. and Y. T. Gu. 2004. Boundary meshfree methods based on the boundary point interpolation methods. *Engineering Analysis with Boundary Elements* 28 (5):475–487. doi:10.1016/S0955-7997(03)00101-2.

Liu, Z., W. Hong, Z. Suo, S. Swaddiwudhipong, and Y. Zhang. 2010. Modeling and simulation of buckling of polymeric membrane thin film gel. *Computational Materials Science* 49 (1, Supplement):S60–S64. doi:10.1016/j.commatsci.2009.12.036.

Liu, G. R. and M. B. Liu. 2003. *Smoothed Particle Hydrodynamics: A Meshfree Particle Method*. World Scientific Publishing Co., Singapore.

Loodts, J., E. Tijskens, C. Wei, E. L. S. Vanstreels, B. Nicolaï, and H. Ramon. 2006. Micromechanics: Simulating the elastic behavior of onion epidermis tissue. *Journal of Texture Studies* 37 (1):16–34. doi:10.1111/j.1745-4603.2006.00036.x.

Lozano, J. E., E. Rotstein, and M. J. Urbicain. 1980. Total porosity and open-pore porosity in the drying of fruits. *Journal of Food Science* 45 (5):1403–1407. doi:10.1111/j.1365-2621.1980.tb06564.x.

Lucy, L. B. 1977. A numerical approach to the testing of the fission hypothesis. *Astronomical Journal* 82:1013–1024.

Maroulis, Z. B., G. D. Saravacos, M. K. Krokida, and N. M. Panagiotou. 2002. Thermal conductivity prediction for foodstuffs: Effect of moisture content and temperature. *International Journal of Food Properties* 5 (1):231–245. doi:10.1081/jfp-120015604.

Martin, O., C. Osvaldo, N. Ganesan, S. Rakesh, and A. Weitnauer. 2006. Food dehydration. In *Handbook of Food Engineering*, 2nd Edition, edited by Heldman, D. R. and D. B. Lund, 601–744. CRC Press, Boca Raton, FL.

Mayor, L., M. A. Silva, and A. M. Sereno. 2005. Microstructural changes during drying of apple slices. *Drying Technology* 23 (9–11):2261–2276. doi:10.1080/07373930500212776.

Mazza, G. and M. Lemaguer. 1980. Dehydration of onion: Some theoretical and practical considerations. *International Journal of Food Science and Technology* 15 (2):181–194. doi:10.1111/j.1365-2621.1980.tb00930.x.

Moreira, R., A. Figueiredo, and A. Sereno. 2000. Shrinkage of apple disks during drying by warm air convection and freeze drying. *Drying Technology* 18 (1–2):279–294. doi:10.1080/07373930008917704.

Morris, J. P., P. J. Fox, and Y. Zhu. 1997. Modeling low Reynolds number incompressible flows using SPH. *Journal of Computational Physics* 136 (1):214–226. doi:10.1006/jcph.1997.5776.

Morris, J. P., Y. Zhu, and P. J. Fox. 1999. Parallel simulations of pore-scale flow through porous media. *Computers and Geotechnics* 25 (4):227–246. doi:10.1016/s0266-352x(99)00026-9.

Murase, H., G. E. Merva, L. J. Segerlind. 1980. Variation of Young's modulus of potato as a function of water potential. *Transactions of the ASAE* 23 (3):794–796.

Nilsson, S. B., C. H. Hertz, and S. Falk. 1958. On the relation between turgor pressure and tissue rigidity. II. *Physiologia Plantarum* 11 (4):818–837. doi:10.1111/j.1399-3054.1958.tb08275.x.

Pabis, S. 2007. Theoretical models of vegetable drying by convection. *Transport in Porous Media* 66 (1):77–87. doi:10.1007/s11242-006-9023-1.

Pabis, S. and M. Jaros. 2002. PH—Postharvest technology: The first period of convection drying of vegetables and the effect of shape-dependent shrinkage. *Biosystems Engineering* 81 (2):201–211. doi:10.1006/bioe.2001.0015.

Pan, T. W. and T. Wang. 2009. Dynamical simulation of red blood cell rheology in microvessels. *International Journal of Numerical Analysis and Modeling* 6 (3):455–473.

Pathmanathan, P., J. Cooper, A. Fletcher, G. Mirams, P. Murray, J. Osborne, J. Pitt-Francis, A. Walter, and S. J. Chapman. 2009. A computational study of discrete mechanical tissue models. *Physical Biology* 6 (3):036001.

Perré, P. 2011. A review of modern computational and experimental tools relevant to the field of drying. *Drying Technology* 29 (13):1529–1541. doi:10.1080/07373937.2011.580872.

Pitt, R. E. and D. C. Davis. 1984. Finite-element analysis of fluid-filled cell response to external loading. *Transactions of the ASABE* 27 (6):1976–1983.

Rahman, M. S. 1994. The accuracy of prediction of the freezing point of meat from general models. *Journal of Food Engineering* 21 (1):127–136. doi:10.1016/0260-8774(94)90099-x.

Rahman, M. S. 2003. A theoretical model to predict the formation of pores in foods during drying. *International Journal of Food Properties* 6 (1):61–72. doi:10.1081/jfp-120016624.

Rahman, M. S. 2008. Dehydration and microstructure. In *Advances in Food Dehydration*, edited by Ratti, C., 97–122. CRC Press, Boca Raton, FL.

Rahman, M. S., I. Al-Zakwani, and N. Guizani. 2005. Pore formation in apple during air-drying as a function of temperature: Porosity and pore-size distribution. *Journal of the Science of Food and Agriculture* 85 (6):979–989. doi:10.1002/jsfa.2056.

Ramos, I. N. 2010. *Integrated Approach on Field Solar Drying, Pilot Convective Drying and Microstructural Changes*. School of Biotechnology, Catholic University of Portugal, Porto, Portugal.

Ramos, I. N., T. R. S. Brandão, and C. L. M. Silva. 2003. Structural changes during air drying of fruits and vegetables. *Food Science and Technology International* 9 (3):201–206. doi:10.1177/1082013030335522.

Ramos, I. N., C. L. M. Silva, A. M. Sereno, and J. M. Aguilera. 2004. Quantification of microstructural changes during first stage air drying of grape tissue. *Journal of Food Engineering* 62 (2):159–164. doi:10.1016/s0260-8774(03)00227-9.

Ratti, C. 1994. Shrinkage during drying of foodstuffs. *Journal of Food Engineering* 23 (1):91–105. doi:10.1016/0260-8774(94)90125-2.

Rosselló, C., J. Canellas, S. Simal, and A. Berna. 1992. Simple mathematical model to predict the drying rates of potatoes. *Journal of Agricultural and Food Chemistry* 40 (12):2374–2378. doi:10.1021/jf00024a010.

Rotstein, E. and A. R. H. Cornish. 1978. Influence of cellular membrane permeability on drying behavior. *Journal of Food Science* 43 (3):926–934. doi:10.1111/j.1365-2621.1978.tb02456.x.

Sansiribhan, S., S. Devahastin, and S. Soponronnarit. 2010. Quantitative evaluation of microstructural changes and their relations with some physical characteristics of food during drying. *Journal of Food Science* 75 (7):E453–E461. doi:10.1111/j.1750-3841.2010.01739.x.

Sansiribhan, S., S. Devahastin, and S. Soponronnarit. 2012. Generalized microstructural change and structure-quality indicators of a food product undergoing different drying methods and conditions. *Journal of Food Engineering* 109 (1):148–154. doi:10.1016/j.jfoodeng.2011.09.019.

Senadeera, W. 2008a. The drying constant and its effect on the shrinkage constant of different-shaped food particulates. *International Journal of Food Engineering* 4 (8):1–6.

Senadeera, W., B. R. Bhandari, G. Young, and B. Wijesinghe. 2003. Influence of shapes of selected vegetable materials on drying kinetics during fluidized bed drying. *Journal of Food Engineering* 58 (3):277–283. doi:10.1016/s0260-8774(02)00386-2.

Senadeera, W., B. R. Bhandari, G. Young, and B. Wijesinghe. 2005. Modeling dimensional shrinkage of shaped foods in fluidized bed drying. *Journal of Food Processing and Preservation* 29 (2):109–119. doi:10.1111/j.1745-4549.2005.00017.x.

Shi, L., T.-W. Pan, and R. Glowinski. 2012. Deformation of a single red blood cell in bounded Poiseuille flows. *Physical Review E* 85 (1):016307.

Simal, S., C. Rosselló, A. Berna, and A. Mulet. 1994. Heat and mass transfer model for potato drying. *Chemical Engineering Science* 49 (22):3739–3744. doi:10.1016/0009-2509(94)00199-5.

Smith, A. E., K. E. Moxham, and A. P. J. Middelberg. 1998. On uniquely determining cell–wall material properties with the compression experiment. *Chemical Engineering Science* 53 (23):3913–3922. doi:10.1016/s0009-2509(98)00198-5.

Stanisław, P. 1999. The initial phase of convection drying of vegetables and mushrooms and the effect of shrinkage. *Journal of Agricultural Engineering Research* 72 (2):187–195. doi:10.1006/jaer.1998.0362.

Sterling, C. 1963. Texture and cell wall polysaccharides in foods. In *Recent Advances in Food Science*, edited by Leitch, J. M., and D. N. Rhodes, 259–279. London: Butterworths and Co.

Stukowski, A. 2010. Visualization and analysis of atomistic simulation data with OVITO—The open visualization tool. *Modelling and Simulation in Materials Science and Engineering* 18 (1):015012.

Suarez, C. and P. E. Viollaz. 1991. Shrinkage effect on drying behavior of potato slabs. *Journal of Food Engineering* 13 (2):103–114. doi:10.1016/0260-8774(91)90013-i.

Taiz, L. and E. Zeiger. 2010. Water and plant cells. In *Plant Physiology*, 73–84. Sinauer Associates, Sunderland, MA.

Tartakovsky, A. M. and P. Meakin. 2005a. Simulation of unsaturated flow in complex fractures using smoothed particle hydrodynamics. *Vadose Zone Journal* 4 (3):848–855. doi:10.2136/vzj2004.0178.

Tartakovsky, A. M. and P. Meakin. 2005b. A smoothed particle hydrodynamics model for miscible flow in three-dimensional fractures and the two-dimensional Rayleigh–Taylor instability. *Journal of Computational Physics* 207 (2):610–624. doi:10.1016/j.jcp.2005.02.001.

Tartakovsky, A. and P. Meakin. 2005c. Modeling of surface tension and contact angles with smoothed particle hydrodynamics. *Physical Review E* 72 (2):026301.

Tartakovsky, A. M. and P. Meakin. 2006. Pore scale modeling of immiscible and miscible fluid flows using smoothed particle hydrodynamics. *Advances in Water Resources* 29 (10):1464–1478. doi:10.1016/j.advwatres.2005.11.014.

Tartakovsky, A. M., P. Meakin, T. D. Scheibe, and R. M. Eichler West. 2007a. Simulations of reactive transport and precipitation with smoothed particle hydrodynamics. *Journal of Computational Physics* 222 (2):654–672. doi:10.1016/j.jcp.2006.08.013.

Tartakovsky, A., P. Meakin, and A. Ward. 2009. Smoothed particle hydrodynamics model of non-aqueous phase liquid flow and dissolution. *Transport in Porous Media* 76 (1):11–34. doi:10.1007/s11242-008-9230-z.

Tartakovsky, A. M., A. L. Ward, and P. Meakin. 2007b. Pore-scale simulations of drainage of heterogeneous and anisotropic porous media. *AIP Physics of Fluids* 19:103301.

Vakilha, M. and M. Manzari. 2008. Modelling of power-law fluid flow through porous media using smoothed particle hydrodynamics. *Transport in Porous Media* 74 (3):331–346. doi:10.1007/s11242-007-9199-z.

Van Liedekerke, P., P. Ghysels, E. Tijskens, G. Samaey, D. Roose, and H. Ramon. 2011. Mechanisms of soft cellular tissue bruising. A particle based simulation approach. *Soft Matter* 7 (7):3580–3591.

Van Liedekerke, P., P. Ghysels, E. Tijskens, G. Samaey, B. Smeedts, D. Roose, and H. Ramon. 2009. A particle based model to simulate plant cells dynamics. In 4th International SPHERIC Workshop. Nantes, France.

Van Liedekerke, P., P. Ghysels, E. Tijskens, G. Samaey, B. Smeedts, D. Roose, and H. Ramon. 2010a. A particle-based model to simulate the micromechanics of single-plant parenchyma cells and aggregates. *Physical Biology* 7 (2):026006.

Van Liedekerke, P., E. Tijskens, H. Ramon, P. Ghysels, G. Samaey, and D. Roose. 2010b. Particle-based model to simulate the micromechanics of biological cells. *Physical Review* E 81 (6):061906.

Vázquez-Quesada, A., M. Ellero, and P. Español. 2009a. Consistent scaling of thermal fluctuations in smoothed dissipative particle dynamics. *The Journal of Chemical Physics* 130 (3):034901–034907.

Vázquez-Quesada, A., M. Ellero, and P. Español. 2009b. Smoothed particle hydrodynamic model for viscoelastic fluids with thermal fluctuations. *Physical Review E* 79 (5):056707.

Verboven, P., G. Kerckhofs, H. K. Mebatsion, Q. T. Ho, K. Temst, M. Wevers, P. Cloetens, and B. M. Nicolaï. 2008. Three-dimensional gas exchange pathways in pome fruit characterized by synchrotron X-ray computed tomography. *Plant Physiology* 147 (2):518–527.

Vesenjak, M., and Z. Ren. 2008. Application of smoothed particle hydrodynamics method in engineering problems. In *Progress on Meshless Methods*, edited by Ferreira, A. J. M. and E. J. Kansa. Springer, New York.

Vishwakarma, V., A. K. Das, and P. K. Das. 2011. Steady state conduction through 2D irregular bodies by smoothed particle hydrodynamics. *International Journal of Heat and Mass Transfer* 54 (1–3):314–325. doi:10.1016/j.ijheatmasstransfer.2010.09.040.

Wang, C. X., L. Wang, and C. R. Thomas. 2004. Modelling the mechanical properties of single suspension-cultured tomato cells. *Annals of Botany* 93 (4):443–453. doi:10.1093/aob/mch062.

Wang, N. and J. G. Brennan. 1995. Changes in structure, density and porosity of potato during dehydration. *Journal of Food Engineering* 24 (1):61–76. doi:10.1016/0260-8774(94)p1608-z.

Weier, T. E., and C. R. Stocking. 1949. Histological changes induced in fruits and vegetables by processing. In *Advances in Food Research*, edited by Mrak, E. M. and F. Stewart George, 297–342. Academic Press, New York.

Wróblewski, P., K. Boryczko, and M. Kope. 2008. Modeling incompressible fluids by means of the SPH method: Surface tension and viscosity. In *Computational Science—ICCS 2008* edited by Bubak, M., G. van Albada, J. Dongarra, and P. Sloot, 600–609. Springer, Berlin and Heidelberg.

Wu, H. I., R. D. Spence, P. J. H. Sharpe, and J. D. Goeschl. 1985. Cell wall elasticity: I. A critique of the bulk elastic modulus approach and an analysis using polymer elastic principles. *Plant, Cell and Environment* 8 (8):563–570. doi:10.1111/j.1365-3040.1985.tb01694.x.

Wu, N. and M. J. Pitts. 1999. Development and validation of a finite element model of an apple fruit cell. *Postharvest Biology and Technology* 16 (1):1–8. doi:10.1016/s0925-5214(98)00095-7.

Wu, Y. L., G. R. Liu, and Y. T. Gu. 2005. Application of meshless local Petrov-Galerkin (MLPG) approach to simulation of incompressible flow. *Numerical Heat Transfer, Part B: Fundamentals* 48 (5):459–475.

Yesilata, B. and M. A. Aktacir. 2009. A simple moisture transfer model for drying of sliced foods. *Applied Thermal Engineering* 29 (4):748–752. doi:10.1016/j.applthermaleng.2008.03.048.

Zhu, Y. and P. J. Fox. 2001. Smoothed particle hydrodynamics model for diffusion through porous media. *Transport in Porous Media* 43 (3):441–471. doi:10.1023/a:1010769915901.

Zhu, H. X. and J. R. Melrose. 2003. A mechanics model for the compression of plant and vegetative tissues. *Journal of Theoretical Biology* 221 (1):89–101. doi:10.1006/jtbi.2003.3173.

Zhu, Y. and P. J. Fox. 2002. Simulation of pore-scale dispersion in periodic porous media using smoothed particle hydrodynamics. *Journal of Computational Physics* 182 (2):622–645. doi:10.1006/jcph.2002.7189.

Zhu, Y., P. J. Fox, and J. P. Morris. 1999. A pore-scale numerical model for flow through porous media. *International Journal for Numerical and Analytical Methods in Geomechanics* 23 (9):881–904. doi:10.1002/(sici)1096-9853(19990810)23:9<881::aid-nag996>3.0.co;2-k.

Zogzas, N. P., M. K. Krokida, P. A. Michailidis, and Z. B. Maroulis. 2002. Literature data of heat transfer coefficients in food processing. *International Journal of Food Properties* 5 (2):391–417. doi:10.1081/jfp-120005794.

17 Nondestructive Measurement of Quality Parameters of Vegetables during Drying by Optical Sensing Technology

Min Huang and Qibing Zhu
Jiangnan University

CONTENTS

17.1 INTRODUCTION

Drying is probably the oldest and most important method of vegetable preservation to prevent deterioration of quality, as it considerably decreases the water activity of the material, reduces the microbiological activity, and minimizes the physical and chemical changes during storage (Mayor and Sereno, 2004). The main advantage of this process is the reduction in weight and volume, which results in decreased packing, storage, and transportation costs and increased storability at room temperature (Mujumdar, 1995). During dehydration of vegetables, chemical (browning and other reactions) and physical (color, texture, shape, porosity, etc.) modifications take place at the same time owing to heating and loss of water. These modifications are the most important parameters in evaluating the drying quality of the final dried product, which can be evaluated by many different methods and techniques.

Expert and consumer panels are widely used to assess the quality of dried vegetables. It is relatively fast but hardly suitable for a large amount of samples, owing to many reasons such as observer's fatigue. Reliability can vary with food item groups and from person to person. The results are typically very subjective, depending on many non-food-related factors, and difficult to evaluate quantitatively. Chemical analysis (destructive measurement) often provides reliable results, but the cost of the analysis per measurement is quite high, and it can produce toxic waste. In addition, chemical analysis cannot be used for monitoring all food units because the sample is destroyed. To overcome the limitations of the aforementioned methods, interest in using innovative and nondestructive measurements for monitoring food quality during drying has been increasing in recent years.

17.2 TRADITIONAL QUALITY MEASUREMENTS OF VEGETABLES DURING DRYING

17.2.1 Color Measurement

Among the properties widely used for the analytical evaluation of qualities of dried products, color is an important quality attribute. This is because of the relation between color, flavor, and aroma of dehydrated products (Morris et al., 1953). The changes in color of thermally treated food materials occur because of chemical changes, such as pigment degradation (especially carotenoids and chlorophyll), and browning reactions, such as Maillard condensation of hexoses, and change in their color is due to the oxidation of highly unsaturated molecules upon exposure to air during processing (Park, 1987). As a visual appearance property, the color of dried products is usually negatively affected by drying and it influences consumer acceptability (Zielińska et al., 2005). Abnormal colors, especially those associated with deterioration in eating quality or with spoilage, cause the dried product to be rejected by the consumer, and color is therefore the first quality judgment made by a consumer on dried vegetables (Maskan, 2001; Zielińska and Markowski, 2011).

Maintenance of satisfactory color in the final product is a challenging aspect of thermal processing of dried vegetables. Tristimulus color coordinates are measured, and this is widely used as a quality indicator for monitoring of color changes during vegetable drying (Koca et al., 2007). Traditional instrumental measurement of color characteristic is performed by a colorimeter in the Hunter lab color system (Hiranuarachat et al., 2011; Zielińska and Markowski, 2011). For each sample, multiple measurements are made at different positions of the sample. Color difference (ΔE) is used to describe the color change in the fresh and dried samples and is calculated as follows:

$$\Delta E = \sqrt{\left(L_0^* - L^*\right)^2 + \left(a_0^* - a^*\right)^2 + \left(b_0^* - b^*\right)^2}, \tag{17.1}$$

where L_0^*, a_0^*, b_0^* are the color readings of a standard white plate at D_{65} illumination. For the measurements of samples during drying, L^*, a^*, and b^* represent the values of the lightness, redness, and yellowness of a dried sample, respectively (Devahastin et al., 2004; Huang et al, 2014b). This measurement is a contact method, which is required to make repeated evaluations of the same sample (Hutchings et al., 2002). Moreover, further error in the analysis might be brought because of undesirable uncertainty from contact measurement (Aguilera, 2003).

17.2.2 Moisture Content Measurement

Moisture content is an important indicator in determining the vegetable quality during the drying process. It also affects the safe storage period of dried vegetables. Therefore, a proper moisture content measurement is crucial in a vegetable-drying process (Huang et al., 2014b). Vegetables are generally dried to a 2%–4% moisture level (Desrosier and Desrosier, 1977) to yield satisfactory storage life and quality retention. The moisture content is expressed as a percentage. The percentage of moisture content was calculated using the following equation:

$$\text{MC}(\%) = \frac{M_{\text{wet}} - M_{\text{dry}}}{M_{\text{wet}}} \times 100, \tag{17.2}$$

where MC is the sample moisture content (kg/kg, d.b.), M_{wet} is the weight of sample before drying (kg), and M_{dry} is the weight of the dried sample (kg).

For moisture content measurement, the gravimetric oven method and the Karl Fischer titration are the commonly used laboratory methods for agricultural products and food products. The Karl Fischer titration can achieve more accurate measurements, but requires toxic chemicals and is expensive. The gravimetric method uses a convection oven at 105° for 7–8 h (or more hours) until a constant weight of samples is reached. Another method for detecting moisture content is based on chemical reaction. By detecting the product of the reaction, such as H_2, the moisture content can be calculated (Xie et al., 2013). However, these methods have limitations in that they are time consuming, inefficient, laborious, and destructive and do not allow the use of the samples for further analysis. Thus, a fast and nondestructive detection method will always be better.

17.2.3 Shrinkage Measurement

During the drying process, structural and physiochemical modifications take place, affecting the dried product quality. Among these modifications, the reduction of the external volume of the food is one of the most important. Changes in shape and dimensions occur because of heating and loss of water that cause stresses in the cellular structure, which are expressed as shrinkage. Shrinkage of food materials has negative consequences on the quality of the dehydrated product. Changes in shape, loss of volume, and increased hardness cause a negative effect on consumer acceptance in many cases (Adiletta et al., 2014). Determination of shrinkage is normally based on the concept of fluid replacement. The volume ratio is used to describe shrinkage, which is defined as

$$\% \text{ Shrinkage} = \frac{V_d}{V_0} \times 100, \quad (17.3)$$

where V_d is the volume of the sample after drying (cm^3) and V_0 is the initial volume of the sample before drying (cm^3).

Drying shrinkage percentage is measured using a displacement method (Thuwapanichayanan et al., 2008). As a simple example, the samples are placed in a 100 mL graduated cylinder containing 50 mL clean oil. The cylinder is held on a vortex vibrator until samples are fully merged into the oil and the volume changes are recorded (Chen et al., 2013a). In another method, the volume of a sample before and after drying is determined by weight and calculated density using a sample coated with wax. This method includes dipping of sample into melted wax for coating. Then the cooled wax-coated sample is weighed and put into a 60-mL density bottle (Wang et al., 2013a). These methods are laborious and only suitable for small samples. Furthermore, they cannot be used for online detection in the vegetable processing industry.

17.2.4 Texture Measurement

Texture is one of the most important quality attributes of dried and semidried foods, because it has been realized that textural behavior is related to the structure of the foods (Martynenko and Janaszek, 2014). In general, the texture quality of foods is negatively affected by conventional drying processes because the cellular structure and underlying tissue of the final product are damaged due to exposure of material to high temperatures (Telis et al., 2005). During the drying process, the mechanical properties change significantly. The fresh vegetable transforms into a dried vegetable or changes from a predominantly plastic behavior to a more elastic behavior. These texture changes are nonreversible. Physical parameters have been made to describe the viscoelastic behavior of dehydrated fruits and vegetables, which are obtained mostly with stress relaxation or compression using Maxwell's or compression models (Nicoleti et al., 2005). Rheological parameters such as maximum stress, maximum strain, and measure of stiffness of solid food are useful information for the textural characterization of the materials. One of the simplest approaches for estimating the rheological parameters is the compression test, which is performed by applying a constant deformation rate and recording force and deformation. Stress–strain compression curves are constructed until the fracture of specimens and the corresponding strain is called maximum strain. The maximum applied force is correlated with hardness or firmness of the material, while the ratio of maximum stress to maximum strain is correlated to crispness (Krokida et al., 2001). In the compression test, texture parameters are measured by using a universal testing machine (TA-XT Plus texture analyzer), which is also a destructive method.

17.2.5 Drying Uniformity Measurement

For a drying process, drying uniformity is one of the important parameters to evaluate the quality of dried products and drying technology. Drying uniformity is defined as the relative standard deviation (RSD, ratio of standard deviation to mean measurement value) of temperature, moisture content (MC), color, and shrinkage (Wang et al., 2013a,b). Among them, moisture content uniformity (MCU) and temperature uniformity during drying are the two important indexes that are of concern to researchers. MC nonuniformity can cause many disadvantages, such as relatively short safe storage life, reduced storage stability, and even mold growth and reduction in product use value at serious nonuniformity (Hashemi and Murray Douglas, 2003). Generally, MCU is calculated by definition as the ratio of MC standard deviation to MC mean measurement value. Temperature uniformity during drying is examined by measuring surface/central temperature of an individual sample from different locations (Wang et al., 2013b). The number of selected samples determines the accuracy of uniformity measurement. The more the selected samples, the higher the uniformity. Because the traditional laboratory methods for MC measurements of agricultural foods—the gravimetric oven method and the Karl Fischer titration (Aguilera, 2003)—are destructive measurements, the same samples cannot be used for further analysis, which limits the number of selected samples for calculating uniformity (Huang et al., 2015a).

17.3 OPTICAL SENSING TECHNIQUES FOR NONDESTRUCTIVE MEASUREMENT

17.3.1 Machine Vision and Image Data Processing

Machine vision, also known as computer vision or computer image processing, is an engineering technology that involves computer, optics, electronics, mathematics, information theory, artificial intelligence, pattern recognition, automation, mathematical morphology, and digital image processing. Machine vision is a branch of artificial intelligence technique that deals with simulating human vision. With a series of advantages that include nondestructiveness, reliability, and rapidity, machine vision has proven to be an effective and powerful technique in safety detection and quality evaluation of food and agricultural products.

A typical machine vision system is shown in Figure 17.1, which generally consists of four basic components: illumination system, sensor or camera, frame grabber or digitizer, and computer. Selecting appropriate light sources and identifying their suitable configurations should be a paramount consideration in the illumination system to obtain target images with the highest quality. A wide variety of light sources and lighting arrangements are available for nondestructive detection, with an extensive spectral region from ultraviolet (UV, 200–400 nm), visible (VIS, 380–780 nm), near-infrared (NIR, 780–2500 nm), and even thermal imaging regions. Generally, most of the applications of machine vision deal with the VIS spectrum (380–780 nm). Sensors or cameras are analogous to the human eyes in a machine vision system. Charge-coupled device and complementary metal-oxide semiconductor cameras are two commercially available and commonly used cameras.

Image processing is an important step in machine vision. To recognize and extract useful features from image data, several steps are performed after image acquisition, including image preprocessing, image segmentation, and feature extraction (Gunasekaran and Irudayaraj, 2001). Noise reduction, geometrical correction, gray-level correction, and correction of defocusing are the general methods used for image preprocessing. The aim of image segmentation is to partition the preprocessed image into multiple parts or objects, which are more meaningful and easier to analyze. The result of image segmentation is directly relevant to further processing and, eventually, to conclusions. Various robust and efficient segmentation algorithms and methods have been proposed in the literature, such as thresholding, histogram-based, clustering, edge detection, region-growing, watershed transformation, compression-based, and split-and-merge. Feature extraction simplifies the description of a large set of image data by transforming the input image into a reduced representation set of features, which is a special form of dimensionality reduction in image processing. Color, size, shape, and texture are the four major image features that have been extensively applied in numerous applications of machine vision. Color is the intensity

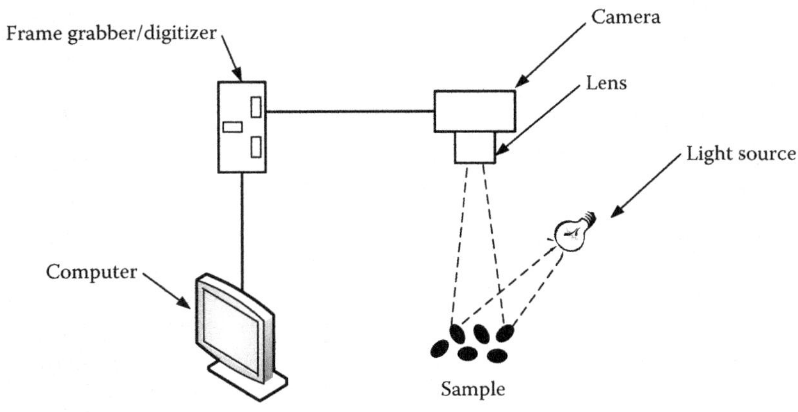

FIGURE 17.1 A typical computer vision system.

of pixels, size reflects the number of pixels, and shape describes the target boundary. Texture is normally the dependency between pixels and their neighboring pixels or the variation of pixel intensities. The superiority, disadvantages, and feasibilities of the different kinds of image features should be considered simultaneously to select the most suitable feature for particular applications (Zheng et al., 2006; Jackman and Sun, 2013). The ultimate goal of the overall image processing is to translate a raw image data into useful information for further analysis. Image or image feature analysis is crucial for a machine vision system to "think" and formulate intelligent decisions (Gunasekaran and Ding, 1994).

17.3.2 Near-Infrared Spectroscopy and Spectral Data Preprocessing

Infrared spectroscopy deals with the infrared region of the electromagnetic spectrum, which has a longer wavelength and lower frequency than visible light. The most commonly used spectroscopy ranges are near-infrared (NIR, 780–2500 nm) and mid-infrared (MIR, 2500–25000 nm) regions. NIR spectroscopy is based on molecular overtone and combined vibration and can typically penetrate deeper into a sample than MIR. Fourier transform infrared (FTIR) spectroscopy, which is used to detect radiation in the MIR region, is a measurement technique capable of recording infrared spectra. As one of the most powerful FTIR methods, attenuated total reflectance spectroscopy has been widely used in biological and liquid sample analysis. Diffuse reflectance infrared Fourier transform spectroscopy (DRIFTS) is a sampling technique that has been developed for infrared analysis of powder materials and turbid liquids (Gunasekaran and Irudayaraj, 2001). A typical NIR spectroscopy system generally consists of four basic components: light source, spectrometer, optical fiber, and computer (Figure 17.2).

The main objectives of the spectral preprocessing are to reduce or remove effects in a raw data with no relevant information, thereby improving the linear relationship between the spectral signals and analyte concentrations. Scatter correction and

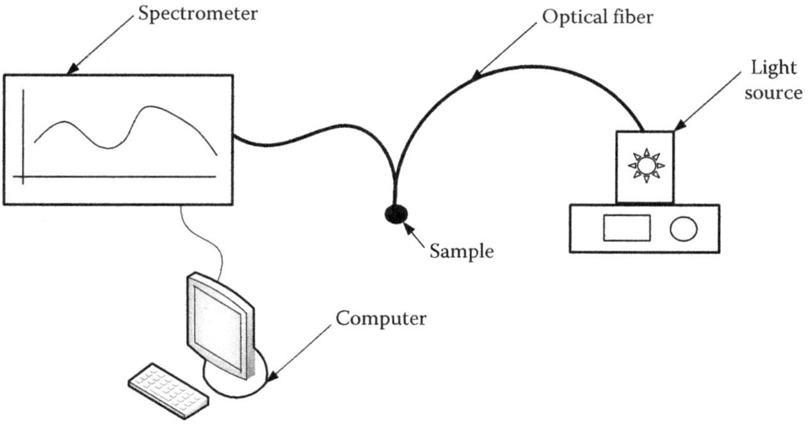

FIGURE 17.2 Schematic of the near-infrared spectral measurement system.

derivation are the most widely used methods. Scatter correction includes multiplicative scatter correction (MSC) or extended MSC, standard normal variate (SNV), detrending, baseline correction (BLC), and normalization. MSC is used to compensate for additive and/or multiplicative effects in spectral data. Detrending is applied to remove nonlinear trends in spectroscopic data. SNV is usually performed to remove scatter effects by centering and scaling each individual spectrum, which is sometimes used in combination with detrending to reduce multicollinearity, baseline shift, and curvature in spectroscopic data. BLC is mainly used to adjust the spectral offset, and normalization is performed to obtain all data on approximately the same scale. Some information hidden in a spectrum may be more easily revealed when using derivation methods, which include finite difference, Savitzky–Golay, and Norris–Williams. The frequently used method is Savitzky–Golay derivative. Some other techniques are also used in data preprocessing, including center and scale, smoothing, compute general, correlation-optimized warping, deresolve, noise, orthogonal signal correction, spectroscopic transformation, quantile normalization, and interpolation.

17.3.3 Hyperspectral Imaging and Spectral Processing

Hyperspectral imaging (also known as imaging spectroscopy or imaging spectrometry) originated from remote sensing, which was introduced into the field of food quality and safety analysis and assessment (Goetz et al., 1985; Lawrence et al., 2003; Mehl et al., 2004). Generally, machine vision and NIR spectroscopy can only provide spatial or spectral information. However, hyperspectral imaging, which integrates the main advantage of machine vision and NIR spectroscopy, can simultaneously obtain spatial and spectral information of the object using only one system. As a consequence, hyperspectral imaging has been widely used by researchers in the evaluation of exterior quality and prediction of internal composition of food and agricultural products.

Hyperspectral images or hyperspectral image cubes, hypercube, or spectral cubes are three-dimensional (3D) in nature, which include two spatial dimensions (x rows × y columns) and one spectral dimension (of λ wavelengths). The hyperspectral image $I(x, y, \lambda)$ can be viewed as an image $I(x, y)$ at each individual wavelength λ or as a spectrum $I(\lambda)$ at each individual pixel (x, y) (Wu and Sun, 2013). The three common sensing modes for hyperspectral imaging are reflectance, transmittance, and interaction. External quality features are typically detected using reflectance mode, whereas internal component concentration and internal defects of relative transparent materials are determined using transmittance mode. Interactive mode can detect more information about the sample and has less surface effects compared with reflectance mode (Qin, 2010).

The basic components for constructing a hyperspectral imaging system include light sources, wavelength dispersion devices, and area detectors. Numerous types of light sources are available for hyperspectral imaging applications, including halogen lamps, light-emitting diodes (LEDs), laser, and tunable sources. As the center of a hyperspectral imaging system, wavelength dispersion devices are used for dispersing broadband light into different wavelengths. The commonly used wavelength dispersion devices include imaging spectrographs, filter wheels, acousto-optic tunable filters (AOTF), liquid crystal-tunable filters (LCTF), Fourier transform imaging spectrometers, and single-shot imagers. The function of an area detector is to collect light from the wavelength dispersion device that carries useful information and then measure the intensity of the light by converting radiation energy into electrical signals. The two major types of solid-state area detectors used in hyperspectral imaging systems are CCD and CMOS cameras. CMOS cameras are popular in the consumer electronics market, but CCD cameras are the dominant devices in the field of technical applications, such as hyperspectral imaging (Wu and Sun, 2013). A VIS/NIR hyperspectral imaging system (Figure 17.3) was developed, with essential components including computer, light source, hyperspectral imager, reflection and transmission device, and transport platform (Huang et al., 2015a).

Hyperspectral images consist of a two-dimensional (2D) spatial information and a one-dimensional spectral information and can be viewed as an image $I(x, y)$ at each individual wavelength λ. Therefore, the methods or techniques used in machine vision for image processing and analysis and in NIR spectroscopy for spectral preprocessing can also be used in hyperspectral image processing and analysis. As a disadvantage in the huge amount of data, selection of optimal wavelengths is usually performed to remove redundant information and simplify calculation. Numerous mathematical algorithms and knowledge-based approaches for selecting optimal wavelengths are available in the literature. Correlation coefficients, loading and regression coefficients, analysis of spectral differences, spectrum derivatives, and stepwise regression are some typical wavelength selection methods. Extensive literature is available for wavelength selection using principal component analysis (PCA) loading (Shahin and Symons, 2011; Zhang et al., 2012; Kaliramesh et al., 2013), regression coefficients (Bhuvaneswari et al., 2011; Kong et al., 2013), and

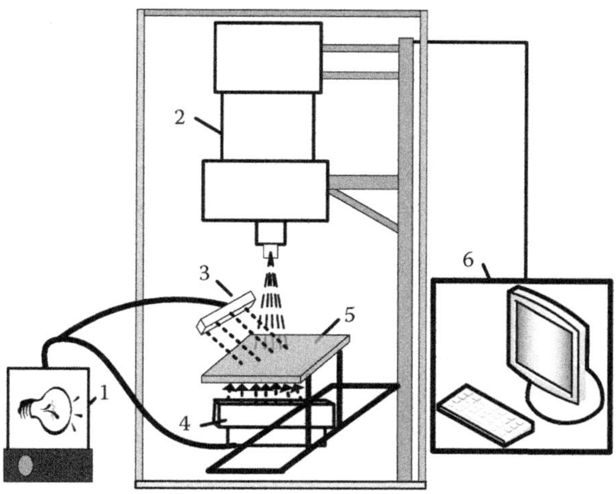

FIGURE 17.3 A VIS/NIR hyperspectral imaging system. Note: light source (1), hyperspectral imager (2), reflection device (3), transmission device (4), transport platform (5), and computer (6).

variable importance in the projection score (Xing et al., 2009; Serranti et al., 2013). Successive projection algorithms, uninformative variable elimination, and genetic algorithms are the three classical and optimal wavelength selection algorithms that have been widely used in hyperspectral imaging for quality and safety evaluation of food and agricultural products.

17.3.4 X-Ray Micro-Computed Tomography and Image Processing

X-ray micro-computed tomography (X-ray CT) is an innovative radiographic imaging technique, which is nondestructive and noninvasive and is used for internal examination of the structural arrangement of products in 3D imaging at a relatively high spatial resolution (Landis and Keane, 2010). Owing to its nondestructive character, X-ray CT allows scanning of the entire sample with its large field of view without any sample preparation. This technique has been applied in the field of food engineering (Léonard et al., 2008).

X-ray CT evaluates the internal structure of a sample using an X-ray source and a detector assembly to accumulate data from a thin projected slice of a sample (Kotwaliwale et al., 2014). The basic principle behind the CT is based on image contrast that is produced by variations in the X-ray attenuation that includes absorption and scattering. Figure 17.4 schematically demonstrates the acquisition principle. During scanning, a sample is mounted on a high-resolution rotary stage and illuminated with X-rays. The X-rays pass through the object in many different directions and along different pathways to create an image illustrating variation in density at numerous points in a 2D slice (Lim and Barigou, 2004). As the sample rotates, a series of 2D radiographs or projection images are acquired with a 2D detector. The detector records the object through which the conical X-ray beam transverses.

Image processing and analysis is required to visualize CT data and to extract suitable information from the image. For microstructural analysis information from the sample volume, density, porosity, object surface-to-volume ratio, particle size, and sample thickness can be obtained. X-ray CT and image analysis are nondestructive tools capable of scanning a whole sample to provide information on pore volume and size distributions and density variations (Schoeman et al., 2016). A typical image processing and analysis procedure is schematically illustrated in Figure 17.5, including X-ray 2D images acquisition, 3D images reconstruction, noise reduction, image

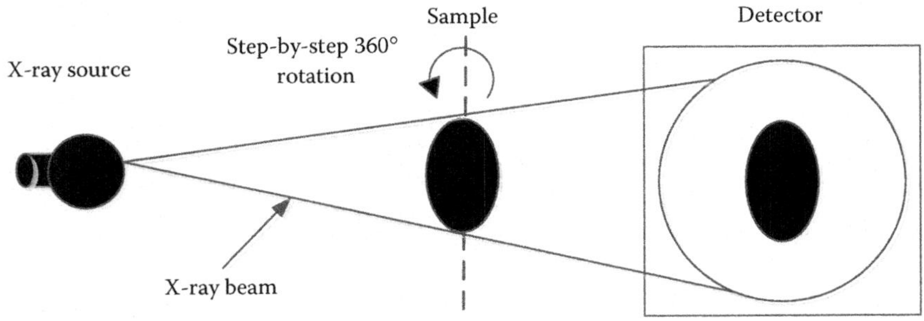

FIGURE 17.4 Schematic illustration of the measurement principle of X-ray CT.

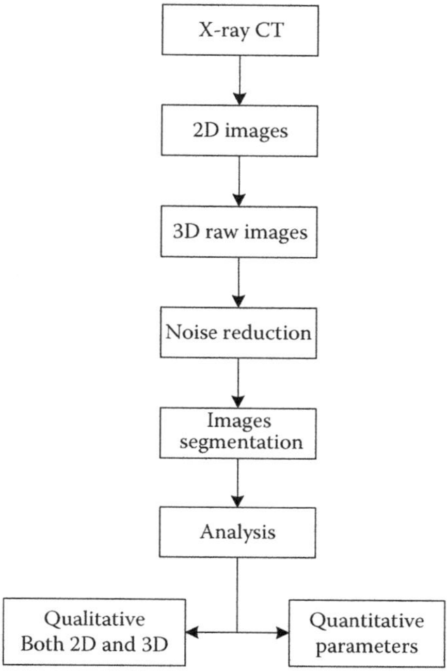

FIGURE 17.5 Schematic illustration of a typical image processing and analysis procedure.

segmentation, and analysis. 3D images are reconstructed by the data from numerous X-ray 2D images processed with a computer. During image processing, the 3D images are initially smoothed using filters (e.g., Gaussian or Median) to reduce random noise. This step is followed by segmentation, where the volume is partitioned into voxel groups of each region of interest (ROI) in the sample. Thus, the gray-scale slices are transformed into a binary layout that consists only of solid (black) and void (white) pixels (Schoeman et al., 2016). Segmentation is usually done using thresholding techniques. Finally, image analysis is used to qualitatively and quantitatively extract visual information and morphometric parameters to characterize the microstructure of a product.

17.3.5 Magnetic Resonance Imaging and Image Processing

Magnetic resonance imaging (MRI) has achieved general acceptance for the diagnosis and assessment of clinical conditions following the widespread introduction of medical imaging systems during the 1980s (Clark et al., 1997). MRI is a unique technology that measures the magnetic properties of spins that can then be related to the physical or chemical properties of subjects. One of the most important advantages of MRI is that the internal images in two or three dimensions can be obtained noninvasively and nondestructively (Schmidt et al., 1996). This advantage makes MRI an attractive technique to research and quantify dynamic phenomena that occur during drying of fruits and vegetables. Numerous publications in the literature have indicated that MRI can be used to noninvasively evaluate important quality attributes of food products (Chen et al., 2013b).

MRI is based on the principles of nuclear magnetic resonance (NMR) spectroscopy. NMR is the physical process in which the nucleus, whose magnetic moment is not zero, resonantly absorbs radiation of a certain frequency under external magnetic field. Detectors detect and receive the NMR signals released as electromagnetic radiation; these signals can then be sent to the computer and be converted into the 2D image through data processing. Generally, the MRI system includes the magnet and power-supply equipment that can produce a wide range of uniform, stable, and constant magnetic fields; a set of gradient magnetic field coil; a controller and power-driven equipment; a radio-frequency (RF) system; a computer system with large storage capacity for data collection and processing; and some auxiliary equipment (Chen et al., 2013b).

A 3D image is compiled from multiple 2D images, which are produced from any plane of view. The 3D image can be rotated and manipulated to be better able to detect tiny changes of structures within the food object. The combination of MRI and image analysis can present decisive capabilities for detection of fruit and vegetables.

17.3.6 Multivariate Data Analysis

Subsequent to image or spectral data processing, multivariate analysis is performed to determine a mathematical relationship between the treated data and the reference chemical and/or physical properties of food products for solving practical application problems. This chapter mainly focused on the classification and prediction problem. Numerous techniques and mathematical algorithms are available for different requirements of applications. The widely used techniques among the classification applications include *K*-nearest neighbor (KNN), linear discriminant analysis (LDA), quadratic discriminant analysis (QDA), partial least squares-discriminant analysis (PLS-DA), soft independent modeling of class analogy (SIMCA), artificial neural network (ANN), and support vector machines (SVMs). Moreover, the following techniques can be found in the literature: nearest mean classifier, stepwise discriminant analysis, regularized discriminant analysis, extended canonical variate analysis, and classification and influence matrix analysis. The mathematical algorithms primarily used in the prediction problems include classical least squares (CLS), inverse least squares (ILS), principal component regression (PCR), partial least squares regression (PLSR), ANN, multiple linear regression (MLR), and least squares support vector machines (LS-SVM). More information about these classification and prediction algorithms can be found in the studies of Ballabio and Todeschini (2009) and Blanco Romía and Alcalà Bernàrdez (2009).

17.3.7 Software for Data Processing and Analysis

Numerous image processing and analysis software are publicly available, such as MATLAB®, SciLab, NumPy, ImageJ, and MaZda. Moreover, a wide variety of commercial software for spectral data preprocessing and multivariate analysis are available, such as Unscrambler (CAMO AS, Oslo, Norway), PLS Toolbox (Eigenvector, Research, Inc., USA), LS-SVM toolbox (Suykens, Leuven, Belgium, www.esat.kuleuven.be/sista/lssvmlab/), WinISI (Infrasoft International, Port Matilda, PA), ParLeS (Sydney, Australia, http://sydney.edu.au/agric/acpa/people/rvrossel/soft01.htm), SAS and JMP (SAS Institute, Inc., Cary, NC), TQ Analyst and OMNIC (Thermo Fisher Scientific Inc., Waltham, MA), OPUS (Bruker Optics, Ettlingen, Germany), and R statistical (www.r-project.org). Some of these methods are specifically developed for the image or spectral processing, which greatly simplify and visualize the processing. In addition, sophisticated image or spectral processing can also be achieved using advanced programming software, such as MATLAB.

17.4 APPLICATIONS IN NONDESTRUCTIVE MEASUREMENT OF QUALITY PARAMETERS OF VEGETABLES

The general methodology of optical detection techniques for evaluating vegetable quality during drying is shown in Figure 17.6. First, optical information is collected from prepared vegetable samples, and different features are extracted using different techniques, where image, spectrum, and combined image–spectrum features are obtained through machine vision, NIR spectroscopy, and hyperspectral imaging, respectively. Second, relational mathematical models, including those for calibration and validation, are developed. Calibration models are implemented using the extracted feature data and known dried vegetable quality parameters, whereas validation models are used to verify the generalizability of calibration models by using unknown dried samples. Finally, the developed models are used to detect dried vegetable quality, particularly the quality of the exterior and internal components (Huang et al., 2015b).

17.4.1 Nondestructive Measurement of Vegetables Using Machine Vision

In recent years, machine vision has been widely studied in variety identification, physical properties detection, and quality grading of many agricultural products including vegetables. Machine vision, in combination with laser diodes emitting at 532 and 635 nm, was investigated to monitor changes in moisture content of red, yellow, and green bell pepper during drying (Romano et al., 2012), and the feasibility of using machine vision alone to analyze color changes during drying is assessed in alternative to the colorimeter. Computer vision was used to analyze the effect of drying on shrinkage, color, and image texture of apple discs during drying (Fernández et al., 2005). Depending on external image features at different stages of drying by a Euclidean distance classifier, the classification model provided 95% classification accuracy coupled with the parameters of shape (area, perimeter, Fourier energy), color (L^*, a^*, b^*), and texture with drying time. A low-cost dual-view computer vision system was used to measure volume and color co-occurrence image textural features of apple slices to find the end of the drying process by comparing physical texture parameters and moisture content (Sampson et al., 2014). Eleven image texture features correlated well with the moisture content, $R^2 > 0.9$ (R is correlation coefficient). Machine vision and ANN coupled with genetic algorithm were used to

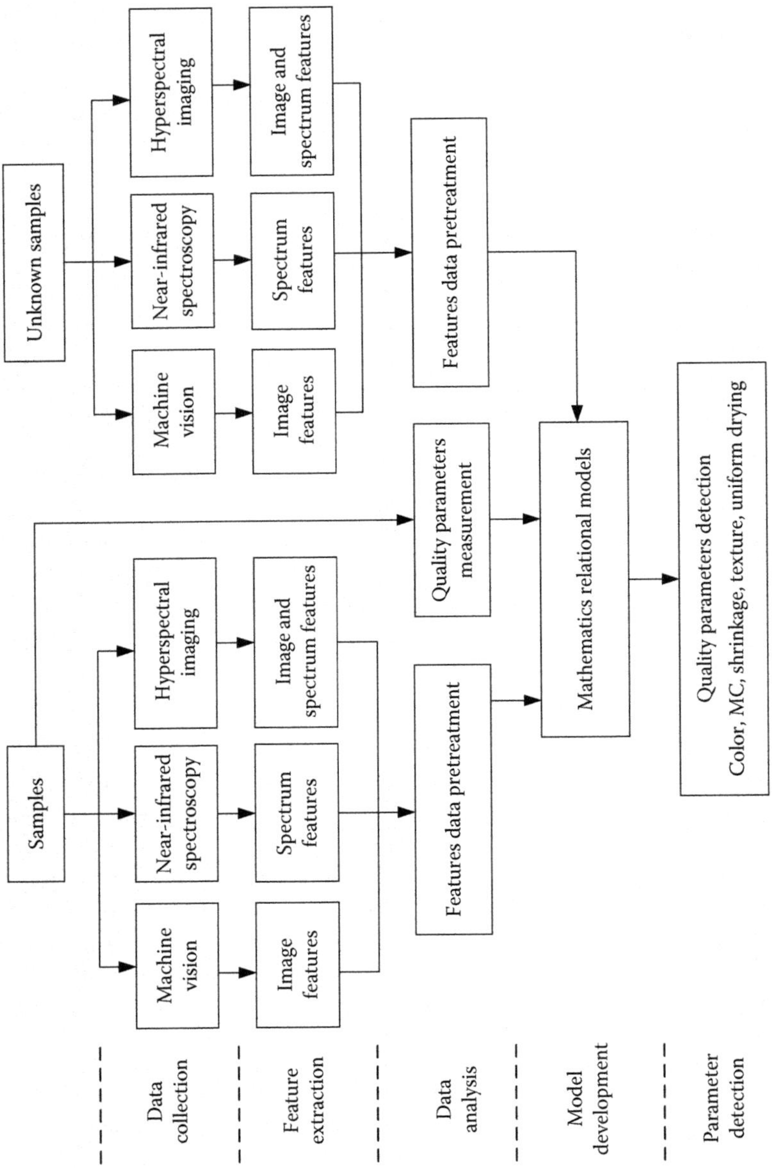

FIGURE 17.6 Block diagram of vegetable quality evaluation by optical detection techniques.

evaluate loss of moisture content and moisture content of apples during drying, and provided a high precision of the predictive model ($R^2 = 0.95$) (Trinca et al., 2014). In addition, digital image processing technology was used to analyze the carotenoid content of carrot powder from different drying techniques with different preprocessing methods (fresh carrot, cured carrot, and freeze thawing carrot) (Gong et al., 2015), which established an estimation model between the carotenoid content and color characteristics. A study used the laser light backscattering imaging technique as a monitoring tool during drying (Romano et al., 2008). LDA coupled with the parameters extracted from images was performed to classify slices according to drying time. The results show that there is significant relationship between changes in backscattering area and moisture content during drying of banana slices, especially at lower temperatures.

With the rapid development of electronics and computers, many commercially and individually customized machine vision systems are available and have been widely used in the food industry. Previous research indicated that machine vision has a great potential in automatic detection during drying. However, previous studies also showed that only some visible characteristics (such as size and color) can be captured by most of the cameras used in the machine vision system. Thus, the machine vision system is not suitable for examining internal quality attributes, even though some linear relationships between the invisible quality attributes and the visible features exist. Therefore, more efforts should be made to improve and extend its function for more advanced applications.

17.4.2 Nondestructive Measurement of Vegetables Using Near-Infrared Spectroscopy

NIR spectroscopy is such a rapid and nondestructive analysis technique for the measurement of quality attributes of fruits and vegetables (Nicolaï et al., 2007). NIR light is irradiated on the products, and the absorbance spectrum is collected and analyzed. This radiation is highly penetrative and thus can be applied to the sample without any preparation. Moreover, the spectra are not very discriminative, but they can give quantitative information on the major organic components of food products. Therefore, NIR spectroscopy has been used successfully as a nondestructive technique to measure the internal quality parameters of vegetables.

NIR reflectance spectra 1108–2490 nm were investigated to quantify the essential oil content and composition parameters for caraway, coriander, carrot, dill, and fennel (Schulz et al., 1998). Boeriu et al. (1998) used NIR spectroscopy over the range 1000–2500 nm for characterization of cell wall pectins in green beans. Roy et al. (1993) studied the determination of the moisture content of intact individual mushrooms by VIS/NIR reflectance spectroscopy (600–2200 nm). Modified partial least square regression with first derivative and scatter correction yielded best performance with a standard error of cross-validation of 0.64%. Steuer and Shulz (2003) investigated the accuracy and transferability of NIR (1100–2498 nm) calibrations for estimating the content and composition of the volatile fraction in fennel fruits. Khuriyati et al. (2004) applied NIR spectroscopy (305–1100 nm) for dry matter calibration in intact hydroponic tomatoes. By using six-position averaged spectra, the calibration equation with the ratio of standard deviation of reference data in validation set to standard error of prediction of 5.05 was obtained. Penchaiya et al. (2009) investigated NIR reflectance spectroscopy over the range of 780–1760 nm and partial least squares to measure the soluble solids content and firmness of bell pepper coupled with Savitzky-Golay second derivative preprocessing and extended multiplicative signal correction. During heating, NIR reflectance spectroscopy over the range of 400–2500 nm was used to predict the change in sensory quality of carrot discs coupled with partial least squares regression method (De Belie et al., 2003). Three different measuring configurations of NIR spectroscopy system were investigated for prediction of dry matter content in whole, unpeeled potatoes, including two off-line measurements and one online measurement, with the online configuration obtaining better prediction performance (Helgerud et al., 2015). According to previous studies, infrared spectroscopy has been proven to be a powerful technique in quality analysis of various fruits and vegetables.

17.4.3 Nondestructive Measurement of Vegetables Using Hyperspectral Imaging

Among the rapid nondestructive technologies for measuring the drying qualities of food products, machine vision and NIR spectroscopy are two important methods. However, the conventional machine vision method can only acquire average image information within the visible range. NIR spectroscopy, although used in a wide range of wavelengths, can only acquire spectral information and cannot obtain the spatial information of the samples. Thus, these methods have limitations in that they cannot provide spectral and spatial information simultaneously, which may result in the loss of useful information (Huang et al., 2014a). As a relatively novel nondestructive technology, hyperspectral imaging integrates the advantages of machine vision and visible infrared spectroscopy, while overcoming the drawbacks of both techniques when used alone. Hyperspectral imaging can provide more detailed or complete information, including internal structure characteristics, morphological information, and chemical composition (Huang et al., 2013). This technology has been applied to the nondestructive measurement of agricultural products for evaluating internal quality (Cheng et al., 2004; Liu et al., 2006; Ariana and Lu, 2008; Huang and Lu, 2010; Li et al., 2012). Thus, this technology may also be used as an alternative for predicting the quality parameters of vegetables during drying.

As a potential nondestructive technique, hyperspectral imaging was investigated to detect moisture loss of carrot samples during storage (Firtha, 2007, 2009). Experiments were carried out to study spectral changes of different cultivars and different tissues of carrot stored under different conditions. Determination of the proper operator on different tissues could help to analyze and model the drying process and to control storage. A hyperspectral transmittance imaging system was developed to detect insect-damaged vegetable soybeans (Huang et al., 2013). Four statistical image features (minimum, maximum, mean, and standard deviation) coupled with support vector data description provided 95.6% classification accuracy. Huang et al. (2014a) used hyperspectral reflectance imaging technique to predict the color and moisture content of vegetables soybean during drying coupled with active contour model. Hyperspectral images of fresh and dried soybeans over the spectral region between 400 and 1000 nm were acquired. Mean reflectance and image entropy parameters were used to predict the color and moisture content of the dried soybeans coupled with partial least squares regression method. Better prediction results for both color and moisture content were achieved using the mean reflectance data (with correlation coefficients or $R_P = 0.862$ and root-mean-square errors of prediction or RMSEP = 1.04 for color, as well as $R_P = 0.971$ and RMSEP = 4.7% for moisture content) than when using entropy data ($R_P = 0.839$ and RMSEP = 1.14 for color, as well as $R_P = 0.901$ and RMSEP = 9.2% for moisture content). Being one of the critical parameters to evaluate the quality of dried products and the drying technique, MCU of maize kernels was investigated by hyperspectral imaging technique during the drying process (Huang et al., 2015a). Two methods, using the prediction value of moisture content to calculate the uniformity (indirect) and predicting the MCU directly, were investigated. Better prediction results were achieved using the direct method (with correlation coefficients $R_P = 0.848$ and root-mean-square error of prediction RMSEP = 2.73) than the indirect method ($R_P = 0.521$ and RMSEP = 10.96). Overall, a simple hyperspectral imaging technique showed significant potential in measuring multiple quality parameters simultaneously during the drying process.

17.4.4 Nondestructive Measurement of Vegetables Using X-Ray Micro-Computed Tomography

Since the first application of X-ray CT in the early 1990s in the detection of maturity in green tomatoes (Brecht et al., 1991), numerous investigations have been performed on vegetables, making it the most prominent field of application of X-ray CT. An X-ray imaging inspection method to detect an internal disorder, spongy tissue, in mangoes was developed in 1993 (Thomas et al., 1993). Differences between the healthy and affected fruit were indicated by variances in the gray values of the X-ray images. Density differences were also used to discriminate between healthy mangoes and fruits infested with weevils (Thomas et al., 1995). X-ray CT was used to predict the salt and water content in dry-cured hams. Prediction models for salt (RMSECV = 0.3% NaCl) and water content (RMSECV = 1.5% water) at the initial stages of the process were achieved, which proved useful for control purposes (Fulladosa et al., 2010). Donis-González et al. (2014) investigated the internal tissue changes of agricultural product, caused by insect damage, disorders, or void presence using X-ray CT imaging. This technique was found to be a powerful technique to explore fruit epidermal and subepidermal structures in 3D at a micro-level through the microstructural characterization of commercial kiwifruit cultivars (Cantre et al., 2014).

X-ray CT is a potential technique for nondestructive online sorting of vegetables, which will detect internal quality characteristics at a relatively early stage and prevent fruits with a short shelf life from entering the supply chain (Donis-González et al., 2014). With the 3D advantage and the ability to visualize the internal structure, improved knowledge of products are obtained that could result in a better understanding of the environmental effects on the vegetable structure. Even though larger sample sets should be used, it is restricted because of the high cost of performing X-ray CT analysis. Nevertheless, X-ray CT can serve as a valuable technique for the development of future prediction models for internal quality (Schoeman et al., 2016).

17.4.5 Nondestructive Measurement of Vegetables Using Magnetic Resonance Imaging

MRI was recently attempted in a laboratory test and introduced commercially into the online food-processing environment after technological and commercial validation in clinical settings. MRI had been extensively used in the past to measure noninvasively the moisture content within different components of food products with good spatial resolution (Du and Sun, 2004). From MRI parameters, many food characteristics can be quantified, including determination of chemical composition and quantification of the structure. Qualitative and quantitative proton MRI techniques have been applied to food quality assessment mainly in the aspect of moisture, carbohydrate, protein, and other quality parameters. The potential of MRI for noninvasively monitoring the subcellular and intercellular redistribution of water in cellular tissue during drying and freezing of parenchyma apple tissue was assessed by Hills and Remigereau (1997). The results showed that freeze-drying apple tissue gave much lower water contents than fluidized bed drying, but the NMR data confirms that it destroys membrane integrity and causes cell wall collapse. In drying process, MRI was used for modeling transient moisture profiles of a drying apple slab (McCarthy and Perez, 1991), studying the rehydration of extruded pasta (Hills et al., 1996) characterizing osmotic dehydrated apple (Cornillon, 2000), and acquiring water mobility and moisture data of strawberry (Evans et al., 2002). Thybo et al. (2003) studied the use of MRI in determination of dry matter content in potatoes and identification of different varieties of potato samples, in which the correlations between dry matter content and the MRI data were investigated using partial least squares regression. Other applications of MRI in food quality inspection have also been reported. Internal structure and sensory analysis in vegetables was investigated (Duce et al., 1992; Martens et al., 2002). The oil and water contents in french fries were determined based on the distribution of relaxation times of MRI parameters (MacMillan et al., 2008). The extent of damage caused by relatively low pressures in strawberry was evaluated by MRI (Otero and Prestamo, 2009). Infestation of apple fruits by the peach fruit moth was studied, and the healthy and infested apples were successfully discriminated (Haishi et al., 2011). The freezing process of sucrose solution was visualized using MRI apparatus (Hindmarsh et al., 2004).

Compared with normal camera images, there are several advantages with NMR images, such as clear contrast and 3D analysis of samples. Because of the principle of MRI, images are converted from electromagnetic signals that represent internal information of samples; so it has great advantages in chemical composition determination and great potential for vegetable quality assessment (Chen et al., 2013b).

17.5 SUMMARY AND FUTURE TRENDS

Machine vision is easier to accomplish than the other two techniques. Numerous in-line detection and grading systems have been put into actual production. Generally, the three steps in a complete detection process include image acquiring, image processing and analysis, and formulating decisions. Considering the increasing development of electronics and microprocessors, the three steps can be accomplished with only one smart camera. Unlike machine vision, infrared spectroscopy is mainly used to determine the internal physical and chemical compositions. Spectral data preprocessing and multivariate statistical analysis methods are crucial for processing spectral data from infrared spectroscopy, and the final results vary with preprocessing and modeling method. As a rapid and nondestructive detection technique, many handheld infrared spectroscopy instruments are available, which make it more practical. Hyperspectral imaging has advantages over conventional machine vision and infrared spectroscopy by providing spectral and spatial information. However, the disadvantage of this technique is the huge amount of data. Therefore, further optimal wavelength selection will help to simplify calculation and improve model performance after the processing of images and spectral data. As a new technology that has been studied for over a decade, hyperspectral imaging has a long way before it can be moved from laboratories to the practical application of vegetable quality and safety detection during drying. X-ray CT is an essential development in imaging technology, which has eliminated some of the shortcomings of traditional imaging by enabling the noninvasive, 3D, and quantitative characterization of food microstructure. Consequently, it has become an increasingly popular device to investigate vegetable microstructure. However, this will remain a challenge as high throughput requirements will have to be met. With improvements in instruments and computational power, it is expected that X-ray CT would become more applicable. MRI is an imaging technique primarily used to obtain high-quality images of the inside of the object in 2D or 3D. Research results in the literature have indicated that MRI can be used to noninvasively evaluate important quality attributes of food products. Although MRI has shown potential to measure food quality parameters, the cost and challenges for in-line use in the sorting line means it is not currently a commercially viable application for high-volume, low-value items such as fruits and vegetables. Future designs of MRI equipment should consider the potential for in-line application.

Machine vision, infrared spectroscopy, hyperspectral imaging, X-ray CT, and MRI as noncontact optical sensing technologies are efficient and reliable in the quality and safety assessment of various vegetables. Machine vision is widely applied in the quality analysis using their visible features (color, shape, texture, and size). Infrared spectroscopy has advantages in quality discrimination and prediction of internal chemical compositions by utilizing spectral information. Hyperspectral imaging integrates the advantages of machine vision and infrared spectroscopy, and has been extensively used in classification and prediction applications using spatial and spectral information of the drying samples. X-ray CT has proved to be a promising device to investigate vegetable microstructure because of its 3D imaging. MRI has advantages that internal images in 2D or 3D can be obtained noninvasively and nondestructively.

Previous studies have indicated that optical sensing technologies can be used effectively as a reliable and accurate tool for variety identification and classification, quality grading, damage detection, and composition prediction. With the urgent need of the industry for advanced testing methods and the rapid development of technology and instruments, optical sensing technologies have great potential to be the dominant method in quality and safety assessment of drying vegetables.

REFERENCES

Adiletta, G., Iannone, G., Russo, P., Patimo, G., Pasquale, D.S., and Matteo, D.M. 2014. Moisture migration by magnetic resonance imaging during eggplant drying: A preliminary study. *International Journal of Food Science and Technology* 49: 2602–2609.

Aguilera, J.M. 2003. Drying and dried products under the microscope. *Food Science and Technology International* 9: 137–143.

Ariana, D.P. and Lu, R. 2008. Quality evaluation of pickling cucumbers using hyperspectral reflectance and transmittance imaging—Part II. Performance of a prototype. *Sensing and Instrumentation for Food Quality and Safety* 2: 152–160.

Ballabio, D. and Todeschini, R. 2009. Multivariate classification for qualitative analysis. In Sun, D.-W. (ed.) *Infrared Spectroscopy for Food Quality Analysis and Control*, 83–104. San Diego, CA: Academic Press.

Bhuvaneswari, K., Fields, P.G., White, N.D.G., Sarkar, A.K., Singh, C.B., and Jayas, D.S. 2011. Image analysis for detecting insect fragments in semolina. *Journal of Stored Products Research* 47: 20–24.

Blanco Romía, M. and Alcalà Bernàrdez, M. 2009. Multivariate calibration for quantitative analysis. In Sun, D.-W. (ed.) *Infrared Spectroscopy for Food Quality Analysis and Control*, 51–82. San Diego, CA: Academic Press.

Boeriu, C.G., Stolle-Smits, T., and van Dijk, C. 1998. Characterisation of cell wall pectins by near infrared spectroscopy. *Journal of Near Infrared Spectroscopy* 6: A299–A301.

Brecht, J., Shewfelt, R., Garner, J., and Tollner, E. 1991. Using X-ray-computed tomography to nondestructively determine maturity of green tomatoes. *HortScience* 26: 45–47.

Cantre, D., East, A., Verboven, P., Trejo Araya, X., Herremans, E., and Nicolaï, B.M. 2014. Microstructural characterisation of commercial kiwifruit cultivars using X-ray micro computed tomography. *Postharvest Biology and Technology* 92: 79–86.

Chen, H.Z., Zhang, M., Fang, Z.X., and Wang, Y.Q. 2013a. Effects of different drying methods on the quality of squid cubes. *Drying Technology* 31: 1911–1918.

Chen, Q., Zhang, C., Zhao, J., and Qin, O. 2013b. Recent advances in emerging imaging techniques for non-destructive detection of food quality and safety. *Trends in Analytical Chemistry* 52: 261–274.

Cheng, X., Chen, Y.R., Tao, C.Y., Kim, M.S., and Lefcourt, A.M. 2004. A novel integrated PCA and FLD method on hyperspectral image feature extraction for cucumber chilling damage inspection. *Transactions of the ASAE* 47: 1313–1320.

Clark, C., Hockings, P., Joyce, D., and Mazucco, R. 1997. Application of magnetic resonance imaging to pre- and post-harvest studies of fruits and vegetables. *Postharvest Biology and Technology* 11: 1–21.

Cornillon, P. 2000. Characterization of osmotic dehydrated apple by NMR and DSC. *LWT-Food Science and Technology* 33: 261–267.

De Belie, N., Pedersen, D.K., Martens, M., Bro, R., Munck, L., and De Baerdemaeker, J. 2003. The use of visible and near-infrared reflectance measurements to assess sensory changes in carrot texture and sweetness during heat treatment. *Biosystems Engineering* 85: 213–225.

Desrosier, N.W. and Desrosier, J.N. 1977. *The Technology of Food Preservation*, 4th edition. Westport, CT: AVT Publishing Company, Inc.

Devahastin, S., Suvarnakuta, P., Soponronnarit, S., and Mujumdar, A.S. 2004. A comparative study of low-pressure superheated steam and vacuum drying of a heat-sensitive material. *Drying Technology* 22: 1845–1867.

Donis-González, I., Guyer, D., Pease, A., and Barthel, F. 2014. Internal characterization of fresh agricultural products using traditional and ultrafast electron beam X-ray computed tomography imaging. *Biosystems Engineering* 117: 104–113.

Du, C., and Sun, D. 2004. Recent developments in the applications of image processing techniques for food quality evaluation. *Trends in Food Science and Technology* 15: 230–249.

Duce, S.L., Carpenter, T.A., and Hall, L.D. 1992. Nuclear magnetic resonance imaging of fresh and frozen courgettes. *Journal of Food Engineering* 16: 165–172.

Evans, S.D., Brambilla, A., Lane, D.M., Torreggiani, D., and Hall, L.D. 2002. Magnetic resonance imaging of strawberry (*Fragaria vesca*) slices during osmotic dehydration and air drying. *LWT-Food Science and Technology* 35: 177–184.

Fernández, L., Castillero, C., and Aguilera, J.M. 2005. An application of image analysis to dehydration of apple discs. *Journal of Food Engineering* 67: 185–193.

Firtha, F. 2007. Development of data reduction function for hyperspectral imaging. *Progress in Agricultural Engineering Sciences* 3: 67–88.

Firtha, F. 2009. Detecting moisture loss of carrot samples during storage by hyperspectral imaging system. *Acta Alimentaria* 38: 22–66.

Fulladosa, E., Santos-Garcés, E., Picouet, P., and Gou, P. 2010. Prediction of salt and water content in dry-cured hams by computed tomography. *Journal of Food Engineering* 96: 80–85.

Goetz, A.F.H., Vane, G., Solomon, J.E., and Rock, B.N. 1985. Imaging spectrometry for earth remote sensing. *Science* 228: 1147–1153.

Gong, Y., Deng, G., Han, C., and Ning, X. 2015. Process optimization based on carrot powder color charateristics. *Engineering in Agriculture, Environment and Food* 8: 137–142.

Gunasekaran, S. and Ding, K.X. 1994. Using computer vision for food quality evaluation. *Food Technology* 48: 151–154.

Gunasekaran, S. and Irudayaraj, J. 2001. Optical methods: Visible, NIR, and FTIR spectroscopy. In Gunasekaran, S. (ed.) *Nondestructive Food Evaluation: Techniques to Analyze Properties and Quality*, 1–37, New York: Marcel Dekker.

Haishi, T., Koizumi, H., Arai, T., Koizumi, M., and Kano, H. 2011. Rapid detection of infestation of apple fruits by the peach fruit moth, *Carposina sasakii* Matsumura, larvae using a 0.2-T dedicated magnetic resonance imaging apparatus. *Applied Magnetic Resonance* 41: 1–18.

Hashemi, S.J. and Murray Douglas, W.J. 2003. Moisture non-uniformity in drying paper: Measurement and relation to process parameters. *Drying Technology* 21: 329–347.

Helgerud, T., Wold, J.P., Pedersen, M.B., Liland, K.H., Balance, S., Knutsen, S.H., Rukke, E.O., and Afseth, N.K. 2015. Towards on-line prediction of dry matter content in whole unpeeled potatoes using near-infrared spectroscopy. *Talanta* 143: 138–144.

Hills, B.P., Babonneau, F., Quantin, V.M., Gaudet, F., and Belton, P.S. 1996. Radial NMR microimaging studies of the rehydration of extruded pasta. *Journal of Food Engineering* 27: 71–86.

Hills, B.P. and Remigereau, B. 1997. NMR studies of changes in subcellular water compartmentation in parenchyma apple tissue during drying and freezing. *International Journal of Food Science and Technology* 32: 51–61.

Hindmarsh, J.P., Buckley, C., Russell, A.B., Chen, X.D., Gladden, L.F., Wilson, D.I., and Johns, M.L. 2004. Imaging droplet freezing using MRI. *Chemical Engineering Science* 59: 2113–2122.

Hiranuarachat, B., Devahastin, S., and Chiewchan, N. 2011. Effects of acid pretreatments on some physicochemical properties of carrot undergoing hot air drying. *Food and Bioproducts Processing* 89: 116–127.

Huang, M. and Lu, R. 2010. Apple mealiness detection using hyperspectral scattering technique. *Postharvest Biology and Technology* 58: 168–175.

Huang, M., Wan, X., Zhang, M., and Zhu, Q. 2013. Detection of insect-damaged vegetable soybean using hyperspectral tansmittance image. *Journal of Food Engineering* 116: 45–49.

Huang, M., Wang, Q., Zhang, M., and Zhu, Q. 2014a. Prediction of color and moisture content for vegetable soybean during drying using hyperspectral imaging technology. *Journal of Food Engineering* 128: 24–30.

Huang, M., Wang, Q., Zhu, Q., Qin, J., and Huang, G. 2015b. Review of seed quality and safety tests using optical sensing technologies. *Seed Science and Technology* 43: 337–366.

Huang, M., Zhao, W., Wang, Q., Zhang, M., and Zhu, Q. 2015a. Prediction of moisture content uniformity using hyperspectral imaging technology during the drying of maize kernel. *International Agrophysics* 29: 39–46.

Huang, W.J., Lu, F.M., and Lee, Y.C. 2014b. GPR system for monitoring paddy moisture content in a drying process. *Drying Technology* 32: 679–685.

Hutchings, J.B., Luo, R., and Ji, W. 2002. Calibrated colour imaging analysis of food. In MacDugall, D. (ed.) *Color in Food*, 352–366. Cambridge: Woodhead Publishing.

Jackman, P. and Sun, D.W. 2013. Recent advances in image processing using image texture features for food quality assessment. *Trends in Food Science and Technology* 29: 35–43.

Kaliramesh, S., Chelladurai, V., Jayas, D.S., Alagusundaram, K., White, N.D.G., and Fields, P.G. 2013. Detection of infestation by *Callosobruchus maculatus* in mungbean using near-infrared hyperspectral imaging. *Journal of Stored Products Research* 52: 107–111.

Khuriyati, N., Matsuoka, T., and Kawano, S. 2004. Precise near infrared spectral acquisition of intact tomatoes in interactance mode. *Journal of Near Infrared Spectroscopy* 12: 391–395.

Koca, N., Burdurlu, H.S., and Karadeniz, F. 2007. Kinetics of color changes in dehydrated carrots. *Journal of Food Engineering* 78: 449–455.

Kong, W., Zhang, C., Liu, F., Nie, P. and He, Y. 2013. Rice seed cultivar identification using near-infrared hyperspectral imaging and multivariate data analysis. *Sensors* 13: 8916–8927.

Kotwaliwale, N., Singh, K., Kalne, A., Jha, S., Seth, N., and Kar, A. 2014. X-ray imaging methods for internal quality evaluation of agricultural produce. *Journal of Food Science and Technology* 51: 1–15.

Krokida, M.K., Oreopoulou, V., Maroulis, Z.B., and Marinos-Kouris, D. 2001. Effect of pretreatment on viscoelastic behavior of potato strips. *Journal of Food Engineering* 50: 11–17.

Landis, E. and Keane, D. 2010. X-ray microtomography. *Materials Characterization* 61: 1305–1316.

Lawrence, K.C., Park, B., Windham, W.R., and Mao, C. 2003. Calibration of a pushbroom hyperspectral imaging system for agricultural inspection. *Transactions of the ASAE* 46: 513–521.

Léonard, A., Blacher, S., Nimmol, C., and Devahastin, S. 2008. Effect of far-infrared radiation assisted drying on microstructure of banana slices: An illustrative use of X-ray microtomography in microstructural evaluation of a food product. *Journal of Food Engineering* 85: 154–162.

Li, J., Rao, X., and Ying, Y. 2012. Development of algorithms for detecting citrus canker based on hyperspectral reflectance imaging. *Journal of the Science of Food and Agriculture* 92: 125–134.

Lim, K. and Barigou, M. 2004. X-ray micro-computed tomography of cellular food products. *Food Research International* 37: 1001–1012.

Liu, Y., Chen, Y., Wang, C., Chan, D., and Kim, M.S. 2006. Development of hyperspectral imaging technique for the detection of chilling injury in cucumbers: spectral and image analysis. *Applied Engineering in Agriculture* 22: 101–111.

MacMillan, B., Hickey, H., Newling, B., Ramesh, M., and Balcom, B. 2008. Magnetic resonance measurements of French fries to determine spatially resolved oil and water content. *Food Research International* 41: 676–681.

Martens, H., Thybo, A.K., Andersen, H.J., Karlsson, A.H., Dønstrup, S., Stødkilde-Jørgensen, H., and Martens, M. 2002. Sensory analysis for magnetic resonance-image analysis: Using human perception and cognition to segment and assess the interior of potatoes. *LWT-Food Science and Technology* 35: 70–79.

Martynenko, A. and Janaszek, M.A. 2014. Texture changes during drying of apple slices. *Drying Technology* 32: 567–577.

Maskan, M. 2001. Kinetics of colour change of kiwifruits during hot air and microwave drying. *Journal of Food Engineering* 48: 169–175.

Mayor, L. and Sereno, A.M. 2004. Modelling shrinkage during convective drying of food materials: A review. *Journal of Food Engineering* 61: 373–386.

McCarthy, M.J., and Perez, E. 1991. Model for transient moisture profiles of a drying apple slab using the data obtained with magnetic resonance imaging. *Biotechnology Progress* 7: 540–543.

Mehl, P.M., Chen, Y.R., Kim, M.S., and Chan, D.E. 2004. Development of hyperspectral imaging technique for the detection of apple surface defects and contaminations. *Journal of Food Engineering* 61: 67–81.

Morris, N.I., Lohmann, LW., O'Conner, R.T., and Freeman, A.F. 1953. Peanut Butter, IV. Determination of color of peanut butter by spectral reflectance method. *Food Technology* 7: 393–396.

Mujumdar, A.S. 1995. *Handbook of Industrial Drying*, 2nd edition. New York: Marcel Dekker.

Nicolaï, B.M., Beullens, K., Bobelyn, E., Peirs, A., Saeys, W., Theron, K., and Lammertyn, J. 2007. Nondestructive measurement of fruit and vegetable quality by means of NIR spectroscopy: A review. *Postharvest Biology and Technology* 46: 99–118.

Nicoleti, J.F., Silveira, J.V., Telis-Romero, J., and Telis, V.R.N. 2005. Viscoelastic behavior of persimmons dried at constant air temperature. *LWT-Food Science and Technology* 38: 143–150.

Otero, L. and Prestamo, G. 2009. Effects of pressure processing on strawberry studied by nuclear magnetic resonance. *Innovative Food Science and Emerging Technologies* 10: 434–440.

Park, Y.W. 1987. Effect of freezing, thawing, drying, and cooking on carotene retention in carrots, broccoli and spinach. *Journal of Food Science* 52: 1022–1025.

Penchaiya, P., Bobelyn, E., Verlinden, B.E., Nicolaï, B.M., and Saeys, W. 2009. Non-destructive measurement of firmness and soluble solids content in bell pepper using NIR spectroscopy. *Journal of Food Engineering* 94: 267–273.

Qin, J. 2010. Hyperspectral imaging instruments. In Sun, D.-W. (ed.) *Hyperspectral Imaging for Food Quality Analysis and Control*, 129–172. San Diego, CA: Academic Press.

Romano, G., Argyropoulos, D., Nagle, M., Khan, M.T., and Müller, J. 2012. Combination of digital images and laser light to predict moisture content and color of bell pepper simultaneously during drying. *Journal of Food Engineering* 109: 438–448.

Romano, G., Baranyai, L., Gottschalk, K., and Zude, M. 2008. An approach for monitoring the moisture content changes of dying banana slices with laser light backscattering imaging. *Food and Bioprocess Technology* 1: 410–414.

Roy, S., Anantheswaran, R., Shenk, J., Westerhaus, M.O., and Beelman, R. 1993. Determination of moisture content of mushrooms by VIS-NIR spectroscopy. *Journal of the Science of Food and Agriculture* 63: 355–360.

Sampson, D.J., Chang, Y.K., Vasantha Rupasinghe, H.P., and Zaman, Q.U. 2014. A dual-view computer-vision system for volume and image texture analysis in multiple apple slices drying. *Journal of Food Engineering* 127: 49–57.

Schmidt, J.S., Sun, X., Bruce Litchfield, J. 1996. Applications of magnetic resonance imaging in food science. *Critical Reviews in Food Science and Nutrition* 36: 357–385.

Schoeman, L., Williams, P., Plessis, A., and Manley, M. 2016. X-ray micro-computed tomography (μCT) for non-destructive characterisation of food microstructure. *Trends in Food Science and Technology* 47: 10–24.

Schulz, H., Drews, H.H., Quilitzsch, R., and Krüger, H. 1998. Application of near infrared spectroscopy for the quantification of quality parameters in selected vegetables and essential oil plants. *Journal of Near Infrared Spectroscopy* 6: A125–A130.

Serranti, S., Cesare, D., Marini, F. and Bonifazi, G. 2013. Classification of oat and groat kernels using NIR hyperspectral imaging. *Talanta* 103: 276–284.

Shahin, M.A. and Symons, S.J. 2011. Detection of *Fusarium* damaged kernels in Canada Western Red Spring wheat using visible/near-infrared hyperspectral imaging and principal component analysis. *Computers and Electronics in Agriculture* 75: 107–112.

Steuer, B. and Shulz, H. 2003. Near-infrared analysis of fennel (*Foeniculum vulgare* Miller) on different spectrometers—Basic considerations for a reliable network. *Phytochemical Analysis* 14: 285–289.

Telis, V.R.N., Telis-Romero, J., and Gabas, A.L. 2005. Solids rheology for dehydrated food and biological materials. *Drying Technology* 23: 759–780.

Thomas, P., Kannan, A., Degwekar, V., and Ramamurthy, M. 1995. Non-destructive detection of seed weevil-infested mango fruits by X-ray imaging. *Postharvest Biology and Technology* 5: 161–165.

Thomas, P., Saxena, S., Chandra, R., Rao, R., and Bhatia, C. 1993. X-ray imaging for detecting spongy tissue, an internal disorder in fruits of Alphonso mango (*Mangifera indica* L.). *Journal of Horticultural Science and Biotechnology* 68: 803–806.

Thuwapanichayanan, R., Prachayawarakorn, S., and Soponronnarit, S. 2008. Drying characteristics and quality of banana foam mat. *Journal of Food Engineering* 86: 573–583.

Thybo, A., Andersen, H., Karlsson, A., Dønstrup, S., and Stødkilde-Jørgensen, H. 2003. Low-field NMR relaxation and NMR-imaging as tools in differentiation between potato sample and determination of dry matter content in potatoes. *Food Science and Technology* 36: 315–322.

Trinca, L.C., Capraru, A.M., Arotaritei, D., Volf, I., and Chiruta, C. 2014. Monitoring methods and predictive models for water status in Jonathan apples. *Food Chemistry* 144: 80–86.

Wang, Y.C., Zhang, M., Mujumdar, A.S., and Mothibe, K.J, 2013a. Microwave-assisted pulse-spouted bed freeze-drying of stem lettuce slices–Effect on product quality. *Food Bioprocess Technology* 6: 3530–3543.

Wang, Y.C., Zhang, M., Mujumdar, A.S., Mothibe, K.J., and Roknul Azam, S.M. 2013b. Study of drying uniformity in pulsed spouted microwave-vacuum drying of stem lettuce slices with regard to product quality. *Drying Technology* 31: 91–101.

Wu, D. and Sun, D. 2013. Advanced applications of hyperspectral imaging technology for food quality and safety analysis and assessment: A review—Part I: Fundamentals. *Innovative Food Science and Emerging Technologies* 19: 1–14.

Xie, C.Q., Li, X.L., Nie, P.C., and He, Y. 2013. Application of time series hyperspectral imaging (TS-HSI) for determining water content within tea leaves during drying. *Transactions of the ASABE* 56: 1431–1440.

Xing, J., Van Hung, P., Symons, S., Shahin, M., and Hatcher, D. 2009. Using a short wavelength infrared (SWIR) hyperspectral imaging system to predict alpha amylase activity in individual Canadian western wheat kernels. *Journal of Food Measurement and Characterization* 3: 211–218.

Zhang, X., Liu, F., He, Y., and Li, X. 2012. Application of hyperspectral imaging and chemometric calibrations for variety discrimination of maize seeds. *Sensors* 12: 17234–17246.

Zheng, C., Sun, D., and Zheng, L. 2006. Recent developments and applications of image features for food quality evaluation and inspection—A review. *Trends in Food Science and Technology* 17: 642–655.

Zielińska, M. and Markowski, M. 2011. Color characteristics of carrots: Effect of drying and rehydration. *International Journal of Food Properties* 15: 450–466.

Zielińska, M., Zapotocany, P., and Markowski, M. 2005. Color standard and homogeneous groups of dried carrots of 34 commercial varieties. *Polish Journal of Food and Nutrition Sciences* 14: 51–56.

18 Computer Vision and Its Applications for Drying of Vegetables

Alex Martynenko
Dalhousie University

CONTENTS

18.1 INTRODUCTION

Computer vision for industrial control has been developed since the early 1980s, mostly for robotic applications. Some of the motivations for using computer vision in the industrial workplace included automated visual inspection and robot vision (Vernon, 1991). Recent advances in computer and digital processing technologies have made it available for food and agricultural applications. The benefits of computer vision for quality assurance of agricultural and food products are widely discussed in excellent reviews of Gunasekaran (1996), Du and Sun (2004), and Zheng et al. (2006). Computer vision made possible noncontact and rapid food quality assessment, which brings benefits of high accuracy and efficiency for the food industry (Gunasekaran, 2000). The applications of computer vision for inspection, sorting, and grading of vegetables can be found in the reviews of Brosnan and Sun (2004), Du and Sun (2004), Cubero et al. (2011), Patel et al. (2012), and Gomes and Leta (2012). However, a concise review of computer vision applications focused on the drying of vegetables with careful analysis of methodology and future trends is currently absent.

It has been widely recognized that drying operations may cause significant loss of vegetable quality from both nutritional and organoleptic standpoints. Vegetable quality is frequently referred to size, shape, color, and texture (Narendra and Hareesh, 2010). Shape distortion, loss of natural flavor and color because of oxidation, and enzymatic and nonenzymatic browning usually occur in drying. Alteration of the redox potentials and pH during drying can affect the structure and functionality of biopolymers (Lewicki, 2006). Vegetables, which are rich sources of vitamins, are mostly sensitive to drying conditions. The extent of these undesirable changes depends on the drying technology and drying conditions. For example, total phenolic, total flavonoid, lycopene, and ascorbic acid contents and antioxidant activity of tomatoes was significantly reduced even at low-temperature drying (Toor and Savage, 2006). Volatile compounds were also reduced in dried parsley (Díaz-Maroto et al., 2002). Shrinkage, density, porosity, and rehydration ability of dried vegetables are among the most studied physical indicators of quality (Lozano et al., 1983; Madamba et al., 1994; Zogzas et al., 1994; Krokida et al., 2003; Martynenko, 2011; López-Ortiz et al., 2013). Color of vegetables is recognized as one of the key indicators of quality for carrot (Lin et al., 1998; Prakash et al., 2004; Zielińska and Markowski, 2012), potato (Krokida et al., 2001; Pedreschi et al., 2006; Wang et al., 2010), onion (Mongpraneet et al., 2002; Arslan and Özcan, 2010), pepper (Vega-Gálvez et al., 2009; Guiné and Barroca, 2012), pumpkin (Guiné and Barroca, 2012), parsley (Soysal, 2004), and dasheen leaves (Maharaj and Sankat, 1996). Table 18.1 shows frequently measured nutritional and organoleptic indicators of quality.

Multiple physicochemical and mechanical changes, such as nonhomogeneous shrinkage, glass transition, case hardening, and surface cracking during drying are difficult to predict. Traditional techniques used for measuring the quality parameters (shape, size, color, and texture) of dried vegetables are labor intensive and not really applicable in the process of drying. For example, bulk volume evaluation requires periodical sampling of the product to be measured with stereopycnometer or volume displacement methodology (López-Ortiz et al., 2013). Traditional color measurements allow to evaluate only cumulative color changes by using *CIE* $L^*a^*b^*$ colorimeter (Krokida et al., 2001; Guine and Barroca, 2012) or spectrophotometer (Zielińska and Markowski, 2012). Existing instrumentation is designed for offline quality control after drying. However, it was found that monitoring of product quality in real-time manner allows optimization of the drying process with respect to quality (Martynenko and Yang, 2007; Jin et al., 2014).

TABLE 18.1
Quality Indicators in Drying of Vegetables

Vegetable	Sample Size	References	Quality Indicators (Color, Texture, Morphology)
Asparagus	Slices	Nindo et al. (2003)	Rehydration ratio, color
Carrot	Slices	Lin et al. (1998)	Color, morphology, texture, rehydration capacity
	Slices	Cui et al. (2004)	Carotene
	Slices	Prakash et al. (2004)	Color, rehydration capacity
	Slices	Doymaz (2004)	Diffusivity
	Slices	Toğrul (2006)	Diffusivity
	Slices	Wang and Xi (2005)	Rehydration ratio, quality
	Slices	Nahimana and Zhang (2011)	Color, shrinkage
	Cubes	Zogzas et al. (1994)	Shrinkage, porosity, density
	Cubes	Zielińska and Markowski (2012)	Color, carotene
	Discs	Lewicki and Duszczyk (1998)	Color
	Cylinders	Krokida et al. (1997)	Color
	Cylinders	Lozano et al. (1983)	Density, shrinkage, porosity
	Cylinders	Marabi et al. (2006)	Rehydration, flavor, texture
	Whole	Krokida et al. (2003)	Diffusivity
Tomato	Slices	Davoodi et al. (2007)	Rehydration, color
	Cuts	Toor and Savage (2006)	Antioxidants, color
	Whole	Doymaz (2007)	Diffusivity
	Whole	Krokida et al. (2003)	Diffusivity
Pumpkin	Slices	Guiné and Barroca (2012)	Color, texture
	Discs	Lewicki and Duszczyk (1998)	Color
	Cubes	Zenoozian et al. (2008)	Color
	Cylinders	Mayor et al. (2011)	Shrinkage, density, porosity, shape
	Whole	Krokida et al. (2003)	Diffusivity
Pumpkin seeds	Whole	Sacilik (2007)	Diffusivity
Mushrooms	Whole	Krokida et al. (2003)	Diffusivity
Green pepper	Slices	Guiné and Barroca (2012)	Color, texture
	Slices	Mendoza et al. (2007)	Color
	Whole	Krokida et al. (2003)	Diffusivity
Red pepper	Slices	Di Scala and Crapiste (2008)	Carotenoids, ascorbic acid
	Slices	Mendoza et al. (2007)	Color
	Slices	Simal et al. (2005)	Color, antioxidants, diffusivity
	Slabs	Vega-Gálvez et al. (2009)	Color, antioxidants, nutrients
	Slices, whole	Sanjuan et al. (2003)	Diffusivity
	Whole	Krokida et al. (2003)	Diffusivity

(Continued)

TABLE 18.1 (*Continued*)
Quality Indicators in Drying of Vegetables

Vegetable	Sample Size	References	Quality Indicators (Color, Texture, Morphology)
Egg plant	Slices	Ertekin and Yaldiz (2004)	Color, rehydration
Potato	Slices	Wang et al. (2004)	Sensory quality, rehydration ratio, energy consumption
	Slices	Wang et al. (2010)	Color, texture, nutrients
	Slices	Romani et al. (2009)	Color
	Slices	Pedreschi et al. (2006)	Color
	Slices	Yadollahinia et al. (2009)	Shrinkage
	Slices	Mogol and Gokmen (2014)	Color
	Slabs	Wang and Brennan (1995)	Microstructure, density, porosity
	Slabs	Azfal and Abe (1998)	Diffusivity
	Slabs	Krokida et al. (2001)	Density, porosity, color
	Slabs	Campos-Mendiola et al. (2007)	Shrinkage, shape
	Cubes	Zogzas et al. (1994)	Density, shrinkage, porosity
	Cubes	Bondaruk et al. (2007)	Sugar, starch, color, microstructure
	Cylinders	Lozano et al. (1983)	Density, shrinkage, porosity
	Cylinders	Krokida et al. (1998)	Color
	Cylinders, cube, slab	Mulet et al. (2000)	Shrinkage
	Cylinders, cube, slab	Senadeera et al. (2003)	Diffusivity
	Discs	Lewicki and Duszczyk (1998)	Color
	Discs	Saravacos (1967)	Water sorption and equilibrium isotherms
Sweet potato	Cylinders	Lozano et al. (1983)	Density, shrinkage, porosity
	Slabs	Orikasa et al. (2010)	Color, texture, sugar, ascorbic acid
Okra	Slices	Inyang and Ike (1998)	Color, texture, nutrients
Onion	Slices	Arslan and Özcan (2010)	Color, minerals, antioxidants
	Slices	Rapusas and Driscoll (1995)	Thermophysical properties
	Slices	Rapusas et al. (1995)	Bulk density
	Slices	Abhayawick et al. (2002)	Density, porosity, diffusivity, thermal conductivity, dielectric permittivity
	Slices	Pathare and Sharma (2005)	Diffusivity
	Slices	Sharma et al. (2005)	Diffusivity, rehydration ability
	Whole	Krokida et al. (2003)	Diffusivity
Welsh onions	Leaves	Mongpraneet et al. (2002)	Color, rehydration ratio, chlorophyll
Green bean	Slices	Doymaz (2005)	Diffusivity
	Cylinders, cube, slab	Senadeera et al. (2003)	Diffusivity
White bean	Whole	Hutchison and Otten (1983)	Diffusivity
	Whole	Adu and Otten (1996)	Diffusivity

(*Continued*)

TABLE 18.1 (*Continued*)
Quality Indicators in Drying of Vegetables

Vegetable	Sample Size	References	Quality Indicators (Color, Texture, Morphology)
Soybean	Whole	Hutchison and Otten (1983)	Diffusivity
	Whole	Huang et al. (2014)	Color, moisture content
Bell pepper	Slabs	Vega-Gálvez et al. (2008)	Color, morphology, microstructure, rehydration, water absorption capacity (WAC)
	Slices	Romano et al. (2012)	Color, moisture content
Cabbage	Strips	Duan et al. (2007)	Diffusivity
Cauliflower	Pieces	Jayaraman et al. (2007)	Sensory, physicochemical properties, shelf life
	Cylinders, cube, slab	Mulet et al. (2000)	Shrinkage
Broccoli	Whole	Mrkìc et al. (2006)	Bioactive, antioxidants
Celery	Whole	Krokida et al. (2003)	Diffusivity
Leek	Whole	Krokida et al. (2003)	Diffusivity
Edamame	Pieces	Hu et al. (2006)	Color, morphology, texture
Dasheen leaves	Pieces (Chopped)	Maharaj and Sankat (1996)	Color, nutrient quality
Garlic	Whole	Sharma and Prasad (2001)	Color, flavor
	Whole	Lozano et al. (1983)	Density, shrinkage, porosity
	Whole	Madamba et al. (1994)	Shrinkage, density, porosity
	Whole	Krokida et al. (2003)	Diffusivity
	Whole	López-Ortiz et al. (2013)	Bulk volume, density
	Slices	Madamba et al. (1994)	Density, shrinkage, porosity
Parsley	Whole	Soysal (2004)	Color, diffusivity
	Leaves	Díaz-Maroto et al. (2002)	Volatile components
Spinach	Whole	Yadav and Sehgal (1995)	Beta-carotene, ascorbic acid
	Whole	Krokida et al. (2003)	Diffusivity
Amaranth	Whole	Yadav and Sehgal (1995)	Beta-carotene, ascorbic acid
Lentils	Whole	Tang and Sokhansanj (1993)	Density, porosity
Ginseng	Pieces	Davidson et al. (2004)	Color, size, ginsenosides
	Whole	Wilhelm (1990)	Color, size, and quality
	Whole	Reynolds (1998)	Color, ginsenosides
	Whole	Sokhansanj et al. (1999)	Shrinkage, density
	Whole	Ren and Chen (2000)	Ginsenosides
	Whole	Davidson et al. (2002)	Shrinkage, color
	Whole	Martynenko (2006)	Shrinkage, moisture, color
	Whole	Davidson et al. (2009)	Shrinkage, moisture
	Whole	Martynenko (2011)	Shrinkage, density, porosity
	Whole	Martynenko (2014)	Shrinkage, density
	Batch	Martynenko (2008)	Shrinkage, moisture, color

To meet the consumer's expectations on quality, the drying process should be made completely observable. One of the alternatives is computer vision technology. Safety, effectiveness, high accuracy, and the ability to establish relationship between visual appearance of foods and quality attributes has made computer vision a promising technique for quality assessment in the process of drying (Aghbashlo et al., 2014). However, the number of applications for vegetable drying are not so numerous. So far it has been successfully used for monitoring of potato and cauliflower shrinkage (Mulet et al., 2000), potato slice shrinkage (Campos-Mendiola et al., 2007; Yadollahinia et al., 2009), color and shrinkage of ginseng roots (Martynenko, 2006), porosity and density of ginseng roots (Martynenko, 2011, 2014), and shrinkage, density, and porosity of pumpkin in osmotic and air drying (Mayor et al., 2011). A computer vision system to measure the color of potato chips was developed by Pedreschi et al. (2006) and Romani et al. (2009). Zenoozian et al. (2008) measured color degradation of pumpkin under osmotic dehydration. Nahimana and Zhang (2011) monitored shrinkage and color change during microwave vacuum drying of carrot. Romano et al. (2012) used computer vision in combination with laser light to predict moisture content and color of bell pepper during drying. Huang et al. (2014) used hyperspectral imaging to predict color and moisture content in soybean drying. Potential applications of computer vision technology for real-time food quality assurance during the drying process has been discussed by Aghbashlo et al. (2014).

18.2 COMPUTER VISION PROCESS

The computer vision process usually includes five steps: image acquisition, image processing, feature extraction, pattern recognition, and decisionmaking. All steps are sequential and can be expressed in the form of a simple flow chart (Figure 18.1).

If computer vision is a part of the robotic or intelligent control system, it might also include elements of supervised or unsupervised learning for pattern recognition, knowledge base, and model development.

18.3 BASIC COMPONENTS OF COMPUTER VISION

The basic components of a computer vision system usually include digital camera(s), the source of illumination, computer hardware and software. Both hardware and software play equally important roles in computer vision. With the development of computer vision systems, the role of software and soft computing is constantly increasing.

18.3.1 Hardware

For drying applications, a set of sensors, monitoring process variables (moisture, pressure, and temperature), and ambient conditions (air temperature, humidity, and velocity) are the ultimate parts of hardware. The data from sensors and computer vision system are combined and recorded in the same time format

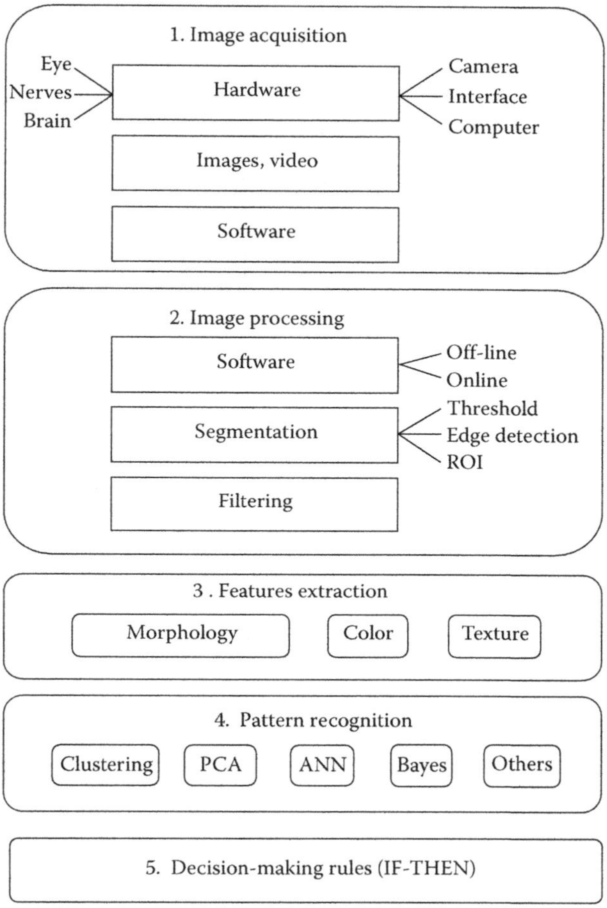

FIGURE 18.1 Computer vision process flow chart.

(sensor fusion). An example of computer vision hardware for ginseng root drying is shown in Figure 18.2.

Computer vision hardware usually consists of a light source, a digital color camera with an optical lens, and a computer. In some cases, for example, multispectral or hyperspectral vision, a set of narrowband (usually 10 nm) optical filters is required. The lighting system is a critical component of the computer vision hardware. The ultimate purpose of lighting design is to provide consistent uniform illumination of the object, avoiding reflections or shadows (Patel et al., 2012). Lighting arrangements and lighting geometry are two important design considerations, determining quality of color reproduction. Proper selection of light sources (incandescent, fluorescent, halogen, xenon, and LED) plays a critical role in appropriate color measurements. Various lighting arrangements, such as directional illumination, diffuse illumination, rear illumination, light-field illumination, dark-field illumination, and

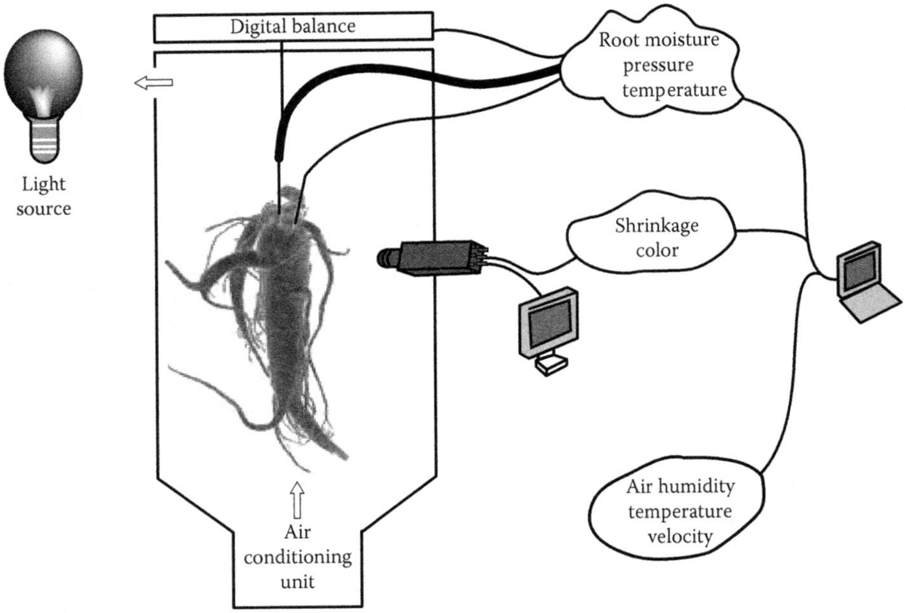

FIGURE 18.2 Computer vision hardware for real-time monitoring of vegetable root drying.

telecentric illumination have been identified (Jähne and Haussecker, 2000). The two commonly used lighting geometries are ring lighting and diffuse lighting (Wu and Sun, 2013).

The geometry of the ring illuminator is simple and is suitable for all-purpose application. On the other hand, the diffuse illuminator delivers 180° of diffuse lighting and is applicable for the illumination of objects with glossy surfaces or those capable of reflection. Since the majority of vegetables are 3-D objects, a diffuse illumination system is more suitable for their imaging.

The illumination system should consider both radiometric properties of the object and lighting source. There may be a need for combined illumination sources depending on the application. It has been suggested that color reproduction for each illumination system should be calibrated using standard color set. Primary factors that influence the selection of light source are: shape of the object (flat or curved), optical properties (absorbing, transmissive, or reflective), and contrast with the background (Zuech, 2003). In a nutshell, the key to successful computer vision is a good contrast, repeatable image, not affected by ambient light or other conditions. Careful design of the controlled lighting system usually simplifies image segmentation and interpretation. Enhancement of image contrast due to good lighting improves feature discrimination, accuracy, and the success of image analysis.

A digital camera with a good optical system captures images with high quality and high resolution, which helps to extract more detailed features and information in time series of images during vegetable drying. Visible light band sensor is useful in quality assessments, but limited to the surface characteristics of vegetables. Analysis of the subsurface qualities monitoring may require sensors that are sensitive to the near infrared, ultraviolet, and X-ray region of the electromagnetic spectrum. Charge-coupled device (CCD) camera is the most commonly used sensor because of the stability of silicon-based semiconductors (Chen et al., 2002). They are available in both color and monochrome modes. The CCD sensor is made up of photodiodes and capacitors in parallel connections. The photodiodes convert light into electrical charge proportional to the intensity of the light. However, CCD color camera requires periodic calibration of gain, brightness, and white balance for the chosen source of illumination. The recently developed complementary metal oxide silicon (CMOS)-based cameras have made image acquisition faster and more efficient. They also allow higher resolution and less power consumption, which is important in field applications.

18.3.2 Software

Recent versions of Windows have a set of drivers, compatible with most of the imaging devices, which allows simple capturing of images or video. However, functionality of standard Windows software is not sufficient to adjust CCD camera settings or time-controlled image acquisition. Therefore, most of the CCD camera manufacturers usually supply specialized software, designed for specific cameras and interface, which often create issues of compatibility. This is probably one of the reasons why most of the researchers are still using two different software packages: one for image capturing (which could be a part of camera software) and another one for offline image processing and analysis. For example, Majumdar and Jayas (2000a) used a Sony video camera with a frame grabber DT2871 (Data Translation Inc., Marlboro, MT) for image acquisition. However, the software for further image processing was written in C language using the Aurora subroutine library (Aurora, Data Translation Inc., Marlboro, MT). Features were extracted by the algorithm developed in the Khoros (Khoral Research Inc., Albuquerque, NM) environment using C++ programing language. Pedreschi et al. (2006) and Mendoza et al. (2007) captured images with Remote Capture Software (v.2.7.0, Canon, Tokyo, Japan) and further processed and analyzed images in MATLAB® code (Mathworks, Natick, MA). Romani et al. (2009) used off-line color scanning of samples with Agfa Snapscan E40 (Agfa-Gevaert, Mortsel, Belgium) with the next processing of images in MATLAB. Romano et al. (2012) used a PAX Cam P1-CMO camera with specialized PAX-IT software (MIS, Villa Park, IL), which allowed sequential capturing and analyzing of images. Yadollahinia et al. (2009) used a Sony TRV-140 (Sony, Japan) camera, controlled by Visual Basic software, with the next image processing in MATLAB. Mayor et al. (2011) used the free software UTHSCSA Image Tool (Health Science Center, University of Texas, TX) to analyze geometrical features of previously captured images. Zenoozian et al. (2008) developed all the algorithms for preprocessing of images, segmentation, and color analysis, using MATLAB 7.1. It turns out that MATLAB is the most popular software for image processing and further analysis. An additional advantage of MATLAB is the flexibility of programming for quantifying of morphological, color, and textural features. However, MATLAB does not allow simultaneous real-time image acquisition, processing, and control. In contrast, LabVIEW software (National Instruments, Austin, TX) sufficiently combines all three functions in real time. Image acquisition and processing with LabVIEW software was successfully used by Martynenko (2008) and Davidson et al. (2009) for temperature control in ginseng drying.

18.4 IMAGE ACQUISITION

Image acquisition is a preliminary and important step in the generation of image data. Image acquisition is the process of image capturing and sending to the computer through an interface. Series of images are acquired in the time-lapse mode, which is controlled by either hardware or software. Usually image acquisition is software-controlled, which allows adjustment of time with respect to monitored feature. Different types of cameras require different interfaces (USB, IEEE-1394 [Firewire], Ethernet, or GigE) and software (Windows, Visual Basic, NI IMAQ, LabVIEW, and MATLAB) for image recording and further processing. The set of drivers are specific for the CCD camera, interface, and software.

Image acquisition can be carried out either off-line or in the online mode. Off-line mode of image acquisition was used in the measurements of color changes of pumpkin slices (Zenoozian et al., 2008), potato chips (Pedreschi et al., 2006; Romani et al., 2009), carrot slices (Nahimana and Zhang, 2011), bell pepper (Romano et al., 2012), and vegetable soybean (Huang et al., 2014) during drying. Off-line image acquisition is more labor-intensive, because it requires conversion of images for further processing and analysis. However, off-line imaging allows better control of illumination and background conditions, which is particularly important for color measurements. The calibration of color reproduction is a part of routine imaging methodology, which usually resulted in comparable high-quality images.

Online mode of image acquisition was not so widely used in vegetable drying. Nonisotropic shrinkage during convective drying of potato slabs was studied by Campos-Mendiola et al. (2007) Yadollahinia et al. (2009). Martynenko (2006, 2008) and Davidson et al. (2009) investigated shrinkage behavior of ginseng roots to maintain temperature on the optimum level on the different stages of ginseng drying. An example of software for online image acquisition is presented in Figure 18.3.

Online mode of image acquisition opened up a range of opportunities for online monitoring and control of drying. For example, time series of images gives additional knowledge about quality of the drying process. The online measurements of ginseng color changes during drying helped to discover the extremely sensitive period of drying, when most browning occurred (Martynenko, 2008). Also, real-time imaging allows to observe porosity evolution during drying (Martynenko, 2011).

It could be concluded that the online mode of image acquisition is justified for vegetables with unique biochemical and medicinal properties, where quality of drying is priority. As a part of the intelligent control system, computer vision was able to prevent undesirable quality degradation (Martynenko and Yang, 2007; Martynenko and Kudra, 2015). Online mode of image acquisition is also desirable for industrial applications to increase the efficiency of quality control without additional personnel.

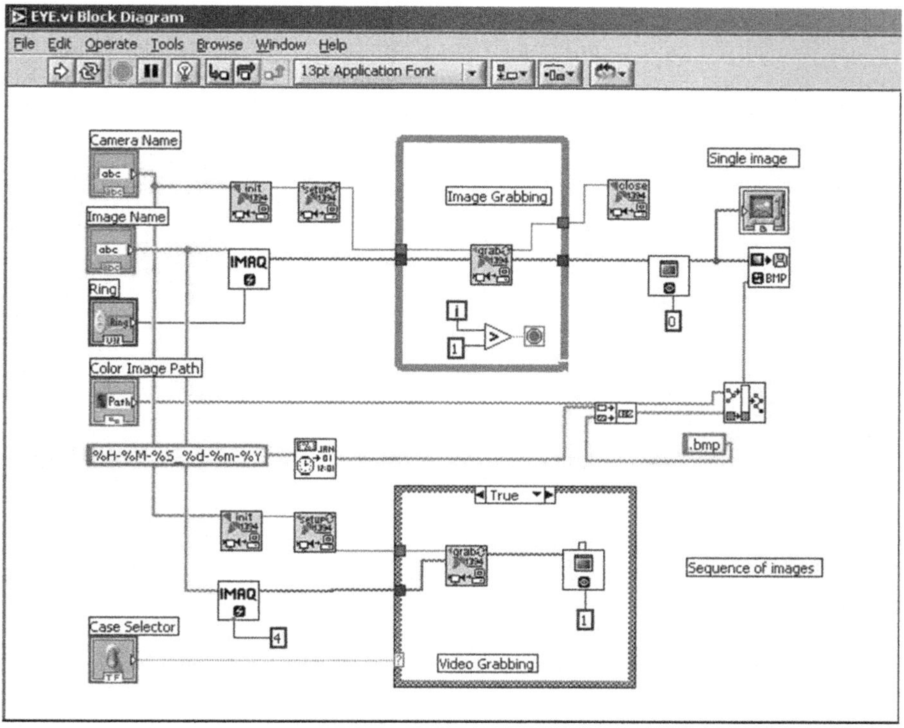

FIGURE 18.3 A functional panel for image and video acquisition, developed in LabVIEW 8.5.

18.5 IMAGE PROCESSING

Image processing could be divided into three major operations: low-level processing or preprocessing, intermediate-level processing, and high-level processing (Figure 18.4).

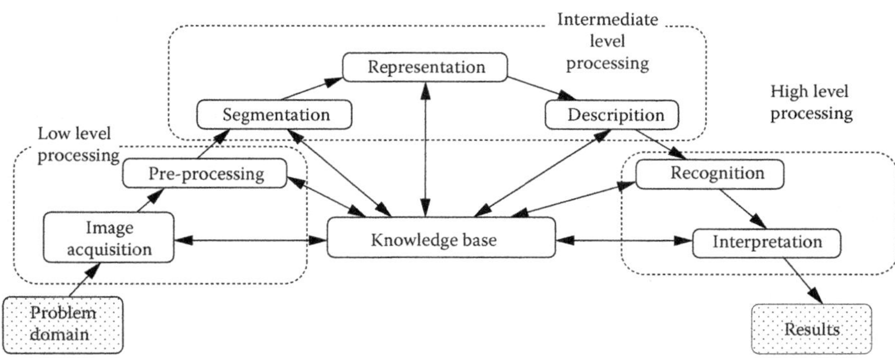

FIGURE 18.4 Image processing illustration. (Adapted from Brosnan, T., Sun, D.-W., *J. Food Eng.*, 61(1), 3–16, 2004. With permission.)

The preprocessing step includes conversion of the video image into numerical format by using digital signal converters (frame grabbers or digitizers). In this format, the image is represented as a two-dimensional grid of pixels intensities. At this step, corrections of gray level, geometrical distortion and defocusing, as well as noise reduction are applied. The choice of a frame grabber is influenced by the mode of camera output, requirements on the spatial and gray image resolution, and speed of the processor (Gunasekaran, 2000). If the computer vision system uses a digital camera, then there is no need for a frame grabber, because the image is already presented in digital form. However, in this case, correction of the gain, brightness, contrast, and white balance of a digital camera with respect to illumination is required. Intermediate-level processing starts with conversion of a color or monochrome image to binary images to distinguish the region of interest (ROI) from the background. The challenge of this step is to choose the most appropriate color plane (R, G, B, H, S, I, L, etc.), which could provide the best contrast of the object from the background. Image segmentation is basically partitioning of the image into clusters based on the gray-scale intensity histogram. The objective of segmentation is to divide the image into regions or areas of interest.

There are three basic techniques for segmentation: thresholding, edge detection, and region detection (Brosnan and Sun, 2004). *Thresholding* is based on extraction objects with similar light reflectivity or absorptivity, *edge detection* is based on detection of discontinuities in gray level, color, or texture, and *region detection* is based on the grouping of similar pixels to form regions, representing single objects. Aggregation of homogenous picture elements to form regions representing single objects within the image forms the region. The basis for the similar picture elements can be gray-level intensity, color, or texture.

Segmented image is usually a binary image and could be used for either extraction of ROI from original color image (image mask) or directly for further image description. Image description is the extraction of quantitative information from the segmented ROI. Established algorithms for the morphological, color, and textural features inside of ROI could be used.

18.5.1 Thresholding

The starting point for thresholding is the histogram of gray-scale intensity distribution. There are a number of techniques to analyze this histogram. *Clustering-based technique* is based on the clustering of gray-level pixels in two parts as either background or foreground objects. A clustering threshold value T is defined automatically or manually. Pixels with gray value higher than T are assigned a gray level value of 255; other pixels are considered as background and assigned a gray level value of 0. In the *shape-based technique*, all peaks, valleys, and curvatures on the histogram are analyzed separately and given physical interpretation. *Entropy-based technique* is based on the difference of entropies of the foreground and background regions, the cross-entropy between the original and binary image, etc. *Object attribute-based technique* analyzes similarities between the gray-level and the binary images, such as fuzzy shape similarity, edge coincidence, etc. Manual thresholding is based on trial and error. It is repeated as many times as possible until the object is finally differentiated from the background. In contrast, *Otsu's method* automatically calculates optimum threshold based on the assumption that the image contains only two classes of pixels (bimodal distribution). This technique is equivalent to Fisher's discriminant analysis.

18.5.2 Edge Detection

Edge detection algorithms are particularly sensitive to noise, so initial filtering for noise reduction is critical. A preprocessing step to edge detection, a smoothing stage, typically Gaussian smoothing, is almost always applied. Most of the methods for edge detection can be grouped into two categories, namely, search-based and zero-crossing-based. The search-based methods detect edges by first computing a measure of edge strength, usually a first-order derivative of the intensity, and then searching for local directional maxima of the gradient using a computed estimate of the local orientation of the edge, usually the gradient direction. The zero-crossing-based methods search for zero crossings in a second-order derivative expression computed from the image to find edges, usually the zero crossings of the Laplacian or the zero crossings of a nonlinear differential expression.

An excellent review of different edge detection methods can be found in Ziou and Tabbone, 1998. The edge detection methods mainly differ in the types of smoothing filters that are applied and the way the measures of edge strength are computed. As many edge detection methods rely on the computation of image gradients, they also differ in the types of filters used for computing gradient estimates in the x- and y-directions.

18.5.3 Filtering

Filtering can be applied before or after segmentation to remove small noise or outliers from the ROI and improve the quality of the image with more detailed information. There are two basic techniques of image filtering: (a) spatial filtering and (b) frequency filtering. Spatial filtering uses a kernel (two-dimensional matrix) for filtering of small objects, blurring, sharpening, or unsharpening of the image. This is accomplished by means of convolution between a kernel and an image. The properties and size of the kernel determine the effects on the output image. Most of the image processing softwares include image filtering subroutines. For example, IMAQ Vision (National Instruments, Austin, TX, USA) include four spatial filter families: smoothing, gradient, Gaussian, and Laplacian (Klinger, 2003). Frequency filtering is based on the fast Fourier transform (FFT) of the original image by converting it to frequency domain, filtering undesirable frequencies, and then returning to spatial domain. It is usually used for either attenuation (low- or high-pass filter) or truncation (bandpass filter) of original image.

18.6 FEATURE EXTRACTION

Feature extraction is a part of high-level processing. The major objective is to extract quantitative information from images for recognition and interpretation of specific features, important for classification or quality control. The main three categories are: morphological, color, and textural features (Majumdar and Jayas, 2000a,b,c,d; Fernández et al., 2005). These features could be extracted from digital images, using streamline image-processing techniques (Martynenko, 2006). In this technique the binary image after segmentation is used for two purposes: (a) estimation of morphological features (area) and (b) masking original color image for extraction of color and textural features. Multiplication of the original image on its binary mask allows conversion of background pixels to zero and elimination of this class from the next calculations. Using streamline technique of feature extraction, an original color image undergoes spatial filtering twice: the first time to extract morphological (area) and textural features, and the second time to extract color features (Martynenko, 2006).

18.6.1 Morphological Features

Image morphology includes geometric structures within an image, either *size-based features*, such as area, perimeter, bounding rectangle, centroid, lower-order moments (normal, central, and invariant), length and width and angle of orientation, or *shape-based features*, such as the degree of roundness, elongation, compactness, particle distribution radius ratio, box ratio, aspect ratio, area ratio, and the coefficient of variation of radii (Majumdar and Jayas, 2000a). Commonly measured morphological features include length, width, diameter, shape factor, and area (Klinger, 2003). Zapotoczny et al. (2008) calculated morphological features based on measurements of linear dimensions, i.e., minimal and maximal length, width, perimeter, and convex perimeter. On the basis of the measurement of basic parameters, shape factors such as aspect ratio, area ratio, circularity, eccentricity, elongation, slenderness, compactness, and corrugation were calculated. The measurements of morphological features also include calculation of center of gravity and geometrical momentum.

Fernández et al. (2005) reported that some specialized software programs, such as Image Pro Plus (Media Cybernetics, Rockville, MD), give more than 30 geometrical parameters of identified objects. In analyzing the features of an image, the approach requires that the image is transformed to another level in which the information represented by the image can be in the form of figures. Blob analysis allows to distinguish the object of interest from isolated small clusters of pixels by using multithreshold filtering with next particle analysis (Klinger, 2003). Area measurements mostly depend on the object geometry. In the case of cubes or parallelepipeds, the area was calculated from the length and width (Mulet et al., 2000). In the case of cylindrical shaped slices, area could be calculated either by direct measurements of pixels with the next multiplication on the conversion coefficient (Martynenko, 2006; Campos-Mendiola et al., 2007) or from diameter measurements (Chen and Martynenko, 2013). In the case of lateral observation of the whole cylindrical shaped root, the surface area was evaluated from two-dimensional projection with the next multiplication by π (Martynenko, 2006). This procedure was successfully applied in thin-layer drying of ginseng slices (Martynenko, 2008). It generally works when pieces of dried material could be separated by background, which allows segmentation and further statistical analysis. In bulk vegetable drying, where roots overlapped with each other, it was difficult to evaluate surface area. However, with preliminary chosen calibrated threshold, it became possible to evaluate relative area changes or root shrinkage (Martynenko, 2008).

Morphological features were used mostly for evaluation of shrinkage, density, porosity, and shape changes during drying (Mulet et al., 2000; Campos-Mendiola, 2007; Yadollahinia et al., 2009; Martynenko, 2011; Mayor et al., 2011; Martynenko, 2014). For example, Mayor et al. (2011) used image analysis for evaluation of surface area (S), perimeter of the contour (p) of cylindrical-shaped pumpkin samples. The length of the major axis (L_m) was determined as the length of the longest line that can be drawn through the object. The length of the minor axis was defined as the length of the longest line that can be drawn through the object perpendicular to the major axis. Elongation was calculated as the ratio of the length of the major axis to the length of the minor axis. Roundness (R) and compactness (C) were calculated from the following equations (Mayor et al., 2011):

$$R = \frac{4\pi \cdot S}{p^2} \tag{18.1}$$

and

$$C = \frac{2}{L_m}\sqrt{\frac{S}{\pi}}. \tag{18.2}$$

The values of R and C determine the circularity of the object. Both shape factors ranged from 0 to 1, with one corresponding to a perfect circle. A decrease in these values indicated deviation from roundness as the result of drying.

Yadollahinia et al. (2009) used similar approach image analysis to calculate area, perimeter, major and minor diameters, roundness, and Feret diameter (FD) of potato slices. For example, FD was calculated from the following equation:

$$\mathrm{FD} = 2 \cdot \sqrt{\frac{S}{\pi}}. \tag{18.3}$$

Calculated values were used to study the effects of temperature, air velocity, and moisture content on the shrinkage of potato slices during drying.

Chen and Martynenko (2013) extended this approach to the evaluation of bulk volume (V), using two (top and side) digital cameras and calculating volume from simultaneous diameter (d) and thickness (h) measurements:

$$V = \pi \cdot (d/2)^2 \cdot h. \tag{18.4}$$

This approach enabled to calculate not only shrinkage but also porosity (Martynenko, 2011) and density (Martynenko, 2014) evolution during drying.

Porosity and density of ginseng root were estimated from real-time imaging with <5% of discrepancy compared to scanning electron microscope (Martynenko, 2011). The high accuracy of porosity and density estimation was due to the uniform shrinkage of ginseng root.

18.6.2 Color Features

Color is considered as a fundamental physical property of vegetables because of high correlation with biochemical and sensorial indicators of quality. However, color is not easy to measure because of complex effects of the light source, the reflectivity of the sample, and the sensitivity of the instrument. An excellent review of calibrated color measurements of agricultural foods with the focus on the methodology of color imaging was presented by Mendoza et al. (2007).

Usually color is represented as a set of three values in one of the color spaces. For the food industry, it is common to measure color in *CIE L*a*b** color space. In computer vision the color is extracted in *RGB NTSC* digital format as intensities of red (R), green (G), and blue (B) channels on 0 to 255 scale. The challenge of this format is high nonlinearity. In contrast, *CIE XYZ* and *L*a*b** outputs are linear and scaled from 0 to 100. To provide compatibility, the output signal from digital camera in *RGB NTSC* has to be converted into *CIE XYZ* and eventually to *L*a*b** color space. The first transformation of *RGB* into *CIE XYZ* requires two steps (Mendoza et al., 2007):

1. Transformation of nonlinear *RGB* into linear *sRGB* values:

   If R or G or $B \le 0.04045$, then

$$sR = 0.0774 \cdot R$$
$$sG = 0.0774 \cdot G \qquad (18.5)$$
$$sB = 0.0774 \cdot B.$$

   Else

$$sR = -\left(\frac{-R+0.055}{1.055}\right)^{2.4}$$
$$sG = -\left(\frac{-G+0.055}{1.055}\right)^{2.4} \qquad (18.6)$$
$$sB = -\left(\frac{-B+0.055}{1.055}\right)^{2.4}.$$

2. Conversion of *sRGB* values into *CIE XYZ* color space coordinates:

$$\begin{bmatrix} X \\ Y \\ Z \end{bmatrix} = \begin{bmatrix} 0.4124 & 0.3576 & 0.1805 \\ 0.2126 & 0.7152 & 0.0722 \\ 0.0193 & 0.1192 & 0.9505 \end{bmatrix} \times \begin{bmatrix} sR \\ sG \\ sB \end{bmatrix}. \tag{18.7}$$

CIE XYZ values could be converted into *CIE L*a*b** color space using the set of Equation 18.8:

$$\begin{aligned} L^* &= 116 \cdot \left(\frac{Y}{Y_n}\right)^{1/3} - 16 \\ a^* &= 500 \cdot \left[\left(\frac{X}{X_n}\right)^{1/3} - \left(\frac{Y}{Y_n}\right)^{1/3}\right] \\ b^* &= 200 \cdot \left[\left(\frac{Y}{Y_n}\right)^{1/3} - \left(\frac{Z}{Z_n}\right)^{1/3}\right], \end{aligned} \tag{18.8}$$

where X_n, Y_n, Z_n correspond to the *XYZ* values of the reference white color.

Due to moisture removal, the surface may shrink during vegetable drying. To reduce the effect of curvature, which can potentially influence the accuracy of color measurements, both *sRGB* and *L*a*b** scales were normalized between 0 and 1 (Mendoza et al., 2007):

$$\begin{aligned} sRGB_{\text{normalized}} &= sRGB/255 \\ L^*_{\text{normalized}} &= L^*/100 \\ a^*_{\text{normalized}} &= (a^* + 120)/240 \\ b^*_{\text{normalized}} &= (b^* + 120)/240. \end{aligned} \tag{18.9}$$

In this normalized format, the total color difference in *L*a*b** can be evaluated using Euclidian distance measure:

$$\Delta E = \sqrt{(L_2^* - L_1^*)^2 + (a_2^* - a_1^*)^2 + (b_2^* - b_1^*)^2}. \tag{18.10}$$

Euclidian measure could be applied to different images to evaluate the effect of treatment or to the time series images to evaluate the effect of drying on the same sample.

An alternative approach to conversion of digital image from *HSI* digital format to *CIE L*a*b** color space was proposed by Sampson et al. (2014), based on the set of the following equations:

$$\text{Hue} = \tan^{-1}\frac{b^*}{a^*}$$

$$\text{Saturation} = \sqrt{(a^*)^2 + (b^*)^2} \tag{18.11}$$

$$\text{Intensity} = kL^*.$$

From this set of equations, it could be easily deduced:

$$a^* = \frac{S}{\sqrt{1+\tan H}}$$

$$b^* = \frac{S}{\sqrt{1+1/\tan H}} \tag{18.12}$$

$$L^* = k_1 I.$$

This approach, however, requires preliminary calibration of the digital camera for intensity (*I*) and chromaticity (*S*) with standard Hunter *L*a*b** values. The chromatic adaptation required for a nonstandard light source could be determined using a known color standard, for example, a ColorChecker (X-rite, Grand Rapids, MI).

To avoid effects of size sampling on color intensity distribution, the number of pixels for each intensity line are normalized with respect to the overall number of pixels in the extracted area. The histogram of color intensity is treated as a fuzzy variable with lightness as a support. Average color intensity is calculated from intensity histogram on the basis of center-of-gravity defuzzification. In time series color measurements means and variances are used to test statistical hypothesis (*F*-test) of color changes on each interval of observation.

18.6.3 Textural Features

The spatial distribution of gray-level intensities is referred to as the textural features of an object. Textural features describe the textural patterns of surface properties of smoothness, coarseness, fineness, and granulation. These textural features, which are related to surface microstructure, correspond to human perception of food surface. The mathematical measurement of texture takes into account the higher moments of the gray-level histograms. The average intensity of an object (*m*) is defined by

$$m = \sum_{i=1}^{k} p(z_i), \tag{18.13}$$

where z_i is a gray level of the image ($i = 1, 2,\ldots, k$), and $p(z_i)$ is a probability density function. The higher order moments of gray-level histogram are defined as

$$\mu_n(z) = \sum_{i=1}^{k} (z_i - m)^n p(z_i). \tag{18.14}$$

For uniform objects, the zero- and the first-order moments (μ_o and μ_1, respectively) are 1 and 0. The variance or the second moment (μ_2) defines the average smoothness or the distribution pattern of the gray-level intensity. Skewness or the third-order moment (μ_3) explains the symmetry of the histogram curve about the mean gray level. Kurtosis, or the fourth-order moment (μ_4) describes the level of peakness or flatness of the histogram of the object (Haralick et al., 1973). Extraction of textural features, based on gray-level co-occurrence matrix (GLCM) or gray-level run length matrix (GLRM), was successfully used by Majumdar and Jayas (2000c).

Texture could be also presented as a geometry-related feature by using empirical mode decomposition or FFT (Xiong et al., 2006). Spectral power density $S(\omega)$ can be calculated by using FFT for one-dimensional color intensity profile as a spatial pattern of intensity pixels. Spectral power density provides information not only about the homogeneity of pixels distribution, but also about some periodical patterns on the vegetable surface, such as surface wrinkling (Martynenko, 2006). First peak of spectral power density corresponds to textural uniformity. Second peak and higher harmonics of spectral power characterizes the development of regular wrinkles on the root surface. Energy is calculated as an integral of spectral power density:

$$E = \int_{0}^{\omega} S(\omega)\,\mathrm{d}\omega, \tag{18.15}$$

where ω (pixels^{-1}) is a parameter, reciprocal to the distance between pixels. Energy is a measure of textural uniformity of food during drying.

Eventually, textural features, such as entropy, spectral density, GLCM, GLRM could eliminate the redundancy of the initial dataset and simplify next pattern recognition or discriminatory analysis.

18.7 PATTERN RECOGNITION

From this survey it follows that the traditional computer vision uses a low number of features and a few number of classifiers, commonly available in software packages. It is unfortunate that considerable research efforts have been focused on developing tailored features and classifiers limited to specific classification problems. Recent computer capabilities make possible simultaneous extraction and analysis of a very large number of features.

The combination of morphological, color, and textural features was used for classification of cereal grains by a computer vision (Majumdar and Jayas, 2000d). Overall

17 morphological features (e.g., area, perimeter, length, width, major axis length, spatial moments, Fourier descriptors, etc.) were extracted. Images were converted from Red-Green-Blue (RGB) into Hue-Saturation-Intensity (HSI) color space and used to optimize the number of morphological and color features that contributes to the classification. Twenty-five textural features (10 GLCM features, 12 GLRM features, and 3 gray-level features) were extracted and used in the discriminant analysis. In terms of texture, GLCM was used. In the combination models, the morphological-texture (MT) model included 23 morphological and 25 textural features. Morphological-color (MC) model included 23 morphological and 18 color features. Texture-color (TC) model included 25 textural and 18 color features. Morphological-texture-color (MTC) model included 23 morphological, 25 textural, and 18 color features. The results showed that all features can be used separately or in combination. When only one feature was used, the accuracy of morphological feature classification was the highest followed by color and texture. When the classification was based on combination of two features, the combination of morphological and color models performed with the highest accuracy followed by morphological combined with texture and texture combined with color. The combination of three features gave the highest accuracy.

Another approach to the pattern recognition problem was proposed by Mery et al. (2013). They formulated their objective as an automated search of the most relevant set of features and classifiers from a list of about 3000 features and about 25 classifiers. Testing of proposed automated optimal computer vision system on eight different food quality evaluation scenarios yielded a classification performance not less than 95%.

18.8 LEARNING TECHNIQUES

It was commonly recognized that pattern recognition often requires supervised or not supervised learning. Development of artificial intelligence and pattern recognition techniques inspired application of learning techniques in the food industry (Goyache et al., 2001; Corney, 2002). Du and Sun (2006) presented an excellent review of learning techniques used in computer vision for food quality analysis. It covers both statistical learning (SL), mostly discriminant analysis, and nonstatistical learning, based on Bayesian classifiers, Artificial Neural Networks (ANN), and Fuzzy Logic (FL). One of the first practical applications of SL by using multivariate image analysis, principal component analysis (PCA), and partial least squares (PLS) for snack classification was reported by Yu and MacGregor (2003). Thybo et al. (2004) used PCA and PLS for learning hidden relationships between image texture and sensory textural attributes in potatoes.

As an example of nonstatistical approach, Martynenko and Yang (2007) used ANN for the prediction of color, shrinkage, and moisture content of ginseng roots from real-time imaging in the process of drying. Zenoozian et al. (2008) successfully used ANN to predict shrinkage and color changes of osmotically dehydrated pumpkin. Next development of this technique was an application of wavelet transform together with ANN that further improved the accuracy of prediction of moisture and color changes in pumpkin dehydration (Zenoozian and Devahastin, 2009).

Further development of learning techniques made possible automated design of a computer vision system for visual food quality evaluation (Mery et al., 2013). This system combines multiple features and multiple classifiers. In the process of training, the system automatically chooses the best problem-related set of features and classifiers. Testing this system in eight different recognition problems with two, three, and six classes showed a high performance of 95% correct classification. Du and Sun (2006) noted that learning techniques can help to learn about meaningful relationships, hidden in the initial dataset, and generalize these relationships to interpret new, unseen data. In this sense, learning techniques is the tool for building the knowledge base of relationships between the image data and the product quality.

18.9 APPLICATIONS

Most of the practical applications of computer vision for drying of vegetables are generalized in Table 18.2.

18.9.1 Dimensional Measurements Using Computer Vision

Over the past decades, computer vision has been extensively used for inspection and quality assessment of vegetables before, after, or during drying (Aghbashlo et al., 2014). In these studies, morphological, color, and textural changes from image analysis were related to physicochemical changes in vegetables during drying.

Early application of computer vision for mushroom shape and whiteness inspection was reported by Heinemann et al. (1994). Images were taken using a Sony black and white video camera and collected using a Data Translation Quick Capture frame grabber installed in an Apple Macintosh computer. Initial calibration of aperture, focus, and zoom settings was performed with a 38.1-mm flat white metal circle disk placed on a black background. After images were captured, gray level and frequency was represented in a histogram of pixels intensity in PASCAL source code. Threshold algorithms to determine the color, size, and shape features were developed in PASCAL. Mushroom color quality was determined by the intensity histograms analysis and it showed that mushrooms grading were performed effectively based on color. Mushroom shape was analyzed from binary images with pixels of mushroom as 1s and background pixels 0s. The boundary-finding subroutine was developed to isolate the mushroom and compute the boundary endpoints. The mushroom cap area was identified as a region within the cap boundary, and veil aria was identified as a region with pixels intensity above 200. An algorithm was developed to find the points where cap and stem are connected. Once the stem was determined another algorithm was written to compute the flare factor as a threshold to detect whether the stem was flared. An average error in classification of individual features was small as compared to individual experts in mushroom quality inspection.

Mulet et al. (2000) used imaging to evaluate shrinkage of potato and cauliflowers of different geometrical shapes (cubes, parallelepipeds, and cylinders) during hot-air-drying. Images were captured by a black and white video camera SSC-M350 CE (Sony, Japan) and processed with SYNOPTICS Ltd. (Version 4.02, 1993) software. Imaging was in good agreement with dimensional measurements for the samples with a regular shape. However, at lower moisture contents the accuracy decreased due to the nonuniform shrinkage. The authors suggested that more accurate shrinkage

TABLE 18.2
Computer Vision Applications in Drying of Vegetables

Feature Extracted	Vegetable	Sample	Quality Parameters	Computer Vision System	Mode of Imaging	References
Morphological	Potato	Cylinders, cube, slab	Shrinkage	SSC-M530 B/W video camera (Sony, Japan)	Off-line	Mulet et al. (2000)
		Slabs	Shrinkage, shape	DC290 digital camera (Kodak, Rochester, NY, USA)	Online	Campos-Mendiola et al. (2007)
		Slices	Shrinkage	TRV140 digital camera (Sony, Japan)	Online	Yadollahinia et al. (2009)
	Carrot	Slices	Shrinkage	i85 CCD digital camera (Samsung, Korea)	Online	Nahimana and Zhang (2011)
	Pumpkin	Cylinders	Shrinkage, density, porosity, shape	SSC-DC50AP digital camera (Sony, Japan)	Off-line	Mayor et al. (2011)
	Ginseng root	Whole	Shrinkage, moisture	Fire-i CCD digital camera (Unibrain Inc.)	Online	Martynenko (2006) and Davidson et al. (2009)
		Whole	Shrinkage, porosity	DFW-SX900 digital camera (Sony, Japan)	Online	Martynenko (2011)
		Whole	Shrinkage, density	DFW-SX900 digital camera (Sony, Japan)	Online	Martynenko (2014)
		Batch	Shrinkage, moisture content	Fire-i CCD digital camera (Unibrain Inc., San Ramon, CA, USA)	Online	Martynenko (2006)
Color	Potato	Slices	Color	E40 Snapscan color scanner (Agfa Gevaert, Belgium)	Off-line	Romani et al. (2009)
		Slices	Color	Canon Power Shot G3 camera (Canon)	Off-line	Pedreschi et al. (2006)
		Slices	Color	N/A	Off-line	Mogol and Gokmen (2014)

(Continued)

TABLE 18.2 (*Continued*)
Computer Vision Applications in Drying of Vegetables

Feature Extracted	Vegetable	Sample	Quality Parameters	Computer Vision System	Mode of Imaging	References
	Carrot	Slices	Color	i85 CCD digital camera (Samsung, Korea)	Off-line	Nahimana and Zhang (2011)
	Pumpkin	Cubes	Color	DSC-40 Cyber Shot (Sony, Japan)	Off-line	Zenoozian et al. (2008)
	Green pepper	Slices	Color	Canon Power Shot A70 (Canon)	Off-line	Mendoza et al. (2007)
	Red pepper	Slices	Color	the same	Off-line	Mendoza et al. (2007)
	Soybean	Whole	Color, moisture content	Hyperspectral imaging with CCD camera Pixelfly QE (Cooke Corp., Romulus, MI, USA) and imaging spectrograph 1003A-10140 Hyperspec VNIR (Headwall Photonics, Bolton, MA, USA)	Off-line	Huang et al. (2014)
	Bell pepper	Slices	Color, moisture content	PAX-Cam P1-CMO (Villa Park, IL, USA)	Off-line	Romano et al. (2012)
	Ginseng	Whole	Color	DFW-SX900 digital camera (Sony, Japan)	Online	Martynenko (2008)
		Batch	Color	DFW-SX900 digital camera (Sony, Japan)	Online	Martynenko (2008)

estimation requires considering concave and irregular shapes by using more sophisticated image processing.

Computer vision for volume estimation of vegetables with noncylindrical shape was developed by Hahn and Sanchez (2000). In their study they used ELECTRIM 1000 monochromatic CCD, which revolved around the carrot. Volume was estimated from two model-based algorithms. One algorithm involved reconstructing the object using finite elements. Periodic turning camera on 1.69° with the next integrating of imaging areas allowed them to calculate volume of a nonperfect cylindrical object from the equation:

$$A_r = x \cdot (r \cdot \sin\theta)/2, \tag{18.16}$$

where x is the distance from the center to the carrot surface, r is the radius, and θ is the angle. To optimize memory use, the erosion and dilation operators were applied on the contour. They found that as the carrot size increased, the estimation error increased due to the optical distortion on the CCD photocell. They also found that white panel with back-illumination of sample allowed to eliminate shadow and to reach better sensor accuracy. The second model-based algorithm used two orthogonal (longitudinal and cross-sectional) projections and an empirical correction factor. Images acquired at 0° and 90° represented the maximum and minimum area, respectively. This approach showed high accuracy in carrot volume estimation.

Igathinathane and Chattopadhyay (1998) modeled food materials as general ellipsoids from measurements of their three principal dimensions under projection and proposed an algorithm to predict surface areas. Gall et al. (1998) described the use of a ring sensor system that scans three-dimensional (3D) objects in real time and creates an enveloping spiral of the convex hull from which volume and other size parameters were calculated using models. The design of many commercial and development sorting machines are particularly suited to obtain axially orientated ("vertical") projections of an object (Crowe and Delwiche, 1996; Gall et al., 1998). Some systems rotate an object in normal, X-ray, infrared (IR) or near infrared (NIR), laser, or spatially modulated (structured) light. Alternatively, several cameras or a mirror viewing system may be positioned around a channel through which the singled product items may pass.

Research carried out in the recent decade has considerably improved imaging techniques and scientific platform for computer vision. Chen and Martynenko (2013) used computer vision for imaging of sample volume and color changes during drying. The computer vision system included two color digital CCD cameras set at 90° angles to each other, capturing images from front side (thickness) and top (diameter). Image processing was performed in real time, using LabVIEW 8.5 software and the library of virtual instruments IMAQ 6.5 (National Instruments). An image analysis includes image segmentation, filtering, and data analysis. ROI was segmented from the original images taken by cameras using image mask. The luminance plane in HSL color space created the best resolution between object and background. The optimal threshold was determined prior to experiment to minimize the differences between real and projected area. Furthermore, filtering with particle filter removed small objects, noises, and outliers from the background. The diameter and thickness of samples are determined via the IMAQ Clamp subroutine configured to IMAQ 6.5. The error of dimensional measurements using computer vision for regular shape objects did not exceed 8%. It was proved that time series imaging has a potential to decrease random error of measurements, while systematic error could be minimized by periodic dimensional measurements and calibration of computer vision system settings.

18.9.2 Color Measurements

Initial application of computer vision for potato color inspection was reported by Tao et al. (1995). Basic image acquisition and processing of color image was performed using frame grabber and Aurora subroutine library software. RGB images were converted to IUV and then to HSI color space. Multivariate discriminant analysis of gray intensity in Hue color plane demonstrated high accuracy (over 90%) in color inspection and sorting of potatoes.

Pedreschi et al. (2006) developed a computer vision system to measure the color of potato chips in the process of frying. Image processing in MATLAB allowed easy conversion of RGB color images into $L^*a^*b^*$ color space. Using MATLAB subroutines, they developed computational software for simultaneous image processing, segmentation, and color conversion. This experimental setup allowed them to investigate kinetics of color changes in potato slices at different temperatures. Due to the high correlation between color of potato chips and acrylamide content, the computer vision system could potentially identify and reject products with undesirably high acrylamide. Imaging of potato chips color during frying with a digital camera was also reported by Romani et al. (2009). The results were comparable with those obtained from a conventional $L^*a^*b^*$ colorimeter.

Mendoza et al. (2007) compared three different color spaces (*sRGB*, *HSV*, and $L^*a^*b^*$) for color quantification of vegetables during drying. $L^*a^*b^*$ color space was suggested as the best color space for quantification of color of vegetables due to high sensitivity to color changes in drying along with robustness to surface curvature.

Romano et al. (2012) measured color changes of bell pepper during drying. It was found that computer vision with CCD camera in combination with laser lighting was useful to predict the moisture content and color of bell pepper (yellow, green, and red) during drying. They converted the original *RGB* image into *XYZ* color space and then calculated the $L^*a^*b^*$ values. Results showed that moisture content can be predicted from the reflected light intensity, normalized to the area of scattering. They also found that light scattering depends on the depth of the surface and tissue structure. Changes of tissue structure result in a different dispersion of photons through the surface. It was concluded that computer vision with laser lighting can be used for online moisture content and color estimation in the process of drying.

Huang et al. (2014) studied sensitivity of hyperspectral imaging in the range from 400 to 1000 nm for quantification of color and moisture content of soybeans during drying. An active contour model (ACM) was used to mitigate the threshold selection problem. Hyperspectral imaging demonstrated potential in simultaneous measurement of soybean color and moisture content during drying.

Fan et al. (2013) used computer vision to study the correlation between texture and color of rice surface. It was found that hardness and gumminess correlated with

L^* and a^* parameters with coefficients of determination more than 0.94. Also, they found that springiness can be calculated from color values (Fan et al., 2013). This study shows the potential of computer vision for both color and texture estimation.

Martynenko (2006) developed a computer vision system for the automatic control of ginseng root drying. Time series of images, captured every minute during the entire process of drying, enabled establishing a relationship between image attributes and physical parameters (shrinkage, moisture content, and quality). A linear relationship between color, measured by colorimeter and digital camera in *XYZ* color space, was established. A strong correlation between color and ginsenosides content makes computer vision a powerful tool for online quality control in ginseng root drying.

Continuous color measurement in the process of drying showed that color of vegetables is highly sensitive to moisture content (Krokida et al., 1998; Martynenko, 2008). In experiments with ginseng it was discovered that most of the quality degradation happens within narrow regions of moisture content in the range from 1.0 to 1.5 g/g (Martynenko, 2008). It was concluded that computer vision can be used for accurate control of temperature in drying vegetables. A prototype control system, based on computer vision, was developed (Martynenko and Yang, 2007).

18.9.3 Texture Measurements

The use of computer vision technology for quality control and assessment of raw and processed vegetables have been reported in the literature. Computer vision inspection of vegetables defects or blemishes for sorting and grading purposes was reviewed by Cubero et al. (2011). Thybo et al. (2004) estimated texture quality attributes of potatoes using MRI in combination with image analysis. Textural features were analyzed, using histogram of pixels distribution, gradients, and GLCM and GLRM matrices. It was found that MR imaging of raw potatoes provides structural/anatomical information, related to textural properties. Computer vision of texture during drying of apple slices considered 17 image texture features; however, only one (uniformity) correlated with sensory texture parameters (Sampson et al., 2014). It should be mentioned that texture measurements require significant mathematical background, which probably limit the number of practical applications. Besides that, computer vision is not able to measure aroma, flavor, and other important organoleptic properties of vegetables. In this case, combination of computer vision with electronic nose and tongue could be beneficial for quality assessment (Ghasemi-Varnamkhasti et al., 2011).

18.10 DISCUSSION AND CONCLUSIONS

It could be concluded that computer vision is a powerful multipurpose technology that has been successfully tested for the evaluation of morphological, textural, and color features of vegetables before, after, or during drying. Many researchers used this capacity of computer vision for the estimation of shrinkage, porosity, density, and color changes of some vegetables in lab-scale drying. Computer vision proved to be economical on an industrial scale for medicinal plants drying (Martynenko and Kudra, 2015). It was found that accuracy of computer vision depends on the quality of illumination and irregularities in the drying material, such as nonisotropic shrinkage and interfaces (Campos-Mendiola et al., 2007). A literature survey provided scarce information about the applications of computer vision for drying of vegetables. The standard methodology for computer vision of fruits and vegetables is still underdeveloped. Further studies are required to determine the suitability of computer vision for different vegetables and different drying methods. Further development of industrial applications of computer vision for vegetable drying is mostly dependent on the progress in the development of pattern recognition, learning techniques, and intelligent control systems.

REFERENCES

Abhayawick, L., Laguerre, J.C., Tauzin, V., Duquenoy, A. (2002). Physical properties of three onion varieties as affected by moisture content. *Journal of Food Engineering*, 55, 253–262.

Adu, B., Otten, L. (1996). Diffusion characteristics of white beans during microwave drying. *Journal of Agricultural Engineering Research*, 64, 61–70.

Aghbashlo, M., Hosseinpour, S., Ghasemi-Varnamkhasti, M. (2014). Computer vision technology for real-time food quality assurance during drying process. *Trends in Food Science and Technology*, 39, 76–84.

Arslan, D., Özcan, M.M. (2010). Study the effect of sun, oven and microwave drying on quality of onion slices. *Food Science and Technology*, 43, 1121–1127.

Azfal, T.M., Abe, T. (1998). Diffusion in potato during far infrared radiation drying. *Journal of Food Engineering*, 37(4), 353–365.

Bondaruk, J., Markowski, M., Błaszczak, W. (2007). Effect of drying conditions on the quality of vacuum-microwave dried potato cubes. *Journal of Food Engineering*, 81, 306–312.

Brosnan, T., Sun, D.-W. (2002). Inspection and grading of agricultural and food products by computer vision systems—A review. *Computers and Electronics in Agriculture*, 36, 193–213.

Brosnan, T., Sun, D.-W. (2004). Improving quality inspection of food products by computer vision—A review. *Journal of Food Engineering*, 61(1), 3–16.

Campos-Mendiola, R., Hernández-Sánchez, H., Chanona-Pérez, J.J., Alamilla-Beltrán, L., Jiménez-Aparicio, A., Fito, P., Gutiérrez-López, G.F. (2007). Non-isotropic shrinkage and interfaces during convective drying of potato slabs within the frame of the systematic approach to food engineering systems (SAFES) methodology. *Journal of Food Engineering*, 83, 285–292.

Chen, Y.-R., Chao, K., Kim, M.S. (2002). Machine vision technology for agricultural applications. *Computers and Electronics in Agriculture*, 36, 173–191.

Chen, Y., Martynenko, A. (2013). Computer vision for real-time measurements of shrinkage and color changes in blueberry convective drying. *Drying Technology*, 31(10), 1114–1123.

Corney, D. (2002). Food bytes: Intelligent systems in the food industry. *British Food Journal*, 104(10–11), 787–805.

Crowe, T.G., Delwiche, M.J. (1996). Real-time defect detection in fruit. Part I: design concepts and development of prototype hardware. *Trans ASAE* 39(6), 2299–2308.

Cubero, S., Aleixos, N., Moltó, E., Gómez-Sanchis, J., Blasco, J. (2011). Advances in machine vision applications for automatic inspection and quality evaluation of fruits and vegetables. *Food and Bioprocess Technology*, 4(4), 487–504.

Cui, Z.-W., Xu, S.-Y., Sun, D.-W. (2004). Effect of microwave-vacuum drying on the carotenoids retention of carrot slices and chlorophyll retention of chinese chive leaves. *Drying Technology*, 22, 563–575.

Davidson, V.J., Li, X., Brown, R.B. (2002). Fuzzy methods for ginseng drying control. The 9th International Conference on Information Processing and Management of Uncertainty in Knowledge-Based Systems, July 1–5, 2002, Annecy, France, 913–918.

Davidson, V.J., Li, X., Brown, R.B. (2004). Forced-air drying of ginseng root: 1. Control strategy for three-stage drying process. *Journal of Food Engineering*, 63, 361–367.

Davidson, V.J., Martynenko, A.I, Parhar, N.K., Sidahmed, M., Brown, R.B. (2009). Forced-air drying of ginseng root: Pilot-scale control system for three-stage process. *Drying Technology*, 27, 451–458.

Davoodi, M.G., Vijayanand, P., Kulkarni, S.G., Ramana, K.V.R. (2007). Effect of different pre-treatments and dehydration methods on quality characteristics and storage stability of tomato powder. *LWT-Food Science and Technology*, 40, 1832–1840.

Díaz-Maroto, M.C., Pérez-Coello, M.S., Cabezudo, M. (2002). Effect of different drying methods on the volatile components of parsley *(Petroselinum crispum* L.). *European Food Research and Technology*, 215(3), 227–230.

Di Scala, K., Crapiste, G. (2008). Drying kinetics and quality changes during drying of red pepper. *LWT-Food Science and Technology*, 41, 789–795.

Doymaz, İ. (2004). Convective air drying characteristics of thin layer carrots. *Journal of Food Engineering*, 61, 359–364.

Doymaz, İ. (2005). Drying behaviour of green beans. *Journal of Food Engineering*, 69, 161–165.

Doymaz, İ. (2007). Air-drying characteristics of tomatoes. *Journal of Food Engineering*, 78, 1291–1297.

Du, C.J., Sun, D.W. (2004). Recent developments in the applications of image processing techniques for food quality evaluation. *Trends in Food Science and Technology*, 15, 230–249.

Du, C.J., Sun, D.W. (2006). Learning techniques used in computer vision for food quality evaluation: A review. *Journal of Food Engineering*, 72, 39–55.

Duan, X., Zhang, M., Mujumdar, A.S. (2007). Studies on the microwave freeze drying technique and sterilization characteristics of cabbage. *Drying Technology*, 25, 1725–1731.

Ertekin, C., Yaldiz, O. (2004). Drying of eggplant and selection of a suitable thin layer drying model. *Journal of Food Engineering*, 63, 349–359.

Fan, F., Ma, Q., Ge, J., Peng, Q., Riley, W.W., Tang, S. (2013). Prediction of texture characteristics from extrusion food surface images using a computer vision system and artificial neural networks. *Journal of Food Engineering*, 118, 426–433.

Fernández, L., Castillero, C., Aguilera, J.M. (2005). An application of image analysis to dehydration of apple discs. *Journal of Food Engineering*, 67, 185–193.

Gall, H., Muir, A., Fleming, J., Pohlmann, R., Gocke, L., Hossack, W. (1998) A ring sensor system for the determination of volume and axis measurements of irregular objects. *Measurement Science and Technology*, 9, 1809–1820.

Ghasemi-Varnamhasti, M., Mohtasebi, S., Rodrigues-Mendez, M., Siadat, M., Ahmadi, H., Razavi, S. (2011). Electronic and bioelectronic tongues, two promising analytical tools for the quality evaluation of non-alcoholic beer. *Trends in Food Science and Technology*, 22, 245–248.

Gomes, G.F.S., Leta, F.R. (2012). Applications of computer vision techniques in the agriculture and food industry: A review. *European Food Research and Technology*, 235(6), 989–1000.

Gonzalez, R.C., Woods, E.E. (2008). *Digital Image Processing*. 3rd Ed. Pearson Education Inc., London.

Goyache, F., Bahamonde, A., Alonso, J., Lopez, S., del Coz, J.J., Quevedo, J.R., Ranilla, J., Luaces, O., Alvarez, I., Royo, L.J., Díez, J. (2001). The usefulness of artificial intelligence techniques to assess subjective quality of products in the food industry. *Trends in Food Science and Technology*, 12(10), 370–381.

Guiné, R.P.F., Barroca, M.J. (2012). Effect of drying treatments on texture and color of vegetables (pumpkin and green pepper). *Food Bioproducts Processing*, 90, 58–63.

Gunasekaran, S. (1996). Computer vision technology for food quality assurance. *Trends in Food Science and Technology*, 7(8), 245–256.

Gunasekaran, S. (2000). *Nondestructive Food Evaluation: Techniques to Analyze Properties and Quality.* CRC Press, Boca Raton, FL.

Hahn, F., Sanchez, S. (2000). Carrot volume evaluation using imaging algorithms. *Journal of Agricultural Engineering Research*, 75, 243–249.

Haralick, R.M., Shanmugam, K., Dinstein, I. (1973). Textural features for image classification. *IEEE Transactions on System, Man and Cybernetics*, 3(6), 610–621.

Heinemann, P., Hughes, R., Morrow, C., Sommer, H., Beelman, R., Wuest, P. (1994). Grading of mushrooms using a machine vision system. *Transactions of the ASAE*, 37, 1671–1677.

Hu, Q.G., Zhang, M., Mujumdar, A.S., Du, W.H., Sun J.C. (2006). Effects of different drying methods on the quality changes of granular edamame. *Drying Technology*, 24, 1025–1032.

Huang, M., Wang, Q., Zhang, M., Zhu, Q. (2014). Prediction of color and moisture content for vegetable soybean during drying using hyperspectral imaging technology. *Journal of Food Engineering*, 128, 24–30.

Hutchison, D., Otten, L. (1983). Thin-layer drying of soybeans and white beans. *Journal of Food Technology*, 18, 507–522.

Igathinathane, C., Chattopadhyay, P.K. (1998). Numerical techniques for estimating the surface areas of ellipsoids representing food materials. *J. Agricultural Engineering Research*, 70, 313–322.

Inyang, U.E., Ike, C.I. (1998). Effect of blanching, dehydration method and temperature on the ascorbic acid, colour, sliminess and other constituents of okra fruit. *International Journal of Food Sciences and Nutrition*, 49, 125–130.

Jähne, B., Haussecker, H. (2000). *Computer Vision and Applications: A Guide for Students and Practitioners.* Academic Press, New York.

Jayaraman, K.S., Gupta, D.K., Rao, N.B. (2007). Effect of pretreatment with salt and sucrose on the quality and stability of dehydrated cauliflower. *International Journal of Food Science and Technology*, 25, 47–60.

Jin, X., van der Sman, R.G.M, van Straten, G., Boom, R.M., van Boxtel, A.J.B. (2014). Energy efficient drying strategies to retain nutritional components in broccoli (*Brassica oleracea var. italica*). *Journal of Food Engineering*, 123, 172–178.

Klinger, T. (2003). *Image Processing with LabVIEW and IMAQ Vision.* Pearson Education, Inc., Upper Saddle River, NJ.

Krokida, M.K., Karathanos, V.T., Maroulis, Z.B., Marinos-Kouris, D. (2003). Drying kinetics of some vegetables. *Journal of Food Engineering*, 59, 391–403.

Krokida, M.K., Maroulis, Z.B. (1997). Effect of drying method on shrinkage and porosity. *Drying Technology*, 15(10), 621–628.

Krokida, M.K., Oreopoulou, V., Maroulis, Z.B., Marinos-Kouris, D. (2001). Effect of predrying on quality of French fries. *Journal of Food Engineering*, 49, 347–354.

Krokida, M.K., Tsami, E., Maroulis, Z.B. (1998). Kinetics on color changes during drying of some fruits and vegetables. *Drying Technology*, 16(3–5), 667–685.

Lewicki, P.P. (2006). Design of hot air drying for better foods. *Trends in Food Science and Technology*, 17(4), 153–163.

Lewicki, P.P., Duszczyk, E. (1998). Color change of selected vegetables during convective drying. *International Journal of Food Properties*, 1(3), 263–273.

Lin, T.M., Durance, T.D., Scaman, C.H. (1998). Characterization of vacuum microwave, air and freeze dried carrot slices. *Food Research International*, 31, 111–117.

López-Ortiz, A, Rodríguez-Ramírez, J., Méndez-Lagunas, L.L. (2013). Effect of drying air temperature on the structural properties of garlic evaluated during drying. *International Journal of Food Properties*, 16, 1516–1529.

Lozano, J.E., Rotstein, E., Urbicain, M.J. (1983). Shrinkage, porosity and bulk density of foodstuffs at changing moisture contents. *Journal of Food Science*, 48(5), 1497–1502.

Madamba, P.S., Driscoll, R.H., Buckle, K.A. (1994). Shrinkage, density and porosity of garlic during drying. *Journal of Food Engineering*, 23(3), 309–319.

Maharaj, V., Sankat, C.K. 1996. Quality changes in dehydrated dasheen leaves: Effects of blanching pre-treatments and drying conditions. *Food Research International*, 29, 563–568.

Majumdar, S., Jayas, D. (2000a). Classification of cereal grains using machine vision: I. Morphology models. *Transactions of the ASAE*, 43, 1669–1675.

Majumdar, S., Jayas, D. (2000b). Classification of cereal grains using machine vision: II. Color models. *Transactions of the ASAE*, 43, 1677–1680.

Majumdar, S., Jayas, D. (2000c). Classification of cereal grains using machine vision: III. Texture models. *Transactions of the ASAE*, 43, 1681–1687.

Majumdar, S., Jayas, D. (2000d). Classification of cereal grains using machine vision: IV. Combined morphology, color, and texture models. *Transactions of the ASAE*, 43, 1689–1694.

Marabi, A., Thieme, U., Jacobson, M., Saguy, I.S. (2006). Influence of drying method and rehydration time on sensory evaluation of rehydrated carrot particulates. *Journal of Food Engineering*, 72, 211–217.

Martynenko, A.I. (2006). Computer-vision system for control of drying processes. *Drying Technology*, 24(7), 879–888.

Martynenko, A. (2008). *Computer Vision System for Ginseng Drying: Remote Sensing, Control and Optimization of Quality in Food Thermal Processing.* ADM Verlag, Saarbrücken, 200 pp.

Martynenko, A.I. (2011). Porosity evaluation from real-time imaging and mass measurements. *Food and Bioprocess Technology*, 4(3), 417–428.

Martynenko, A. (2014). True, particle and bulk density of shrinkable biomaterials: Evaluation from drying experiments. *Drying Technology*, 32(11), 1319–1325.

Martynenko, A., Kudra, T. (2015). Non-isothermal drying of medicinal plants. *Drying Technology*, 33(13), 1550–1539.

Martynenko, A.I., Yang, S.X. (2007). An intelligent control system for thermal processing of biomaterials. *IEEE Conference on Networking, Sensing and Control*, 2007, 93–98.

Mayor, L., Moreira, R., Sereno, A.M. (2011). Shrinkage, density, porosity and shape changes during dehydration of pumpkin. *Journal of Food Engineering*, 103(1), 29–37.

Mendoza, F., Dejmek, P., Aguilera, J.M. (2007). Colour and image texture analysis in classification of commercial potato. *Food Research International*, 40, 1146–1154.

Mery, D., Pedreschi, F., Soto, A. (2013). Automated design of a computer vision system for visual food quality evaluation. *Food and Bioprocess Technology* 6(8), 2093–2108.

Mogol, B.A., Gokmen, V. (2014). Computer vision-based analysis of foods: A non-destructive colour measurement tool to monitor quality and safety. *Journal of the Science of Food and Agriculture*, 94, 1259–1263.

Mongpraneet, S., Abe, T., Tsurusaki, T. (2002). Accelerated drying of Welsh onion by far infrared radiation under vacuum conditions. *Journal of Food Engineering*, 55, 147–156.

Mrkìc, V., Cocci, E., Rosa, M.D., Sacchetti, G. (2006). Effect of drying conditions on bioactive compounds and antioxidant activity of broccoli (*Brassica oleracea* L.). *Journal of the Science of Food and Agriculture*, 86, 1559–1566.

Mulet, A., García-Reverter, J., Bon, J., Berna, A. (2000). Effect of shape on potato and cauliflower shrinkage during drying. *Drying Technology*, 18, 1201–1219.

Nahimana, H., Zhang, M. (2011). Shrinkage and color change during microwave vacuum drying of carrot. *Drying Technology*, 29, 836–847.

Narendra, V.G., Hareesh, K.S. (2010). Prospects of computer vision automated grading and sorting systems in agricultural and food products for quality evaluation. *International Journal of Computer Applications*, 1(4), 1–9.

Nindo, C., Sun, T., Wang, S., Tang, J., Powers, J. (2003). Evaluation of drying technologies for retention of physical quality and antioxidants in asparagus (*Asparagus officinalis* L.). *LWT-Food Science and Technology*, 36, 507–516.

Orikasa, T., Wu, L., Ando, Y., Maramatsu, Y., Roy, P., Yano, T., Shiina, T., Tagawa, A. (2010). Hot air drying characteristics of sweet potato using moisture sorption isotherm analysis and its quality changes during drying. *International Journal of Food Engineering*, 6(2), 12.

Patel, K.K., Kar, A., Jha, S.N., Khan, M.A. (2012). Machine vision system: A tool for quality inspection of food and agricultural products. *Journal of Food Science and Technology*, 49(2), 123–141.

Pathare, P.B., Sharma, G.P. (2006). Effective moisture diffusivity of onion slices undergoing infrared convective drying. *Biosystems Engineering*, 93(3), 285–291.

Pedreschi, F., León, J., Mery, D., Moyano P. (2006). Development of a computer vision system to measure the color of potato chips. *Food Research International*, 39(10), 1092–1098.

Prakash, S., Jha, S., Datta, N. (2004). Performance evaluation of blanched carrots dried by three different driers. *Journal of Food Engineering*, 62, 305–313.

Rapusas, R.S., Driscoll, R.H. (1995). Kinetics of non-enzymatic browning in onion slices during isothermal heating. *Journal of Food Engineering*, 24, 417–429.

Rapusas, R.S., Driscoll, R.H., Schrednicki, G.S. (1995). Bulk density and resistance to airflow of sliced onions. *Journal of Food Engineering*, 26(1), 67–80.

Ren, G., Chen, F. (2000). Drying of steamed Asian ginseng (*Panax ginseng*) roots by microwave-hot air combination. *Pharmazie*, 55(2), 124–128.

Reynolds, L.B. (1998). Effects of drying on chemical and physical characteristics of American Ginseng (*Panax guinquefolius* L.). *Journal of Herbs, Spices and Medicinal Plants*, 6(2), 9–21.

Romani, S., Rocculi, P., Mendoza, F., Dalla Rosa, M. (2009). Image characterization of potato chip appearance during frying. *Journal of Food Engineering*, 93(4), 487–494.

Romano, G., Argyropoulos, D., Nagle, M., Khan, M.T., Müller, J. (2012). Combination of digital images and laser light to predict moisture content and color of bell pepper simultaneously during drying. *Journal of Food Engineering*, 109, 438–448.

Sacilik, K. (2007). Effect of drying on thin-layer drying characteristics of hull-less seed pumpkin (*Crcurbita pepo* L.). *Journal of Food Engineering*, 79, 23–30.

Sampson, D.J., Chang, Y.K., Rupasinghe, H.V., Zaman, Q.U. (2014). A dual-view computer-vision system for volume and image texture analysis in multiple apple slices drying. *Journal of Food Engineering*, 127, 49–57.

Sanjuan, N., Lozano, M., García-Pascual, P., Mulet, A. (2003). Dehydration kinetics of red pepper (*Capsicum annuum* L var. Jaranda). *Journal of the Science of Food and Agriculture* 83, 697–701.

Saravacos, G.D. (1967). Effect of the drying method on the water sorption of dehydrated apple and potato. *Journal of Food Science*, 32, 81–84.

Sharma, G., Prasad, S. (2001). Drying of garlic (*Allium sativum*) cloves by microwave–hot air combination. *Journal of Food Engineering*, 50, 99–105.

Sharma, G.P., Verma, R.C., Pathare, P.B. (2005). Thin-layer infrared radiation drying of onion slices. *Journal of Food Engineering*, 67, 361–366.

Simal, S., Garau, C., Femenia, A., Rosselló, C. (2005). Drying of red pepper (*Capsicum annuum*): Water desorption and quality. *International Journal of Food Engineering*, 1(4), 1022.

Sokhansanj, S., Bailey, W.G., Guo, Y.P., van Dalfsen, K.B. (1999). Drying of North American ginseng roots (*Panax quinquefolius* L.). *Drying Technology*, 17(10), 2293–2308.

Soysal, Y., 2004. Microwave drying characteristics of parsley. *Biosystems Engineering*, 89, 167–173.

Tang, J., Sokhansanj S. (1993). Moisture diffusivity in laird lentil seed components. *Transactions of the ASAE*, 36(6), 1791–1798.

Tao, Y., Heinemann, P.H., Varghese, Z., Morrow, C.T., Sommer, H.J. III. (1995). Machine vision for colour inspection of potatoes and apples. *Transactions of the ASAE*, 38(5), 1555–1561.

Thybo, A.K., Szczypiński, P.M., Karlsson, A.H., Dønstrup, S., Stødkilde-Jørgensen, H.S., Andersen, H.J. (2004). Prediction of sensory texture quality attributes of cooked potatoes by NMR-imaging (MRI) of raw potatoes in combination with different imaging methods. *Journal of Food Engineering*, 61, 91–100.

Toğrul, H. (2006). Suitable drying model for infrared drying of carrot. *Journal of Food Engineering*, 77, 610–619.

Toor, R.K., Savage, G.P. (2006). Effect of semi-drying on the antioxidant components of tomatoes. *Food Chemistry*, 94, 90–97.

Vega-Gálvez, A., Di Scala, K., Rodríguez, K., Lemus-Mondaca, R., Miranda, M., López, J., Perez-Won, M. (2009). Effect of air-drying temperature on physico-chemical properties, antioxidant capacity, colour and total phenolic content of red pepper (*Capsicum annuum*, L. var. Hungarian). *Food Chemistry*, 117, 647–653.

Vega-Gálvez, A., Lemus-Mondaca, R., Bilbao-Sáinz, C., Fito, P., Andrés, A. (2008). Effect of air drying temperature on the quality of rehydrated dried red bell pepper (var. Lamuyo). *Journal of Food Engineering*, 85, 42–50.

Vernon, D. (1991). *Machine Vision: Automated Visual Inspection and Robot Vision*, Prentice-Hall International (UK) Ltd., London, 260 pp.

Wang, J., Xi, Y.S. (2005). Drying characteristics and drying quality of carrot using a two-stage microwave process. *Journal of Food Engineering*, 68, 505–511.

Wang, J., Xiong, Y.-S., Yu, Y. (2004). Microwave drying characteristics of potato and the effect of different microwave powers on the dried quality of potato. *European Food Research and Technology*, 219, 500–506.

Wang, N., Brennan, J.G. (1995). Changes in structure, density and porosity of potato during dehydration. *Journal of Food Engineering*, 24, 61–76.

Wang, R., Zhang, M., Mujumdar, A.S. (2010). Effects of vacuum and microwave freeze drying on microstructure and quality of potato slices. *Journal of Food Engineering*, 101, 131–139.

Wilhelm, L.R. (1990). Ginseng drying-the effect of drying characteristics of *Panax notoginseng Panax quinguefolium*. *Journal of Chinese Agriculture Engineering*, 7(1), 62–69.

Wu, D., Sun, D.W. (2013). Colour measurements by computer vision for food quality control. *Trends in Food Science and Technology*, 29(1), 5–20.

Xiong, C.Z., Xu, J.Y., Zou, J.C., Qi, D.X. (2006). Texture classification based on EMD and FFT. *Science A*, 7(9), 1516–1521.

Yadav, S.K., Sehgal, S. (1995). Effect of home processing on ascorbic acid and beta-carotene content of spinach (*Spinacia oleracia*) and amaranth (*Amaranthus tricolor*) leaves. *Plant Foods for Human Nutrition*, 47, 125–131.

Yadollahinia, A., Latifi, A., Mahdavi, R. (2009). New method for determination of potato slice shrinkage during drying. *Computers and Electronics in Agriculture*, 65(2), 268–274.

Yu, H., MacGregor, J.F. (2003). Multivariate image analysis and regression for prediction of coating content and distribution in the production of snack foods. *Chemometrics and Intelligent Laboratory Systems*, 67, 125–144.

Zapotoczny, P., Zielińska, M., Nita, Z. (2008). Application of image analysis for the varietal classification of barley. *Journal of Cereal Science*, 48(1), 104–110.

Zenoozian, M.S., Devahastin, S. (2009). Application of wavelet transform coupled with artificial neural network for predicting physicochemical properties of osmotically dehydrated pumpkin. *Journal of Food Engineering*, 90(3), 219–227.

Zenoozian, M.S., Devahastin, S., Razavi, M.A., Shahidi, F., Poreza, H.R. (2008). Use of artificial neural network and image analysis to predict physical properties of osmotically dehydrated pumpkin. *Drying Technology*, 26(1), 132–144.

Zheng, C., Sun, D.W., Zheng L. (2006). Recent developments and applications of image features for food quality evaluation and inspection—A review. *Trends in Food Science and Technology*, 17, 642–655.

Zielińska, M., Markowski, M. (2012). Color characteristics of carrots: Effect of drying and rehydration. *International Journal of Food Properties*, 15(2), 450–466.

Ziou, D., Tabbone, S. (1998). Edge detection techniques—An overview. *International Jornal of Pattern Recognition and Image Analysis*, 8, 537–559.

Zogzas, N.P., Maroulis, Z.B., Marinos-Kouris, D. (1994). Densities, shrinkage and porosity of some vegetables during air drying. *Drying Technology*, 12(7), 1653–1666.

Zuech, N. (2003). Machine vision and lighting. www.visiononline.org/vision-resources-details.cfm/vision-resources/Machine-Vision-and-Lighting/content_id/1269 (Accessed April 6, 2016).

19 Novel Packaging of Dried Vegetable Products

Huizhi Chen and Min Zhang
Jiangnan University

CONTENTS

19.1 INTRODUCTION

Appropriate packaging of dried vegetable products can prolong shelf life and ensure food safety and quality. In the circulation of dried food from producers to consumers, they may be contaminated or damaged. Dried food can be contaminated by humans, rodents, insects, and other animals. Damage of dried food often occurs during picking, stacking, transport, display, etc. Thus, packaging of dried vegetable products plays a vital role in protection of foods against contamination and damage. In addition, packaging provides an important medium for manufacturers and retailers to communicate with the consumers, both at the "point of sale" and through shelf life of a product over time (Westerman et al., 2013). On-package graphics have the potential to influence consumers' product-related attitudes and behaviors, and a good design with a beautiful pattern and detailed instruction is also very important for consumers' acceptance.

Processes and types of packaging are dependent on the dried foods. Generally, there are two basic principles of packaging of dried products: (1) robust sealability of internal packaging to prevent moisture absorption of dried foods; (2) good resistance to pressure and shock to prevent crushing of dried foods.

19.2 FOOD-PACKAGING INTERACTIONS

Food-packaging interactions refer to the exchange of mass and energy between the food, the packaging, and the environment. These interactions may be briefly classified into three phenomena: permeability, sorption, and migration (Figure 19.1). Permeability is the transfer of molecules through the package from the environment to the packaged food or from the packaged food to external storage environment.

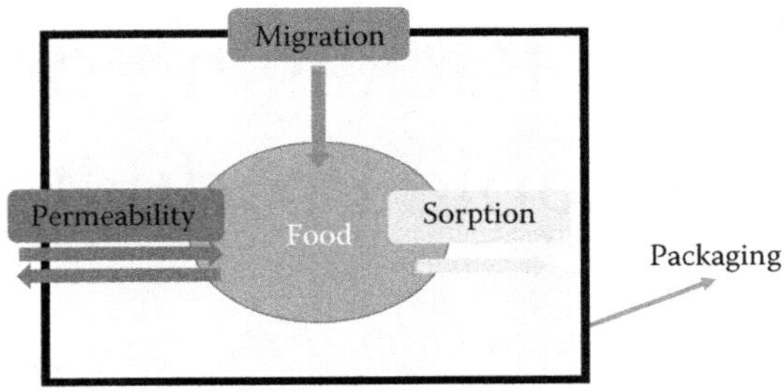

FIGURE 19.1 Interactions between food, packaging, and environment.

Sorption includes the take-up of molecules from the packaged food into, but not through, the package. Migration involves the passage of molecules originally contained in the packaging material into the packaged food (Sadaka et al., 2013). Safety issues arise mainly from the primary packaging for direct contact with the atmosphere surrounding the food or with the food itself, owing to potential undesirable migration of chemical components. The main parameters evaluating food-packaging interactions in a certain food/packaging system are: diffusion coefficient D (kinetics of diffusion in the polymer and foodstuff), partition coefficient K (thermodynamics of partition equilibrium between packaging and food), and geometric dimensions (Bart, 2006). An optimum package should balance the packaging material properties, product protection requirements, environmental and transport conditions, and cost. Two phenomena should be considered in calculating the moisture uptake by a packaged dried food product: (1) the transfer of water vapor through the package, and (2) the kinetic uptake of water by the food product (Macedo et al., 2013).

19.3 PACKAGING MATERIALS AND THEIR PROPERTIES

The property of packaging materials determines the environmental conditions (e.g., composition of air) inside the package, which in turn is known to affect the rate and extent of nutrient loss and microbial activity. In other words, the shelf life of food is highly dependent on the permeability characteristics of the packaging materials, which emphasizes the importance of packaging design. For instance, oxygen-barrier plastic films can limit the growth of aerobic spoilage microorganisms to extend the shelf life of packaged foods. Packages can be generally divided into two major groups, rigid and flexible, with each group having distinct advantages and limitations.

1. *Rigid packaging*
   Metal cans and glass jars are two typical rigid containers. These materials have good barrier properties and can offer physical protection against vapor and gas transfer between inside and outside of the packaging, which is beneficial to the dried foods. Some freeze-dried foods are currently packaged in this way by backflushing with nitrogen to an oxygen concentration of <2% (to reduce oxidative rancidity and browning and to maintain high

concentration of nutritional ingredients like ascorbic acid) (Salunkhe et al., 1973). Compared to metal cans, glass jars have transparency, chemical stability, and good compressive strength and rigidity. However, most glass are fragile and bulky and have poor thermal stability and high energy consumption for container production.

2. *Flexible packaging*

   Flexible packaging is commonly used for dried food products, which can maximize the use of available packaging space with cheaper cost than rigid types (Salunkhe et al., 1973). For dried vegetable products, the flexible packaging needs to be impervious to moisture, light, and oxygen.

   a. *Paper*

      Paper is one of the oldest packaging materials and is also a major material used in external packing of dried foods. Its advantages include low cost, good protective performance, high production flexibility, convenient transportation, easy to shape and decorate, safe, recycling, and so on. For food packaging, paper can be divided into three categories, i.e., plain paper (thickness of 250–300 μm), cardboard (thickness of 500–2500 μm), and fiberboard (generally thickness of fiberboard is not clearly defined). Paper is produced to different grades, according to specific purposes of consumers. Polyethylene (PE) sheet could be incorporated in paper package as a moisture barrier.

   b. *Thermoforming films*

      Plastic polymers are commonly used for food packaging owing to their low cost and good physical properties for industrial applications, such as lightweight, excellent formability, and remarkable diversity. It can be made in various forms of packaging and containers such as bags, bottles, cans, hoses, casings, shrink film, stretch film, special functional film, foam containers, thermoformed cups, plates, pots and containers, and skin packaging blister pack containers. However, the plastics also have disadvantages that their temperature resistance and barrier properties are not as good as metal and glass containers, and they also have an impact on the environment. These thermoplastic synthetic polymers can be melt-processed by simply applying heat and shear (Mensitieri et al., 2011). The majority of packaging films are made of polyamides, polyesters, and polyolefins (Arvanitoyannis and Stratakos, 2012). Some examples of thermoforming films are as follows:

      Polyamides (PAs), i.e., nylons, consist of large carbon chains (>6), which contribute to their high strength and resistance to puncture, abrasion, and tearing (Mullan and McDowell, 2003). PA films are thermally stable and flexible at low temperatures and have dilute acid and alkali resistance. Monolayer or multilayer films based on PAs (like PA-6) are widely applied for packaging food products under vacuum or modified atmosphere conditions (Félix et al., 2014). In addition, multilayer films based on PA-6 are usually used for heat treatment during food processing.

Polyester is a category of polymers containing an ester functional group in their main chain and most commonly refers to polyethylene terephthalate (PET), although there are many other types of polyester such as polylactic acid (PLA) and polybutylene terephthalate (PBT). The PET film provides good resistance to heat, cold, has high mechanical strength, forms a good barrier to oxygen (O_2 permeability of 6–8 nmol m^{-1} s^{-1} GPa^{-1}), and is easy to print on with ink (Coles et al., 2003); therefore it is widely used in food packaging.

Polyolefins are the major plastic materials for the development of thermoforming films, such as polypropylene (PP), polystyrene (PS), polyvinyl chloride (PVC), and various grades of polyethylene (high-density polyethylene: HDPE; low-density polyethylene: LDPE, etc.) (Duncan, 2011).

PVC film is relatively inexpensive and easy to use due to its good heat sealability. However, PVC film also has some disadvantages in dried food packaging, such as being highly susceptible to tearing and punctures, thereby leading to a frequency of "leaky" packages, and bacterial growth and browning due to relatively high oxygen permeability, which leads to a short shelf life (Cornforth and Hunt, 2008).

Polyethylene (PE) film has high elasticity, good heat sealablility, sufficient water vapor barrier properties, and resistance to cold temperature, acid, and alkali except for nitric acid (Piringer and Baner, 2000). HDPE provides a relatively low O_2 permeability (200–400 nmol m^{-1} s^{-1} GPa^{-1}) and significantly good water vapor barrier, while LDPE offers a low permeability to water vapor but a high permeability to gases. For example, LDPE could be easily penetrated by essential oils (Greengrass, 1999). Prakash et al. (2004) investigated the storage stability of dried carrot products during storage at room temperature in LDPE of different gauges (200, 300, and 400), and the 400 gauge was found to retain β-carotene content and rehydration potential of dried carrot better than the thinner gauges. Therefore, lower stability toward water-activity-dependent deteriorative processes could be expected for products packaged in LDPE compared with those packaged with HDPE due to its lower oxygen and water vapor permeability (Almeida-Dominguez et al., 1992). Uadal and Sagar (2008) showed that HDPE was better than LDPE and PP for dehydrated leafy vegetables (viz., amaranth, fenugreek, and spinach) for storage of up to 3 months. Dried vegetable products packaged by HDPE retained higher β-carotene, ascorbic acid, and total chlorophyll content and had less moisture, non-enzymatic browning, and rehydration ratio.

The experimental results of Adom et al. (1996) showed that the packaging material significantly affected food quality parameters of dried okra, and a polyethylene package was more suitable than cotton package. The 9-month-storage study of dehydrated carrots within polyethylene packaging showed that retention of β-carotene and ascorbic acid was better in samples within double layers of HDPE bags stored in cold temperature, and it also showed minimum browning during storage (Negi and Roy, 2000).

Other thermoforming film materials that can be used for food packaging include ethylene vinyl alcohol (EVOH), polystyrene (PS), linear low-density polyethylene (LLDPE), polypropylene-oriented polypropylene (OPP), ethylene vinyl acetate (EVA), and polyvinylidene chloride (PVdC), etc., with each one having different mechanical and barrier properties. However, pure simple polymer generally does not exhibit all the desired mechanical and barrier properties for food packaging application.

c. *Composite materials*

One of the major trends for the thermoforming film packaging is development of flexible multilayer plastic films, which can meet more technical requirements. Composite materials in packaging are made of two or more materials after one or more complex processes. Generally, it is composed of substrate layer, functional layer, and heat-sealing layer. The substrate layer plays a part in printing, beautiful appearance, moisture barrier, and so on. The functional layer is primarily a barrier layer and provides protection from light, gas, and moisture. The heat-sealing layer is in direct contact with the packaged foods and plays a role in permeation resistance, good heat-sealing properties, and transparency.

For example, two multilayered plastic films include a medium barrier, which comprises PP/tie/PA6/tie/PA6/tie/LDPE and a high barrier where the central adhesive (tie) layer is replaced by a layer of EVOH. These multilayer films were designed to improve the film oxygen-barrier property, while the film's mechanical and optical properties were also improved (Crippa et al., 2007). Films used for dried food packaging where low vapor transfer is demanded are customarily three-ply laminates, like a polyolefin-foil-mylar (Salunkhe et al., 1973). It was noted that the strength of polyethylene could be improved by coextrusion with polyamide (Cornforth and Hunt, 2008). Coextruded polyamide-polyethylene films are typically used for high oxygen-modified atmosphere packaging (Sørheim et al., 1999). Aluminized composite material has advantages of shading and anti-ultraviolet radiation, which not only prolongs the shelf life of the contents, but also increases the thickness of the film. Because of its low cost and useful properties, such as being a good barrier, this composite material can replace aluminum foil to a certain extent. For packaging of dried products, the use of high-density polypropylene (HDPP) is currently in practice. Being transparent, it allows consumers to assess the product quality easily. Dak et al. (2014) found that aluminum-laminated polyethylene (ALP) has even better protective effect than HDPP for dried pomegranate arils according to the magnitude of the quality change during storage. This is due to lower permeability to water vapor, thus extending the shelf life of the dried product.

19.4 TYPE OF PACKAGING

According to the packaging form, there are three basic types of packaging for dried food products, i.e., primary packaging, secondary packaging, and tertiary packaging. Generally, primary packaging is in direct contact with the dried foods. Secondary packaging refers to packaging several single packages together. Tertiary packaging is always used for bulk transport and store distribution. According to packaging methods, there are five types of packaging, i.e., conventional packaging, vacuum packaging (VP), modified atmosphere packaging, active packaging, and intelligent packaging.

1. *Conventional packaging*
   Conventional packaging is a packaging in which dried products are stored into packaging containers under normal atmospheric pressure. Bag packaging is widely used in a variety of dried foods. According to the material used, bag packaging can be divided into paper bags and plastic bags. Paper bags can be divided into single bag, double bag, and multibag. A plastic layer can be added to multipaper bag to improve the air tightness and moisture resistance. Cartons are usually made with multiple layers of cardboard, which is lightweight, strong, economical, practical, and easily recyclable, etc. It facilitates mechanization during loading and unloading and is commonly used in domestic and export of the dried products. Plastic bags, owing to their airtight ability and moisture resistance, are also widely used in dried food packaging.
2. *VP*
   VP for dried vegetable products are generally low-O_2 packaging systems that put the material in shrinkable barrier films (like barrier styrene or PE films), followed by vacuum sealing the heat-shrinkable barrier films to conform to the shape of the product (Belcher, 2006). Common materials for VP are PA, EVA, EVOH, PET-PVdC, etc. It should be noted that the reduced thickness at the corners significantly affects the gas-barrier properties of the vacuum package. Oliveira et al. (2006) suggested the use of EVOH in VP, because this material does not affect the gas-barrier properties of the packaging corners. A typical VP material is usually three-layered coextrusions of EVA/PVdC/EVA with O_2 permeability of $<15.5\,ml \cdot m^{-2} \cdot (24\,h)^{-1}$ at 1 atmosphere (Jenkins and Harrington, 1991).
3. *Modified atmosphere packaging*
   Modified atmosphere packaging (MAP) is a very important and practical preservation technique used for extending the shelf life of food. In this technique, the air in the package is removed, and the normal composition of atmospheric air is exchanged to another gas composition, which can inhibit growth of microorganisms that cause spoilage, and thus assist in maintaining high quality (Rao and Sachindra, 2002). The most commonly used gases in MAP are oxygen (O_2), carbon dioxide (CO_2), and nitrogen (N_2). Nitrogen is a relatively inert gas that can neither support the growth of aerobic microorganisms nor inhibit the growth of anaerobic bacteria. The

noble gases such as helium (He), argon (Ar), xenon (Xe), and neon (Ne) have been investigated in some food MAP applications due to their lack of reactivity. However, it seems to be difficult to find advantages of noble gases compared with N_2 used in the dried vegetable packaging field because of their high costs.

It was reported that package with an inert atmosphere appeared to be an advantageous packaging method according to the stability of vacuum-dried tomato juice powder under a variety of packaging conditions (Wong et al., 1956). Nitrogen packaging had little effect on anthocyanin degradation in freeze-dried strawberry puree, but it did delay the degree of browning (Erlandson and Wrolstad, 1972). It was reported that nitrogen packaging can maintain the flavor of potato flakes during 12-month storage at 23°C (Sapers et al., 1973). Freeze-dried mushrooms packed in aluminum–film combination pouches under nitrogen had better quality retention (Luh and Eidels, 1969).

In general, both VP and MAP can reduce the oxygen content (<2%) in the packaging environment and extend the shelf life of dried product by delaying the oxidative deterioration.

4. *Active packaging*
The new European Regulation 450/2009/EC formulates specific rules for active and intelligent packaging materials and articles as well as the general requirements established in Regulation 1935/2004/EC for their safe application (Restuccia et al., 2010). Active packaging involves interactions between package or package components and food or internal gas atmosphere and meets consumers' requirements for high-quality and safe dried vegetable products (Ozdemir and Floros, 2004). Active packaging has many functions, like scavenging of oxygen, ethylene, moisture, odor, etc. and releasing of antimicrobial agents (such as silver, alcohol, sulfur dioxide, and bacteriocins) and antioxidants (e.g., butylated hydroxyanisole [BHA], butylated hydroxytoluene [BHT], and tertiary butylhydroquinone [TBHQ]). Active agents may either be imbedded into a solid, dispersed in the plastic, or introduced into various layers of the package, including adhesive, lacquer, or enamel layers (Figure 19.2). One example is adding a sachet into the package. The sachet material is always highly permeable. Sachets

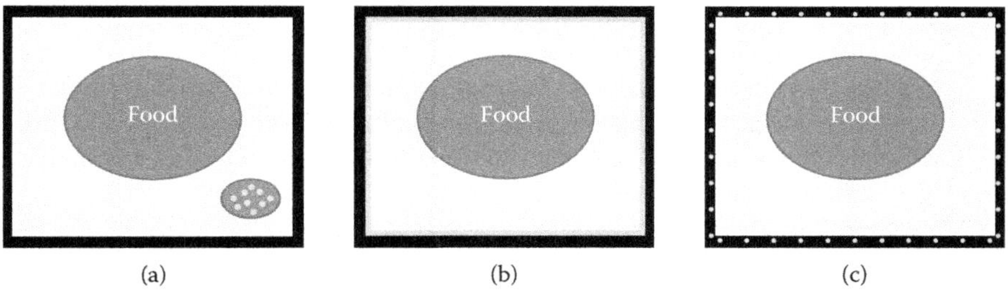

FIGURE 19.2 Three forms of active packaging. (a) Adding a sachet into the package. (b) Coating active agents on the surface of film. (c) Incorporating active agents in the film.

need to be labeled "Do not eat" for safety reasons and regulatory purposes. Although sachets can be concealed using secondary packages, this practice increases packaging costs. The other methods are coating active agents on the surface of film and incorporating active agents in packaging films (Figure 19.2).

Important examples of active agents for packaging dried vegetable products are oxygen scavenger, carbon dioxide scavenger/emitter, and moisture scavenger.

1. *Oxygen scavenger*
   Control of oxygen levels in food packaging is important to limit the rate of oxygen-related deteriorative and spoilage reactions in foods, thus improving product quality and extending shelf life. Oxygen scavenger system provides an alternative to vacuum and gas flushing packaging.
   The materials of oxygen scavenging sachets need to be highly permeable to oxygen. The type and amount of oxygen absorbent is determined by the initial oxygen level in the package, the amount of dissolved oxygen present in the food, the permeability of the packaging material, the nature (size, shape, weight, etc.), and the water activity of the food. For scavengers incorporated in packaging films, these allow the absorption of oxygen from the surfaces of the food that are in contact with the film. According to the mode of action, substances used in oxygen scavengers systems can be divided into chemical systems (such as powdered iron oxide, sulfite salt–copper sulfate, photosensitive dye oxidation, antioxidant oxidation, catalytic conversion of oxygen by platinum catalyst, etc.) and enzymatic systems (such as glucose oxidase-glucose, alcohol oxidase, etc.) (Restuccia et al., 2010). Multilayer oxygen scavengers absorb oxygen more effectively than single-layer scavenging systems. The structure of a typical multilayer oxygen scavenging system is composed of inner layer, oxygen-absorbing layer, barrier layer, and outer layer (Ozdemir and Floros, 2004).
2. *Carbon dioxide scavenger/emitter*
   Carbon dioxide emitting system may be used to reduce the rate of microbial growth of dried products. A CO_2 scavenging sachet can be composed of a porous envelope containing calcium oxide (CaO) and a hydrating agent, like silica gel, on which water is emitted. In this system, H_2O reacts with CaO and produces $Ca(OH)_2$, which then reacts with CO_2 to form $CaCO_3$. Other substances like ferrous carbonate–metal halide and iron powder–calcium hydroxide can also be used in a CO_2 scavenging system. The use of a dual function system consisting of an oxygen scavenger and a carbon dioxide emitter is the usual practice for increasing the shelf life of semidried foods (Ozdemir and Floros, 2004).
3. *Moisture scavenger*
   For dried food packages, moisture scavengers are commonly used in sachet forms to maintain low levels of moisture. The accumulation of excess water inside the package contributes to microbial growth, leading to quality deterioration and shelf life reduction. Using moisture scavengers is an effective

way for removing excess water in a package that has a high barrier to water vapor. The common moisture-absorbing substances are silica gel, calcium oxide, calcium chloride molecular sieves, propylene glycol, polyvinyl alcohol, natural clays (e.g., montmorillonite), and modified starch. Among these desiccants, silica gel is the most widely used, because it is nontoxic and noncorrosive (Ozdemir and Floros, 2004). A typical moisture-absorbing system comprises a superabsorbent polymer located between two layers of a microporous or nonwoven polymer (Realini and Marcos, 2014).

4. *Others*

Active packaging changes the condition of packaged food, and the main advantage of active packaging is to extend food shelf life and maintain product quality (Puligundla et al., 2012). Antimicrobial packaging is a form of active packaging, particularly for perishable products. Antimicrobial agents (e.g., sorbates, benzoates, propionates, ethanol, etc.) incorporated into packaging systems can inhibit microbial contamination by delaying the growth of population or suppressing the activity of microorganisms (Quintavalla and Vicini, 2002). Using the dry phase inversion technique, Uz and Altınkaya (2011) developed cellulose acetate-based monolayer and multilayer antimicrobial food packaging materials for controlled release of potassium sorbate (an antimicrobial agent). In this technique, drying induces crystal formation of potassium sorbate in packaging materials and produces packaging materials with structures ranging from asymmetric and porous to dense by changing the drying temperature, initial casting composition, and wet casting thickness, thereby forming crystal dissolution diffusion-controlled release systems. On the other hand, low level of antioxidants can also be incorporated into thermoforming films, which improves antioxidant ability of the packaging films. For example, synthetic antioxidants like BHA and BHT and natural antioxidants such as rosemary extract and tocopherols were incorporated into LDPE at the 0.1% level and thermoformed into 25-mm-diameter, 0.25-mm-thick discs (Moore et al., 2003).

Currently used active packaging technologies are mainly based on sachet technologies. However, sachets suffer from inadequate consumer acceptance due to fears of ingestion by children and accidental consumption with package contents. The development and use of active packaging systems in the form of thin films can be expected to increase in the next decade. Continuous innovations in active packaging are expected to lead to further improvements in food quality, safety, and stability.

5. *Intelligent Packaging*

Intelligent or smart packaging is emerging as a new branch of packaging science and technology that uses the communication function of the package to facilitate decision making to achieve enhanced food quality and safety (Yam et al., 2005). Intelligent packaging systems are always attached as labels printed onto, or incorporated into, a food packaging material. The primary advantage of intelligent packaging is the ability to (1) monitor storage conditions, such as temperature, time, oxygen, or carbon dioxide content inside the package; (2) indicate food quality by monitoring changes of pathogenic bacteria, biogenic amines, toxins, or volatile compounds (like organic acids, ethanol, CO_2, nitrogen compounds, sulfuric compounds), etc.; and (3) give more detailed information throughout the supply chain (Dainelli et al., 2008). Three principal types of this packaging technology are sensors, indicators, and radio frequency identification (Table 19.1).

Intelligent indicators/sensors were first developed to provide information on packaging conditions and thus indirectly indicate the quality of packed food. Puligundla et al. (2012) classified CO_2 sensors for intelligent food packaging applications as follows: conventional CO_2 sensors (like nondistributive infrared) and innovative optical CO_2 sensors (like pH-based wet optical CO_2 indicators, fluorescent CO_2 sensors, dry optical CO_2 sensors, sol–gel-based optical CO_2 sensor, photonic crystal sensors, and color changing metals). A number of oxygen indicators are designed to show color changes due to leaking or tampered packages. Alginate polymer has been used to prevent dyes from leaching out of colorimetric oxygen indicator films. Vu and Won (2013) used alginate as a coating polymer to make a water-resistant oxygen indicator. The indicators/sensors have been designed to provide some information on quality attributes (Gontard, 2004), such as fish freshness indicator (Chan et al., 2006), chicken freshness indicator (Rukchon et al., 2014), and apple ripeness indicator (Lang and Hübert, 2011). However, commercially, direct indicators of quality attributes are not reported for packaging of dried vegetable products yet.

On the other hand, radio frequency identification (RFID) tags, placed outside the primary packaging, use radio frequency electromagnetic fields to efficiently store and communicate real-time information of the product for automatic identification and traceability of product (Realini and Marcos, 2014). High frequencies enable higher data transmission rates and larger reading ranges but are more expensive. Compared to high RF (10–15 MHz) and ultra-high RF (850–950 MHz, 2.4–2.5 GHz,

TABLE 19.1
Three Types of Intelligent Packaging Technology

Type of Intelligent Packaging	Measures Applied in Intelligent Packaging Systems
Sensors	Printed electronics, carbon nanotechnology (e.g., carbon nanotubes and carbon nanofibers), silicon photonics, biotechnology, and nose systems
Indicators	Gas indicators (e.g., leaking seals or gas concentrations indicators for water vapor, carbon dioxide, ethanol, hydrogen sulfide, and other gases), freshness indicators, time–temperature indicators (critical temperature indicators, partial history indicators, and full history indicators), and thermochromic ink
Radio frequency identification	Sensor-enabled RFID tags, passive RFID sensors

Source: Vanderroost, M. et al., *Trends Food Sci. Technol.*, 39 (1), 47–62, 2014. With permission.
RFID, radio frequency identification.

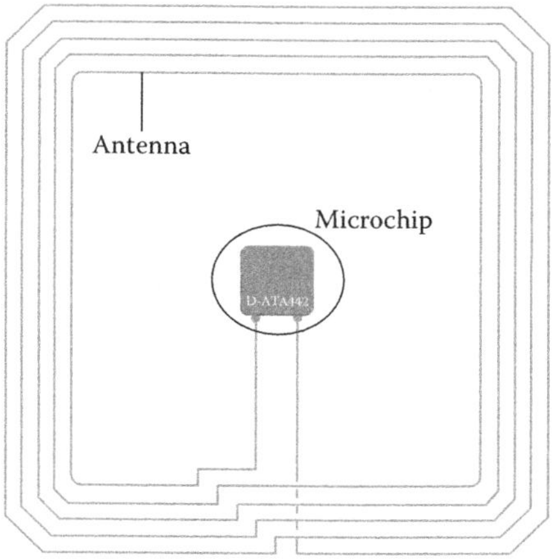

FIGURE 19.3 Schematic diagram of an RFID tag. (From Vanderroost, M. et al., *Trends Food Sci. Technol.*, 39 (1), 47–62, 2014. With permission.)

508 GHz), low RF (30–500 kHz) is fast enough for most applications of food packaging. In a typical RFID system, a read/write device composed of a transmitter and/or a receiver uses electromagnetic waves to communicate with an RFID tag through antennas (Vanderroost et al., 2014). Figure 19.3 shows an RFID tag that is composed of a microchip attached to an antenna.

Despite active research in the area, safety issues of intelligent systems come mainly from direct indicators/sensors with potential undesirable migration of chemical components, because these intelligent systems are required to be put into the packaging for direct contact with the atmosphere surrounding the food or with the food itself (Dainelli et al., 2008).

19.5 DEVELOPMENT OF PACKAGING

Environmental concerns and safety issues enhance and stimulate the use of renewable bioresources as novel packaging materials to maintain or improve the quality of packaged foods. Meanwhile, the application of the nanocomposite concept has proven to be a promising option to improve mechanical and barrier properties of packaging (Avella et al., 2005).

1. *Biocomposite materials*
   a. *Biopolymers for packaging film*
      Thermoplastic bioderived polysaccharides (such as cellulose, gum, starch, chitosan, etc.) as blends with other polymers have garnered attention due to their biodegradability (Lagarón and Sanchez-García, 2008). Cellulose acetate has high toughness and excellent optical clarity; so, it can be used to prepare films by either solvent-casting (Paul et al., 2005) or melting techniques (Mohanty et al., 2003). Different ratios of k-carrageenan and locust bean gum have been used to

tailor edible films with enhanced barrier and mechanical properties. For all k-carrageenan/locust bean gum-based films, the 40/60 ratio films showed the best mechanical properties and water vapor barrier (Martins et al., 2012). The study by Al-Hassan and Norziah (2012) showed that starch/gelatin solutions with ratios of 3:1, 4:1, and 5:1 appear to form good flexible films.

Several techniques have been used to prepare edible chitosan films. Compared to ambient air drying, hot air drying, and vacuum drying, low-pressure superheated steam drying at 70°C was proposed as the most favorable conditions for drying chitosan films (Mayachiew and Devahastin, 2008). Chitosan films have good mechanical properties and selective permeability to CO_2 and O_2 gases, but have poor barrier to moisture due to its hydrophilic nature. To control moisture transfer, several approaches were conducted as follows: incorporation of hydrophobic compounds such as lipids, optimization of the interaction between polymers, and cross-linking or functionalization through physical, chemical, or enzymatic treatments. For example, Pereda et al. (2012) investigated the properties of plasticized chitosan–olive oil emulsion films. It showed that the emulsifying property of chitosan was able to stabilize olive oil droplets in the film-forming emulsions.

Proteins are also important biopolymers with excellent film-forming ability. Thermoplastic proteins have been investigated recently, including gluten (Pommet et al., 2005), gelatin (Salerno et al., 2007) and zein (Mensitieri et al., 2011; Selling and Sessa, 2007). Similar to chitosan-based films, protein-based films also have high water absorptivity owing to hydrophilicity of amino acids in protein molecules. Therefore, hydrophobic substances such as fats and oils can be added to improve the water vapor barrier properties (Artharn et al., 2009). Whey protein can be used to make edible films, which act as barriers to oxygen, aroma, and/or moisture and improve mechanical integrity. Javanmard (2008) suggested that a whey protein–olive oil composite film is a viable alternative packaging process for peanuts and could improve the shelf life and sensory properties. Huang et al. (2011) also used whey protein isolate as edible coatings to improve the rehydration behavior of freeze-dried strawberry pieces.

In fact, high crystallinity and strong intermolecular interactions in proteins and polysaccharides lead to thermal degradation of the biomaterial before achieving melt flow during thermoforming process of packaging the film. Therefore, a suitable plasticizer is necessary for biopolymers as an internal lubricant to increase molecular mobility, thus promoting the melt flow (Mensitieri et al., 2011). Biomaterials blended with different synthetic polymers (e.g., blends of PLA with polyethylene glycol PEG, polyhydroalkanoates PHA, and PCL) can improve the water-barrier properties of biopolymers but results in less O_2 transmission capacity (Wu et al., 2014). Other approaches are available to improve the barrier properties of biopolymers (Petersen et al., 1999); Mensitieri et al. (2011) suggested as follows: (1) lamination (coextrusion) of two or more biopolymers; (2) chemical and/or physical modification of biopolymers; for example, micro and nanocomposites; and (3) use of coating with materials that would add hydrophobicity to the packaging material.

b. *Bioactive agent*
The application of bioactive packaging would increase significantly in the near future due to consumers' preference for naturally preserved food. Selection of bioactive agent is often restricted by (1) its heat instability during the manufacturing process, (2) the incompatibility of active agent with the packaging material, and (3) the organoleptic effect on the packaged food (Sadaka et al., 2013). Many natural antimicrobial/antioxidant agents could be incorporated into edible films. The natural antibacterial agent allyl isothiocyanate encapsulated in cyclodextrin has previously been evaluated as a slow-release additive in polylactide-co-polycaprolactone (PLA-PCL) films (Plackett et al., 2007). Optimization of drying methods and conditions on antimicrobial activity of edible chitosan films incorporated with galangal extract will be useful for making active packaging films, which could help prevent surface growth of microorganisms like *Staphylococcus aureus* (Mayachiew et al., 2010). Natural antioxidants (like Indian gooseberry extract that contains high concentrations of phenolic compounds) can also be added to edible chitosan films, which release active agents and reduce oxidation in packed foods (Mayachiew and Devahastin, 2010).

c. *Biointelligent agent*
Chitosan is a natural polymer with good biodegradability and ability to retard the transport of moisture, gas, flavor, and lipids (Yoshida et al., 2010). Anthocyanins are a group of natural, water-soluble, and nontoxic pigments that can change color depending on pH. Yoshida et al. (2014) developed a chitosan film containing anthocyanin as a natural pH colorimetric indicator, in which it presented a pink color at acid pH, bluish-green at neutral pH, and yellow at basic pH. Pereira et al. (2015) used an active chitosan/PVA film with anthocyanins from *Brassica oleracea* (red cabbage) as time–temperature indicators in intelligent food packaging.

2 *Nanocomposite material*
When particle size is reduced to 1–100 nm, the packaging material exhibits physical and chemical properties different from those in macroscales. Three main applications of nanomaterials as packaging films were proposed by Duncan (2011), including high barrier packaging materials (e.g., polymer/clay nanocomposites, see Figure 19.4), antimicrobial agents (e.g., silver nanoparticles) for active packaging materials, and nanosensors for intelligent packaging materials. In Figure 19.4, compared to polymer (a), polymer (b) with nanocomposite theoretically increases the average gas diffusion length due to its tortuous pathway, thus forming the high gas-barrier property of packaging material.

Nanocomposites based on biopolymers have been considered as potential materials for the development of new eco-friendly food packaging. Dispersion of clays in

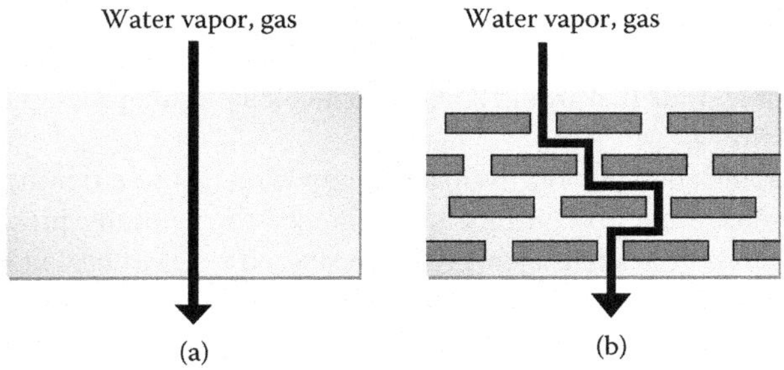

FIGURE 19.4 (a) Schematic diagram of a regular polymer film. (b) Polymer film incorporated with exfoliated clay nanoplatelets. (Adapted from Duncan, T. V., *J. Colloid Interface Sci.*, 363 (1), 1–24, 2011. With permission.)

polymer matrices has shown important changes in the gas permeability of different nanocomposites (Rodríguez et al., 2012). Rodríguez et al. (2014) developed an antimicrobial material based on a nanocomposite cellulose acetate film for active food packaging, including cellulose acetate (CA), commercial organoclay Cloisite30B (C30B), thymol (T) as a natural antimicrobial component and triethyl citrate (TEC) as a plasticizer, and the antimicrobial activity of active film was increased with the presence of C30B in the exfoliated nanocomposite films. Nanocomposites based on CA, C30B, TEC, and variable content of antimicrobial agents (thymol and cinnamaldehyde) were also obtained using a solution casting technique. The cellulose eco-nanocomposites were developed for food packaging, which exhibited antimicrobial properties (Rodríguez et al., 2013).

19.6 CONCLUSIONS

A properly designed package can help the packed dried vegetable products alleviate quality loss, such as mechanical damage caused by friction, collision and extrusion, contamination from dust, pests and microorganisms, and moisture evaporation. Innovative food packaging technologies such as active packaging, intelligent packaging, and application of nanomaterials must comply with strict national and international regulations. Packaging materials should be harmless to humans and should not affect the quality of the packed foodstuff in an unacceptable way. Continued innovations in packaging design and manufacture are expected to lead to further improvements in safety, quality, and stability of dried vegetable products.

REFERENCES

Adom, K., Dzogbefia, V., Ellis, W., and Simpson, B. (1996). Solar drying of Okra—Effects of selected package materials on storage stability. *Food Research International*. 29 (7):589–593.

Al-Hassan, A. A. and Norziah, M. H. (2012). Starch–gelatin edible films: Water vapor permeability and mechanical properties as affected by plasticizers. *Food Hydrocolloids*. 26 (1):108–117.

Almeida-Dominguez, N. G., Higuera-Ciapara, I., Goycoolea, F. M., and Valencia, M. E. (1992). Package, temperature and TBHQ effects on oxidative deterioration of corn-based snacks. *Journal of Food Science*. 57 (1):112–117.

Artharn, A., Prodpran, T., and Benjakul, S. (2009). Round scad protein-based film: Storage stability and its effectiveness for shelf-life extension of dried fish powder. *LWT-Food Science and Technology*. 42 (7):1238–1244.

Arvanitoyannis, I. S. and Stratakos, A. C. (2012). Application of modified atmosphere packaging and active/smart technologies to red meat and poultry: A review. *Food and Bioprocess Technology*. 5 (5):1423–1446.

Avella, M., De Vlieger, J. J., Errico, M. E., Fischer, S., Vacca, P., and Volpe, M. G. (2005). Biodegradable starch/clay nanocomposite films for food packaging applications. *Food Chemistry*. 93 (3):467–474.

Bart, J. C. J. (2006). Quality of food contact polymers. In Press, F. U. (Ed.), *Polymer Additive Analytics: Industrial Practice and Case Studies*. Florence, Italy: Firenze University Press.

Belcher, J. N. (2006). Industrial packaging developments for the global meat market. *Meat Science*. 74 (1):143–148.

Chan, S. T., Yao, M. W. Y., Wong, Y. C., Wong, T., Mok, C. S., and Sin, D. W. M. (2006). Evaluation of chemical indicators for monitoring freshness of food and determination of volatile amines in fish by headspace solid-phase microextraction and gas chromatography-mass spectrometry. *European Food Research and Technology*. 224 (1):67–74.

Coles, R., McDowell, D., and Kirwan, M. J. (2003). *Food Packaging Technology*. Oxford, UK: Blackwell.

Cornforth, D. and Hunt, M. (2008). Low-oxygen packaging of fresh meat with carbon monoxide. *AMSA White Paper Series*. 2 (10):1–10.

Crippa, A., Sydenstricker, T. H. D., and Amico, S. C. (2007). Performance of multilayer films of thermoformed packages. *Polímeros*. 17 (3):188–193.

Dainelli, D., Gontard, N., Spyropoulos, D., Zondervan-van den Beuken, E., and Tobback, P. (2008). Active and intelligent food packaging: Legal aspects and safety concerns. *Trends in Food Science and Technology*. 19:S103–S112.

Dak, M., Sagar, V. R., and Jha, S. K. (2014). Shelf-life and kinetics of quality change of dried pomegranate arils in flexible packaging. *Food Packaging and Shelf Life*. 2 (1):1–6.

Duncan, T. V. (2011). Applications of nanotechnology in food packaging and food safety: Barrier materials, antimicrobials and sensors. *Journal of Colloid and Interface Science*. 363 (1):1–24.

Erlandson, J. and Wrolstad, R. (1972). Degradation of anthocyanins at limited water concentration. *Journal of Food Science*. 37 (4):592–595.

Félix, J. S., Manzoli, J. E., Padula, M., and Monteiro, M. (2014). Evaluation of different conditions of contact for caprolactam migration from multilayer polyamide films into food simulants. *Packaging Technology and Science*. 27 (6):457–466.

Gontard, N. (2004). Active packaging for food processing and preservation. In: International Congress on Engineering and Food: ICEF 9, March 7–11, 2004, Montpellier, France.

Greengrass, J. (1999). *Packaging Materials for MAP of Foods. Principles and Applications of Modified Atmosphere Packaging of Food*. Gaithersburg, MD: Aspen.

Huang, L. L., Zhang, M., Yan, W. Q., Mujumdar, A. S., and Sun, D. F. (2011). Rehydration characteristics of freeze-dried strawberry pieces as affected by whey protein edible coatings. *International Journal of Food Science and Technology*. 46 (4):671–677.

Javanmard, M. (2008). Effect of whey protein edible film packaging on the quality and moisture uptake of dried peanuts. *Journal of Food Process Engineering*. 31 (4):503–516.

Jenkins, W. A. and Harrington, J. P. (1991). *Packaging Foods with Plastics*. Lancaster, PA: Technomic Pub. Co.

Lagarón, J. M. and Sanchez-García, M. (2008) *Thermoplastic Nanobiocomposites for Rigid and Flexible Food Packaging Applications. Environmentally Friendly Food Packaging*. Boca Raton, FL: Woodhead Publishers.

Lang, C. and Hübert, T. (2011). A colour ripeness indicator for apples. *Food and Bioprocess Technology*. 5 (8):3244–3249.

Luh, B. and Eidels, L. (1969). Chemical changes in freeze-dried mushrooms (*Agaricus bisporus*). *Confructa*. 14 (1):8.

Macedo, I. S. M., Sousa-Gallagher, M. J., Oliveira, J. C., and Byrne, E. P. (2013). Quality by design for packaging of granola breakfast product. *Food Control*. 29 (2):438–443.

Martins, J. T., Cerqueira, M. A., Bourbon, A. I., Pinheiro, A. C., Souza, B. W. S., and Vicente, A. A. (2012). Synergistic effects between κ-carrageenan and locust bean gum on physicochemical properties of edible films made thereof. *Food Hydrocolloids*. 29 (2):280–289.

Mayachiew, P. and Devahastin, S. (2008). Comparative evaluation of physical properties of edible chitosan films prepared by different drying methods. *Drying Technology*. 26 (2):176–185.

Mayachiew, P. and Devahastin, S. (2010). Effects of drying methods and conditions on release characteristics of edible chitosan films enriched with Indian gooseberry extract. *Food Chemistry*. 118 (3):594–601.

Mayachiew, P., Devahastin, S., Mackey, B. M., and Niranjan, K. (2010). Effects of drying methods and conditions on antimicrobial activity of edible chitosan films enriched with galangal extract. *Food Research International*. 43 (1):125–132.

Mensitieri, G., Di Maio, E., Buonocore, G. G., Nedi, I., Oliviero, M., Sansone, L., and Iannace, S. (2011). Processing and shelf life issues of selected food packaging materials and structures from renewable resources. *Trends in Food Science and Technology*. 22 (2–3):72–80.

Mohanty, A., Wibowo, A., Misra, M., and Drzal, L. (2003). Development of renewable resource-based cellulose acetate bioplastic: Effect of process engineering on the performance of cellulosic plastics. *Polymer Engineering and Science*. 43 (5):1151–1161.

Moore, M. E., Han, I. Y., Acton, J. C., Ogale, A. A., Barmore, C. R., and Dawson, P. L. (2003). Effects of antioxidants in polyethylene film on fresh beef color. *Journal of Food Science*. 68 (1):99–104.

Mullan, M. and McDowell, D. (2003). *Modified Atmosphere Packaging. Food Packaging Technology*. Boca Raton, FL: CRC Press.

Negi, P. S. and Roy, S. K. (2000). Effect of low-cost storage and packaging on quality and nutritive value of fresh and dehydrated carrots. *Journal of the Science of Food and Agriculture*. 80 (15):2169–2175.

Oliveira, L. M., Sarantópoulos, C. I. G. L., Cunha, D. G., and Lemos, A. B. (2006). Thermoformed and thermoprocessable plastic packages for processed meat. *Polímeros*. 16 (3):202–210.

Ozdemir, M. and Floros, J. D. (2004). Active food packaging technologies. *Critical Reviews in Food Science and Nutrition*. 44 (3):185–193.

Paul, D., Zeng, Q., Yu, A., and Lu, G. (2005). The interlayer swelling and molecular packing in organoclays. *Journal of Colloid and Interface Science*. 292 (2):462–468.

Pereda, M., Amica, G., and Marcovich, N. E. (2012). Development and characterization of edible chitosan/olive oil emulsion films. *Carbohydrate Polymers*. 87 (2):1318–1325.

Pereira, V. A., de Arruda, I. N. Q., and Stefani, R. (2015). Active chitosan/PVA films with anthocyanins from *Brassica oleraceae* (Red Cabbage) as time–temperature indicators for application in intelligent food packaging. *Food Hydrocolloids*. 43:180–188.

Petersen, K., Væggemose Nielsen, P., Bertelsen, G., Lawther, M., Olsen, M. B., Nilsson, N. H., and Mortensen, G. (1999). Potential of biobased materials for food packaging. *Trends in Food Science and Technology*. 10 (2):52–68.

Piringer, O. G. and Baner, A. L. (2000). *Plastic Packaging Materials for Food.* Weinheim, Germany: Wiley-VCH.

Plackett, D., Ghanbari-Siahkali, A., and Szente, L. (2007). Behavior of α- and β-cyclodextrin-encapsulated allyl isothiocyanate as slow-release additives in polylactide-co-polycaprolactone films. *Journal of Applied Polymer Science.* 105 (5):2850–2857.

Pommet, M., Redl, A., Guilbert, S., and Morel, M.-H. (2005). Intrinsic influence of various plasticizers on functional properties and reactivity of wheat gluten thermoplastic materials. *Journal of Cereal Science.* 42 (1):81–91.

Prakash, S., Jha, S. K., and Datta, N. (2004). Performance evaluation of blanched carrots dried by three different driers. *Journal of Food Engineering.* 62 (3):305–313.

Puligundla, P., Jung, J., and Ko, S. (2012). Carbon dioxide sensors for intelligent food packaging applications. *Food Control.* 25 (1):328–333.

Quintavalla, S. and Vicini, L. (2002). Antimicrobial food packaging in meat industry. *Meat Science.* 62 (3):373–380.

Rao, D. N. and Sachindra, N. M. (2002). Modified atmosphere and vacuum packaging of meat and poultry products. *Food Reviews International.* 18 (4):263–293.

Realini, C. E. and Marcos, B. (2014). Active and intelligent packaging systems for a modern society. *Meat Science.* 98 (3):404–419.

Restuccia, D., Spizzirri, U. G., Parisi, O. I., Cirillo, G., Curcio, M., Iemma, F., Puoci, F., Vinci, G., and Picci, N. (2010). New EU regulation aspects and global market of active and intelligent packaging for food industry applications. *Food Control.* 21 (11):1425–1435.

Rodríguez, F. J., Galotto, M. J., Guarda, A., and Bruna, J. E. (2012). Modification of cellulose acetate films using nanofillers based on organoclays. *Journal of Food Engineering.* 110 (2):262–268.

Rodríguez, F., Sepulveda, H. M., Bruna, J., Guarda, A., and Galotto, M. J. (2013). Development of cellulose eco-nanocomposites with antimicrobial properties oriented for food packaging. *Packaging Technology and Science.* 26 (3):149–160.

Rodríguez, F. J., Torres, A., Penaloza, A., Sepulveda, H., Galotto, M. J., Guarda, A., and Bruna, J. (2014). Development of an antimicrobial material based on a nanocomposite cellulose acetate film for active food packaging. *Food Additives and Contaminants Part A: Chemistry, Analysis, Control, Exposure and Risk Assessment.* 31 (3):342–353.

Rukchon, C., Nopwinyuwong, A., Trevanich, S., Jinkarn, T., and Suppakul, P. (2014). Development of a food spoilage indicator for monitoring freshness of skinless chicken breast. *Talanta.* 130:547–554.

Sadaka, F., Nguimjeu, C., Brachais, C.-H., Vroman, I., Tighzert, L., and Couvercelle, J.-P. (2013). Review on antimicrobial packaging containing essential oils and their active biomolecules. *Innovative Food Science and Emerging Technologies.* 20:350.

Salerno, A., Oliviero, M., Di Maio, E., and Iannace, S. (2007). Thermoplastic foams from zein and gelatin. *International Polymer Processing.* 22 (5):480–488.

Salunkhe, D. K., Do, J. Y., and Bolin, H. R. (1973). Developments in technology and nutritive value of dehydrated fruits, vegetables, and their products. *Critical Reviews in Food Science and Nutrition.* 4 (2):153–192.

Sapers, G., Panasiuk, O., Talley, F., and Shaw, R. (1973). Flavor quality and stability of potato flakes. Effects of raw material and processing. *Journal of Food Science.* 38 (4):586–589.

Selling, G. W. and Sessa, D. J. (2007). Sample preparation and testing methods affect the physical properties and evaluation of plasticized zein. *Industrial Crops and Products.* 25 (3):266–273.

Sørheim, O., Nissen, H., and Nesbakken, T. (1999). The storage life of beef and pork packaged in an atmosphere with low carbon monoxide and high carbon dioxide. *Meat Science.* 52 (2):157–164.

Uadal, S. and Sagar, V. (2008). Influence of packaging and storage temperature on the quality of dehydrated selected leafy vegetables. *Journal of Food Science and Technology-Mysore*. 45 (5):450–453.

Uz, M. and Altınkaya, S. A. (2011). Development of mono and multilayer antimicrobial food packaging materials for controlled release of potassium sorbate. *LWT-Food Science and Technology*. 44 (10):2302–2309.

Vanderroost, M., Ragaert, P., Devlieghere, F., and De Meulenaer, B. (2014). Intelligent food packaging: The next generation. *Trends in Food Science and Technology*. 39 (1):47–62.

Vu, C. H. and Won, K. (2013). Novel water-resistant UV-activated oxygen indicator for intelligent food packaging. *Food Chemistry*. 140 (1–2):52–56.

Westerman, S. J., Sutherland, E. J., Gardner, P. H., Baig, N., Critchley, C., Hickey, C., Mehigan, S., Solway, A., and Zervos, Z. (2013). The design of consumer packaging: Effects of manipulations of shape, orientation, and alignment of graphical forms on consumers' assessments. *Food Quality and Preference*. 27 (1):8–17.

Wong, F., Dietrich, W., Harris, J. G., and Lindquist, F. (1956). Effect of temperature and moisture on storage stability of vacuum-dried tomato juice powder. *Food Technology*. 10 (2):96–100.

Wu, Y., Qin, Y., Yuan, M., Li, L., Chen, H., Cao, J., and Yang, J. (2014). Characterization of an antimicrobial poly(lactic acid) film prepared with poly(ε-caprolactone) and thymol for active packaging. *Polymers for Advanced Technologies*. 25 (9):948–954.

Yam, K. L., Takhistov, P. T., and Miltz, J. (2005). Intelligent packaging: Concepts and applications. *Journal of Food Science*. 70 (1):R1–R10.

Yoshida, C. M. P., Bastos, C. E. N., and Franco, T. T. (2010). Modeling of potassium sorbate diffusion through chitosan films. *LWT-Food Science and Technology*. 43 (4):584–589.

Yoshida, C. M. P., Maciel, V. B. V., Mendonça, M. E. D., and Franco, T. T. (2014). Chitosan biobased and intelligent films: Monitoring pH variations. *LWT-Food Science and Technology*. 55 (1):83–89.

20 Microbiology and Safety of Dried Vegetables

Wunwisa Krasaekoopt
Assumption University

CONTENTS

20.1 INTRODUCTION

Dehydration of vegetables is a traditional preservation technique used for the production of artificially dried vegetables, for instance, garlic, onion, pea, carrot, potato, etc. Dried vegetables are generally stable after processing and during storage due to low water activity, which is reduced during the drying process to a level that does not support the growth of microorganisms. Therefore, these products rarely cause food-borne diseases. After drying, the microbiological stability of these products depends on how these products are kept, including product packaging. This chapter describes the effects of drying process on microorganisms, microflora, and spoilage of microorganisms, pathogen and safety, as well as its control.

20.2 EFFECTS OF PROCESSING ON MICROORGANISMS

The drying process of vegetables comprises of grading, washing, blanching, and drying. Effects of blanching of vegetables for drying are similar to those of vegetables for freezing. Therefore, the microflora of dried vegetables is influenced by the destruction degree of blanching. A significant reduction of vegetable cells of microorganisms is found, whereas less marked reduction of spore-forming bacteria is seen after blanching. Vegetables for drying may be favorable for the growth of microorganisms when they are kept at room temperature too long or at improper cooling conditions. Contamination may be also caused by unclean equipment (Vaughn 1951; Goresline 1963). Moreover, before the drying process, vegetables may be subjected to some antibrowning treatments such as sulfiting, which can reduce the number of microorganisms (Wedzicha 1984). On the other hand, some vegetables are not blanched before drying, for example, onion and garlic.

For the drying process, most of the vegetables are subjected to heated air by using cabinet tray or belt tunnel dryers. Therefore, the loading of trays or belts is an important factor that needs to be taken into consideration due to uneven loading. This condition can lead to improper air circulation and temporary wet spots that can favor the growth of microorganisms. Due to the water evaporation occurring during the drying process, the temperature of the products is normally lower than 50°C, while the heated air temperature is approximately higher than 80°C. Destruction of microorganisms by conventional drying process is rarely found, and in some cases the number may be increased due to the increased concentration in the smaller amount of the products (Murphy 1973). In some conditions, long-exposure drying is required in vegetables such as edible mushroom; therefore, microorganism growth can be observed as being at the same level as before drying.

Moreover, drying method is another factor that has to be taken into consideration. The method of drying affects not only the number but also the type of microorganisms. Hot air drying has adverse effects on the quality of dried vegetables, especially sensory and nutrient qualities, due to high temperature and long drying time used (Vadivambal and Jayas 2007; Head et al. 2008). Compared to hot air drying, vacuum drying uses lower drying temperature in a lower oxygen condition, resulting in better qualities (Suvarnakuta et al. 2005; Lewicki 2006; Wu et al. 2007; Alibas 2009). Recently, a low-pressure superheated steam drying (LPSSD) method has been reported to be an alternative for drying of heat-sensitive food products. Many studies reported that LPSSD performed better than hot air and vacuum drying, providing better quality of products (Devahastin et al. 2004; Methakhup et al. 2005; Suvarnakuta et al. 2005; Leeratanarak et al. 2006; Thomkapanish et al. 2007). For microbial inactivation, it is well recognized that dry heat is less effective than moist heat due to proteins, which are involved in maintaining cell viability and are more stable in low-moisture conditions (Archer et al. 1998; Doyle and Mazzotta 2000). Furthermore, cells attached to a tissue are more heat resistant than unattached cells (Murphy et al. 2002; Chiewchan et al. 2007). The changes of the food surface characteristics during drying also resulted in microbial cell entrapment, leading to protection from direct heating and longer drying time (Chiewchan and Morakotjinda 2009; Hawaree et al. 2009). LPSSD is demonstrated to be effective in mycotoxin and microbial decontamination due to its high-temperature environment (Cenkowski et al. 2007; Nygaard and Hostmark 2008). In addition, the type of machine used for drying influences the microbial quality of the dried vegetables, even though the same hot air drying is used. Belted-dried onion has lower microbial count than tray-dried onion. Majority of the belted-dried onions contain spore-forming bacteria, whereas high number of lactic acid bacteria are found in tray-dried ones (Sheneman 1973).

20.3 POSTDRYING TREATMENT OF DRIED VEGETABLES

Dried vegetables can be contaminated by air- and soil-borne bacteria, fungi, and insects, especially during open air drying. Pathogens such as *Samonella*, *Escherichia coli*, *Clostridium perfringens*, *Bacillus cereus*, and toxigenic molds can also be present (Bendini et al. 1998). Good manufacturing practices (GMPs) during harvesting and processing could improve their microbiological quality, but not enough to reach standard acceptable level (WHO 1999). Therefore, postdrying treatments are required, such as conditioning, fumigation, and irradiation.

20.3.1 Conditioning

When the drying process is over, conditioning may be required due to different moisture levels of the products caused by size and location in the oven during drying. Conditioning distributes residual moisture evenly in dried vegetables, hence the chance of spoilage is less. For conditioning, the cooled, dried vegetables are kept in a warm, dry, and well-ventilated place for 4–10 days with stirring daily. The conditioning is not required for the product containing the same size due to its very low level of moisture. If there is evidence of moisture, the products are returned to the drying trays and heated in a 65°C oven for 30 min, then cooled and packaged.

20.3.2 Fumigation

Many commercial food processing units of dried vegetables use fumigation with methyl bromide and ethylene oxide (EO) to eliminate the microbiological problems of the products, especially bacteria and molds. Nevertheless, these compounds have been recognized as extremely harmful substances for the consumers and the environment. Methyl bromide has the potential to deplete the atmospheric ozone layer, resulting in greenhouse effects. EO has been used for fumigation for more than 60 years; it uses alkylation as the reaction mechanism. EO fumigation could significantly reduce spoilage due to microorganisms in spices and dried vegetables. For an effective result, the EO gas must be diffused freely throughout the product being treated, making breathable packaging a necessity. Aeration is required because nonvolatile residues remain after fumigation. Generally, the aeration is completed within 24 h. EO is normally used with inert gases (carbon dioxide or a chlorinated hydrocarbon), due to its extreme flammability, in specially designed vacuum chambers. The concentration used is in the range of 400–1000 mg L^{-1}. The aerobic plate count can be reduced by 10^1 to 10^4, depending on the types of products, microorganisms, and condition of treatments. Mold count is reduced by 10^2 to 10^3. The microbial destruction rates depend on the concentration of EO, temperature, relative humidity, and moisture content of the products as well as the porosity of products and permeability of packaging used. EO was prohibited in Europe due to its safety and environmental friendly consideration. The use of this compound has been also banned in USA for the treatment of ground spices (Loaharanu 1994). The U.S. Clean Air Act and the Montreal Protocol of the Vienna Convention required that any substance listed as ozone depleting must be withdrawn from production and use by the year 2001.

20.3.3 Irradiation

Dried vegetables, herbs, and spices are currently treated with ionizing radiation to eliminate microbial contamination. It has been definitely confirmed that the treatment with ionizing energy is more effective against bacteria than thermal treatments such as sterilization. It also does not have chemical residues in the food product (Thayer et al. 1996; Olson 1998). Heat treatment can cause the loss of thermolabile aromatic volatiles and/or cause additional thermally induced changes (e.g., thermal decomposition or production of thermally induced radicals). Pezzutti et al. (2005) reported the use of gamma ray doses between 5 and 25 kGy to reduce the microbial counts in dried garlic and onion. Mesophilic aerobic bacterial spores were 1.4–2.7 and 3.5–4.6 log CFU g^{-1} in garlic and onion, respectively. For onion, the dose of 10 kGy reduced the spore counts to nondetectable levels, whereas the spore counts were lower than 30 CFU g^{-1} in garlic. Effects of irradiation on dried vegetables and spices are reported in Table 20.1.

TABLE 20.1
Irradiation of Dried Vegetables and Spices

Dried Vegetables and Spices	Dose (kGy)	Remarkable Results
Black pepper, white pepper, turmeric, rosemary, and basil	4–10	Elimination of coliform
Coriander, cumin, turmeric, chili, rosemary, and basil	5	Elimination of coliform Reduction of mold by 2–6 logs
Garlic powder	5	Elimination of coliform
White pepper	5	Reduction of total plate count by 3 logs
Nutmeg	5	Reduction of total plate count by 4 logs
Ginger	5	Reduction of total plate count by 2 logs
Paprika and crushed red pepper	6.5	Elimination of coliform and mold Reduction of total plate count to $<3 \times 10^3$
Cinnamon, cloves, coriander, nutmeg, white pepper, and black pepper	7	Reduction of total plate count by 2.5–4 logs
Garlic powder	7	Reduction of total plate count by 4 logs
Pepper, cardamom, mace, cinnamon, marjoram, cloves, caraway, coriander, charlock, juniper, paprika, black pepper, pimento, and commercial spice blends	7.7–10	Virtual sterility
Cinnamon, cloves, coriander, nutmeg, white pepper, and black pepper	10	Reduction of total plate count to 1.7 log CFU/g
Chili	10	Reduction of total plate counts to below detectable level
Herbs, garlic powder, and onion powder	10	Elimination of coliform and mold Reduction of total plate count to $<3 \times 10^3$
Black pepper, white pepper, turmeric, rosemary, and basil	10	Reduction of spore-forming bacteria to $<10^3$
Paprika and caraway seeds	10	Reduction of total plate count to $<3 \times 10^3$
Black pepper, white pepper, turmeric, rosemary, and basil	12–15	Reduction of total plate counts to below detectable level

20.4 PACKING AND STORING OF DRIED VEGETABLES

Dried vegetables can be significantly deteriorated during storage. It is an important consideration for storage of dried vegetables—both the type of packaging material including the method used and the storage conditions—in order to avoid insect infestation and moisture reabsorption as well as to reduce microbiological degradation of the products. After the drying process, the dried vegetable must be packed when it is completely cooled to avoid the sweating that can lead to the favorable moisture condition for mold growth. The moisture content of dried vegetables is not constant because of their hygroscopicity and is always in equilibrium with relative humidity of air in storage rooms. To maintain low moisture content of dried vegetables, the products should be stored at relative humidity lower than 78% in water vapor-proof containers, such as waterproof plywood drums.

Tin boxes or drums are the most effective packages mainly used for long-term storage of dried vegetables as are laminated packages (carton with metallic sheet, plastic materials, etc.), which are mainly for consumer packages. With regard to plastic bag, Singh and Sagar (2010) reported that packing dried leafy vegetables in high-density polyethylene (HDPE) at 7°C provided better quality of products than those packed in low-density polyethylene (LDPE) and polypropylene (PP). Higher retention nutrients such as β-carotene and ascorbic acid as well as chlorophyll content and sensory scores were found. Moreover, this condition maintained low moisture in products, which resulted in less microbial deterioration. In addition, Seevaratnam et al. (2012) evaluated packing of dried leafy vegetables in HDPE, PP, and metalized polypropylene (MPP) and storing at 28°C–36°C. MPP retained higher levels of chlorophyll, ascorbic acid, and β-carotene than the others as well as lower moisture content during 3 months storage.

In addition, moisture absorbers or desiccants are introduced into the containers, hermetically closed, not only to maintain but also reduce moisture of the product. The most popular desiccant is calcium oxide, which is packed in small bags made from materials permeable to water vapor but which do not permit the desiccant to escape into the products. Packing with desiccant can reduce moisture of the product to as low as <4%, inhibiting or reducing microbial growth. Vacuum packing and inert gas packing (carbon dioxide or nitrogen) are also good options for dried vegetables, mainly to reduce chemical and biochemical deterioration rather than microbial spoilage such as β-carotene oxidation, leading to foul smell and discoloration.

Storage temperature also plays an important role in reducing or inhibiting microbial spoilage. The storage temperature should be below 25°C (and preferably 15°C); lower temperatures (0°C–10°C) help to maintain taste, color, water rehydration ratio, and, also to an extent, vitamin C.

20.5 MICROFLORA AND SPOILAGE MICROORGANISMS OF DRIED VEGETABLES

Dried vegetables can be infected with fungi and other contaminants either already present on the primary product or during the drying process that takes place under unhygienic conditions; further spoilage can occur during storage, handling, and transportation till sale (Mandeel 2005). Thus, the number of microorganisms is exactly variable. Dried celery, onion, and garlic, for instance, tend to have a high microbial count because they are grown in soil. For vegetables requiring blanching before drying, the number of microorganisms is remarkably reduced; therefore, the microflora of these products is composed of the microorganisms that grow well on the product-contact surfaces of the equipment. In addition, the dried vegetables that are not subjected to blanching have microflora approximately close to that seen in raw materials (Sheneman 1973).

The number and types of microflora also depend upon the places where these products are produced. Akeredolu and Adebajo (2013) reported the microbiological property of three dried vegetables (bitter leaf, bell pepper, and okra) produced in Nigeria. Nine bacteria, namely, *Staphylococcus aureus*, *Acinetobacter iwoffi*, *Corynebacterium* sp., *Bacillus pumilus*, *Micrococcus* sp., *Pseudomonas aeruginosa*, *Flavobacterium* sp., *Bacillus* sp., and *Micrococcus kristianae* and six fungi, namely, *Aspergillus niger*, *A. flavus*, *Aspergillus* spp., *Penicillium* spp., *Cladosporium* sp., *Fusarium* spp. were found. It was indicated that proper handling of vegetables during processing and storage is required to minimize microbial contamination. Kudjawu et al. (2011) also reported on the microbiota of dried vegetables produced in Ghana. Molds and *Bacillus* spp. were the dominant microflora of all the dried vegetables although lactic acid bacteria and coliforms were also isolated in most of these products. It was suggested that primarily sanitizing the fresh vegetables prior to drying in solar dryers is required. These microorganisms originally come from the raw material, equipment in the processing line, and storage.

In addition, improper storage of dried vegetables makes these products easily susceptible to the growth of fungi, leading to mycotoxin contamination (Garbutt 1997; Bankole and Mabekoje 2004). Mycotoxins are hazardous compounds to consumers' health and affect food quality, resulting in economic loss as well as loss of commercial value (Bhat and Vashanti 1999; Otzuki et al. 2001). Natural occurrence of mycotoxins and fungal contamination on dried vegetables was reported in many countries on the African continent (Alghalibi and Shater 2004; Magnoli et al. 2004; Nutsugah et al. 2004; Begum et al. 2005; Mandeel 2005; Hell et al. 2009).

20.6 PATHOGENS AND SAFETY OF DRIED VEGETABLES

In dried vegetables, vegetative cells of pathogens are rarely found due to lower water activity. However, spores of some spore-forming bacteria such as *Bacillus cereus*, *Clostridium botulinum*, and *Cl. perfringens* may be found in the products due to their heat resistance to the blanching process. These microorganisms are found in soil previously contaminated with human or animal wastes and can carry through onto the finished products and remain harmless in the products until the products are rehydrated or in the improper conditions that allow these microorganisms to grow. Blanching can eliminate most of the vegetative cells, nevertheless, if proper hygienic practices are not followed, the pathogens can be reintroduced. Dried vegetables possibly contain low number of pathogens such as *S. aureus*, *Listeria monocytogenes*, and *Salmonella* sp. However, these appear to be uncommon in a well-controlled process.

In addition, Banerjee and Sarkar (2003) noticed that *B. cereus*, *C. perfringens*, *S. aureus*, and members of *Enterobacteriaceae* occurred in 85%, 59%, 11%, and 85% of 154 samples of 27 kinds of dried spices from the retail shops, respectively. Bacterial and mold counts were also in the unacceptable level ($>10^6$ cfu g^{-1}). *B. cereus* was not found in black pepper powder, caraway, garlic, and red chili, but was found in all the samples of ajmud, small cardamom, and cumin powder. Coliforms and fecal coliforms were found in 33% and 15% of the samples, respectively. Sagoo et al. (2009) studied the microbiological properties of dried spices and herbs from retail and production in the UK during 2004. *Salmonella* spp. were detected in 1.5% and 1.1% of dried spices and herbs sampled at production and retail, respectively. Overall, 3.0% of herbs and spices contained high counts of *B. cereus*. (1%, $\geq 10^5$ cfu g^{-1}), *C. perfringens* (0.4%, $\geq 10^3$ cfu g^{-1}), and/or *Escherichia coli* (2.1%, $\geq 10^2$ cfu g^{-1}). Van Doren et al. (2013) reported that there were 14 pathogenic outbreaks due to the consumption of dried spices during 1973–2010. The affected countries included Canada, Denmark, England and Wales, France, Germany, New Zealand, Norway, Serbia, and the United States. They found that there were 128 hospitalizations and two deaths. Infants/children were the primary population segments impacted by 36% (5/14). Four outbreaks were associated with multiple organisms. *Salmonella enterica* ssp. *enterica* was identified as the causative agent in 71% (10/14) of outbreaks, followed by *Bacillus* sp. (29%).

Salmonella-related outbreak caused by dried spice and herbs in Europe and North America was also reported (Zweifel and Stephan 2012). A broad diversity of serovars was found in many studies with Salmonella-positive samples and ranged from 0% to 8.4%. Salmonella can survive longer in spices and dried herbs due to its high desiccation resistance. Therefore, the use of untreated spice and herbs for seasoning of ready-to-eat foods or production of unheated foods can lead to food poisoning to consumers.

Moreover, Phungamngoen et al. (2011) studied the effects of drying methods on the heat resistance of *Sal. anatum* attached to the surface of cabbage by using hot air drying, vacuum drying, and LPSSD. The results showed that drying methods had a significant effect on destruction of *Sal. anatum*. The higher drying temperatures resulted in greater destruction. LPSSD provided the highest degree of Salmonella inactivation. DiPersio et al. (2005, 2007) also evaluated the influence of traditional home-type drying treatments recommended by U.S. Cooperative Extension Services on the inactivation of Salmonella in dried carrot slices. Predrying treatments, such as steam or water blanching and immersion in NaCl solution, as well as post-oven drying were investigated (Brennand 1994; Dinstel 1997; Kendall and Allen 1998; Andress and Harrison 1999; Hughes and Willenberg 1999; Roberts and Cox 1999). They found that blanching in acidic solution, such as citric acid, prior to drying appeared to be the most effective way to enhance the inactivation of Salmonella on dried vegetables during home-type dehydration and storage.

Recently, Jernberg et al. (2015) reported that from 24 December to 24 July 2015, 174 cases of a nationwide salmonellosis outbreak were reported in Sweden. There were 108 cases connected to a single restaurant. A spice mix, containing dried vegetables from the restaurant tested positive for the outbreak strain. In addition, spice mixes with similar content from different suppliers also gave positive results. The outbreak investigation suggested the possible risk of contaminated products in the market of other countries.

For microbiological standard of dried vegetables and spices including herbs, the criteria of microorganisms concerned are different based on the regulation of each country and standard organization as well as types of products investigated. The common maximum levels of microorganisms are as follows: *Salmonella* must be not detectable (ND) in 25 g, *E. coli* 1.1×10^2–10^3 MPN/g, molds 10^2–10^4 cfu/g, standard plate counts 10^4–10^6 cfu/g, coliforms 10^2–10^3 cfu/g, and *S. aureus* (coagulase +) 10^2–10^4 cfu/g (Food Standards Australia New Zealand 2003; International Commission on Microbiological Specifications for Food 1986; International Organization for Standardization 1981; World Health Organization 2000).

20.7 MYCOFLORA AND OCCURRENCE OF MYCOTOXINS

Natural occurrence of fungal contamination including mycotoxins in dried vegetables was investigated in many countries (Alghalibi and Shater 2004; Magnoli et al. 2004; Nutsugah et al. 2004; Begum et al. 2005; Mandeel 2005). The presence of mold in dried vegetables can be either already present on the primary product or get contamination during the drying process under improper hygienic conditions, which can lead to further spoilage during storage, handling, and transportation (Mandeel 2005). High fungal contamination was found above 3 logs CFU g^{-1} in red chili and black pepper in Bahrain. Simultaneously, *Aspergillus*, *Cladosporium*, *Penicillium*, and *Fusarium* were the most commonly isolated molds from smoked paprika in Spain, with the level above 4 logs CFU g^{-1} (Martin et al. 2005). As a consequence, these products can be contaminated with mycotoxins (Bankole and Mabekoje 2004). Mycotoxins have effects on consumer health and quality of food, resulting in loss of commercial value of the products (Otzuki et al. 2001). The summary of mold and mycotoxins contamination in dried vegetables and spices is presented in Table 20.2.

Spices have been used for flavor, color, aroma, and preservation of food or beverages for thousands of years. The most important spice crops in the world trade are black pepper (*Piper nigrum* L.), capsicum (*Capsicum annuum* L.), nutmeg (*Myristica fragrans*), cumin (*Cuminum cyminum*), cinnamon (*Cinnamomum*), ginger (*Zingiber officinale*), turmeric (*Curcuma longa*), clove (*Syzygium aromaticum*), and coriander (*Coriandrum sativum*) (UNIDO/FAO 2006). Depending on their processing and environmental conditions, spices are highly susceptible to mycotoxin development (McKee 1995). There are two groups of mycotoxins concerned, which are aflatoxins (AFs) and ochratoxin A (OTA).

AFs are the most toxic group produced by some *Aspergillus* species (*Aspergillus flsper*, *A. parasiticus*, and the rare *A. nomius*). Many types of AFs produced naturally belong to a group called the difuranocoumarins, but only four, aflatoxin B_1 (AFB_1), aflatoxin B_2 (AFB_2), aflatoxin G_1 (AFG_1), and aflatoxin G_2 (AFG_2) are

TABLE 20.2
Mold and Mycotoxin Contamination in Dried Vegetables and Spices

Country	Product	Microorganisms	Mycotoxins	References
Spain	Smoked paprika and red chili	*Aspergillus, Cladosporium, Penicillium,* and *Fusarium*	Presence	Martin et al. (2005), Hierro et al. (2008), and Santos et al. (2010)
Bahrain	Red chili and black pepper	Mold	NA	Mandeel (2005)
Brazil	Red chili	Mold	Presence	Shundo et al. (2009)
Hungary	Red chili	Mold	Presence	Fazekas et al. (2005)
India	Dried red chili	Mold	AFB_1	Ravi Kiran et al. (2005), Zinedine et al. (2006), Saha et al. (2007)
Malaysia	Red chili	Mold	Presence	Jalili and Jinap (2012)
Benin, Togo, Mali	Dried okra, hot chili, tomato, melon seeds, onion, baobab leaves, ginger, garlic, and black pepper	*Aspergillus flavus*	AFB_1 AFB_2	Hell et al. (2009) Gnonlonfin et al. (2013)
Italy	Chili peppers, nutmeg, and cinnamon	NA	AFB_1 AFB_2 AFG_1 OTA	Romagnoli et al. (2007), Prelle et al. (2014)
Turkey	Red chili flake, red chili powder, black pepper powder, and cumin	NA	AFB_1 OTA	Ozby and Kabak (2012)
Egypt	Sun-dried jew's mallow leaves Sun-dried okra	*A. flavus*, *A. niger*, *A. fumigatus*, *A. awamori*, *A. foetidus*, and *A. ficuum*	Presence	Youssef (2008)
Nigeria	RCP, dried *Adansonia digitata* leaves, dried *Hibiscus sabdariffa* leaves, and dried baobab leaves powder	*A. flavus*, *A. niger*, *A. fumigatus*, *Mucor racemosus*, and *Rhizopus oryzae*	AF	Anthony et al. (2012), Aliero and Ibrahim (2014)
China	Pepper, chili, prickly ash	NA	AFB_1	Zhao et al. (2013)
Qatar	Chili powder, black pepper, tandoori masala, turmeric, and garam masala	*A. flavus*, *A. nomius*, and *A. niger* *Penicillium*	AFB_1AFB_2 AFG_1	Hammami et al. (2014)

NA, not available

naturally found in foods. AFB_1 is the most potent genotoxic and carcinogenic compound most commonly found in agricultural products (Sweeney and Dobson, 1998). The International Agency for Research on Cancer has classified AFB_1 and naturally occurring mixtures of AFs as Group I (carcinogenic to humans) (IARC 1993). While OTA is a kidney toxin produced mainly by *Penicillium verrucosum* in temperate climates, it is produced by *A. ochraceus* and the rare *A. carbonarius* in warm and tropical countries and is contaminated prior to harvest or more commonly during storage. This compound has nephrotoxic effects on all mammalian species and has been associated with fatal human kidney disease, referred to as Balkan endemic nephropathy and with an increased incidence of tumors of the upper urinary track. The IARC has classified OTA as a probable human carcinogen (Group 2B) (IARC 1993). These mycotoxins are able to contaminate spices in the field, during drying, and storage, which are mostly found in red chili (capsicums). The investigation was performed in many countries such as Brazil (Shundo et al. 2009), Hungary (Fazekas et al. 2005), India (Saha et al. 2007), Malaysia (Jalili and Jinap 2012), and Spain (Hierro et al. 2008; Santos et al. 2010).

Ravi Kiran et al. (2005) reported AFB_1 production in dried chilies (*Capsicum annum* L.) with the mean level of 5.5 μg kg^{-1} in India; Zinedine et al. (2006) reported high level of AFB_1 (9.68 μg kg^{-1}) in red paprika. Hell et al. (2009) investigated fungal infection and aflatoxin contamination of dried vegetables such as okra, hot chili, tomato, melon seeds, onion, and baobab leaves from Benin, Togo, and Mali. Baobab leaves, followed by hot chili and okra had high fungal contamination compared with other species of *Aspergillus* as dominant. Only okra and hot chili were naturally contaminated with AFB_1 and AFB_2, at concentrations of 6.0 μg kg^{-1} on okra and 3.2 μg kg^{-1} on hot pepper when the mycotoxin investigation was performed.

The presence of AFs in 103 sample spices and herbs in Italy was investigated (Romagnoli et al. 2007). Only 7 spices were found to have positive results (5 chili peppers, 1 nutmeg, and 1 cinnamon). Two samples had the toxin at nonpermissible levels. None of the aromatic herbs, herb tea, and medicinal plant samples analyzed were contaminated, even if they were from tropical countries. The co-occurrence of AFs and OTA in 105 samples of spices marketed in Turkey was also studied using high-performance liquid chromatography (HPLC) coupled with fluorescence detection (HPLC-FD) after immunoaffinity column (IAC) cleanup (Ozby and Kabak 2012). AFs were detected in 79.2% of red chili flake samples, 63.6% of red chili powder, 30.4% of black pepper powder, and 21.1% of cumin samples, while none was found in cinnamon powder samples at detectable levels. Four red chili flake samples and three red chili powder samples were above the EU regulatory limit of 5 mg kg^{-1} for AFB_1. OTA was found in 75% of red chili flake, 54.5% of red chili powder, 17.4% of black pepper powder, with four red chili flake and three red chili powder samples exceeding EU limit of 30 mg kg^{-1}. No OTA was found in cinnamon powder samples, while detectable levels of OTA were found in only one cumin sample. The co-occurrence of AFB and OTA was detected in 62.5% of red chili flake, 40.9% of red chili powder, and 4.3% of black pepper powder samples. Moreover, sun-dried jew's mallow leaves and okra fruits collected from Egypt were investigated for mold and mycotoxin contamination in Egypt (Youssef 2008). Mycological investigation revealed that dried okra fruit samples were more highly contaminated with fungal spore than jew's mallow leaves samples. *Aspergillus* was the highest occurrence (100% of the samples) and was represented by 13 species one variety of, *A. flavus*, *A. niger*, *A. fumigatus*, *A. awamori*, *A. foetidus*, and *A. ficuum* were the predominant ones. *Mucor*, *Rhizopus*, *Fusarium*, *Myrothecium*, *Emericella*, and *Cochliobolus* were fungal genera isolated with different occurrences. Mycotoxin analysis proved that jew's mallow leaf samples were free from any detectable mycotoxins, while five samples of dried okra fruits out of 30 tested (16.7%) proved to be toxic.

The dried vegetable samples composed of baobab leaves (*Adansonia digitata*) (20), okra (*Abelmoschus esculentus*) (20) and red hot chili pepper (RCP) (*Capsicum annuum*) (20) from Minna and Nigeria were evaluated for AFs (Anthony et al. 2012). The mycotoxins were not found in baobab leaves and okra but were detected in 60% of the hot chili pepper samples at high concentration of 19.45 $\mu g\ kg^{-1}$, which was above the European Union maximum tolerance level of 4 $\mu g\ kg^{-1}$ for total AF, and thus raised public health concern. In Asia, Zhao et al. (2013) also estimated the risk of illness posed by AFB_1 resulting from consumption of Chinese spices. The 480 samples comprising of pepper, chili, prickly ash, cinnamon, aniseed, fennel, curry powder, cumin, and ginger were analyzed using HPLC. The results showed that approximately 11% of samples contained detectible levels of aflatoxin. The highest concentrations were found in chili, prickly ash, and pepper. In Africa, fungal infection and aflatoxin contamination were evaluated on 114 samples of dried and milled spices such as ginger, garlic, and black pepper from southern Benin and Togo (Gnonlonfin et al. 2013). These products were dried to preserve them for lean periods available throughout the year. Higher mold contamination was found in ginger and pepper compared to garlic that had lower levels of mold count. Species of *Aspergillus* were dominant on all dried and milled spices. After gene characterization and amplification analysis, in most of the isolation *A. flavus* was found. AFB_1 were naturally contaminated with AFB_1ranging from 390 to 1045 $mg\ kg^{-1}$ only in garlic and ginger.

Recently, Hammami et al. (2014) also surveyed for the presence of potentially harmful mycoflora and for contamination with AFB and AFG in Doha, Qatar. Among the tested spice samples, chili powder had the highest fungal contamination, while ginger, curry, and garlic samples did not have any fungal contamination. A total of 120 isolates, mostly belonging to *Aspergillus* and *Penicillium* genera, were collected. *A. flavus*, *A. nomius*, and *A. niger* were the most dominant. AFs were detected in five spices (black pepper, chilli, tandoori masala, turmeric, and garam masala), and with the exception of garam masala, the tested samples of turmeric, black pepper, tandoori masala, and chili powder exceeded B_1, B_2, G_1, and/or total aflatoxin maximum levels. In addition, spices from India, China, South America, USA, Northern Africa, Europe, and Sub-Saharan Africa marketed in Italy were investigated for AFs (AFs: AFB_1, AFB_2, AFG_1, AFG_2) and OTA (Prelle et al. 2014). They reported that 15.4% and 23.8% out of 130 samples were contaminated with AFs and OTA, respectively. The average concentration was 0.64 $ng\ g^{-1}$, far below the maximum threshold admitted by the European legislation [5 $ng\ g^{-1}$ for AFB_1, and 10 $ng\ g^{-1}$ for total AFs (AFB and AFG)]. A higher incidence of OTA was found in chili (60.0%) more than in pepper (13.3%), ranging from 2.16 to 16.35 $ng\ g^{-1}$ and from 1.61 to 15.85 ng g^{-1}, respectively. The co-occurrence of OTA and AFs in spices was detected as 4.6% of samples, ranging from 1.61 to 15.85 ng g^{-1} and from 0.57 to 3.19 ng g^{-1} for AFB_1 and AFG_1, respectively.

Moreover, Aliero and Ibrahim (2014) evaluated the aflatoxigenic fungi and AF content of dried vegetables sold in Sokoto metropolis, Nigeria. The fungi associated with the dried vegetables were *A. flavus*, *A. niger*, *A. fumigatus*, and *Mucor racemosus*. Dried *Adansonia digitata* leaves had high AF content of 62.70 μg kg^{-1} and dried *Hibiscus sabdariffa* leaves had the lowest value of 11.60 μg kg^{-1}. The fungi associated with the dried baobab leaves powder were *A. niger*, *A. flavus*, and *Rhizopus oryzae*. *A. niger* had the highest frequency of occurrence of 46.67%. The highest AF was 812 μg kg^{-1} and was obtained from Kwamberu sample, and the lowest value was 18.17 μg kg^{-1}.

The maximum AF in dried vegetables, herbs, and spices varied depending on the regulations of each country (FAO, 2003), such as the United States of America (20 μg kg^{-1}), the United Kingdom (10 μg kg^{-1}), the European Union (10 μg kg^{-1}), and Thailand (20 μg kg^{-1}).

20.8 MICROBIOLOGICAL CONTROL

Microbiological control of dried vegetables composes of confidence in process, ingredients, and hygiene program. Therefore, verification of routine checking and testing are focused. Useful testing for dried vegetable is summarized in Table 20.3. Microbiological testing can be classified into three steps as raw materials, in-process products, and finished products.

20.8.1 Raw Materials

The microbiological and safety of dried vegetables are mainly influenced by the quality of raw materials used and hygienic practice, especially when blanching method is not used in the production of dried vegetables. It is beneficial to perform microbiological analysis to build assurance in suppliers, particularly the test for index microorganisms. Nevertheless, due to the perishable characteristics of the raw materials as well as the stability characteristics of the finished products, verification testing of finished products is more appropriately focused for the effectiveness. Increased analysis may lead to more warranty of the suppliers regarding the ability to provide quality consistency of the raw materials. It should be taken into consideration that the drying process is not an effective method of reducing the number of microorganisms and pathogens, especially when blanching is not involved. Therefore, pathogens present in raw materials will frequently survive.

20.8.2 In-Process and Processing Environment

In-process microbial testing is normally of limited value and is not recommended. Since dried vegetable contamination depends on hygienic practices before and after drying, environmental processing sampling may be useful to verify the effectiveness of sanitation and hygienic programs. For control, if blanching is applicable, time and temperature must be controlled to ensure effectiveness of blanching. On the other

TABLE 20.3
Microbiological Safety and Quality Testing and Controls for Dried Vegetable Production

Step in Production	Risk	Control	Useful Analysis
Raw material	Low	Fresh vegetable control	Index microorganism analysis
In-process and processing environment	Medium	Blanching time and temperature when applicable Prompt drying to $a_w < 0.6$ Cleaning of equipment Even drying loading Moisture control in the processing environment	Hygienic testing Yeast and mold count Coliform and *E. coli* count Salmonella determination
Finished products	Low	$aw < 0.6$	Periodic testing for specific indicators including their levels that are product dependent Pathogen testing in case production condition indicates potential contamination

hand, it is generally of concern that the primary purpose of blanching is to inactivate enzymes; therefore, not all vegetables can be blanched because of significant effects on quality of the products.

Drying is the process in which the majority of water is removed, leading to an increase in microbial numbers. The principle of drying is to reduce water activity to below 0.6 to inhibit the growth of microorganisms, extending the shelf life of the products. Similar to other vegetable processing, proper and frequent equipment cleaning should be done to prevent contamination as well as to ensure hygienic handling of the finished products. Therefore, hygienic testing of the equipment is recommended. Proper loading of the product during drying is also of concern to assure even drying. Pockets of moist products can lead to retention of spoilage moisture; hence moisture control of the processing environment is required to minimize the risk of recontamination of dried vegetables.

20.8.3 Finished Products

Due to the nonperishable characteristics of dried vegetables, the variation of microbial population in dried vegetables is based on the type of vegetables and the conditions of growing and processing; the microbial analysis for finished products is not recommended, although aerobic plate count may be useful to measure hygienic and process control. It is possible that specific pathogens for specific products might be required. For example, dried vegetables for infants could be tested in a similar manner as that for infant formulae. Coliform count may not be useful as a fecal contamination indicator; however, the presence of *E. coli* may indicate cause of consideration. Periodic testing may be used for verification of the integrated effectiveness of process controls. Specific indicators vary based on the individual product that may include lactic acid bacteria, yeast and mold, and spore-forming bacteria.

For food safety concern, potential spoilage controlling is based on the condition of the storage of finished products. Products should be stored in the condition that can maintain water activity at below 0.6 for effective microbial growth controlling. Storage under preclusion of moisture may lead to subsequent fungal growth and mycotoxin contamination.

20.9 CONCLUSION

Dried vegetables are important foods consumed by people in most countries. Drying of vegetables is a traditional preservation technique that reduces the water activity to levels that do not encourage the growth of microorganisms, resulting in nonperishable and self-stable characteristics of products. The microbiological stability of the products depends on how to maintain the dry condition of products after the drying process throughout storage via proper storage conditions and proper packaging of the products to assure safety for the consumers. Due to the hastening development of the food industry, dried vegetables are used not only for cooking but also for making snacks. Vegetables can be made into snacks using different drying methods. Development of new pre- and posttreatments to improve the microbiological and sensory qualities of dried vegetables as well as to produce novel dried vegetable snacks are of interest; for instance, novel dried vegetable snacks of different shapes, sizes, flavors, textures, tastes, and colors. In addition, further research and development of microwave-assisted drying is considered to resolve the nonuniformities during microwave heating for large-scale application along with improvement in capital and operating costs.

REFERENCES

Akeredolu, A. A. and L. O. Adebajo. 2013. Microflora of three dehydrated vegetables. *British Microbiology Research Journal* 3(3):295–308.

Alghalibi, S. M. S. and A. R. M. Shater. 2004. Mycoflora and mycotoxin contamination of some dried fruits in Yemen Republic. *Association of University Bulletin and Environmental Research* 7:19–27.

Alibas, I. 2009. Microwave, vacuum and air drying characteristics of collard leaves. *Drying Technology* 27:1266–1273.

Aliero, A. A. and A. D. Ibrahim. 2014. Aflatogenic and aflatoxin contents of dried vegetables sold in Sokoto Metropolis, Nigeria. *Journal of Emerging Trends in Engineering and Applied Sciences* 5(7):116–120.

Andress, E. L. and J. A. Harrison. 1999. *So Easy to Preserve*. 4th edition. Cooperative Extension Service, The University of Georgia, Athens, GA. pp 344.

Anthony, M. H., Simeon, M. C., Abdulramoni, S. A., Chimeririm, O. B., and O. M. Uchenna. 2012. A preliminary survey of aflatoxin in fresh and dried vegetables in Minna, Nigeria. *African Journal of Food Science and Technology* 3(10):268–272.

Archer, J., Jervis, E. T., Bird, J., and J. E. Gaze. 1998. Heat resistance of *Salmonella weltevreden* in low-moisture environments. *Journal of Food Protection* 61:969–973.

Banerjee, M. and P. K. Sarkar. 2003. Microbiological quality of some retail spices in India. *Food Research International* 36:469–474.

Bankole, S. A. and O. O. Mabekoje. 2004. Mycoflora and occurrence of aflatoxin B_1 in dried yam chips from markets in Ogun and Oyo States, Nigeria. *Mycopathologia* 157:111–115.

Begum, M., Lokesh, S., and V. T. Kumar. 2005. Pathogenicity of *Macrophomina phaseolina* and *Fusarium verticillioides* in okra. *Integrative Bioscience* 9:37–40.

Bendini, A., Galina Toschi, T., and G. Lercker. 1998. Influence of gamma irradiation and microwaves on the linear unsaturated hydrocarbon fraction in spices. *Zeitschrift für Lebensmittel Untersuchung und-Forschung A* 207:214–218.

Bhat, R. V. and S. Vashanti. 1999. Occurrence of aflatoxins and its economic impact on human nutrition and animal feed. The new regulation. *Agricultural Development* 23:50–56.

Brennand, C. P. 1994. Home drying of food. Utah State University Cooperative Extension. https://extension.usu.edu/files/publications/publication/FN-330.pdf (Accessed July 2, 2015).

Cenkowski, S., Pronyk, C., Zmidzinska, D., and W. E. Muir. 2007. Decontamination of food products with superheated steam. *Journal of Food Engineering* 83:68–75.

Chiewchan, N. and P. Morakotjinda., 2009. Effects of acetic acid pretreatment and hot air drying on resistance of *Salmonella* on cabbage slices. *Drying Technology* 27:955–961.

Chiewchan, N., Pakdee, W. and S. Devahastin. 2007. Effect of water activity on thermal resistance of Salmonella krefeld in liquid medium and on rawhide surface. *International Journal of Food Microbiology* 114:43–49.

Devahastin, S., Suvarnakuta, P., Soponronnarit, S., and A. S. Mujumdar. 2004. A comparative study of low-pressure superheated steam and vacuum drying of a heat-sensitive material. *Drying Technology* 22:1845–1867.

Dinstel, R. R. 1997. *Food Preservation-Drying Fruits and Vegetables*. Alaska Cooperative Extension, University of Alaska, Fairbanks, AK. www.uaf.edu/ces/publications/freepubsFdPresDrying.pdf (Accessed May 15, 2015).

DiPersio, P. A., Kendall, P. A., Yoon, Y., and J. N. Sofos. 2007. Influence of modified blanching treatments on inactivation of *Salmonella* during drying and storage of carrot slices. *Food Microbiology* 24:500–507.

DiPersio, P. A., Yoon, Y., Sofos, J. N., and P. A. Kendall. 2005. Inactivation of *Salmonella* during drying and storage of carrot slices prepared using commonly recommended methods. *Journal of Food Science* 70:M230–M235.

Doyle, M. E. and A. S. Mazzotta. 2000. Review of studies on the thermal resistance of Salmonellae. *Journal of Food Protection* 63(6):779–795.

Fazekas, B., Tar, A. and M. Kovácz. 2005. Aflatoxin and ochratoxin A content of spices in Hungary. *Food Additives and Contaminants* 22:856863.

Food and Agricultural Organization (FAO). 2003. Worldwide regulations for mycotoxins in food and feed in 2003. www.fao.org/docrep/007/y5499e/y5499e00.htm (Accessed March 7, 2016).

Food Standards Australia New Zealand. 2003. Standard 1.6.1: Microbiological limits for food. Online. Available at www.foodstandards.gov.au/_srcfiles/Standard161_Micro_v63.pdf. Accessed March 11, 2016.

Garbutt, J. 1997. *Essentials of Food Microbiology*. 1st edition. Hodder Arnold, London.

Gnonlonfin, G. J. B., Adjovi, Tokpo, Y. C., Agbekponou, E. D., Ameyapoh, Y., de Souza, C., Brimer, L., and A. Sanni. 2013. Mycobiota and identification of aflatoxin gene cluster in marketed spices in West Africa. *Food Control* 34:115–120.

Goresline, H. E. 1963. A discussion of the microbiology of various dehydrated foods. In *Microbiological Quality of Foods*, eds Slanetz, L. W., Chicheste, C. O., Gaufin, A. R., and Z. J. Ordal, 179–192. Academic Press, New York.

Hammami, W., Fiori, S., Thani, R. A., Kali, N. A., Balmas, V., Migheli, Q., and S. Jaoua. 2014. Fungal and aflatoxin contamination of marketed spices. *Food Control* 37:177–181.

Hawaree, N., Chiewchan, N., and S. Devahastin. 2009. Effects of drying temperature and surface characteristics of vegetable on the survival of *Salmonella. Journal of Food Science* 74:E16–E22.

Head, D. S., Cenkowski, S., Holley, R., and G. Blank. 2008. Effects of superheated steam on *Geobacillus stearothermophilus* spore viability. *Journal of Applied Microbiology* 104:1213–1220.

Hell, K., Gnonlonfin, B. G. J., Kodjogbe, G., Lamboni, Y., and I. K. Abdourhamane. 2009. Mycoflora and occurrence of aflatoxin in dried vegetables in Benin, Mali and Togo, West Africa. *International Journal of Food Microbiology* 135:99–104.

Hierro, J. M. H., García-Villanova, R. J., Torrero, P. R., and I. M. T. Fonseca. 2008. Aflatoxins and ochratoxin A in red paprika for retail sale in Spain: Occurrence and evaluation of a simultaneous analytical method. *Journal of Agricultural and Food Chemistry* 56:751–756.

Hughes, K. V. and B. J. Willenberg. 1999. Quality for Keeps-Drying Foods, Publication GH1562, University Extension, University of Missouri, Columbia, Missouri. http://extension.missouri.edu/explore/hesguide/foodnut/gh1562.htm.

International Agency for Research on Cancer (IARC). (1993). Some naturally occurring substances, food items and constituents, heterocyclic aromatic amines and mycotoxin, Vol. 56, World Health Organization, Lyon, France, 489–521.

International Commission on Microbiological Specifications for Food (ICMSF). 1986. The ICMSF sampling plans and recommended microbiological limits for vegetables, fruits, nuts and yeast 3. In: *Microorganisms in Foods 2, Sampling for Microbial Analysis: Principles and Specific Applications*, 2nd edition. ICMSF Blackwell Scientific Publications, Palo Alto, CA, 197–202. http://www.icmsf.org/pdf/icmsf2.pdf.

International Organization for Standardization (ISO). 1981. General Guidance for Detection of Salmonella. ISO 6579. ISO, Washington, DC.

Jalili, M. and S. Jinap. 2012. Natural occurrence of aflatoxins and ochratoxin A in commercial dried chili. *Food Control* 24:160–164.

Jernberg, C., Hjertqvist, M., Sundborger, C., Castro, E., Lofdahl, G., Pááj́ärvi, A., Sundqvist, L., and E. Lóf. 2015. Outbreak of *Salmonella enteritidis* phage type 13a infection in Sweden linked to imported dried-vegetable spice mixes, December 2014 to July 2015. *Euro Surveillance.* 20(30):21194. www.eurosurveillance.org/ViewArticle.aspx?ArticleId=21194 (Accessed November 3, 2015).

Kendall, P. A. and L. Allen. 1998. Drying Vegetables. Service in Action 9.308. Cooperative Extension, Colorado State University, Fort Collins, CO.

Kudjawu, B. D., Sakyi-Dawson, E., and W. K. Amoa-Awua. 2011. The microbiota of dried traditional vegetables produced in the Sudan Savannah and Guinea Savannah agroecological zones of Ghana. *International Food Research Journal* 18:101–108.

Leeratanarak, N., Devahastin, S., and N. Chiewchan. 2006. Drying kinetics and quality of potato chips undergoing different drying techniques. *Journal of Food Engineering* 77:635–643.

Lewicki, P. P. 2006. Design of hot air drying for better foods. *Trends in Food Science and Technology* 17:153–163.

Loaharanu, P. 1994. Status and prospects of food irradiation. *Food Technology* 52:124–131.

Magnoli, C., Astoreca, A., Ponsone, L., Combina, M., Palacio, G., Rosa, C. A., and A. M. Dalcero. 2004. Survey of mycoflora and ochratoxin A in dried vine fruits from Argentina markets. *Letter of Applied Mycobiology* 39:326–331.

Mandeel, Q. A. 2005. Fungal contamination of some imported spices. *Mycopathologia* 59:291–298.

Martin, A., Aranda, E., Benito, M. J., Pérez-Nevado, F., and M. G. Córdoba. 2005. Identification of fungal contamination and determination of mycotoxigenic molds by micellar electrokinetic capillary chromatography in smoked paprika. *Journal of Food Protection* 68:815–822.

McKee, L. H. 1995. Microbial contamination of spices and herbs: A review. *LWT-Food Science Technology* 28:1–11.

Methakhup, S., Chiewchan, N., and S. Devahastin. 2005. Effects of drying methods and conditions on drying kinetics and quality of Indian gooseberry flake. *LWT-Food Science and Technology* 38:579–587.

Murphy, R. P. 1973. Microbiological contamination of dried vegetables. *Processing Biochemistry* 8:17–19.

Murphy, R. Y., Duncan, L. K., Johnson, E. R., Davis, M. D., and J. A. Marcy. 2002. Thermal inactivation of *Salmonella senftenberg* and *Listeria innocua* in beef/turkey blended patties cooked via fryer and/or air convection oven. *Journal of Food Science* 67:1879–1885.

Nutsugah, S. K., Vibeke, L., Atokple, I. D. K., and F. K. Ayensu. 2004. Seed-borne mycoflora of major food crops in Ghana. *Journal of Science and Technology* 24:22–31.

Nygaard, H. and O. Hostmark. 2008. Microbial inactivation during superheated steam drying of fish meal. *Drying Technology* 26:222–230.

Olson, D. G. 1998. Irradiation of food. *Food Technology* 52:56–62.

Otzuki, T., Wilson, J. S., and M. Sewadeh. 2001. What price precaution? European harmonisation of aflatoxin regulations and African groundnut exports. *European Review of Agricultural Economy* 28:263–283.

Ozby, F. and B. Kabak. 2012. Natural co-occurrence of aflatoxins and ochratoxin A in spices. *Food Control* 28:354–361.

Pezzutti, A., Marucci, P. L., Sica, M. G., Matzkin, M. R., and C. A. Croci. 2005. Gamma-ray sanitization of Argentinean dehydrated garlic (*Allium sativum* L.) and onion (*Allium cepa* L.) products. *Food Research International* 38:797–802.

Phungamngoen, C., Chiewchan, N., and S. Devahastin. 2011. Thermal resistance of Salmonella enterica serovar Anatum on cabbage surfaces during drying: Effects of drying methods and conditions. *International Journal of Food Microbiology* 147:127–133.

Prelle, A., Spadaro, D., Garibaldi, A., and M. L. Gullino. 2014. Co-occurrence of aflatoxins and ochratoxin A in spices commercialized in Italy. *Food Control* 39:192–197.

Ravi Kiran, D., Narayana, K. J. P., and M. Vijayalakshmi. 2005. Aflatoxin B_1 production in chillies (*Capsicum annum* L.) kept in cold stores. *African Journal of Biotechnology* 4:791–795.

Roberts, T. and R. Cox. 1999. Drying Fruits and Vegetables, Document 348597. Virginia Cooperative Extension, Virginia Polytechnic Institute, Blacksburg, VA. www.ext.vt.edu/pubs/foods/348–597/348–597.html (Accessed August 3, 2015).

Romagnoli, B., Menna, V., Gruppioni, N., and C. Bergamini. 2007. Aflatoxins in spices, aromatic herbs, herb-teas and medicinal plants marketed in Italy. *Food Control* 18:697–701.

Sagoo, S. K., Little, C. L., Greenwood, M., Mithani, V., Grant, K. A., McLauchlin, J., de Pinna, E., and E. J. Threlfall. 2009. Assessment of the microbiological safety of dried spices and herbs from production and retail premises in the United Kingdom. *Food Microbiology* 26:39–43.

Saha, D., Acharya, D., Roy, D., Shrestha, D., and T. K. Dhar. 2007. Simultaneous enzyme immunoassay for the screening of aflatoxin B_1 and ochratoxin A in chili samples. *Analytica Chimica Acta* 584:343–349.

Santos, L., Marín, S., Sanchis, V., and A. J Ramos. 2010. Co-occurrence of aflatoxins, ochratoxin A and zearalenone in capsicum powder samples available on the Spanish market. *Food Chemistry* 122:826–830.

Seevaratnam, V., Banumathi, P., Premalatha, M. R., Sundaram, S. P., and T. Arumugam. 2012. Effect of packaging materials on retention of quality characteristics of selected dehydrated green leafy vegetables during storage. *World Journal of Dairy and Food Sciences* 7:190–194.

Sheneman, J. M. (1973). Survey of aerobic mesophilic bacteria in dehydrated onion products. *Journal of Food Science* 38:206–209.

Shundo, L., De Almeida, A. P., Alaburda, J., Lamardo, L. C. A., Navas, S. A., Ruvieri, V., and M. Sabino. 2009. Aflatoxins and ochratoxin A in Brazilian paprika. *Food Control* 20:1099–1102.

Singh, U. and V. R. Sagar. 2010. Quality characteristics of dehydrated leafy vegetables influenced by packaging materials and storage temperature. *Journal of Scientific and Industrial Research* 69:785–789.

Suvarnakuta, P., Devahastin, S., and A. S. Mujumdar. 2005. Drying kinetics and β-carotene degradation in carrot undergoing different drying processes. *Journal of Food Science* 70:S520–S526.

Sweeney, M. J. and A. D. W. Dobson. 1998. Mycotoxin production by *Aspergillus*, *Fusarium* and *Penicillium* species. *International Journal of Food Microbiology* 43:141–158.

Thayer, D. W., Josephson, E. S., Brynjolfsson, A., and G. G. Giddings. 1996. Radiation pasteurization of food. Council for Agricultural Science and Technology—CAST, Ames, IA, April 7.

Thomkapanish, O., Suvarnakuta, P., and S. Devahastin. 2007. Study of intermittent low-pressure superheated steam and vacuum drying of banana. *Drying Technology* 25:205–223.

United Nations Industrial Development Organization/Food and Agriculture Organization of the United Nations (UNIDO/FAO). 2006. *Herbs, Spices and Essential Oils: Post-harvest Operations in Developing Countries*. UNIDO/FAO, Vienna, Austria.

Vadivambal, R. and D. S. Jayas. 2007. Changes in quality of microwave-treated agricultural products—A review. *Biosystems Engineering* 98:1–16.

Van Doren, J. M., Neil, K. P., Parish, M., Gieraltowski, L., Gould, L. H., and K. L. Gombas. 2013. Foodborne illness outbreaks from microbial contaminants in spices, 1973–2010. *Food Microbiology* 36:456–464.

Vaughn, R. H. 1951. The microbiology of dehydrated vegetables. *Food Research* 16:429–438.

Wedzicha, B. L. 1984. *Chemistry of Sulfur Dioxide in Foods*. Elsevier Applied Science Publishers, Barking, UK.

World Health Organization (WHO). 1999. High-dose irradiation: Wholesomeness of food irradiated with doses above 10 kGy. Report of a joint FAO/IAEA/WHO study group. WHO Technical Report Series 890. World Health Organization, Geneva.

World Health Organization (WHO). 2000. Database of Microbiological Specifications for Selected Countries. www.who.dk/foodsafety/Microbiological/20020418_8 (Accessed March 7, 2016).

Wu, L., Orikasa, T., Ogawa, Y., and A. Tagawa. 2007. Vacuum drying characteristics of eggplants. *Journal of Food Engineering* 83:422–429.

Youssef, M. S. 2008. Studies on mycological status of sundried jew's mallow leaves and okra fruits in Egypt. *Research Journal of Microbiology* 3:375–385.

Zhao, X., Schaffner, D. W. and T. Yue. 2013. Quantification of aflatoxin risk associated with Chinese spices: Point and probability risk assessments for aflatoxin B_1. *Food Control* 33:366–377.

Zinedine, A., Brera, C., Elakhdari, S. et al. 2006. Natural occurrence of mycotoxins in cereals and spices commercialized in Morocco. *Food Control* 17:868–874.

Zweifel, C. and R. Stephan. 2012. Spices and herbs as source of *Salmonella*-related foodborne diseases. *Food Research International* 45:765–769.

Index

D

I

J

K

L

N

O

T